Proceedings

Ein stetig steigender Fundus an Informationen ist heute notwendig, um die immer komplexer werdende Technik heutiger Kraftfahrzeuge zu verstehen. Funktionen, Arbeitsweise, Komponenten und Systeme entwickeln sich rasant. In immer schnelleren Zyklen verbreitet sich aktuelles Wissen gerade aus Konferenzen, Tagungen und Symposien in die Fachwelt. Den raschen Zugriff auf diese Informationen bietet diese Reihe Proceedings, die sich zur Aufgabe gestellt hat, das zum Verständnis topaktueller Technik rund um das Automobil erforderliche spezielle Wissen in der Systematik aus Konferenzen und Tagungen zusammen zu stellen und als Buch in Springer.com wie auch elektronisch in Springer Link und Springer Professional bereit zu stellen. Die Reihe wendet sich an Fahrzeug- und Motoreningenieure sowie Studierende, die aktuelles Fachwissen im Zusammenhang mit Fragestellungen ihres Arbeitsfeldes suchen. Professoren und Dozenten an Universitäten und Hochschulen mit Schwerpunkt Kraftfahrzeug- und Motorentechnik finden hier die Zusammenstellung von Veranstaltungen, die sie selber nicht besuchen konnten. Gutachtern, Forschern und Entwicklungsingenieuren in der Automobil- und Zulieferindustrie sowie Dienstleistern können die Proceedings wertvolle Antworten auf topaktuelle Fragen geben.

Today, a steadily growing store of information is called for in order to understand the increasingly complex technologies used in modern automobiles. Functions, modes of operation, components and systems are rapidly evolving, while at the same time the latest expertise is disseminated directly from conferences, congresses and symposia to the professional world in ever-faster cycles. This series of proceedings offers rapid access to this information, gathering the specific knowledge needed to keep up with cutting-edge advances in automotive technologies, employing the same systematic approach used at conferences and congresses and presenting it in print (available at Springer.com) and electronic (at Springer Link and Springer Professional) formats. The series addresses the needs of automotive engineers, motor design engineers and students looking for the latest expertise in connection with key questions in their field, while professors and instructors working in the areas of automotive and motor design engineering will also find summaries of industry events they weren't able to attend. The proceedings also offer valuable answers to the topical questions that concern assessors, researchers and developmental engineers in the automotive and supplier industry, as well as service providers.

More information about this series at http://www.springer.com/series/13360

Peter E. Pfeffer
Editor

11th International Munich Chassis Symposium 2020

chassis.tech plus

Springer Vieweg

Editor
Peter E. Pfeffer
Munich University of Applied Science
Munich, Germany

ISSN 2198-7432 ISSN 2198-7440 (electronic)
Proceedings
ISBN 978-3-662-63192-8 ISBN 978-3-662-63193-5 (eBook)
https://doi.org/10.1007/978-3-662-63193-5

Responsible Editor: Markus Braun
This Springer Vieweg imprint is published by the registered company Springer-Verlag GmbH, DE part of Springer Nature.
The registered company address is: Heidelberger Platz 3, 14197 Berlin, Germany

Preface

The chassis will play a key role in the future of the automobile. End customers want a dynamic driving experience – without any compromise on safety. At the same time, automated and autonomous functions are generating higher demands for comfort. The future lies in platforms that can be combined as flexibly as possible. The aim is to optimize the interaction between chassis systems to achieve functional integration.

chassis.tech plus, which was being held in June 2020 for the eleventh time, is a unique event throughout Europe that will bring together numerous leading experts in chassis systems, steering, brakes, and tires/wheels from all over the world for an industry-wide exchange of ideas and a transfer of knowledge. Renowned speakers from industry and research discuss the very latest developments in the chassis, steering systems, brakes, and wheels/tires in four parallel strands of lectures. This conference proceedings documents these advances in chassis technology.

Dr. Peter E. Pfeffer
Munich University of Applied Sciences
Scientific Director of the Symposium

Contents

Autorenverzeichnis

Masato Abe Kanagawa Institute of Technology, Atsugi, Kanagawa, Japan

D. Allgeier University of Applied Science Kempten, Kempten, Germany

Daniel Alt Joyson Safety Systems, Berlin, Germany

Lars Altseimer University of Stuttgart, Stuttgart, Germany

Matthias Becker MdynamiX AG, Munich, Germany

Adrian Bernhard Research Institute of Automotive Engineering and Vehicle Engines Stuttgart (FKFS), Stuttgart, Germany

Frédéric Biesse Michelin, Research Center Ladoux, Clermont-Ferrand, France

Maximilian Böhle formerly municHMotorsport e.V., Munich, Germany

Robert Buchmann MAN Truck & Bus SE, Munich, Germany

Stefan Büchner BMW Group, Garching, Germany

P. Camocardi McLaren Automotive Ltd, Woking, England

Francesco Camozzi Brembo S.p.A., Stezzano, Italy

Salvatore Danise FCA Emea Product Development, Torino, Italy

S. De Pinto McLaren Automotive Ltd, Woking, England

Norbert Deixler BMW Group, Garching, Germany

Antonino DI GERLANDO Politecnico di Milano, Milano, Italy

A. Doria Cerezo Universidad Politécnica de Cataluña, Barcelona, Spain

Eberhard Drechsel formerly Munich University of Applied Sciences, Munich, Germany

Toimothy Drotar Ford Research and Innovation Center, Dearborn, USA

Lutz Eckstein Institute for Automotive Engineering (ika) RWTH Aachen University, Aachen, Germany

Alexander Ein Waldt Ford-Werke GmbH, Cologne, Germany

Cristiano Fissore Brembo S.p.A., Stezzano, Italy

F. Fuhr Dr.-Ing. h.c. F. Porsche AG, Weissach, Germany

Masato Fujiyama Toyota Motor Corporation, Susono Shizuoka, Japan

Joachim Funke HELLA GmbH & Co. KGaA, Lippstadt, Germany

Bernhard Gerster BFH-TI and DTC AG, Bern, Switzerland

Alberto Girelli Consolaro Ford Research and Innovation Center Aachen, Aachen, Germany

Massimiliano Gobbi Politecnico di Milano, Milano, Italy

Florian Goy Hyundai Motor Technical Center Europe GmbH, Rüsselsheim, Germany

Ingo Greue HELLA GmbH & Co. KGaA, Lippstadt, Germany

Roland Grimm Joyson Safety Systems, Berlin, Germany

P. Gruber University of Surrey, Guildford, England

Takashi Gunji NSK Ltd., Kanagawa, Japan

Mark Harris FKA GmbH, Aachen, Germany

Ralf Hintzen Ford Research and Innovation Center, Aachen, Germany

Manuel Höfer Dr. Ing. h.c. F. Porsche AG, Weissach, Germany

Jens Hunecke HELLA GmbH & Co. KGaA, Lippstadt, Germany

Mitsunori Ishibashi Kanagawa Institute of Technology, Atsugi, Kanagawa, Japan

Norio Ishihara NTN Corporation, Iwata-shi, Shizuoka, Japan

Francesco Italiano Ford Research and Innovation Center Aachen, Aachen, Germany

Atsushi Ito NTN Corporation, Iwata-shi, Shizuoka, Japan

D. Ivone Pirelli & C. S.p.A., Milan, Italy

Jaffrey Immanuel Jeyakumar BMW Group, Munich, Germany

Benedikt Jooß Dr. Ing. h.c. F. Porsche AG, Weissach, Germany

Jongmin Jung Hyundai Motor Company, Hwaseong, South Korea

Niels Justen BMW Group, Munich, Germany

T. Kaneko Toyota Motor Corporation, Susono Shizuoka, Japan

Yoshio Kano Kanagawa Institute of Technology, Atsugi, Kanagawa, Japan

Shinji Kato KYB Corporation, Dota, Japan

Thomas Kersten Volkswagen AG, Wolfsburg, Germany

Kei Kitahara JTEKT Corporation, Kashihara-city, Nara, Japan

Joe Klesing Nexteer Automotive Corporation, Auburn Hills, US

Benjamin Koller TÜV SÜD Auto Service GmbH, Munich, Germany

T. Kondo Toyota Motor Cooperation, Susono Shizuoka, Japan

Ioannis Konstantinou Siemens Digital Industries, Helmond, The Netherlands

Roland Krivachy MAN Truck & Bus SE, Munich, Germany

Carsten Kügeler HELLA GmbH & Co. KGaA, Lippstadt, Germany

Martin Kundla BMW Group, Munich, Germany

Timo Kuthada Research Institute of Automotive Engineering and Vehicle Engines Stuttgart (FKFS), Stuttgart, Germany

Isao Kuwayama Bridgestone Corporation, Kodaira city, Japan

Miso Kvesic formerly municHMotorsport e.V., Munich, Germany

Seungmin Kwon Hyundai Motor Company, Hwaseong, South Korea

Olav Lange Ford Werke GmbH, Köln, Germany

Jun-Mo Lee Hyundai-Kia America Tech Center Inc., Irvine, US

S. Lefèvre Dr.-Ing. h.c. F. Porsche AG, Weissach, Germany

Tom Lehmkuhl fka GmbH, Aachen, Germany

Markus Lemmen Bochum University of Applied Sciences Campus Velbert/ Heiligenhaus , Heiligenhaus, Germany

Markus Lienkamp Technical University of Munich, Munich, Germany

Tim Lindner BMW Group, Munich, Germany

Alexander Lion Bundeswehr University, Neubiberg, Germany

Christian Ludwig Hyundai Motor Technical Center Europe GmbH, Rüsselsheim, Germany

Ian MacLachlan Ford Motor Company, Dearborn, USA

Daniel Mainz Ford Research and Innovation Center, Aachen, Germany

Mauro Martino FCA Emea Product Development, Torino, Italy

Gianpiero Mastinu Politecnico di Milano, Milano, Italy

Robert Matawa TÜV SÜD Auto Service GmbH, Munich, Germany

Victor Mazzilli University of Surrey , Guildford, England

Michael Moczala ZF Automotive Germany GmbH, Düsseldorf, Germany

Hiroshi Mouri Tokyo University of Agriculture and Technology, Tokio, Japan

V. Vidal Muñoz University of Surrey , Guildford, England

Oliver Nehls Ford-Werke GmbH, Cologne, Germany

Jonas Nesensohn University of Applied Science Kempten, Kempten, Germany

Hirokazu Ohba NTN Corporation, Iwata-shi, Shizuoka, Japan

Ryosuke Okubo Toyota Motor Corporation, Susono Shizuoka, Japan

Patrick Pallot Manufacture Française des Pneumatiques MICHELIN, Clermont-Ferrand, France

Ho Joon Park Hyundai Motor Company, Hwaseong, South Korea

L. Pascali McLaren Automotive Ltd, Woking, England

Michael Peperhowe dSPACE GmbH, Paderborn, Germany

Peter E. Pfeffer Munich University of Applied Sciences, Munich, Germany

Antonino Pizzuto Hyundai Motor Technical Center Europe GmbH, Rüsselsheim, Germany

Florian Rath Ford-Werke GmbH, Cologne, Germany

Martin Reichenecker Dr. Ing. h. c. F. Porsche AG, Weissach, Germany

Paolo RIGHETTINI University of Bergamo, Bergamo, Italy

Thomas Rotter Hochschule Esslingen, Esslingen, Germany

Andreas Saathoff Volkswagen AG, Wolfsburg, Germany

Alessandro Salgarello Hyundai Motor Technical Center Europe GmbH, Rüsselsheim, Germany

Navid Samadi KYB Technical Center Europe, Munich, Germany

Florian Scheirl BMW Group, Munich, Germany

B. Schick University of Applied Science Kempten, Kempten, Germany

Dieter Schramm University Duisburg-Essen, Duisburg, Germany

Domenik Schramm Research Institute of Automotive Engineering and Vehicle Engines Stuttgart (FKFS), Stuttgart, Germany

Bernhard Schweizer Dr. Ing. h. c. F. Porsche AG, Weissach, Germany

Dominic Scopacasa HELLA Corporate Center USA, Northville Township, USA

Konstantin Sedlan Volkswagen AG, Wolfsburg, Germany

Ludwig Seethaler BMW Group, Garching, Germany

Yousuke Sekino Showa Corporation, Gyoda, Japan

Stefan Sempach DTC AG, Bern, Switzerland

Osamu Shimizu The University of Tokyo, Tokyo, Japan

Jieun Song Hyundai Motor Company, Hwaseong, South Korea

A. Sorniotti University of Surrey, Guildford, England

Roberto Strada University of Bergamo, Bergamo, Italy

Ralph Stroph BMW Group, Garching, Germany

Christian Strümpler Joyson Safety Systems, Berlin, Germany

Marcel Strup DTC AG, Bern, Switzerland

G. Tarquinio McLaren Automotive Ltd, Woking, England

Adolfo Tee Hyundai-Kia America Tech Center Inc., Irvine, US

Florian Till BMW Group, Munich, Germany

Damiano Todisco Brembo S.p.A., Stezzano, Italy

Violaine Todoroff Michelin, Research Center Ladoux, Clermont-Ferrand, France

Stefan Uhlar Audi AG, Ingolstadt, Germany

Willem Versteden Siemens Digital Industries, Helmond, The Netherlands

Andreas Wagner Research Institute of Automotive Engineering and Vehicle Engines Stuttgart (FKFS), Stuttgart, Germany

Nick Watson Ford Dunton Technical Centre, Laindon Basildon, Great Britain

Jochen Wiedemann Research Institute of Automotive Engineering and Vehicle Engines Stuttgart (FKFS), Stuttgart, Germany

Friedrich Wolf-Monheim Ford Research and Innovation Center, Aachen, Germany

Makoto Yamakado Kanagawa Institute of Technology, Atsugi, Kanagawa, Japan

Yong-Sub Yi Hyundai Motor Company, Hwaseong, South Korea

Ventseslav Yordanov Institute for Automotive Engineering (ika) RWTH Aachen University, Aachen, Germany

Paul Zandbergen Ford Research and Innovation Center, Aachen, Germany

Markus Zimmermann Technical University of Munich, Munich, Germany

The new chassis of Volkswagen ID.3 within MEB platform – Front Axle, Rear Axle

Dr. Andreas Saathoff and Dr. Thomas Kersten

Volkswagen AG

P. E. Pfeffer (ed.), *11th International Munich Chassis Symposium 2020*, Proceedings, https://doi.org/10.1007/978-3-662-63193-5_1

The new chassis of Volkswagen ID.3 within MEB platform

Front Axle, Rear Axle

Dr. Andreas Saathoff
Dr. Thomas Kersten
Volkswagen AG

Agenda

1) The new MEB e-mobility Matrix of VW ID.3
2) Axles: Concept, Boundary Conditions, Simulation
3) The new MEB McPherson Front Axle
4) The new MEB Multilink Rear Axle
5) Axle Testing and Simulation of the MEB
6) Summary

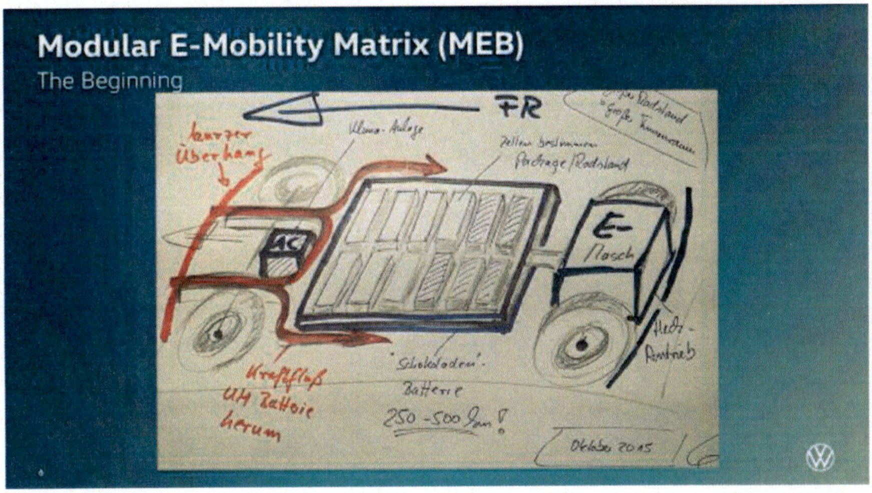
Modular E-Mobility Matrix (MEB)
The Beginning
FR
kurzer Überhang
AC
Batterie
250 -500 km!
Oktober 2015

Modular E-Mobility Matrix (MEB)
The Concept
100% Electric | Highest Flexibility | Scalable Matrix | Long Wheelbase | Best Weight Distribution

Design and Package
From the outside like a Golf.
From the inside like a Passat.
Short Overhang
Long Wheel Base
Δ Lenght -23 mm
In comparison to Golf 8
Short Overhang
5

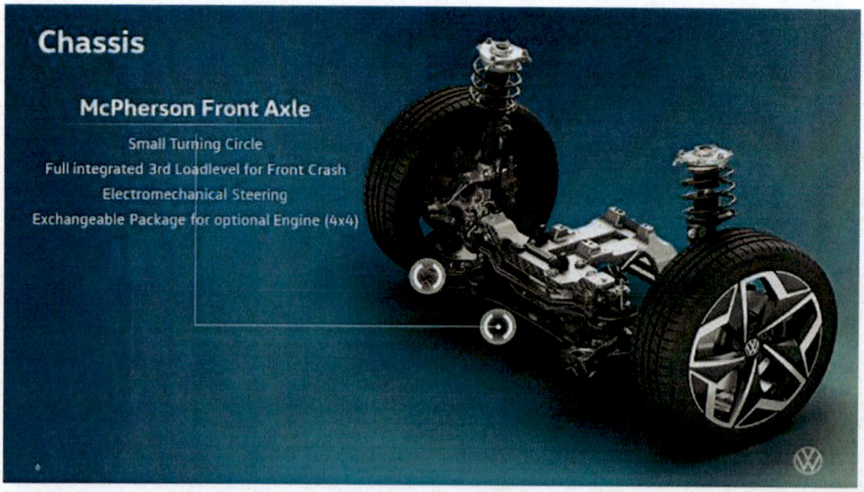
Chassis
McPherson Front Axle
Small Turning Circle
Full integrated 3rd Loadlevel for Front Crash
Electromechanical Steering
Exchangeable Package for optional Engine (4x4)
6

Chassis
Multilink Rear Axle
Subframe with hydro formed Tubes carrying the Engine
Subframe with Bushings for Acoustic Decoupling
Optimized Rear Axle Package to realize maximum amount of Battery Cells

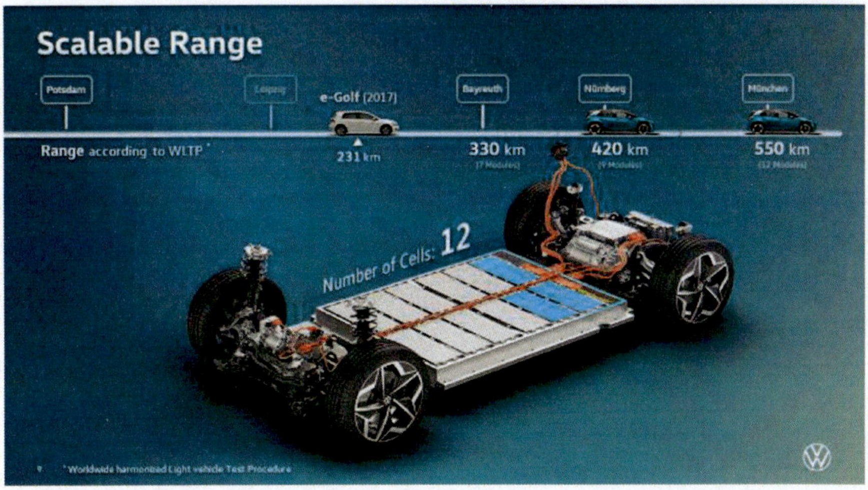
Scalable Range
Potsdam
Bayreuth
Nürnberg
München
e-Golf (2017)
Range according to WLTP *
231 km
330 km
420 km
550 km
Number of Cells: 12
* Worldwide harmonized Light vehicle Test Procedure

e-Mobility to make all customers happy

MEB-Lineup

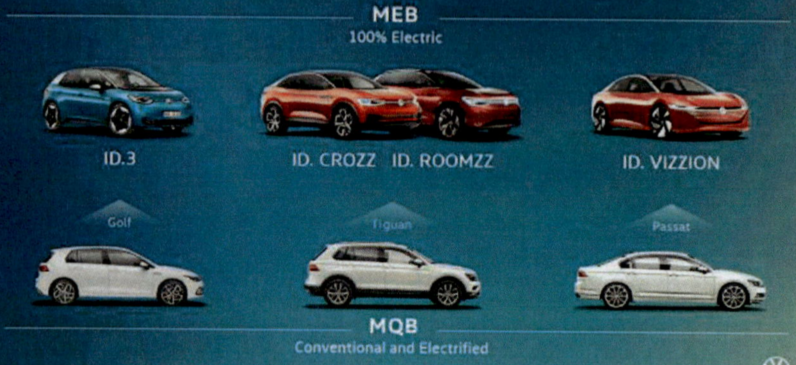

10

Agenda

1) The new MEB e-mobility Matrix of VW ID.3
2) Axles: Concept, Boundary Conditions, Simulation
3) The new MEB McPherson Front Axle
4) The new MEB Multilink Rear Axle
5) Axle Testing and [illegible]
6) Summary

11

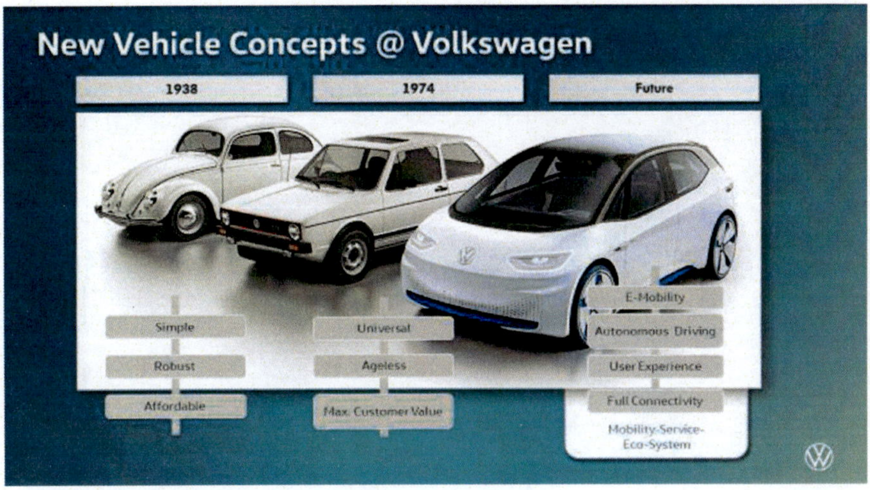
New Vehicle Concepts @ Volkswagen
1938
1974
Future
Simple
Robust
Affordable
Universal
Ageless
Max. Customer Value
E-Mobility
Autonomous Driving
User Experience
Full Connectivity
Mobility-Service-Eco-System

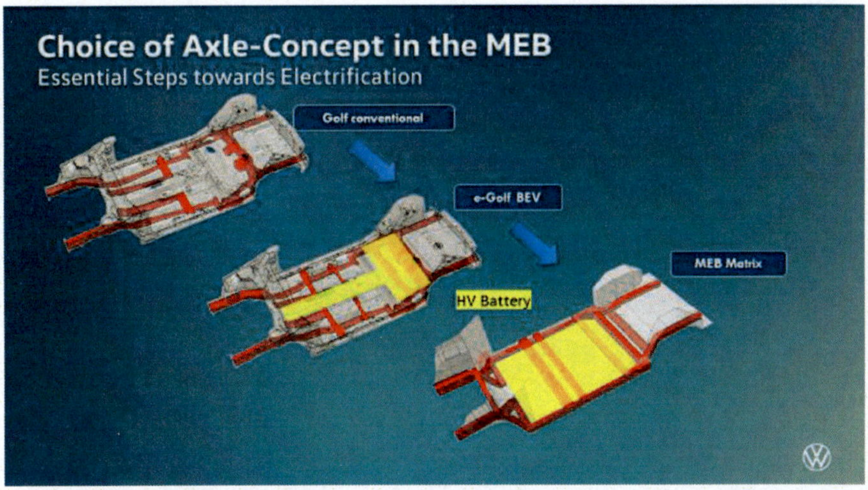
Choice of Axle-Concept in the MEB
Essential Steps towards Electrification
Golf conventional
e-Golf BEV
MEB Matrix
HV Battery

Choice of Axle-Concept in the MEB

Matrix Overview

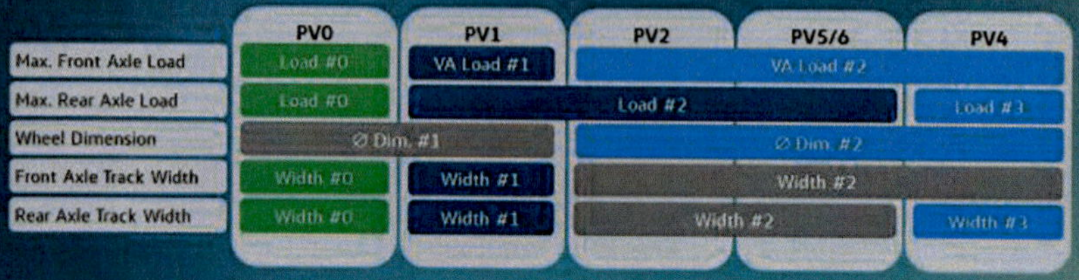

Summary Concept – Rear Drive

+ Very good Traction
+ Large Wheel Base = Huge Passenger Compartment
+ Optimized Turning Circle
+ Steering Location in front of Wheel Center
− High Sensitivity of Cross Winds
− Small Rear Axle Package
− High Rear Axle Loads must be realised (mixed tire dimensions necessary)

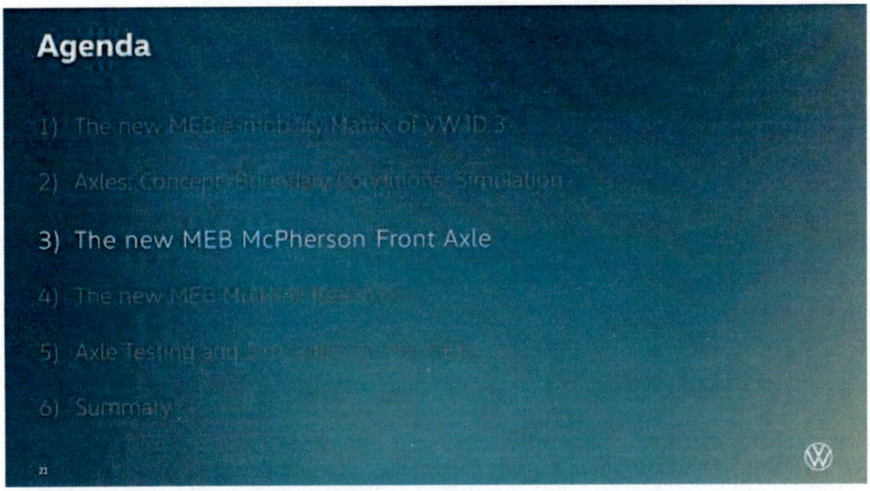
Agenda
1) The new MEB e-mobility Matrix of VW ID.3
2) Axles: Concept, Boundary Conditions, Simulation
3) The new MEB McPherson Front Axle
4) The new MEB
5) Axle Testing and
6) Summary
22

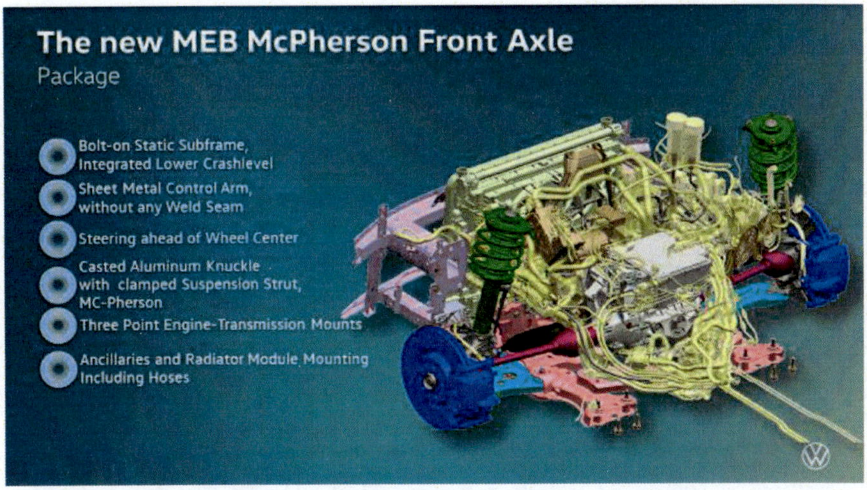
The new MEB McPherson Front Axle
Package
Bolt-on Static Subframe, Integrated Lower Crashlevel
Sheet Metal Control Arm, without any Weld Seam
Steering ahead of Wheel Center
Casted Aluminum Knuckle with clamped Suspension Strut, MC-Pherson
Three Point Engine-Transmission Mounts
Ancillaries and Radiator Module Mounting Including Hoses

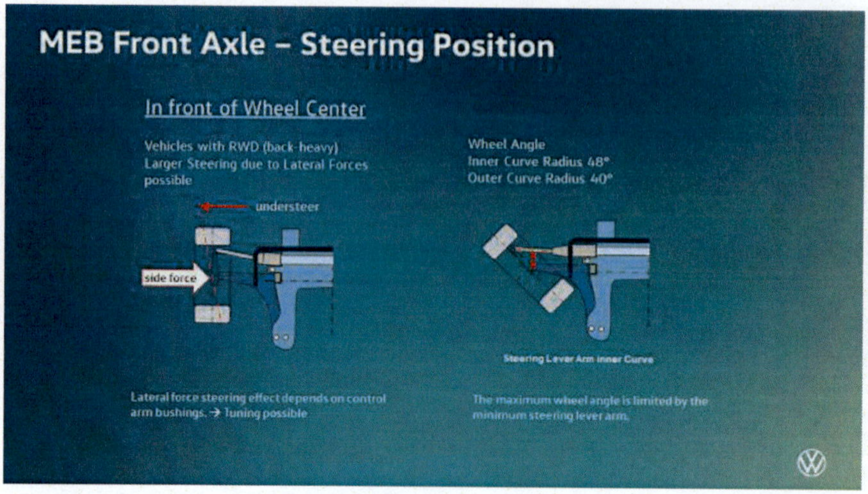
MEB Front Axle – Steering Position
In front of Wheel Center
Vehicles with RWD (back-heavy)
Larger Steering due to Lateral Forces possible
understeer
side force
Lateral force steering effect depends on control arm bushings. → Tuning possible
Wheel Angle
Inner Curve Radius 48°
Outer Curve Radius 40°
Steering Lever Arm inner Curve
The maximum wheel angle is limited by the minimum steering lever arm.

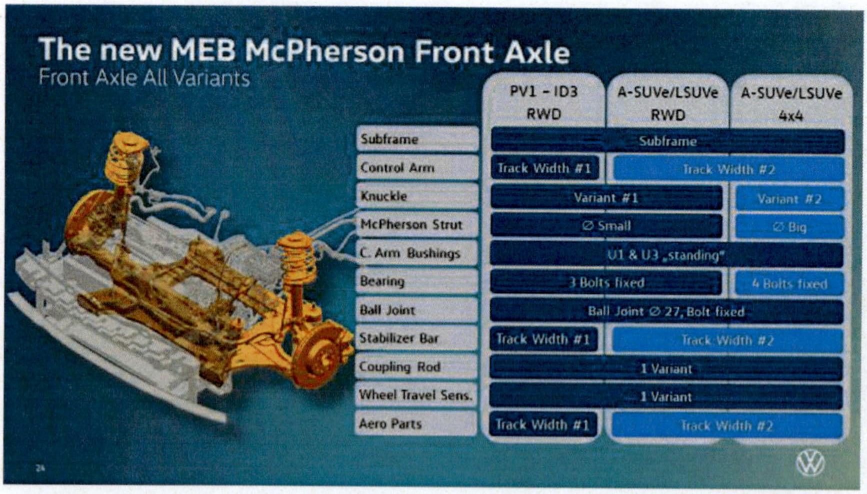
The new MEB McPherson Front Axle
Front Axle All Variants
PV1 – ID3 RWD
A-SUVe/LSUVe RWD
A-SUVe/LSUVe 4x4
Subframe
Subframe
Control Arm
Track Width #1
Track Width #2
Knuckle
Variant #1
Variant #2
McPherson Strut
∅ Small
∅ Big
C. Arm Bushings
U1 & U3 „standing"
Bearing
3 Bolts fixed
4 Bolts fixed
Ball Joint
Ball Joint ∅ 27, Bolt fixed
Stabilizer Bar
Track Width #1
Track Width #2
Coupling Rod
1 Variant
Wheel Travel Sens.
1 Variant
Aero Parts
Track Width #1
Track Width #2
24

The new MEB McPherson Front Axle

MEB31 Front Axle Variants

RWD Track Width #1 and #2 — 4x4 Track Width #2

25

Agenda

1) The new MEB e-mobility Matrix of VW ID.3
2) Axles: Concept, Boundary Conditions, Simulation
3) The new MEB McPherson Front Axle
4) The new MEB Multilink Rear Axle
5) Axle Testing and Simulation of the MEB
6) Zusammenfasung

26

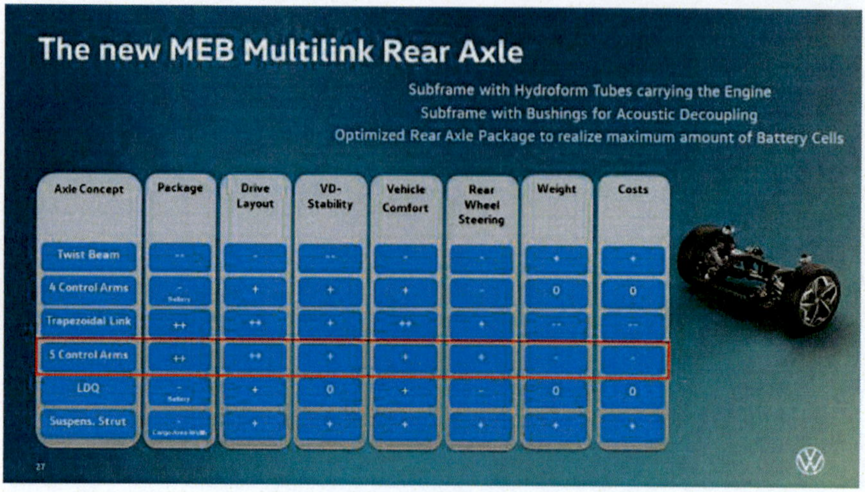

Axle Concept	Package	Drive Layout	VD-Stability	Vehicle Comfort	Rear Wheel Steering	Weight	Costs
Twist Beam	--	-	--	-	-	+	+
4 Control Arms	- Battery	+	+	+	-	0	0
Trapezoidal Link	++	++	+	++	+	--	--
5 Control Arms	++	++	+	+	+	-	-
LDQ	- Battery	+	0	+	-	0	0
Suspens. Strut	- Cargo Area Width	+	+	+	+	+	+

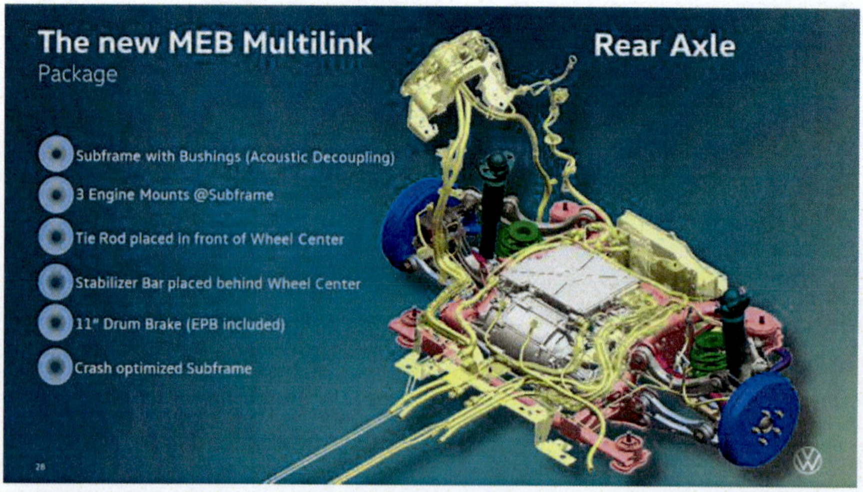

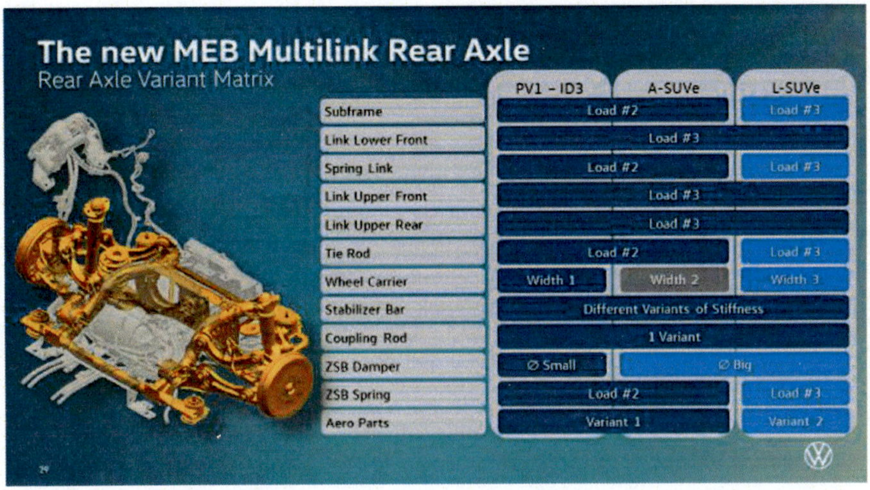
The new MEB Multilink Rear Axle
Rear Axle Variant Matrix
PV1 – ID3
A-SUVe
L-SUVe
Subframe
Load #2
Load #3
Link Lower Front
Load #3
Spring Link
Load #2
Load #3
Link Upper Front
Load #3
Link Upper Rear
Load #3
Tie Rod
Load #2
Load #3
Wheel Carrier
Width 1
Width 2
Width 3
Stabilizer Bar
Different Variants of Stiffness
Coupling Rod
1 Variant
ZSB Damper
Ø Small
Ø Big
ZSB Spring
Load #2
Load #3
Aero Parts
Variant 1
Variant 2
29

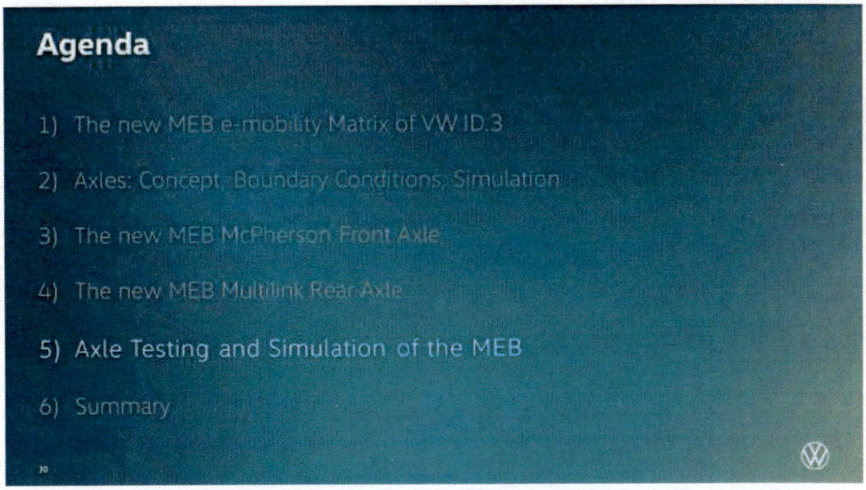
Agenda
1) The new MEB e-mobility Matrix of VW ID.3
2) Axles: Concept, Boundary Conditions, Simulation
3) The new MEB McPherson Front Axle
4) The new MEB Multilink Rear Axle
5) Axle Testing and Simulation of the MEB
6) Summary
30

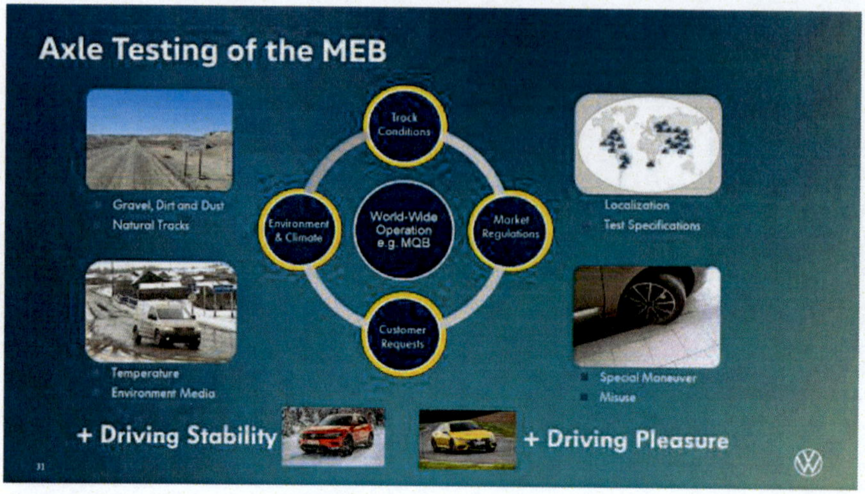

Axle Testing of the MEB
Gravel, Dirt and Dust
Natural Tracks
Temperature
Environment Media
Track Conditions
Environment & Climate
World-Wide Operation e.g. MQB
Market Regulations
Customer Requests
Localization
Test Specifications
Special Maneuver
Misuse
+ Driving Stability
+ Driving Pleasure
31

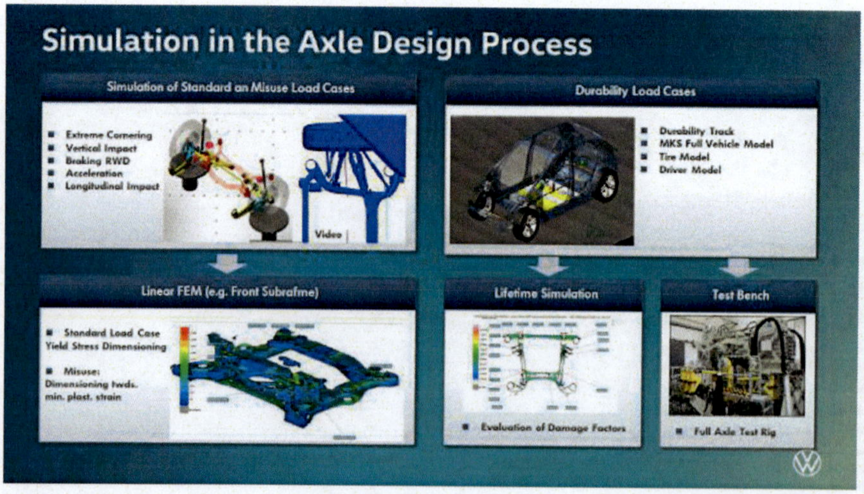

Simulation in the Axle Design Process
Simulation of Standard an Misuse Load Cases
Extreme Cornering
Vertical Impact
Braking RWD
Acceleration
Longitudinal Impact
Video
Durability Load Cases
Durability Track
MKS Full Vehicle Model
Tire Model
Driver Model
Linear FEM (e.g. Front Subfraime)
Standard Load Case
Yield Stress Dimensioning
Misuse:
Dimensioning twds. min. plast. strain
Lifetime Simulation
Evaluation of Damage Factors
Test Bench
Full Axle Test Rig

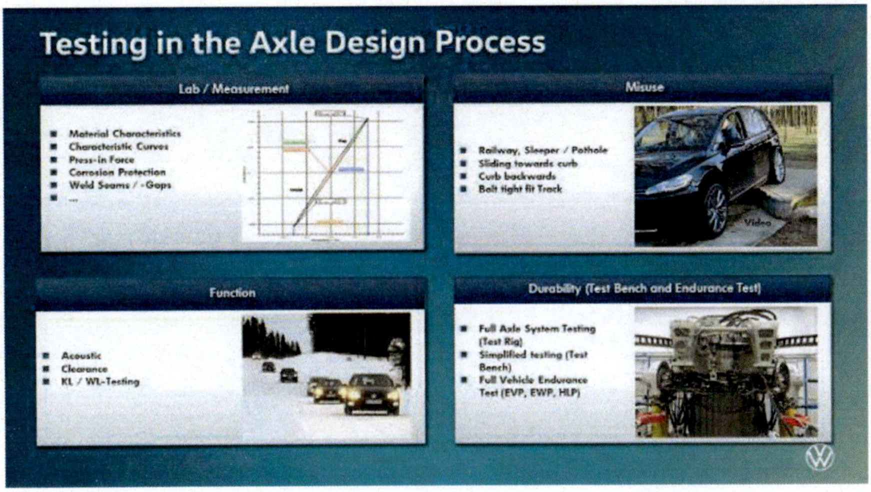
Testing in the Axle Design Process
Lab / Measurement
Material Characteristics
Characteristic Curves
Press-in Force
Corrosion Protection
Weld Seams / -Gaps
...
Misuse
Railway, Sleeper / Pothole
Sliding towards curb
Curb backwards
Bolt tight fit Track
Video
Function
Acoustic
Clearance
KL / WL-Testing
Durability (Test Bench and Endurance Test)
Full Axle System Testing (Test Rig)
Simplified testing (Test Bench)
Full Vehicle Endurance Test (EVP, EWP, HLP)

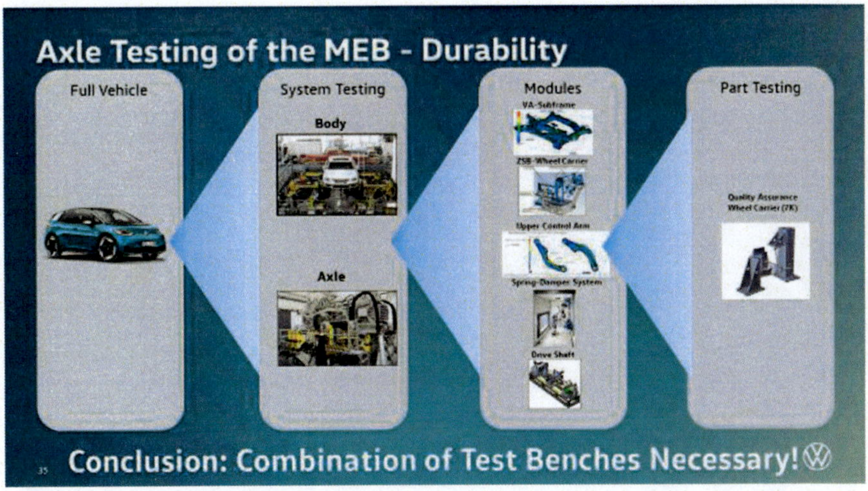
Axle Testing of the MEB - Durability
Full Vehicle
System Testing
Body
Axle
Modules
VA-Subframe
ZSB-Wheel Carrier
Upper Control Arm
Spring-Damper System
Drive Shaft
Part Testing
Quality Assurance Wheel Carrier (7K)
35
Conclusion: Combination of Test Benches Necessary!

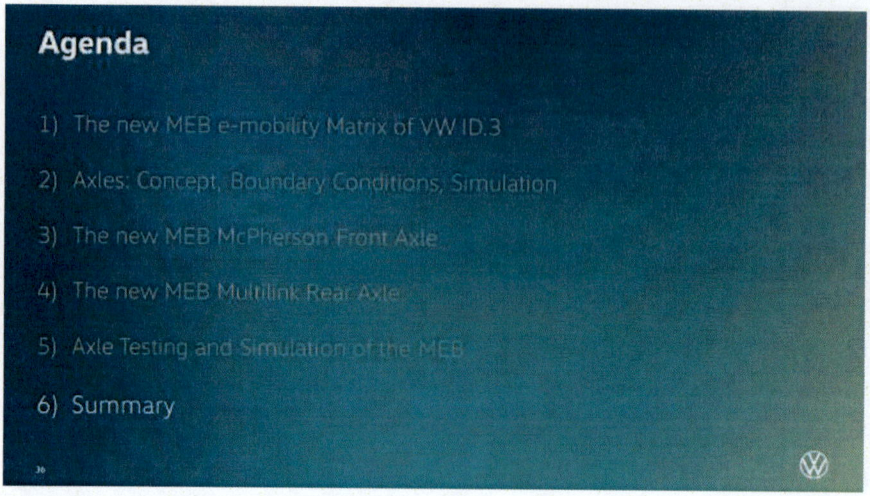
Agenda
1) The new MEB e-mobility Matrix of VW ID.3
2) Axles: Concept, Boundary Conditions, Simulation
3) The new MEB McPherson Front Axle
4) The new MEB Multilink Rear Axle
5) Axle Testing and Simulation of the MEB
6) Summary
36

Summary Chassis ID.3
McPherson Front Axle
Small Turning Circle
Full integrated 3rd Load-Level for Front Crash
Electromechanical Steering
Exchangeable Package for optional engine (4x4)
Mulitlink Rear Axle
Subframe with hydro formed Tubes carrying the Engine
Subframe with Bushings for Acoustic Decoupling
Optimized Rear Axle Package to realize
maximum amount of Battery Cells
37

e-Mobility
to make all customers happy
Thank you.

Without A Chassis We Will Go Nowhere

Yousuke SEKINO

SHOWA Corporation, Japan

The Desire to Move...

All great Empires in history were looking for solutions to achieve expansion. To give you some examples: The Romans built roads, the Ottomans used light-weight chain armor and the Mongolians used a dynamic cavalry.

What did these solutions have in common? They supported "speedy and secure movement" of personnel.

Basically, humans always had the urge to change their location and therefore have kept moving throughout history.

What has changed drastically over the centuries, are the distances and required times. Let me give you an example: two and a half centuries ago, on June 6, 1763, Wolfgang Amadeus Mozart and his family left Salzburg to travel to Paris. In their horse-drawn carriage, it took them two days to arrive here in Munich. And it was torture. Nowadays, this trip takes one and a half hours in the comfort of your car.

...Will Always Be in Us

People's need or desire for changing location, which means "moving", will never end. No matter how widespread Virtual Reality will ever be.

And the Covid-19 crisis has brought back to our minds how important "moving" is for us, and how much we miss it, when we cannot do it.

This means that also vehicles which enable such "movement" will remain and evolve, and so will their chassis technology.

You and I – who are dealing with important components of vehicles – we are responsible to "turn the necessity of moving into the pleasure of moving", as far as possible.

Especially with the future vision of autonomous driving.

Showa History

Let me briefly introduce Showa as one of the companies that strives to further improve the pleasure of driving:

Showa started in 1938 as a company that developed and built landing gear for aircraft. After the War, we started to develop and produce automotive components.

P. E. Pfeffer (ed.), *11th International Munich Chassis Symposium 2020*, Proceedings, https://doi.org/10.1007/978-3-662-63193-5_2

Up to now we have grown to about 13,000 people in 39 facilities across 12 countries worldwide.
We develop and manufacture products for cars, motor bikes and even marine applications.
The amount of customers, or people who benefitted from our products, is difficult to count, because one of our products used to be the pantograph of Shinkansen. We estimate hundreds of millions.

Product Highlights

Next, I would like to boast a bit with some of the highlights in our history with automotive components:
We were the first one to launch EPS on a production car – the 1991 Honda NSX
On the Honda S2000 we launched the first variable steering system VGS. The steering ratio varied over steering angle AND vehicle speed!
The Showa Progressive Valve 'SPV' enabled a linear damping force. It initially contributed to huge success in racing and was later transferred into mass production

From Now into The Future

Of course, the target of each new development we, or our competitors make, is always to improve performance and function in all regards and aspects.
We have now electronically controlled shock absorbers in production, which don't require additional sensors – under the trade name IECAS.
For EPS we are strong in DPA and BRA.
Some of the technologies we are now working on for the future include:
New technologies for ADAS
Coordinated control between IECAS and EPS…
And also, an elaborated Steer-by-wire system – for that we would have had a simulator on our booth in the foyer

At the beginning of Disruption to Our Industry

Talking about the future leads me to an important point.
For the past few years, we have witnessed changes in our industry.
chassis.tech_plus has been one of the forums in which these changes have been discussed from the beginnings. People here recognised early the significance of these changes and the disruption they represent to our industry.
We have to open totally new departments full of engineers with skills that 15 years ago nobody would have linked to the automotive industry.

Changes Are Happening

Some of the traditional departments have to reduce their manpower or might even not be needed at all in the future.
The new market requirements and technologies are forcing us to think in new ways. We also have to deal with the increasing speed of change and development – they

require new management structures and methods, as well as a new approach to budget allocation.
I have a feeling that even we in this forum have not yet fully recognised and thought through all the consequences of the upcoming change of our situation. Nobody has. It is probably not even possible currently.

Consequences

For companies that due to their size cannot allocate the anticipated budgets and manpower to new technologies, a consequence might be that they will have to close their doors.
The same will happen to companies that are big enough, however not quick enough to adapt to the new situation.
In the case of Showa, the first point potentially applied.
We had to look at our options and produce a plan that will allow our company to adapt and stay sustainably in business.

How Showa Copes with This New Era

The plan is to team up with other Honda affiliates and Hitachi Automotive Systems.
We are in the process of merging three Honda affiliate companies with Hitachi Automotive Systems into one new company.
The SHOWA brand will be kept also under the new roof.
The merger is not just the simple addition of the four.
Rather, the strong points of each company will be amplified, and weaknesses will be eliminated by maximising the synergies. The result will be a sustainable big player in the development of advanced chassis technology, as well as other automotive areas.

愛 – Respectful Enthusiasm

Some people are worried that chassis technology will lose importance in a future with autonomous vehicles.
But as I said earlier, I have no doubt that engineers will keep on improving chassis technology to evolve along with vehicles.
However, for chassis technology to stay relevant, we have to take an active approach.
From my personal experience as leader of large car development project teams, leaders in R&D must express "愛". Let me interpret this with "Respectful Enthusiasm".
"愛" for team members and colleagues,
"愛" for product and technology and
"愛" for customers – I think the order depends on your daily mood…

Future Relevance of Chassis Technology Depends on Us

On the other hand, engineers sometimes must be ready to deal with "almost impossible" requests from their superiors or from other departments.
The combination of "愛" and readiness to deal with demanding requests will make sure that chassis technology will keep up with the pace of car development in general and remain relevant also for autonomous vehicles. It is in our hands.
Every vehicle that is in contact with the ground, or at least touches it every now and then, needs a chassis.
Even if air taxis such as Volocopter replace cars in the future.

The History of Improvement Also Has A Future

From the beginning of cars, the majority of drivers were men. Male physiognomy on average is taller, heavier and stronger than female. Therefore, seating positions, view angles, size of steering wheels, operating forces, and many more features were optimised for men. For decades, cars were designed with an inherent disadvantage for female drivers.
Ergonomics, driver support systems, operational forces, etc. have reached levels at which the physiognomy of any adult driving person is irrelevant. We have come so far, that also drivers with special needs can now participate in traffic with few limitations.
What the industry did in the past, will continue into the future and cars will become even more performing, comfortable and safe.

Conclusion and Outlook

The operation of vehicles will be further optimized for all users and become largely autonomous.
Automotive products of the future – which are the results of our work in response to the market needs – will be very different and very much better.
For the benefit of the acceptance of autonomous driving, the focus should not be taken off chassis technology. For a bright future of the automotive industry – now it is our time to shine.
We should keep in mind:
Without a chassis you have no grip.
You'll even find that you have lost touch…

The All-New YARIS

Toyota New Global Architecture (TNGA)

Masato Fujiyama, T. Kondo, T. Kaneko[1]

[1] Toyota Motor Cooperation, Japan

1 Introduction

P. E. Pfeffer (ed.), *11th International Munich Chassis Symposium 2020*, Proceedings, https://doi.org/10.1007/978-3-662-63193-5_3

Toyota New Global Architeture (TNGA) is the latest company-wide program to design "ever better cars" by enhancing the vehicle structure and the development process.

The principle aim of TNGA is to enable the development and production of more attractive vehicles by relying on essential improvements of both basic performances and product value. To achieve this, a virtuous cycle of creating ever better cars is set up by sharing high quality components strategically across the platform, leading to cost reduction and thus more appealing products.

TNGA genealogy is illustrated in Figure 1. The GA-C TNGA platform, dedicated to mid-size vehicles, was the first to be rolled-out with the launch of the New Prius in 2015. Later on, the GA-K platform for large size vehicles and the GA-L platform for luxury vehicles were developed. Nowadays, many vehicles have already been designed based on these platforms. This roll-out is expanding with continuous improvements. Indeed, the gained knowledge from existing vehicles is properly feedbacked to next models. As a result, TNGA development has been sequentially refined and the GA-B platform is the forth variant of TNGA platform. The latter platform is dedicated to compact size vehicles with the first application on the New Yaris.

Fig.1. TNGA Genealogy

GA-B platform is equipped with brand new style & vehicle packaging, body, chassis and powertrain (Fig.2). These key features were thoroughly designed to increase their functions and performances. In addition to these individual developments, an integrated approach was also adopted to optimize the entire vehicle performances and to realize "ever better car".

Fig.2. Key features of GA-B platform

Concerning packaging, "Small" is an invariant characteristic of utmost importance. While respecting this constraint, "Styling flexibility" and "high dynamic performance" were targeted. To meet these objectives, driving position, powertrain layout, lengthen wheelbase and shorten overhang were considered from the initial planning phase in order to achieve of low hood height, low center of gravity and small moment of inertia. The size of each part, its impact on gravity center height and inertia, as well as its risk to implement were deeply investigated. With such efforts, the final package is very closed to the initial ideal one (Fig.3).

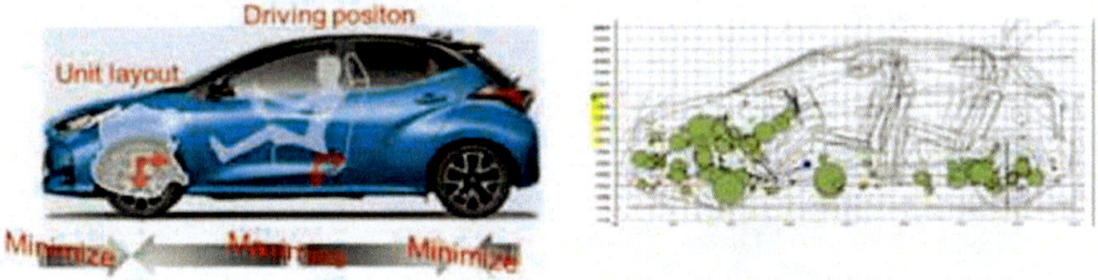

Fig.3. Redesigned vehicle package and parts management

2 Versatility of the new GA-B platform

When developing the GA-B platform, a lot of discussions was needed to clarify the targeted vehicle group and features to be included in the integrated planning. For instance, the vehicle size and mass, the powertrain power, tire size as well as the new

4WD system were discussed (Fig.4). Application range was enlarged compared to previous models and thus the complexity of interconnections between various systems. On the other hand, compact segment vehicles are expected to be lightweight (fuel consumption and environmental impact) and affordable. These are key items for this segment and to be consistent with them, structure and composition were deeply considered.

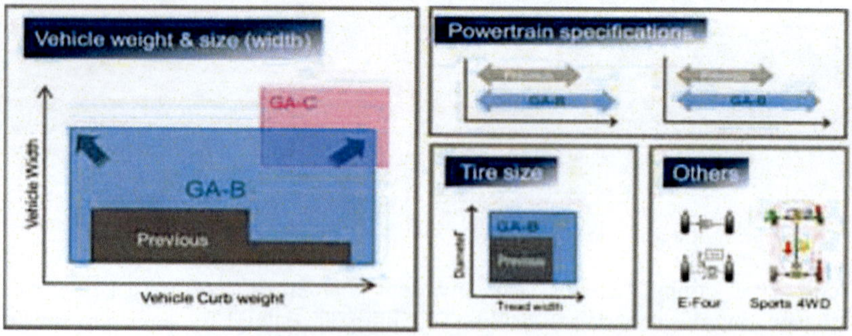

Fig.4. Example of GA-B application range

To be applicable on wider application range and vehicle group, GA-B platform is equipped with variable wheelbase, track and height (Fig.5). The wheelbase is adjusted by the body structure, the track by the chassis structure and height by both, the body and the chassis structures.

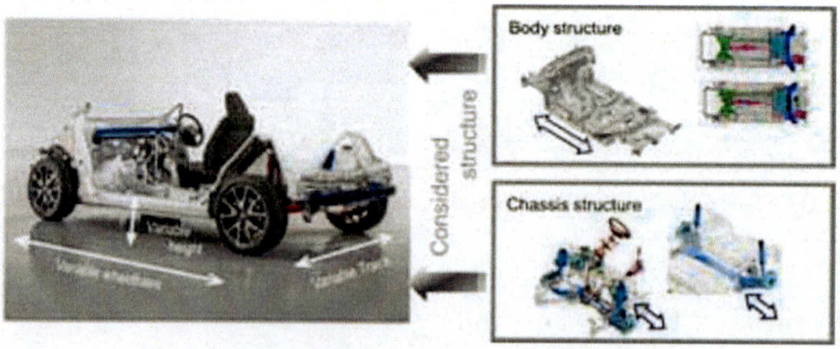

Fig.5. Variable dimensions

For major component layout, especially engine and driver operation area, it was decided based on commonization purposes in order to be applicable to all vehicles of the platform. Main components determining vehicle basic performances were designed following the TNGA thinking way. i.e. "Develop competitive components and share strategically". Only decoration parts that are visible and can be touched by our customers are individualized for each model to create a distinctive and attractive vehicle.

Besides these commonization, the GA-B platform was designed to exchange modules and to accommodate already existing units. For instance, the rear suspension was developed while considering the possibility of installing e-4WD unit. As a result, e-4WD unit can be installed within the same body structure. Moreover, the body structure allows exchanging the rear section with another platform in order to obtain specialized model, like 4WD-sport.

These combinations generate a wide range of characteristics (performance, axle load, space, etc.) that can be efficiently implemented by using existing units and modules (Fig.6).

Fig.6. Commonization & Accommodation of GA-B

3 Common principles of TNGA chassis

Performance concept of TNGA chassis is given in Figure 7. The so-called "Confident and Natural" concept of refined vehicle development is used for vehicle performance. At the bottom of the pyramid lie fundamental chassis performances, which are defined by stability and body control. Once these aspects are fixed, performance aspects that are directly related to the perceived quality, such as ride comfort, steering feel and braking feel are developed.

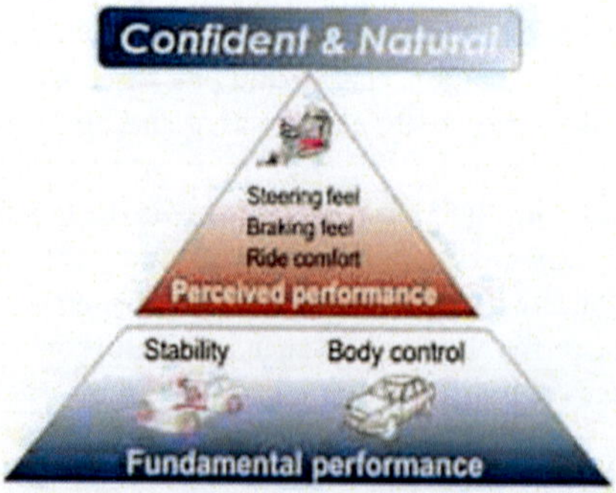

Fig.7. Performance concept of TNGA Chassis

To apply this performance concept on TNGA platform, the development process illustrated in Figure 8 is used. The left side shows the technology development phase. The objectives of development and performances are harmonized through a cycle including phenomenon analysis, mechanism generalization and component development. The right side shows the vehicle development phase. Developed technology is fully integrated in the V-shape development process. Simulations and HILS are employed to design and evaluate vehicles, systems and components. Through the process, conflicting performances are balanced to achieve the target. Furthermore, technical concerns faced during the vehicle development, market feedback and latest benchmark information are reflected to the technology development. These continuous improvement cycles aim at reinforcing TNGA development and render it more efficient and more robust.

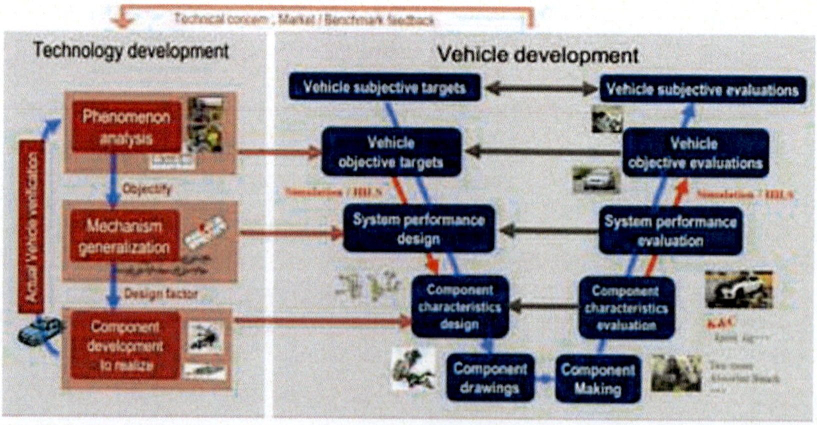

Fig.8. Development process

Figure 9 illustrates the crucial aspects and their characteristics for both fundamental and perceived performances, which are invariant independently of the considered platform. During GA-B platform development, each aspect was considered with latest knowledge and improvements that were reflected from other platform developments. Moreover, the latest technology developments were incorporated after a brash-up in order to match features of compact segment vehicle and/or system restrictions for GA-B.

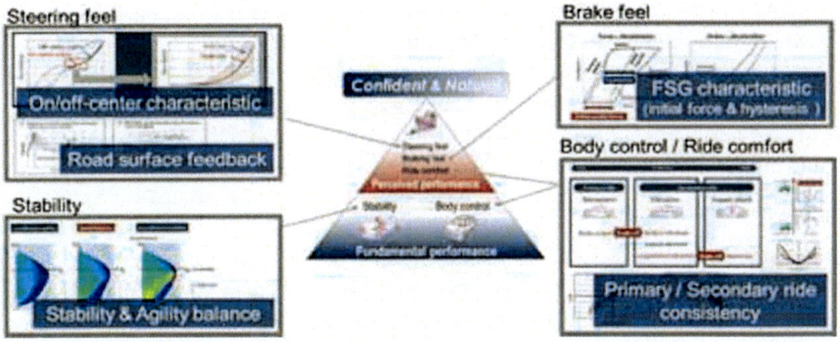

Fig.9. Key performances of TNGA chassis

4 Main aspects of GA-B development

Figure 10 compares compact car features with respect to higher vehicle segments. The track is narrower, the wheelbase is shorter, the vehicle weight is lighter and moments of inertia, especially yaw, is smaller and the center of gravity is relatively higher.

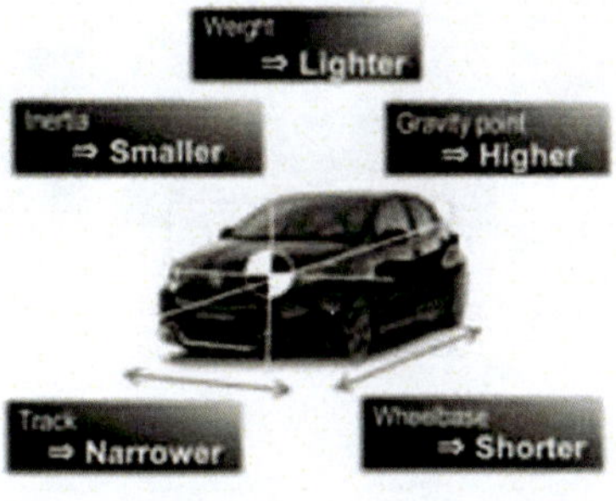

Fig.10. Features of compact size car for dynamic performances

These features are considered as advantages for dynamic performances since they promote agile and sharp vehicle maneuvers as well as steering effectiveness including emergency handling situation. However, due the lighter weight, the vehicle can be easily affected by driving disturbances such as road input, unexpected driver operation, side wind, etc. Therefore, to achieve confident and natural driving quality, the following two main aspects are carefully treated during development: the stability-handling balance and the ride comfort, arising from vertical vibrations.

To go further, stability involves for instance straight stability and rollover, even if the driver suddenly brakes or if the vehicle is hit by a strong lateral wind, the vehicle should not deviate from its lane. Also, the vehicle must not have a tendency to rollover when the driver quickly steers in case of an emergency situation. On the contrary, for handling purpose, an agile and linear response matching driver's intentions is expected. Even in case of an emergency situation, steering must be effective and the vehicle must be able to avoid an obstacle with confidence. It is very important to balance these performances at a higher level.

Concerning ride comfort, since the vehicle sprung mass is lighter compared to higher vehicle segments, damping of high frequency vibration tends to get worse despite a fine turning of coil spring rate and shock absorber damping. To prevent this tendency, suspension friction, spring rate at small inputs as well as the unsprung mass have to be properly reduced. As a consequence, body control performance (roll speed & roll damping) under steering input must be adjusted accordingly. In addition, a damping force at idling speed needs to be secured (Fig. 11).

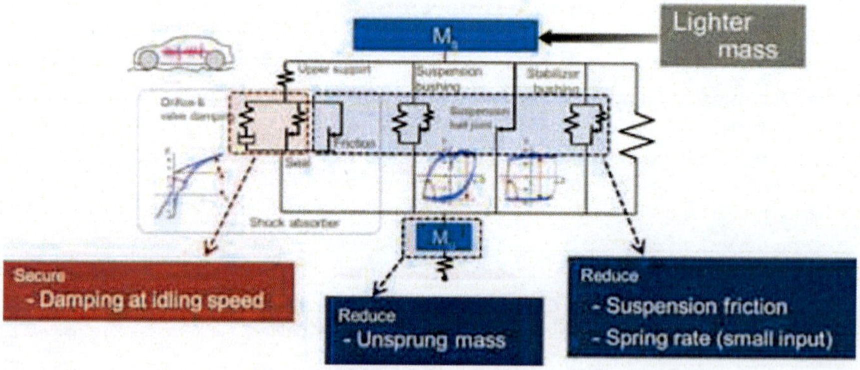

Fig.11. Improvements of ride comfort (Vertical vibration)

5 Approaches of GA-B development

Key activities of GA-B development related to vehicle development process are shown in Figure 12. As already mentioned, platform, powertrain, etc. are renewed, simultaneous studies of performance prediction and vehicle packaging are possible during the initial planning phase. Therefore, essential improvements such as vehicle specification, system geometry and so on are possible by refining part layout and rebalancing performances.

The use of simulations is also expanded. A closed loop simulation involving chassis control was developed. At the design stage, it is used to predict performances and set the targets for sub-system design. At the evaluation stage, it is employed to confirm exhaustively performances of each vehicle variation. These efforts contributed to shorten the lead time and reduce development costs, which enables to secure an appropriate period of time for vehicle tuning in order to reach the ideal driving quality. Concrete examples are hereafter described.

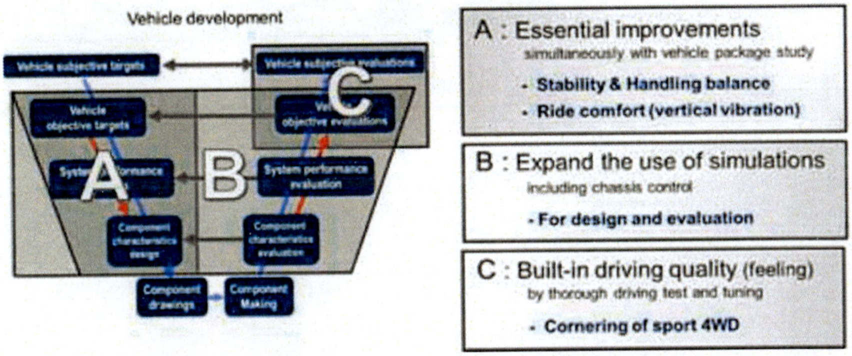

Fig.12. Key activities of GA-B development

A. Essential improvements

- Stability & Handling balance

When investigating thoroughly the stability and handling balance, a fundamental question is to what extend we can take advantage of tire characteristics. For example, in order to ensure braking stability, if the vehicle does not have enough potential, we may have to sacrifice tire cornering power at high load to secure safety performances. However, the steering effectiveness is thus getting worse, which is not an optimal balance.

During GA-B platform development, every part layout such as powertrain, brake unit, battery, air control unit, ECU's, wire harness and up to spare tire was deeply studied. Basic specifications such as lateral & longitudinal vehicle weight difference (gravity offset) and gravity center height were improved. At the same time, all chassis characteristics such as suspension geometry (which had to be consistent with the vehicle packaging), steering stiffness, load transfer, brake force distribution were re-engineered. Thanks to this approach, appropriate tire characteristics were ensured (Fig.13).

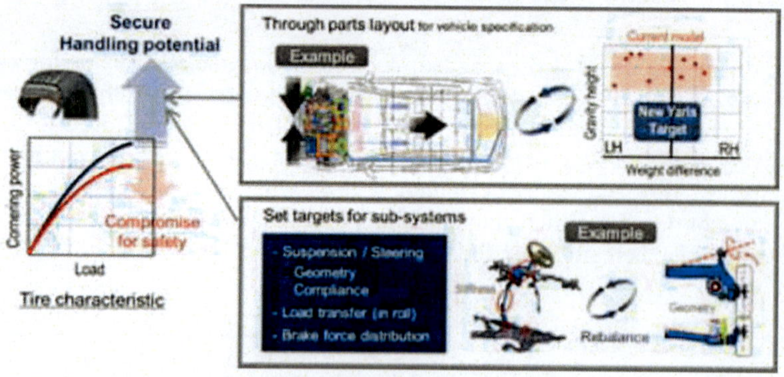

Fig.13. Improvement of stability & handling balance

Figure 14 shows G-G diagrams resulting from simulations. Understeer & oversteer behaviors are visualized in colors. The horizontal and vertical axes represent cornering G (lateral acceleration) and longitudinal G (acceleration and deceleration) respectively. One can observe that the maximum cornering G, steering effectiveness at high cornering G and braking stability are improved.

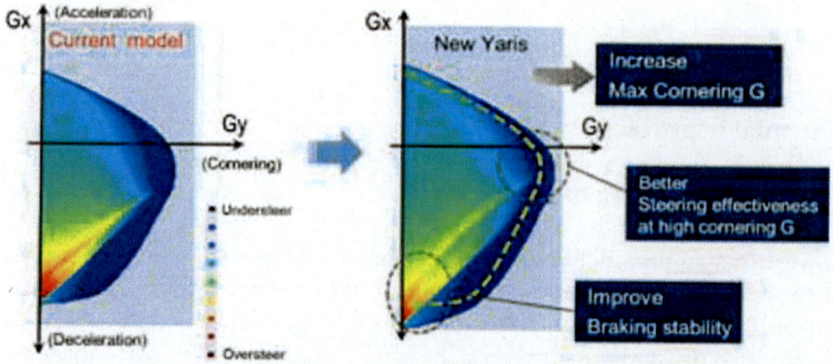

Fig.14. G-G diagram

- **Ride comfort (Vertical vibration)**

Figure 15 illustrates the improvements related to ride comfort, i.e. vertical vibration. To reduce suspension friction and spring rate at small input, the strut inclination and the angle of the coil spring reaction force were increased to minimize the lateral force acting on shock absorber. Many difficulties were faced on realized these changes, for instance keeping the packaging consistency with the engine and brake boosters but also guaranteeing collision performances, by conducting simultaneous investigations from the initial planning, ideal angles were achieved. In addition, bonded stabilizer bushes were adopted, the angle of the trailing bush was optimized, and toe-correct bushes were implemented in order to further improve ride comforts.

To reduce unsprung mass, high strength materials were adopted and new production technologies were simultaneously developed to maximize the benefits of these materials. These efforts contributed to reduce the overall vehicle mass.

In order to secure damping force at idling speed, sliding elements of the shock absorber were focused on and guided the developments to achieve the desired characteristics. As a result, vertical vibrations are better controlled without negatively impacting body control.

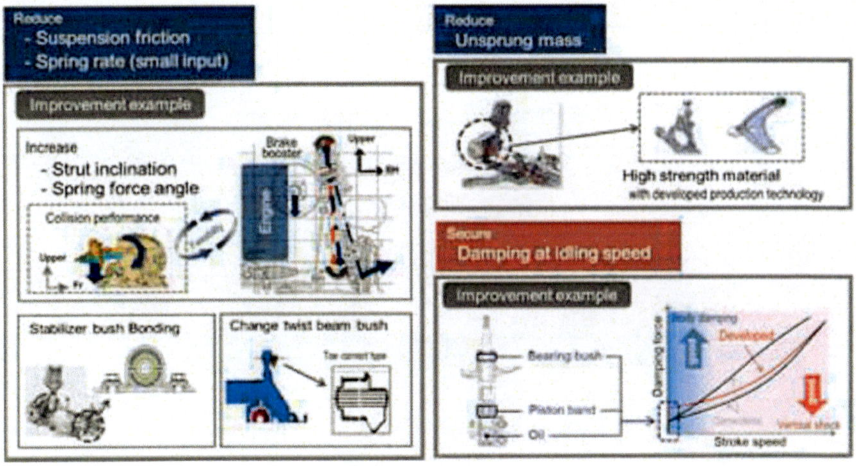

Fig.15. Improvement of ride comfort

On figure 16, one can observe vehicle measurement results regarding suspension friction (front suspension) and vehicle vertical vibration. Compared with the current model, friction is significantly reduced as well as the spring rate at small inputs. Consequently, the vehicle vertical vibration is improved as expected.

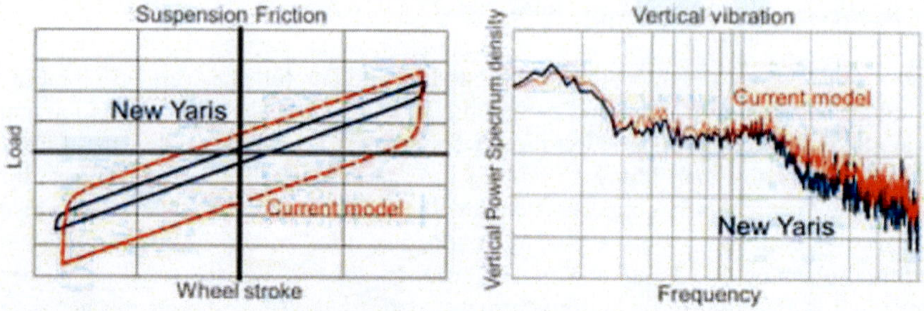

Fig.16. Vehicle measurement result

B. Expand the use of simulations

The use of simulations is pro-actively expanded into both design and evaluation stages for built-in quality and shortened lead time (Fig.17). In the design stage, accurate performance prediction and target setting for sub-systems are performed. In the evaluation stage, the use of simulations enables to evaluate many vehicle variations and to confirm the design virtually.

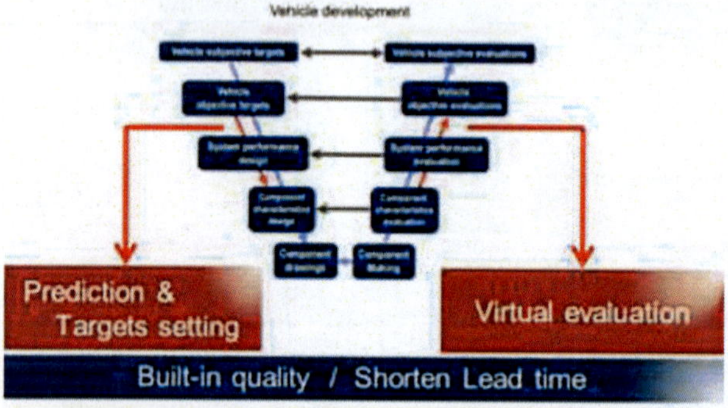

Fig.17. Simulation usage in vehicle development

Of course, simulations were already employed for vehicle development but it was mainly applied to the chassis structure without control. For chassis control, the usage was to rely on vehicle tunings during the evaluation stage. However, with the new technological developments of closed-loop simulations for chassis control, the simulation environment is now mature enough and simulations can be used to design the full vehicle including chassis control at once.

An overview of the simulation process is given in Figure 18. It consists of vehicle model, maneuver mode and controller module embedding individual control models. Database reinforcement and individual model refinement were carefully conducted. Simulation enabled to evaluate exhaustively all vehicle variations (Fig.19).

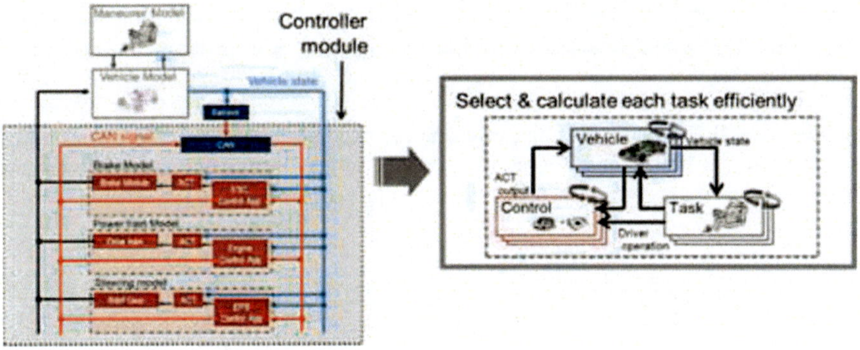

Fig.18. Overview of the simulation process

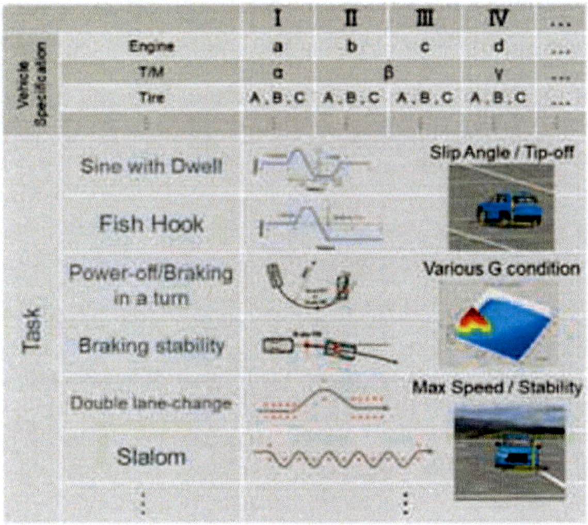

Fig.19. Vehicle variation and task matrix for virtual evaluation

The following example illustrates a concrete design case. For models with a relatively higher gravity center (e.g. SUV models), it is more difficult to achieve roll-over and handling performances. Moreover, it is usually not possible to meet the targets by only improving vehicle basic specifications and chassis structure. If chassis control items are not considered during the performance prediction, too many adjustments are required in the evaluation stage. In some cases, due to schedule limitations, the target setting of some control items was compromised. To avoid these issues, simulation was intensively used in the design stage to confirm the potential of each target setting and to realize improvements.

To achieve rollover and handling performances of GA-B platform, high boost brake actuator and new brake control logic were developed (Fig.20). In addition, clear targets for tire characteristics were set in early design stage (Fig.21).

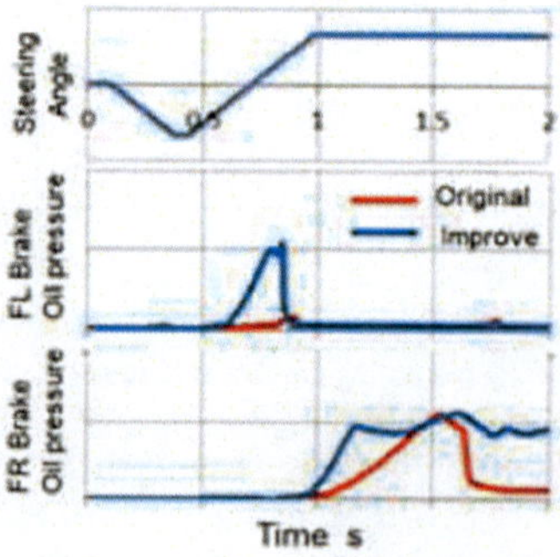

Fig.20. New brake control logic

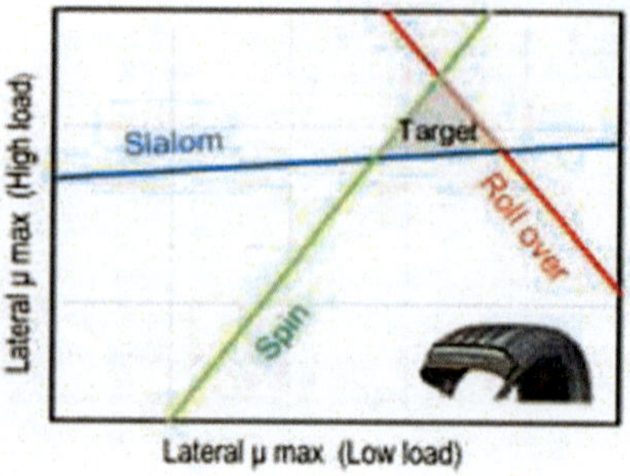

Fig.21. Target of tire characteristic

C. Built-in driving quality (feeling)

Performance targets were achieved by realizing essential improvements on vehicle and components, and by using simulations for vehicle development. Nonetheless, to reach the performance concept ("confident and natural"), energy must be invested into built-in driving quality. It aims at maximizing vehicle potential by iterating, in an effective manner, between driving test and tuning of component parameters. This approach is exemplified by a case of the GR Yaris.

The GR Yaris is a sport 4WD vehicle and it has various tuning factors for vehicle dynamics (Fig. 22). Therefore, by taking advantage of all these factors, excellent sport driving feeling and accurate controllability can be expected even in the grip limit areas. By using simulations, these factors were tuned and high performances of the vehicle were confirmed. However, driving feeling and controllability should be adjusted with a driver-oriented approach.

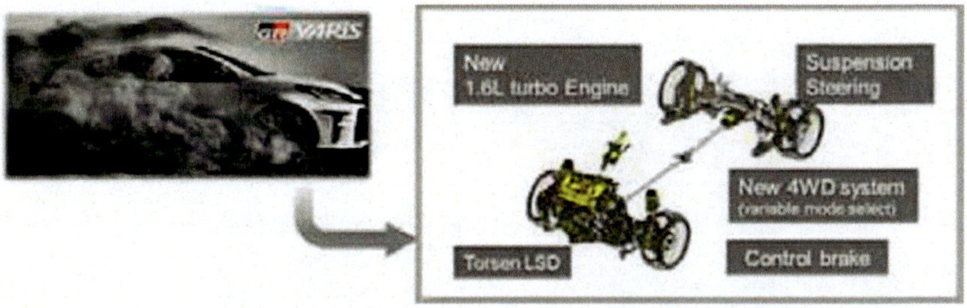

Fig.22. Various turning factors

To confirm the controllability in the grip limit areas, driving skills to accurately drive under these conditions are required. Therefore, a tight collaboration was established with skilled drivers who conducted the driving tests under various situations (including circuit, gravel, dirt, snow, etc).

During the driving tests, driver operations and vehicle data were acquired simultaneously (Fig. 23). Then, by consistently analyzing them and linking them with simulation results, tuning parameters were further refined.

Fig.23. Driver operation & vehicle data

In Figure 24, a G-G diagram that is used for vehicle turning is illustrated. The transition phases of vehicle behavior such as braking to turning or turning to acceleration are complex zones where lie tuning factors between engine, drivetrain, chassis structure and brake control. To achieve driving feel and accurate vehicle controllability, detailed studies were carried out based on measurement data and drivers' feedback, and results were used to tune the parameters. By repeating these steps, the built-in driving quality with a driver (human)-centric approach was enhanced.

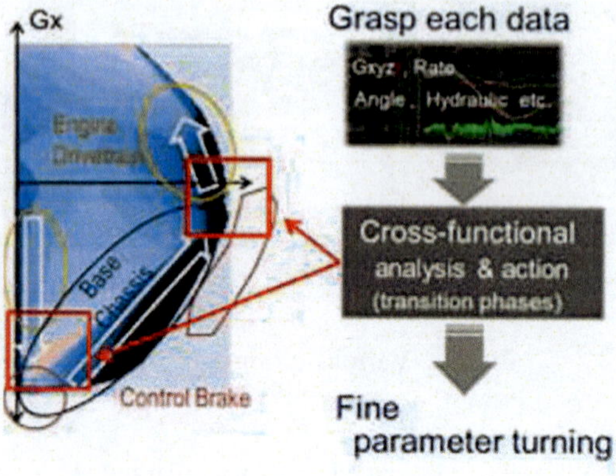

Fig.24. Cross-functional vehicle tuning

6 Conclusions

The GA-B platform is the fourth of the TNGA series and is dedicated to compact size vehicles. Its first application is on the all-new Yaris and its series. It has been developed with the aim of a strategic share of components, flexible dimensions and accommodation of various vehicle types.

Chassis performances were developed based on the performance concept of TNGA chassis. Furthermore, in order to maximize the advantages of a compact car, essential improvements were carried out from the initial planning phase. Also, in the development process, efficient designs and evaluations were ensured by expanding the use of simulations. As a result, target performances were reached and built-in driving quality was enhanced.

By combining all efforts, the production cycle of creating "ever better cars" was carried out more efficiently and more thoroughly, leading to an overall considerable progress.

7 Acknowledgement

The author would like to express his sincere gratitude to all members who contributed to this development for their invaluable teamwork and assistance.

Automated Driving Requires International Regulations

Benjamin Koller and Robert Matawa

TÜV SÜD Auto Service GmbH, Germany

Abstract. The rapid launch of automated vehicles on global markets requires uniform global regulations. To date, however, the regulations have varied greatly from region to region, and the development of uniform, internationally binding guidelines has only just begun. The first UNECE regulation for an automated driving system at SAE level 3 is expected in 2020—a pioneering achievement and a template for further internationally binding regulations.

What is the status of global regulations for automated driving? Who has done what? What does a detailed look at the first recommendation for the UNECE regulation for ALKS reveal? What are the key issues? Which requirements are defined for manufacturers? How is testing done?

This article takes the UNECE regulation for ALKS as a basis to examine the development and creation of uniform legal foundations for international partners. It also takes a close look at efforts to approve automated driving functions in the various automotive markets of Europe, the USA, and Asia.

The analysis features a detailed description of the upcoming, new UNECE regulation for ALKS in relation to different subject areas. Focus is placed on the system's functionality and requirements for type approval, which form the context for an examination of cybersecurity, software update processes, data storage systems (DSSAD), functional safety, SOTIF, and simulations. The article also explores the issue of physical testing in connection with the type-approval procedure for the automated driving function ALKS.

P. E. Pfeffer (ed.), *11th International Munich Chassis Symposium 2020*, Proceedings, https://doi.org/10.1007/978-3-662-63193-5_4

A look at the current state of developments

Mobility as we know it is facing disruptive change in industrialized countries. The key word in tomorrow's mobility is: intermodal mobility. From A to B on the roads, the water, the rails, and through the air—algorithms will play a decisive role. Vehicles will no longer be driven exclusively by people; they will drive autonomously as well. And against the backdrop of the current COVID-19 crisis and social distancing, individual mobility will maintain its place alongside public transport and sharing models in an intermodal context.

Assisted driving functions with a high degree of automated functionality have ceased to be a vision of the future. Emergency braking and traffic jam assist systems as well as automatic distance control and lane-keeping assistance have been installed in many vehicles. And yet, the step toward automated driving—that is, without the driver continuously monitoring or delegating the responsibility for driving to the system—has many hurdles to overcome, especially with regards to regulations, because technically a great deal is now fully functional. This becomes clear with the leap to SAE level 3 [1], when driving responsibility is temporarily transferred to the automated system and the vehicle must be able master critical situations without the driver as a direct fallback solution [2].

As it stands, the requirements for the approval of automated driving functions have been regulated very differently worldwide. The requirements are frequently defined only in terms of special permits for automated driving on public roads. One important step on the path to global harmonization is the first UNECE regulation for approval of an automated lane-keeping system (ALKS), which is expected to be adopted in the first half of 2020. Its stipulations on monitoring, functional requirements, and tests will become the template for further regulations [3].

Forcing ahead with new, uniform global regulations for approval is the precondition for rapid market introduction of automated vehicles—initially level 3—and thus for joining the field of new mobility. Only if the safety of vehicles around the world is uniformly regulated will drivers feel confident in handing the computer the wheel. And here, of course, change also means looking at security issues from a completely new perspective. This particularly applies to passive and active safety. The focus automated for driving functions has turned to such key safety and security issues as cybersecurity, data storage systems, and software updates—for example, over-the-air updates [4].

For the rapid introduction of automated driving, product safety and OEM safety and security concepts also play an important role in approval procedures. Uniform regulations, such as those of the UNECE, offer clear guidelines and ensure that the approval process for vehicles runs smoothly in global markets.

In summary, it is safe to say that a few more years are needed before automated driving at level 4/5 is accepted.

Questions

What is the status of global regulations for automated driving? Who has done what? What does a detailed look at the first recommendation for the UNECE regulation for ALKS reveal? What are the key issues? Which requirements are defined for manufacturers, for example? How is testing done? A look at the status of the international regulations for automated driving.

Overview of current developments

Automated vehicles are being developed and tested around the world. However, the requirements for the vehicles as well as the knowledge of and approaches to traffic rules, certification regulations, and test standards vary from country to country. There are still no uniform or established testing methods and standards for the reliable validation and approval of automated driving functions. This presents a real hurdle for the testing and market launch of automated vehicles, despite OEMs having made considerable technological progress: since 2018 various manufacturers have advertised level-3 vehicles as individually approved vehicles undergoing testing or even their market launch. As early as this year, test vehicles will be launched with functionalities beyond SAE level 3 [1].

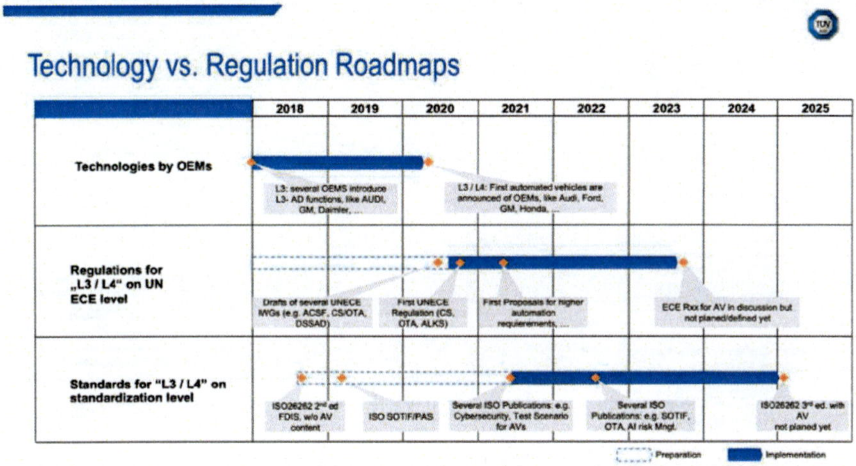

By comparison, a look at the status of the UNECE regulations reveals a clear deficit: the foundations for automated driving were laid and obstacles removed with the adaptation of the Vienna Convention on Road Traffic in 2016. The UNECE's World Forum for the harmonization of vehicle regulations created a basis with its framework document [4]; this spring saw various working groups present their first drafts for the future UNECE regulations. These were the results of international working groups on topics including Automated Commanded Steering Function (ACSF) or cybersecurity and software updates (over the air, CS/OTA). They have led to the first UNECE regulation ALKS mentioned earlier and two further UNECE regulations for cybersecurity and software update procedures; their adoption is expected in the first half of 2020.

This is also the case for standards. Standards such as ISO 24089 for software updates or ISO 21434 for cybersecurity are still in the detailing phase. The first publications are not expected until the end of 2020.

Despite developers and OEMs already driving innovation in these areas based on these standards and pending regulations, an overview of regulation development reveals that uniform and binding UNECE rules and standards for developing automated vehicles of the levels 3 and 4 can be expected in 2025 at the earliest.

Advancements around the world

Europe

The individual European member states have also taken different paths to implementing the amendments to the Vienna Convention. The approach of individual countries differs primarily in terms of practical implementation. France, the Netherlands, Austria, and Great Britain are leading the way, having implemented the Vienna model primarily in guidelines and regulations for testing automated vehicles (AV) on public roads. Their main focus is on safety, functional security, and cybersecurity.

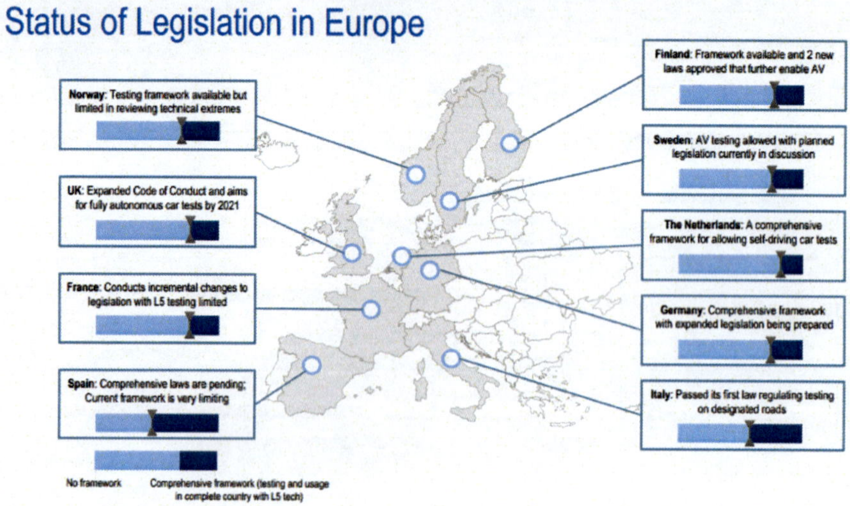

Although Germany reacted very quickly to Vienna by adopting the Vienna convention's provisions in national law by means of corresponding changes to the Road Traffic Act, technical requirements have not yet been defined. In various funded projects, such as Germany's successfully completed project PEGASUS [5] or its follow-up research consortium VVM (research association for verification/validation of autonomous systems) [6], experts from highly diverse areas of industry and supplier industries are developing legally secure and efficient verification and validation methods for automated driving functions.
Overall, other member states present a heterogeneous picture. Norway, for example, has already implemented the framework conditions for tests, while Spain is still in the early stages of developing specifications. Italy, on the other hand, has already passed its first law for tests on certain roads. Finland is also well advanced in this area, having stipulated the framework conditions for tests and passed the first laws that form the basis for later AV approval.

USA

The situation in the United States seems particularly confusing at first glance but follows a familiar pattern. Federalism is particularly pronounced in the United States. The states have considerable independence in matters of transport. To some extent, more variation in traffic regulations exists among individual states than has ever been the case in past decades among European countries. A striking example illustrates this: the various American states interpret a red traffic signal in very different ways. A red light can have the same significance as a stop sign in certain states; in others, it is comparable to Germany's green arrow sign, which allows drivers to turn right on red when there are no drivers with the right-of-way at an intersection. In still other states, a red light means the same as it does in Europe. For this reason, it is only logical that the rules for automated test-drives are created at the state level.

Production vehicles equipped for level 3 and above will have to be designed for all regions. However, it is not at all uncertain that UNECE rules and regulations will find their way into US legislation. This country has always followed its own course with regards to vehicle approval. The reverse is also true. Across the Atlantic, regulations like the Federal Motor Vehicle Safety Standards (FMVSS) of the Department of Transportation (DOT, or frequently USDOT) stipulate the requirements for vehicles and components. There is no homologation comparable to the procedure in countries within the UNECE region. Manufacturers confirm that their products comply with the DOT rules, for example, by printing this on a component. A check by technical service providers or the like is not required.

States on the West Coast, such as California and Nevada, began to create legal frameworks for automated driving and testing over ten years ago. The Department of Transportation has stated its intention to create federal rules.

Asia

Asia is more advanced in developing regulations, standards, and tests than the rest of the world. China is rapidly adapting its legislation, albeit with major differences in the various regions. South Korea and Singapore are at the forefront of development and can boast the first standards and regulations concerning testing and approval in the field of automated driving. One example of this is the four-part TR68 standard from Singapore, developed in connection with activities at CETRAN (Center of Excellence for Testing and Research of AVs) at Nanyang Technology University in Singapore[7]. By contrast, Japan, one of the most significant OEM locations, has only adapted their laws for testing, and efforts are being made to include the first automation functions, such as ALKS, in the national approval process.

Many of the world's national platforms and initiatives that strive for standardization and uniform requirements, such as the creation of testing methods for automated and connected driving, have formed the International Alliance for Mobility Testing and Standardization (IAMTS) to develop practicable and harmonized standards around automated driving [8].

The work of the UNECE in automated and connected driving

For years, the UNECE has been involved in all regional and national efforts to get AV on the road by drafting uniform regulatory framework for global requirements to approve such vehicles. The reasons:

- Vehicles are global products whose components are produced throughout the world.
- Uniform specifications are a basic requirement for the rapid introduction of AV and its approval in core markets.
- Uniform global regulations engender trust in consumers.

The first steps toward an internationally uniform regulation were the corresponding adjustments to the Geneva and Vienna Conventions. The various contracting states and their relevant licensing authorities, such as partner organizations NHTSA; the European Commission and CATARC are collaborating on this, supported by international expert associations including CITA, OICA, and CLEPA in the definition of requirements and design.

The World Forum for the Harmonization of Vehicle Regulations (WP.29) is responsible for the development and formulation of a uniform set of rules for the approval of AV. The international agreements of 1958 and 1998 involve the participation of nearly all countries. The understanding that automated driving would increase overall safety and improve mobility has led to the definition of thirteen key issues for the global harmonization of regulations and requirements [4].

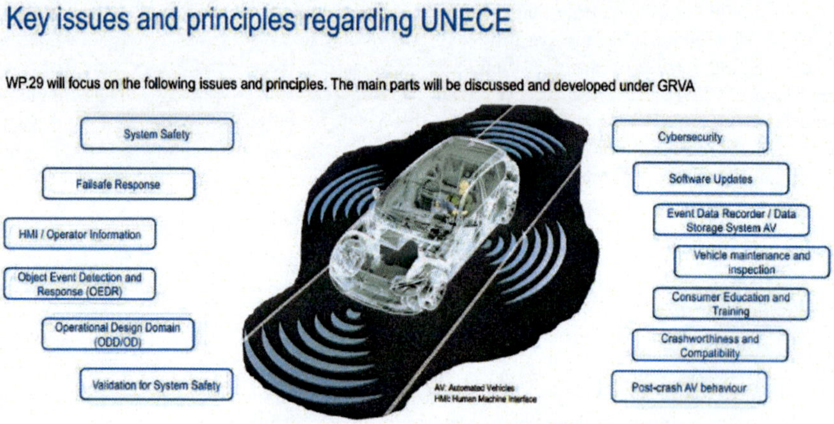

A total of five working groups, the Working Party on Automated/Autonomous and Connected Vehicles (GRVA), were designated to draft the relevant regulations.

1. Informal Working Group (IWG) on Automatically Commanded Steering Function (ACSF)
2. IWG on Functional Requirements for Automated and Autonomous Vehicles (FRAV)
3. IWG on Validation Method for Automated Driving (VMAD)
4. IWG on Data Storage System for Automated Driving / Event Data Recorder (DSSAD/EDR)
5. Task Force on Cybersecurity and Over-the-Air issues (CS/OTA)

Their focus is on the functional safety of automated driving functions as well as new validation methods and systems for storing driving and accident data. Other key issues are cybersecurity and software updates (including over the air).

Another participant in the development of uniform global regulations relating to autonomous driving is the Global Forum for Road Safety WP.1. National authorities are working closely with teams of experts from OEM, industry, technical monitoring, and other specialist areas on international requirements for AV's global approval.
Accordingly, in the spring of 2020 the aforementioned GRVA expert bodies were able to submit drafts for future new regulations on the following topics:

- Automatic steering function
- Functional safety requirements
- Validation methods for automated driving

- Cybersecurity and software updates (over the air)
- Data storage system for automated driving

This has resulted in templates for three new UNECE regulations:

- UNECE regulation for uniform testing of software update processes
- UNECE regulation for the uniform testing of vehicles for cybersecurity and cybersecurity management systems
- UN regulation for the first automated driving function: Automated Lane Keeping System (ALKS)

Pioneering achievement: First UNECE regulation for a level-3 lane-keeping system

The WP.29 will address the aforementioned drafts in the upcoming 181st session. After the adopted UN regulations are approved and conveyed to the contracting states of the 1958 UNECE Agreement and an accompanying six-month phase for comments has passed, the individual countries can incorporate the regulations, such as those for automated lane-keeping systems (ALKS), into applicable law.

The UNECE regulation in detail

The ALKS is the first set of rules and regulations for automated driving and as such is initially limited to vehicles of the category M1—that is, passenger vehicles with up to eight seats plus driver. The regulation is also limited to systems that function at speeds of up to sixty kilometers per hour. In this area, the automated driving function controls the vehicle in the longitudinal and transverse directions, thereby keeping it in the lane without the driver's involvement.

An important aspect of the planned regulation deals with monitoring the driver's condition, also known as driver monitoring or driver awareness monitoring. It must ensure that the driver can follow a request to take over the lane-keeping system within a defined period of time. As long as the system autonomously takes over driving, drivers at level 3 no longer have to keep their hands on the steering wheel or otherwise prove their attentiveness. Conversely, the driver must be able to intervene at any time and take control again. UNECE is thus following the Vienna Agreement, which always gives the driver priority over an automatic steering function. The same applies to the operation of the brake or accelerator pedal; like manual steering, it switches off the ALKS.

The human-machine interface and the demands placed on the system to monitor the human driver are a core element of the new regulation and a real innovation. The regulation requires the system to monitor whether, during system activation, the driver conforms to requirements by sitting in the driver's seat and wearing the seatbelt. Or in the case of an event, such as reaching the system limits, the system can record the situation and take over the driving task responsibly [2].

The draft gives examples of criteria to be detected:

- blinking
- closed eyelids
- deliberate head or body movements
- pressing switches and other control elements only within reach of the driver can also serve as a positive indication of attention

At least two independent criteria must be met for the driver to be considered available. The time between checks cannot be longer than thirty seconds. If the human driver needs to take over, the design requires the system to instruct the driver intuitively and unambiguously to take over manual control of the vehicle.

An automatic lane-keeping system for level 3 is a complex system that must first be defined. This can be found in the drafted regulation. Naturally, this kind of system contains a great deal of electronics, software, and interfaces, and their safeguarding and inspection are of central importance. In addition to the requirements for complex electronic systems in Appendix 4 of the regulation, the ALKS regulation also requires the fulfillment of the two new UNECE regulations for cybersecurity and software updates.

Protection against attacks
One of the three upcoming regulations is concerned with cybersecurity and the requisite management systems at the manufacturer and its supply chain. This is to ensure that the vehicle's systems and software cannot be compromised. The requirements of the cybersecurity regulation intervene in the three phases of a vehicle's life cycle: during development, in production, and afterwards.

The future core element that will check for compliant cybersecurity management systems is a Certificate of Compliance, issued by an accredited organization like TÜV SÜD [9]. Until July 2024, the vehicle manufacturer could alternatively demonstrate in a suitable manner that its cybersecurity management systems meet the requirements.

Security during an update
Another UNECE regulation is concerned with software update processes. It specifies how updates should be handled and which management processes must be established. The software used in highly automated vehicles will most likely require regular updates. The reasons for this are manifold. On the one hand, new threats are possible due to previously unknown security vulnerabilities; an update can eliminate them. Improvements to ease of use, for example, are also possible. Changes to legal requirements can be implemented via software updates as well. Until now, these have been installed almost exclusively in workshops. But manufacturers and suppliers have begun working on a so-called over-the-air method (OTA). The new program code is imported, for example, via connected Wi-Fi or a cellular connection. For this reason, the Software Update Management System (SUMS) required by UNECE places high demands on the quality of the procedure. For example, the exclusive transfer of software approved by the vehicle manufacturer must be guaranteed. A check is provided before and after the update to determine whether the program code fits not only the vehicle model but the individual type. The SUMS also requires a Certificate of Compliance [10].

Basic functions of a lane-keeping system for level 3
Based on the regulations under consideration, there are three operational states once the driver has activated the system, which must always be switched on manually. Furthermore, the process must be repeated after each engine start and is accompanied by a self-test. An exception is made for operation using a start-stop system; in this case, activation at the start of the journey is sufficient.

Generic Function Diagram of the Level 3 ALKS system

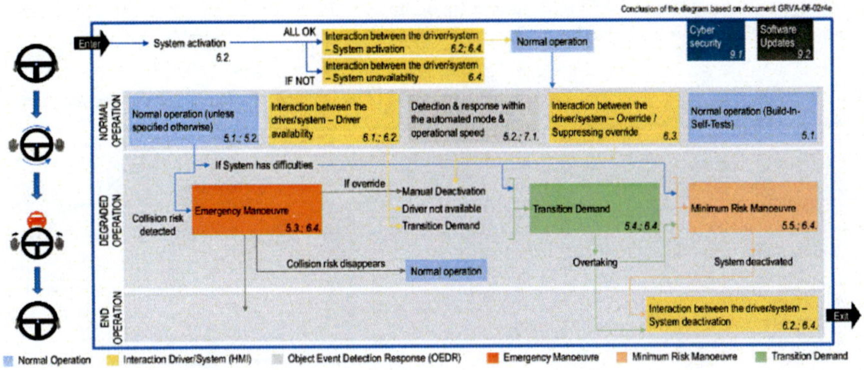

Afterwards, the system is running in one of three states:

Normal operation. The automated driving function records all necessary data and information via the sensors and carries out the driving task independently by controlling lateral and longitudinal movements. The interaction between the system and the driver is running undisturbed. The driver-condition monitoring system reports that the requirements for the driver to operate the ALKS system are met, and he/she could intervene if necessary. Deliberate corrections—for example a brief change in speed—are also possible.

Reduced operation. A problem has occurred, and the driver is warned. At the same time, the system tries independent countermeasures such as evasive action within its own lane. If this succeeds, it will go back to normal operation. Otherwise, the driver receives a takeover request. If he/she does not intervene, a driving maneuver will be initiated to minimize risk, which ends, for example, when the vehicle comes to a stop. The system always follows the primary goal of avoiding a collision.

Deactivated. The system is ready for operation, but the driver is steering or has perhaps assumed responsibility for driving following a request for system takeover.

Definition of test criteria

The basic requirement for a complex electronic system is initially easy to describe: it must follow the prompts it receives. In the case of an automatic lane-keeping system, this means that after explicit activation by the driver, the system processes the longitudinal and lateral functions while driving and monitors the driver. This includes checking via condition monitoring whether the driver is available. The system then performs its normal task via sensors and actuators and keeps the vehicle in the selected lane. As soon as an intervention becomes necessary or the driver deliberately intervenes, the automated driving function must switch to emergency mode.

Test scenarios for the automated lane-keeping system (ALKS)

To verify the safety of the complex overall system, the draft of the ALKS regulation requires an audit of the safety concept for its functional and operational safety; verification is achieved using simulations and real tests. The documentation needed for the audit is provided by the manufacturer, the responsible licensing authorities, or the designated technical service providers. Further requirements include a series of physical tests, listed in Appendix 5, which are carried out by a designated technical service provider or in cooperation with the manufacturer. In addition to normal lane guidance in longitudinal and transverse directions, they primarily deal with disturbances caused by stationary objects on the road and by other vehicles, for example, moving into or out of the vehicle's lane. It will also be important to assess the area covered by the system.

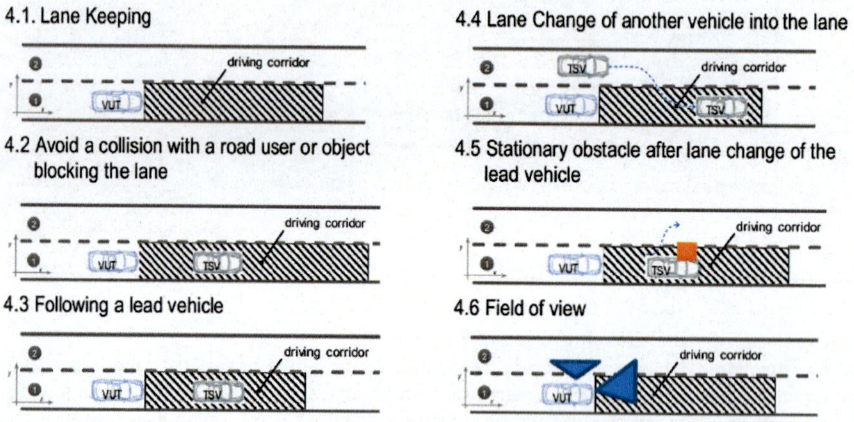

The validation requirements relate to the main criteria human-machine interface, normal function, dealing with emergency situations, and the identification of objects.

These are first checked during the audit according to Annex 4 of the draft regulation [3]. Audit results serve as the basis for the actual tests. The function itself and the manufacturer's safety concept must be confirmed. The methods used are tried and tested: driving on test tracks and operation in real traffic. The design also opens the possibility of using simulations in the type-approval process, for example, to verify valid system behavior in a critical situation, or to test a very broad range of parameters, see [11].

Merging physical and virtual tests
Some scenarios are too complex to test on a test track or even on public roads. The simulation can provide a valuable service in these cases. This calls for the creation of appropriate tools and the guarantee that the simulation models used sufficiently adhere to reality [12]. The results of real test-drives are combined with those from simulations to allow experts to make statements about the overall performance of the systems. This enables them to conclude from correct reactions in physical and simulated test operation that other requirements are also correctly fulfilled [12]. Detailed documentation is included in the expert's report. Hence, the requirements, the range of functions, the manufacturer's security and management concept, and information for the end user are thoroughly appraised. The overall aim is to determine the suitability of the system for use on public roads. Approval by an approval authority then takes place via the documentation of the technical service.

Dynamic tests
Six basic scenarios are used to check the automated driving function's system behavior in reaction to defined traffic situations, from common to critical, according to Annex 5 of the draft

regulation. This includes dealing with motorcycles but also awareness for traffic light systems, emergency vehicles, construction sites as well as correctly reacting to people on the road.

Validation on the road

Regarding the issue of what, where, and under what conditions is being tested, strict security conditions apply. Critical driving maneuvers prohibited in public traffic areas are possible on closed test tracks. It is also possible to predict the behavior of a vehicle and its systems in other environments.

However, only public roads provide a range of realistic conditions, without which a comprehensive assessment is not possible. The appropriate procedure is to validate the knowledge gained on the test track later on the road.

Scope of tests on road and test site

The steps for testing on the test track are far more extensive than those in public traffic. On their premises, the experts work through more than seventy points of a checklist. Data read or recorded from seven categories must deliver results across 117 scenarios, whereas only around twenty-four criteria from a checklist need be checked on general roads.

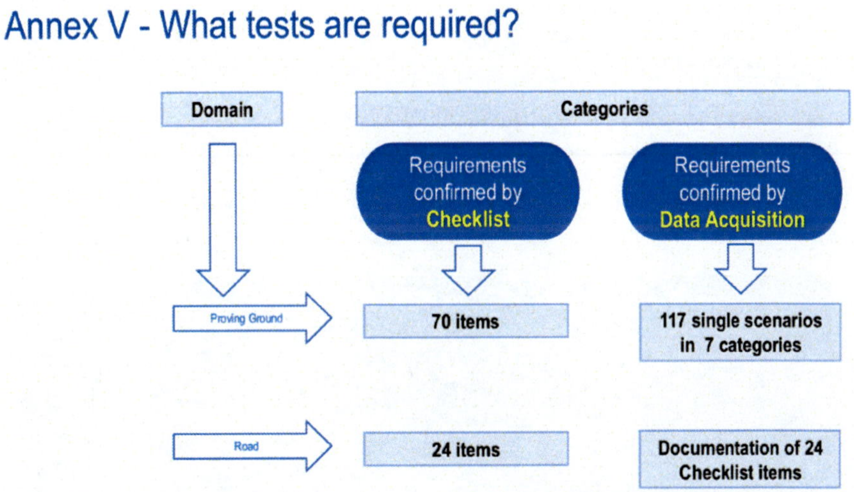

The drafted regulations define criteria not only for an automated lane-keeping system itself but for the test procedures and their environment. The challenge inherent in the regulation's new draft: at first glance, it does not define any explicit tests and requirements. However, ninety-six points for verification in testing can be derived from the requirements in the regulation's main text and the corresponding appendices that address the requirements for the complex electronic system as well as the test specifications. Some points do not require a complex test setup, with test drivers simply buckling up and the vehicle registering this. Even this must be considered and documented in a test plan. The validation of the test environment has a larger scope. Is it safe to use? Are the required speeds and distances drivable? This includes observing the weather and road conditions.

Chapter	Number	Excerpt from the regulations text	Number of test points	Comment
Annex 5 Test specifications	6.1.3.1	Criteria for deeming driver available	5	Single points described in 6.1.3.1
Annex 5 Test specifications	5.1.3	Driver support system active	3	Single points described in 5.1.3 like wiper, ...
Annex 5 Test specifications	6.3.1.1	Driver attentiveness	3	Points redundant
Annex 5 Test specifications	5.5.	System behavior during a Minimal Risk Manoeuvre [...] Driver take over [...] Standstill (hazard lights) [...] Re-activation disabled if reached standstill	3	Sum point for tests required in Annex 4 and can be transferred to checklist tests
Annex 5 Test specifications	5.1.4	Transition demand and behavior/escalation	3	Sum valuation per observation

Test-drives in reality

The seven main categories to be assessed are:

1. Staying in the lane
2. Detection of static objects
3. Detection of moving objects
4. Following a leading vehicle
5. Entering a lane
6. Leaving a lane
7. Areas monitored by the system (field of view)

In the 117 scenarios, individual points are checked at different speeds, distances, and environments. Different static and dynamic obstacles to which the automated driving function must accurately react are also used.

Targets, soft vehicle-dummies

Enabling deterministic statements
It is not possible to check all events and conditions that are conceivable and inconceivable in real traffic. This would require many millions of kilometers of test-drives. A deterministic approach is required to conclude that an automated-driving system will master all practical requirements.

To achieve this, a number of requirements must be met in the real tests. These include 100 percent precision, guaranteed repeatability, and absolute, secure control during execution. In this case, precision means that dynamic objects on the test premises achieve an accuracy within ten centimeters for measurements in the longitudinal direction and three centimeters for the transverse direction, which places special demands on the localization and control of dynamic objects. The right safety concept is crucial for the safe execution of the tests. At a certain level of criticality, the tests are carried out with controlled self-driving platforms, thus reducing the risk for security drivers in the event of a collision. This is the only way to test automated driving functions safely up to the system's limit.

The appropriate high-quality equipment for the test site and laboratories is naturally of the utmost importance.

Summary

With this set of rules, the UNECE has not only created the first mandatory requirements for an automated lanekeeping system. It is also calling for measures to prevent cyber threats or faulty software updates as well as a data-storage system that ascertains who was responsible in each respective driving situation. These points thus become an essential part of the homologation.

The upcoming UNECE regulation for ALKS is the important first step in rapid implementation of automated driving and in enabling AVs internationally uniform and simplified market access. This development also demonstrates that it will take some time for automated vehicles to become a ubiquitous presence on our roads. One reason for this has to do the upcoming regulations' adoption into applicable law—such as the EU regulation coming into effect in the autumn of 2020. Another reason is that regulations for automated driving functions beyond the scope of ALKS have yet to be drafted.
Until then, an increasing number of individual permits for automated vehicles will ensure that people will be able to experience tomorrow's mobility, even in this country. Trust in the new technology is another fundamental requirement for market success.

An example of this can be found in the spa town of Bad Birnbach in Lower Bavaria: the first fully automated public buses have shuttled spa guests between the market square and the spa for the last two years, and since last year, between the train station and the city center. This means of public transport conveys more than 130 people safely through a wide variety of traffic situations. To be able to operate on public roads, a single vehicle approval according to §21 StVZO (Germany's road traffic licensing act) in connection with exceptions according to §70 StVZO was necessary for every bus shuttle. The exceptions according to the StVZO are necessary because the automated vehicles do not fully meet the requirements of the StVZO. To evaluate the underlying safety concept of the vehicles and to justify the necessary exceptions (e.g., for the braking system) the TÜV SÜD technical inspection body in Bavaria accompanied the approval process.

Vehicle safety, functional safety SOTIF, and cybersecurity—these are key topics alongside the UNECE regulations. In addition to preliminary examinations and vehicle tests, the program also includes in-depth document reviews. These are supplemented by comprehensive dynamic tests in which driving functions are tested in a variety of real and simulated situations. This ultimately ensures that the vehicle always reacts safely and reliably. A detailed safety report from the TÜV on the vehicle's use and functional safety, along with the required expert reports, closes the regulatory gap for the recommended approval for road use.

With the existing national possibilities for automated vehicles for testing purposes on public roads and the upcoming UNECE regulation for the first automated driving function in production vehicles, the legislators have taken a major step toward closing the gap between technology and legislation. This presents an opportunity for society to accept and increasingly trust the technology, thus allowing automated driving to establish itself and continue to develop, step by step.

[1] SAE J3016
[2] Gnandt, Christian and Düser, Tobias, "Homologation and Validation of Automated Driving Functions: It's all about an efficient method and process," in Eighth International Symposium on Development Methodology (Wiesbaden, 2019),
[3] UNECE, "Proposal for a new UN Regulation on Uniform provisions concerning the approval of vehicles with regard to Automated Lane Keeping System," GRVA-06-02-Rev.4, Sixth GRVA Session (March 2020).
[4] UNECE, "Revised Framework document on automated/autonomous vehicles," ECE/TRANS/WP.29/2019/34/Rev.2, in 180th Session of World Forum for Harmonization of Vehicle Regulations (WP.29), (2019).
[5] Pegasus, URL: https://www.pegasusprojekt.de/en/, accessed May 18, 2020.
[6] "VVMethoden: Verifikations- und Validierungsmethoden automatisierter Fahrzeuge Level 4 und 5," TÜV Rheinland Consulting, URL: http://www.tuvpt.de/index.php?id=vvmethoden, accessed May 18, 2020.
[7] "TÜV SÜD entwickelt vorläufige nationale Standards in Singapur mit – zur Konstruktion und zum Einsatz autonomer Fahrzeuge," TÜV SÜD, April 2019, URL: https://www.tuvsud.com/de-de/presse-und-medien/2019/april/tuev-sued-entwickelt-vorlaeufige-nationale-standards-in-singapur-mit.
[8] IAMTS, LinkedIn, URL: https://www.linkedin.com/company/autonomous-vehicle-testing/, accessed May 18, 2020.
[9] UNECE, Proposal: Draft new UN Regulation on uniform provisions concerning the approval of vehicles with regard to cyber security and of their cybersecurity management systems, GRVA-06-19-Rev.1, Sixth GRVA Session (March 2020).
[10] UNECE, Proposal: Draft new UN Regulation on uniform provisions concerning the approval of software update processes, GRVA-05-06, Fifth GRVA Session (January 2020).
[11] Koller, Benjamin and Düser, Tobias, "Homologation und Validierung von automatisierten Fahrfunktionen," ATZextra, March 2020, pp. 22–26.
[12] Düser, Tobias, Abdellatif, Houssem, Gutenkunst Christian, Gnandt Christian, "Ansätze für die Homologation automatisierter Fahrfunktionen," ATZelektronik 14, no. 10 (2019): 54–60.

Dipl.-Ing (FH) **Benjamin Koller** has been head of technical regulations and knowledge management of automated driving at TÜV SÜD Auto Service GmbH since 2019. The trained mechanical engineer and officially recognized expert has been with the company since 2010 and has held various positions there, including as assistant to the board member for mobility at TÜV SÜD

AG, and as PMO automated driving and driver assistance systems at TÜV SÜD Auto Service GmbH.

Dipl.-Ing. (FH) **Robert Matawa** has headed the department of testing automated driving functions at TÜV SÜD Auto Service GmbH since 2016. Before joining TÜV SÜD in 2002, the qualified vehicle technician was a motorsport engineer for Audi in the STW and at the 24 Hours of Le Mans. At TÜV SÜD, he started as a test and development engineer for chassis and steering systems and headed the department for analytical reports, among other divisions.

Porsche Taycan - recuperative brake system for combined braking

Martin Reichenecker[1] and Bernhard Schweizer[1]

[1] Dr. Ing. h.c. F. Porsche AG, Weissach

Abstract. The requirements imposed on the brake system of a **B**attery **E**lectric **V**ehicle (**BEV**); above all the requirements imposed on the recuperative capacity of the system for maximum energy recovery, are constantly increasing. Energy recovery, not just in the usual form as a major contribution to energy efficiency and increasing the range in day-to-day operation, but rather in this case – which is typical of Porsche, using the first electric Porsche sports car, the Taycan as the example – this also extends to the limits of driving dynamics, under extremely high lateral acceleration and during ABS control. Thus recuperation is a very large and important contributor to thermal offload of the wheel brake on the one hand, but also towards increasing the drivable number of laps until the next charging procedure, on the other hand. The demand is to allow the interactions of the various deceleration actuators (friction brake and electric motor) required in this regard, to run in the best manner possible, unnoticed by the driver. This applies equally in the context of "requirements for racing circuit driving", and to the objective of a brake pedal feel that is as natural as possible, such as retention of the haptic ABS pressure point relative to pedal force and pedal travel, as well as to the challenge of minimal influence on front-axle to rear-axle braking force distribution, so as not to disadvantageously influence the driving dynamics that are customary and typical for Porsche. In the article below – after a brief description of drive system and chassis – the recuperative brake system for combined braking of the Porsche Taycan will be presented in detail. This includes all participating components, from the pedal system to the brake actuation system and the recuperation strategy and on through to the wheel brake. In this regard a technical description of the lightweight design concepts and technologies that are used will be provided.

Keywords: Recuperation, brake system, sports car, Porsche Taycan.

1 Brief description of the drive system and complete chassis

The drive system consists of two permanently excited synchronous motors (PSM) on the front axle and rear axle with a total system power of 625 hp, which is available – as continuous capacity without degradation – over a longer period of time. These motors are fed by an underbody HV battery with 800 V system voltage rating, which through its arrangement ensures a particularly low overall vehicle centre of gravity (again 80 mm lower than in the current 911 and which is lower than that of all current series

P. E. Pfeffer (ed.), *11th International Munich Chassis Symposium 2020*, Proceedings, https://doi.org/10.1007/978-3-662-63193-5_5

production Porsche models). The 800 V technology enables reduced amperage ratings at the same power output, enabling reduction of all HV cable cross-sections – this saves weight. The low centre of gravity, paired with the broad track width and high-performance tyres with optimised roll-resistance, guarantee the level of maximum lateral dynamics, i.e. driving dynamics that is typical for the brand. Electromechanical anti-roll stabilisation, rear axle steering, as well as fully variable drive torque distribution between front and rear axle are additional highlights where driving dynamics are concerned. The combination with three-chamber air suspension with level adjustment and lift function, guarantees the maximum spread between driving dynamics and comfort, i.e. everyday usability

Fig. 1. Complete chassis and drive system of the Porsche Taycan

2 Brake system and components

2.1 Lightweight design brake pedal

A lightweight design brake pedal with organic sheet technology and plastic mounting saddle. The brake pedal meets the rigorous load requirements, for example, thanks to a multi-axial fibre layer structure (glass fibre) in the inlay encapsulated in "one shot". The continuous fibre-reinforced thermoplastic composites in the top layer ensure the excellent pull and bending resilience and the inner layers ensure a high torsional resilience of the brake pedal. With this laminate we succeeded in meeting the extremely high technical specifications imposed on the mechanical performance of the safety component, and at the same time exploiting the optimum potential of lightweight design. The weight of the complete pedal system, incl. mounting saddle is approximately one kilogram, which is virtually half the weight of a steel pedal. Use of this technology is the consistent extension of the lightweight design strategy of organic sheet pedal systems, introduced for the first time in the 918 project, and which is now found in multiple series production Porsche model ranges.

Fig. 2. Pedal system – Porsche Taycan

2.2 Brake actuation and recuperation system

A vacuum-free brake actuation and control system is used, consisting of a second generation electromechanical brake booster, combined with an ESC premium hydraulic unit, six-piston pump and open rear axle circuit, i.e. as distinguished from a conventional ESC in the rear axle circuit, it is equipped with an isolating valve, a pressure regulating valve, as well as a direct suction connection to the brake fluid reservoir. On the other hand, the low pressure accumulator in the rear axle circuit – see illustration or circuit diagram Fig. 3 and 4, is dispensed with. The main brake cylinder on the EBB is dimensioned with a reduced diameter of just 25.4 mm in such a manner that in terms of volume, only approximately two-thirds of the total volume capacity of the wheel brake is covered, the rest of the volume is "pulled-in" as needed by the ESC hydraulic unit, directly via the rear axle suction connection and thus the brake system is completely supplied with fluid. Consequently due to these technical modifications the overall system bears the name EBB + ESD HEV II.
During a recuperative brake procedure without intervention and pressure build-up on the wheel brake, i.e. exclusively via the electric motors as generators, the EBB and ESC HEV II components interact in the job split as follows:

1. The volume that is not shifted into the wheel brake is discharged directly into the brake fluid reservoir without pressure, via the pressure regulating valve and the suction line (volume blending via ESC HEV II)
2. In this case the counterforce of the non-existent pressure column (since it is pressureless) is simulated via a reduced EBB support and its suspension characteristic curve (force blending via EBB)
3. At a required pressure build-up during a recuperative braking procedure the sequences occur inversely, in other words the non-existent volume is pulled-in from the brake fluid reservoir via the suction line connection and the ESC pump; thus the support force in the EBB is successively increased.

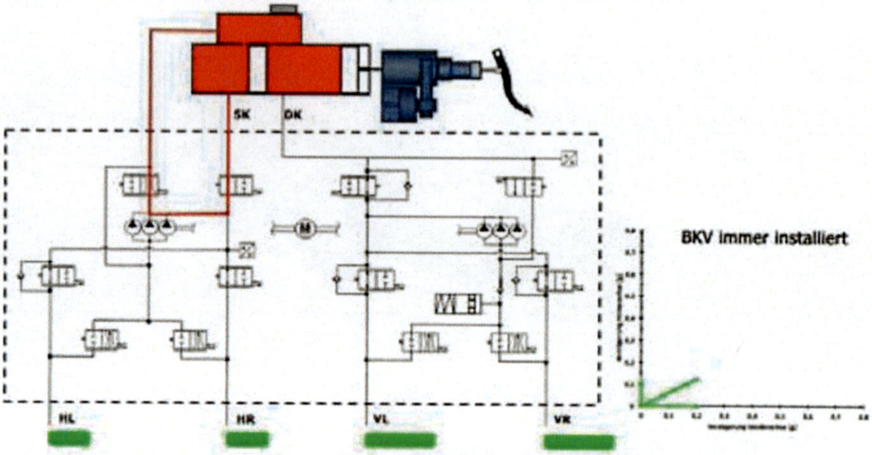

Fig. 3. 100% recuperative braking: The volume shifted by the driver is shifted into the reservoir without pressure, the corresponding brake torque is made available via the generators.

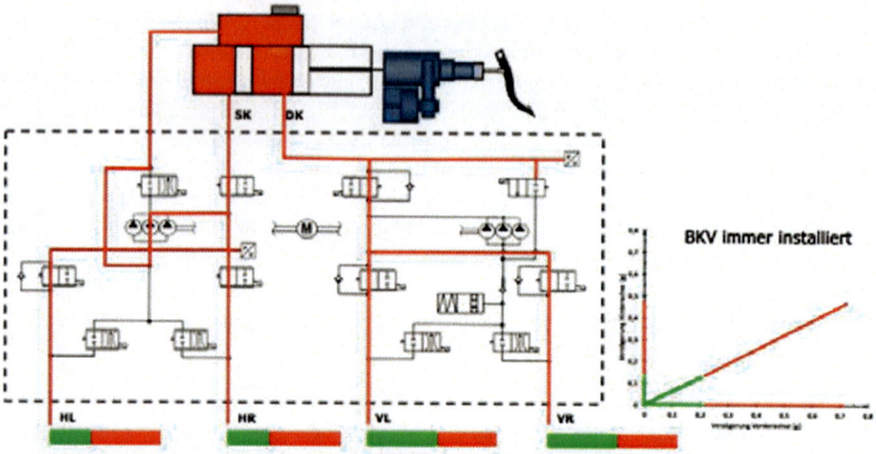

Fig. 4. Proportional recuperative braking if, in the course of a recuperative braking procedure, a hydraulic/friction proportion is required, the volume will be pumped into both circuits of the brake system via the suction line connection and the ESC pump.

This blending strategy is referred to as system-level blending, as pressure build-up and dissipation always occurs simultaneously in both circuits. This strategy is enabled because on the rear axle, as well as the front axle, electric motors as generators can provide pure electrical brake torque. Through the pressure equilibrium on front axle and rear axle in conjunction with the electrical braking torque distributed in the same ratio, the constant, installed braking force distribution relevant for driving dynamics, is ensured. The performance of the system can be expressed in numbers, as follows:

Brake recuperation

- Pure electric decelerations via the brake pedal to maximum 3.0 $^{m}/_{s^2}$
- With a maximum recuperation power of 265 kW extending in to the limit area of driving dynamics.
- approximately 90 % of day-to-day braking can be executed on a purely recuperative basis

Moreover the Porsche Taycan has the possibility of an optional, 3-stage accelerator lift off recuperation, operable via steering wheel button.

Accelerator pedal lift off recuperation

- Accelerator lift off recuperation OFF, pure sailing (maximum efficiency)
- Accelerator lift off recuperation ON [maximum 0.8 $^{m}/_{s^2}$]
- Auto-recuperation via camera [maximum 1.3 $^{m}/_{s^2}$]

The values of the accelerator lift off recuperation must be considered as additive to the values of the brake recuperation and can be combined at any time, in accordance with the desired behaviour. The Taycan's available driver assistance systems have also been optimised for use of the maximum recuperation and in this regard utilise accelerator lift off recuperation, as well as brake recuperation in full scope – also combined, as needed.

2.3 Wheel brake and brake cooling

The challenge in configuring and dimensioning the wheel brake was primarily how to master the extremely high medium braking power due to the high turnover of kinetic energy in use on the racing circuit. The current relevant key data in this regard are:

- 0-100 km/h (62 mph) in 2.8 s
- 0-200 km/h (124 mph) in 9.8 s, and this more than 25 times, in other words, there is no derating worth mentioning
- v_{max} 260 km/h (162 mph)
- at a weight of approximately 2,300 kilograms
- Nürburgring Nordschleife in 7:42 minutes.

In addition, at start of development it was not clear which thermal offload of the wheel brake through recuperation could be calculated on what driving profile, or on which circuit. Given the massive conflict of objectives with the range-relevant coefficient of air resistance, cW – the brake cooling in the front axle is designed to be switchable in the form of a closed channel (see Fig. 1). This channel routes the cooling air, including downstream spoiler, on through to the inlet of the brake disc chamber and is opened or closed as needed via temperature model. A permanently installed non-switchable, air guidance channel for brake cooling is used on the rear axle.

For the wheel brake technology, such as we relied on in the lightweight design high-performance brake technology that has been proven in multiple model ranges. Solid aluminium, fixed calliper with drawn pad pin guide on front axle and rear axle, likewise paired with cooling-optimised composite brake discs. These fixed callipers on the rear axle are enabled by use of the directly actuated electric Duo Servo parking brake, which likewise is used across model ranges.

As brake disc technology, we used the ceramic brake (PCCB – series production in the Turbo S), which has been familiar for some time, as well as the PSCB technology, which was introduced for the first time with the current Cayenne Turbo (grey cast iron disc coated with tungsten carbide – series production in the Turbo). As basic brake technology we used grey cast iron integral discs.

Thanks to the high efficiency of the recuperation, extending to the limits of driving dynamics, it was possible to lower the maximum temperature values of the brake disc on the front axle on the circuit by > 80 Kelvin, and to lower the temperature of the brake fluid by > 10 Kelvin, so that currently the following dimensions are allocated to the wheel brake (see modular brake system, Fig. 5). The brake variants PCCB and PSCB are on the front axle, due to the extremely large pad surfaces, paired with a ten-piston calliper whereas the cast iron disc uses six-piston callipers. Four-piston callipers are used on the rear axle. All callipers are clearance-optimised for reduction of the residual torque.

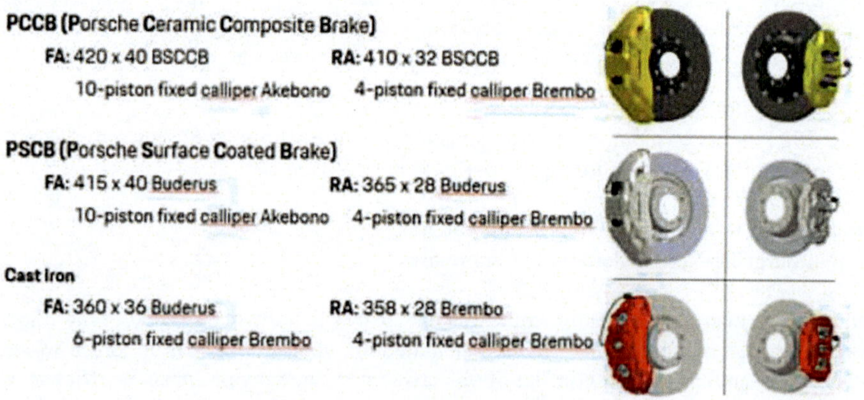

Fig. 5. Modular brake system of the Porsche Taycan

Thanks to the extremely high proportion of recuperation (90 % of braking operations in everyday use) it was necessary to develop a brake bedding function, as well as a brake refresh function. The bedding function, as the name indicates, for the initial brake bedding of the pad and disc (in new condition or after replacing pads) and the refresh function for maintaining the coefficient of friction in operation. The core of these two functions is to deactivate the recuperation for a defined energy conversion and to cover all decelerations in terms of technical friction solutions. The challenge is to determine the best compromise between cleaning/refreshing effect and lost recuperative energy.

3 Summary

Through numerous lightweight design features, and components, sophisticated brake cooling, as well as an extremely efficient and high-performance recuperation, extending into the limit area of driving dynamics, the recuperative brake system of the Porsche Taycan substantially contributes to the fact that Porches's first purely electric sports car continues the tradition and brand promise of superior driving dynamics at maximum spread in the direction of comfort and day-to-day usability.

References

1. Fig. 1, 2 and 5 PAG internal presentations
2. Fig. 3 and Fig. 4. Bosch presentation – EBB + ESC HEV II System.

The Integral Bushing Rear Suspension Architecture of the New Lincoln Corsair

Dr. Friedrich Wolf-Monheim[1], Dr. Paul Zandbergen[1], Daniel Mainz[1], Ralf Hintzen[1], Timothy Drotar[2]

[1] Ford Research and Innovation Center, Germany
[2] Ford Research and Innovation Center, United States of America

Abstract. The Lincoln Corsair is a compact premium crossover vehicle on Ford's global C (compact class) platform, replacing the Lincoln MKC. The prime focus of the research and development workstreams of the integral bushing rear suspension architecture for the new Lincoln Corsair was to significantly improve the performance in terms of impact harshness as well as noise, vibration and harshness (NVH) behavior compared to the Lincoln MKC featuring a control blade rear suspension architecture. The new integral bushing rear suspension architecture is an evolution of the integral link rear suspension architecture developed for Ford's global CD (midsize class) platform. Compared to the integral link rear suspension architecture the integral bushing rear suspension architecture is significantly lighter and the cost structure could be improved without compromising the performance level of the integral link rear suspension architecture. With regards to the interior noise attribute the performance level could be significantly improved. The integral bushing rear suspension architecture development included benchmarking activities, definition of attributes and targets, suspension concept design as well as system evaluation and selection. The concept development work was conducted with special attention to key requirements implied by a compact premium crossover product like the Lincoln Corsair as well as design constraints introduced by the platform and the sister product Ford Kuga deploying a control blade rear suspension architecture. This paper focuses on the design aspects and the fundamental layout principles of the integral bushing rear suspension architecture by presenting the journey starting from the integral bushing rear suspension prototype development towards the final series production design of the premium rear suspension architecture of the new Lincoln Corsair.

Keywords: Premium Suspension, Suspension Architecture, Lincoln.

P. E. Pfeffer (ed.), *11th International Munich Chassis Symposium 2020*, Proceedings, https://doi.org/10.1007/978-3-662-63193-5_6

1 Introduction

The Lincoln Corsair is based on Ford's global C (compact class) platform, a front-wheel drive / all-wheel drive architecture with east-west powertrain layout. The premium rear suspension research and development workstreams included benchmarking activities, definition of attributes and targets, suspension concept design as well as system evaluation and selection. The paper at hand focuses on the technical design aspects of the rear suspension architecture.

Various key architectural principles enabled the successful implementation in the new Lincoln Corsair. Both the springs as well as the shock absorbers are directly connected to the wheel carriers to optimize the motion ratios to finally maximize the vehicle dynamics performance and subsequently reduce cost and weight of the spring and damper system. The novel lower control arm design is driven by compactness and simplicity to enable a robust and cost-efficient sheet metal construction. To optimize material utilization in the subframe design only a single longitudinal load path was implemented per side. The camber link is realized as a planar structure to minimize cost and weight. In contrast to the integral link rear suspension architecture the integral link is superseded by an advanced integral bushing integrated into the structure of the wheel carrier. From a functional perspective the integral bushing mimics the integral link.

2 Suspension Prototype Design and Development

2.1 Base Suspension Layout – Computer-Aided Design (CAD)

Fig. 1 shows the prototype CAD design of the integral bushing rear suspension system in isometric view (left) and the right-hand side of the suspension in top view (right).

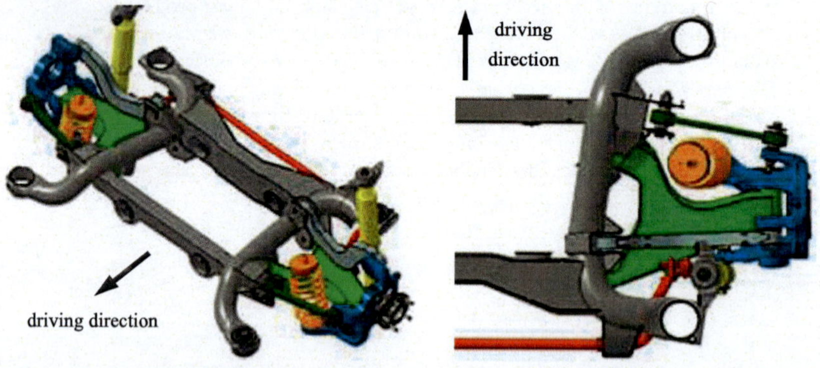

Fig. 1. Integral bushing rear suspension layout in isometric view (left) and top view of the right-hand side of the suspension (right)

Fig. 2 shows the left-hand side of the prototype CAD design of the integral bushing rear suspension system in front view (left) and side view (right).

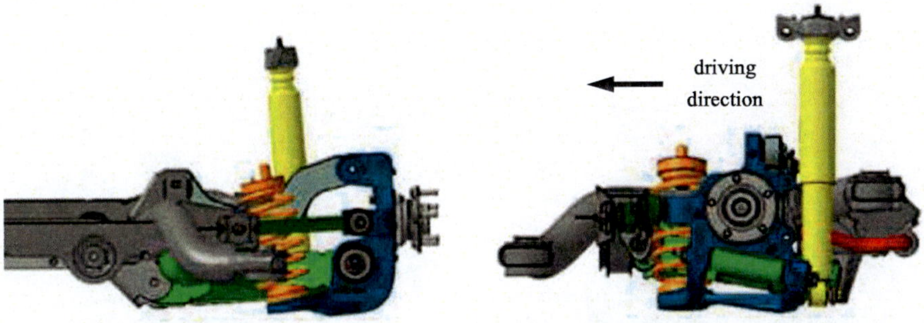

Fig. 2. Left-hand side of integral bushing rear suspension system layout in front view (left) and side view (right)

2.2 Kinematic Points

In Fig. 3 the suspension system kinematic points (A – Q) and the wheel center (WC) location are shown.

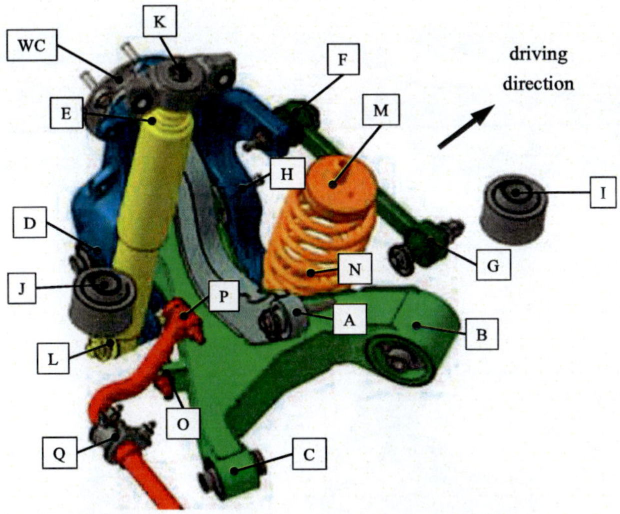

Fig. 3. Integral bushing rear suspension system kinematic points (A – Q) and wheel center (WC) location (left-hand side suspension corner)

2.3 Suspension Component Design

In the following the individual components of the integral bushing rear suspension system are shown. The prototype design and the key functions of the individual components are shown starting with the wheel carrier. The basic development process for the individual suspension components follows similar principles, considering relevant procedures and guidelines for stiffness and strength. A general overview of the left rear suspension corner is shown in Fig. 4.

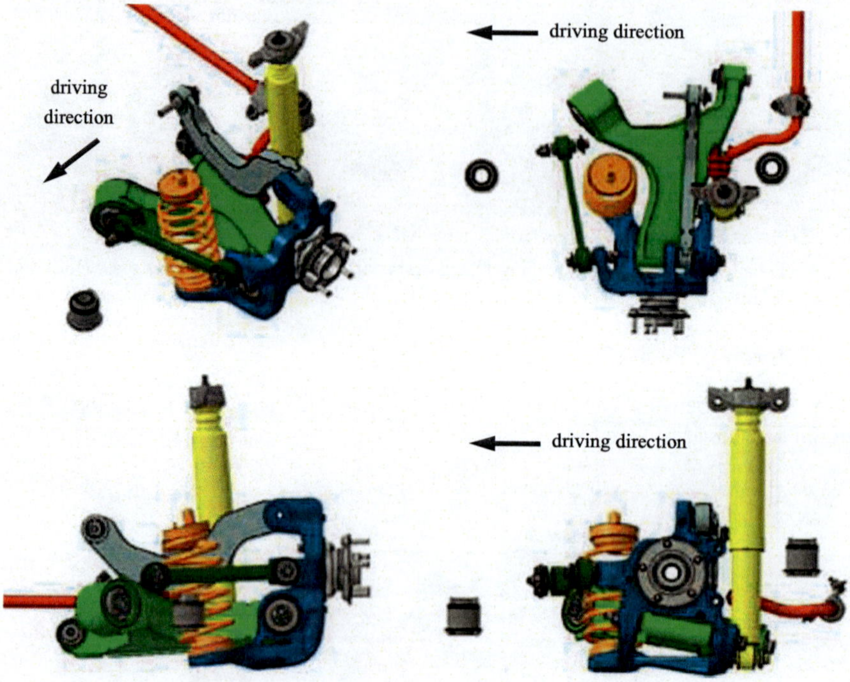

Fig. 4. Left-hand side suspension corner of integral bushing rear suspension (top left: isometric view, top right: top view, bottom left: front view, bottom right; side view)

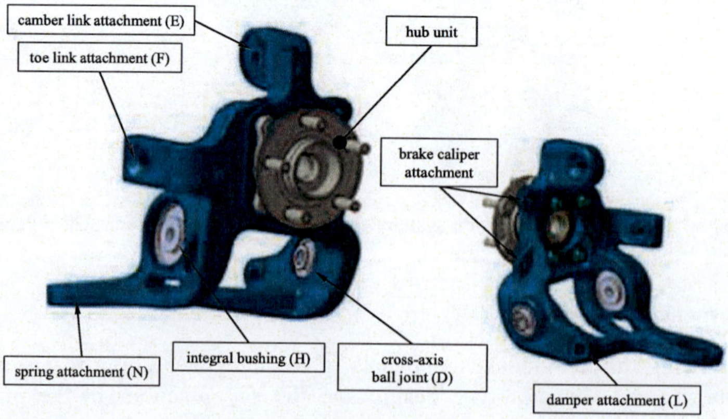

Fig. 5. Integral bushing rear suspension wheel carrier

Wheel Carrier. The wheel carrier shown in Fig. 5 is made from aluminum. A key function of the wheel carrier is to carry the wheel hub and bearing unit which is directly connected to the wheel and tire assembly.

On the inner side the wheel carrier is attached to the subframe via an upper lateral link, called camber link, a lower lateral link, called lower control arm or Y-arm and one additional lateral link, called toe link. The wheel carrier also carries the lower attachments of the damper and the coil spring.

The connection between the wheel carrier and the lower lateral link is realized via a rubber metal bushing in kinematic point H and a cross-axis ball joint in kinematic point D. Kinematic point H is the so-called integral bushing. The integral bushing is mainly loaded in global z-direction (vertical direction) whereas the camber link bushing (kinematic point E) and the toe link bushing (kinematic point F) are mainly loaded in global y-direction (lateral direction) or more precisely in the direction of the particular center line of the control arm. The ball joint connection between the wheel carrier and the lower control arm (kinematic point D) is loaded in x-, y- and z-direction. The toe link, the camber link and the damper are connected to the wheel carrier via rubber metal bushings. The wheel hub and bearing unit is connected to the wheel carrier by means of four bolts.

The wheel carrier development followed several steps to meet strength and stiffness requirements. The package space for the wheel carrier is limited by neighboring systems and parts like wheel and tire assembly, control arms, brake system components, vehicle body structure as well as ground clearance requirements.

Lower Control Arm. Fig. 6 shows the lower control arm (Y-arm). The main function of the Y-arm is to guide the wheel carrier and to transfer lateral loads between the wheel carrier and the subframe.

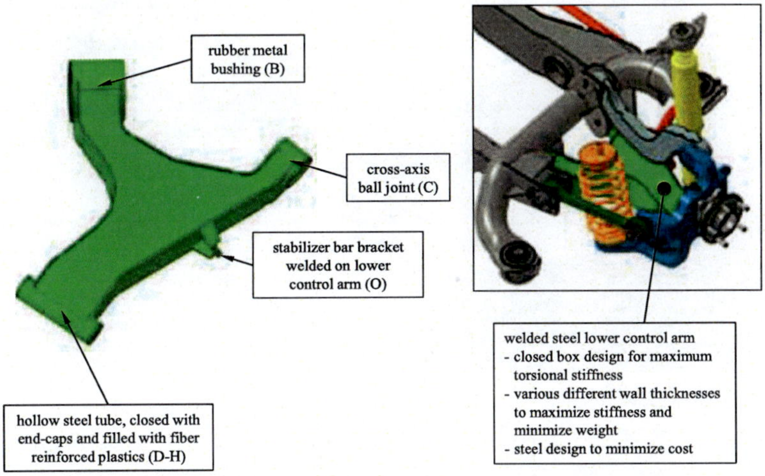

Fig. 6. Integral bushing rear suspension system lower control arm (Y-arm) design

On the inboard side the Y-arm is connected to the subframe in kinematic points B and C. Both interfaces are equipped with rubber metal bushings which are pressed into the metal sleeves of the Y-arm. The structure between kinematic points C and D is designed to transfer lateral loads coming from the wheel and tire assembly via the wheel carrier and the Y-arm into the subframe. Therefore, the path between kinematic points C and D is targeted to be very stiff. To aid this the rubber metal bushing in kinematic point C has a high radial stiffness. To support the kinematics of the suspension system the arm can rotate around the axis B-C to allow the wheel and tire assembly to move up and down as well as forward and rearward in a defined manner. Therefore, torsional and conical stiffness in kinematic points B and C is low.

The bushing in kinematic point B is the comfort bushing. The main task of the bushing in kinematic point B is to provide a low longitudinal wheel center compliance. Therefore, the bushing in kinematic point B is soft in terms of the radial stiffness in the Y-arm plane. The stabilizer bar drop link is connected to the Y-arm with the aid of a bolt-on rubber metal bushing.

The connection between kinematic points D and H is designed as a metal tube in which a connection bolt is integrated. To deliver a high caster stiffness on suspension level the vertical stiffness in kinematic point H and the torsional stiffness of the Y-arm around the global Y-axis of the vehicle need to be very high.

To realize the required clamping forces in the connection between kinematic points D and H the load path between these points is supported by a hollow sleeve. The sleeve allows the connection bolt to go through and supports the whole assembly structure to avoid clamping force losses.

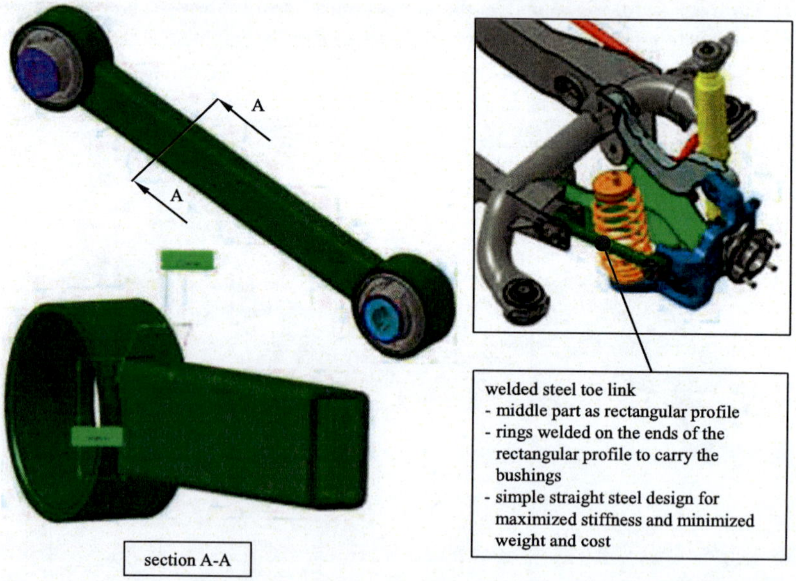

Fig. 7. Integral bushing rear suspension system toe link design

Toe Link. Fig. 7 is showing the toe link design of the integral bushing rear suspension system. The toe link is made from a rectangular profile and two metal cylinders. The two metal cylinders are welded to the central rectangular profile. The metal cylinders are designed to press in rubber metal bushings. The toe link can only take loads along its center line.

Camber Link. Fig. 8 shows the upper control arm (camber link) design of the integral bushing rear suspension system. The upper control arm is based on a center section with a rectangular shape made from four sheet metal elements and two metal cylinders welded on to this center section to press in rubber metal bushings. The link is mainly loaded along a line through the kinematic points A and E. The camber link is designed to meet all package, strength and stiffness requirements.

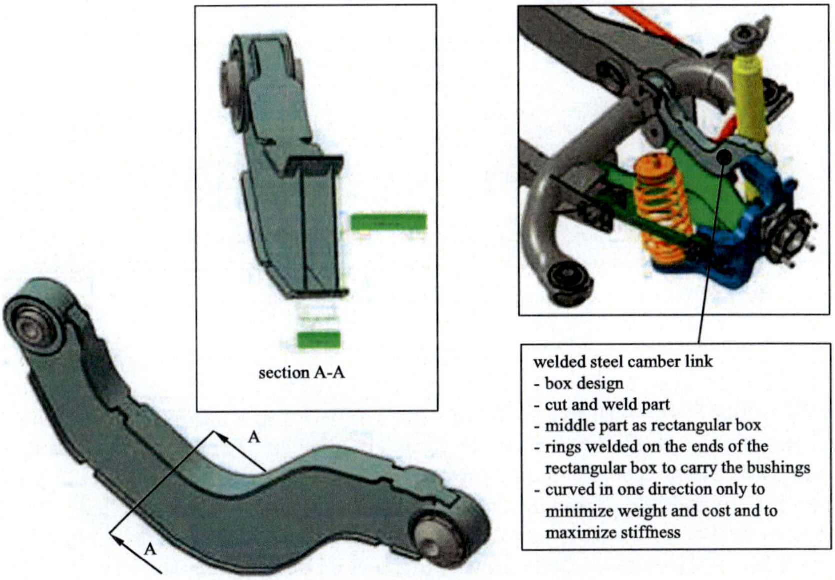

Fig. 8. Integral bushing rear suspension system upper control arm design

Subframe. Fig. 9 shows the subframe design of the integral bushing rear suspension system. The subframe is made from two longitudinal tubes, two transverse cross members and several brackets. The longitudinal tubes are composed of several tube sections with different wall thicknesses welded together. This approach has the advantage that the wall thicknesses can be tailored to the specific local stiffness requirements to finally save weight. Additionally, it is difficult to manufacture the longitudinal tubes from a single tubular semi-finished product due to the tight bending radii required.

At the end of the longitudinal members metal cylinders are integrated to press in the subframe mounts. The front and rear transverse cross members are made from laser-cut steel plates welded together. The wall thicknesses of the steel plates were chosen based on the stiffness as well as the strength requirements. Brackets are welded to the longitudinal members and to the cross members to connect the control arms to the subframe.

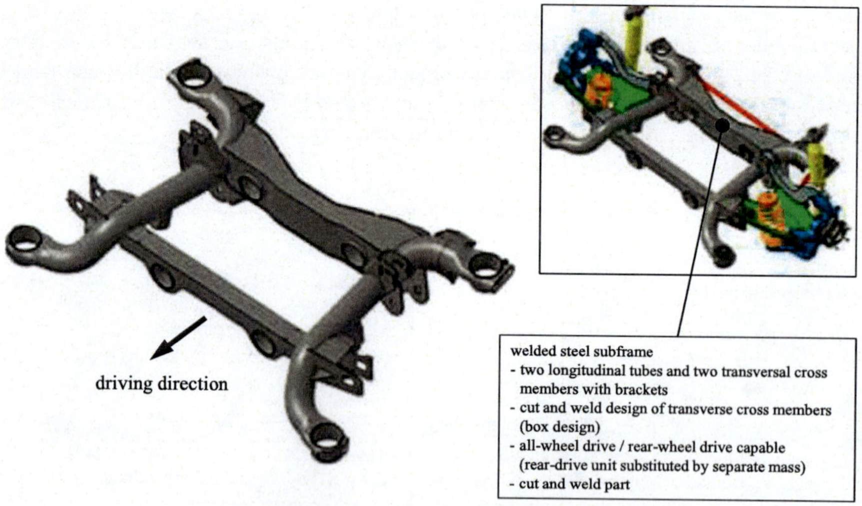

Fig. 9. Integral bushing rear suspension system subframe design

Stabilizer Bar System. During the development of the integral bushing rear suspension system various stabilizer bar routings were investigated. The key targets were to maximize the stabilizer bar motion ratio (vertical travel of the attachment point between the stabilizer bar drop link and the lower control arm relative to the total wheel travel), to minimize the component weight and to minimize the loads in the stabilizer bar mounts. The final stabilizer bar design of the integral bushing rear suspension system delivers a high motion ratio.

Fig. 10 and Fig. 11 show the stabilizer bar routing behind the wheel center. The stabilizer bar is connected to the rear cross member of the subframe with the aid of rubber metal bushings placed far outboard close to the subframe mounts. The stabilizer bar drop links are attached to the lower control arms. By optimizing the positions of the stabilizer bar mounts and the drop links the best balance between the motion ratio, the stabilizer bar length and the reaction forces in the stabilizer bar mounts was found.

2.4 Suspension Design Principles

Roll Pole. The roll pole can be approximated with the aid of a simplified geometric approach as shown in Fig. 12.

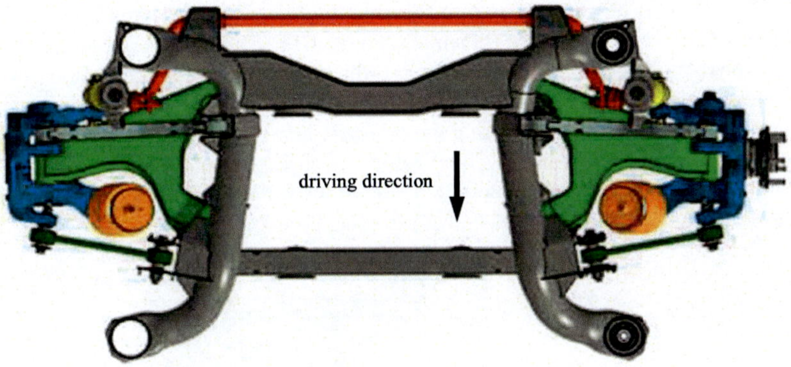

Fig. 10. Integral bushing rear suspension system stabilizer bar routing (bottom view)

Fig. 11. Integral bushing rear suspension system stabilizer bar routing (rear view)

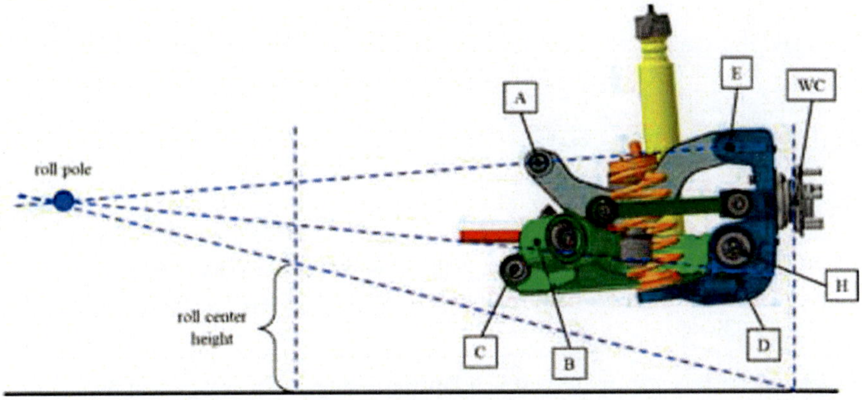

Fig. 12. Integral bushing rear suspension system (left-hand side suspension corner) – roll pole and roll center height (front view)

A first line parallel to the camber link (kinematic points A and E) and a second line parallel to the lower control arm (kinematic points B, C, D, and H) are both projected onto a global YZ-plane in global vehicle coordinates through the wheel center (WC). The intersection of theses two lines is the role pole. The vertical distance of the intersection point of a line connecting the role pole and the tire contact patch with the vehicle center plane (XZ-plane in global vehicle coordinates) and the ground is the roll center height (see Fig. 12). The lateral position of the roll pole defines the roll camber.

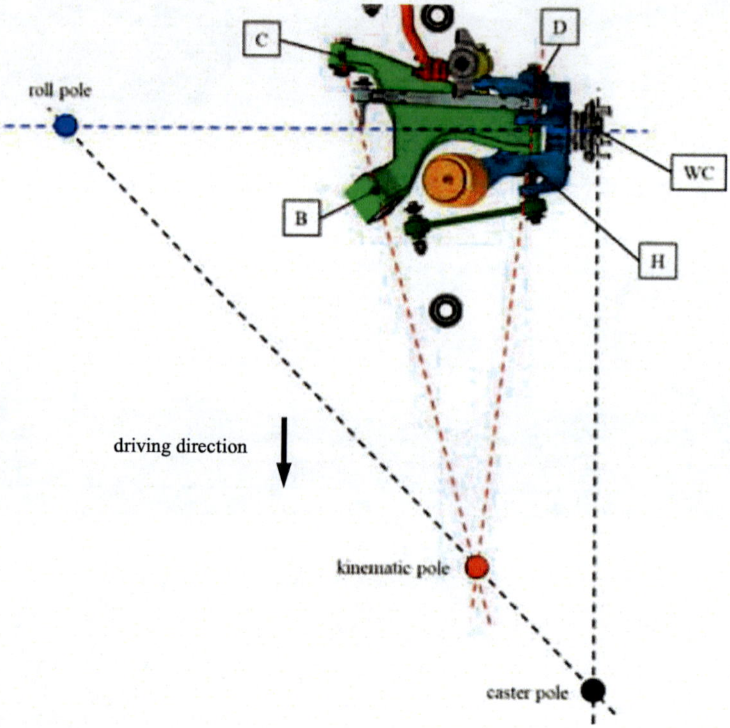

Fig. 13. Integral bushing rear suspension system (left-hand side suspension corner) – roll, kinematic, and caster pole (top view)

Kinematic Pole and Caster Pole. In Fig. 13 the top view of the integral bushing rear suspension system is shown. The kinematic points B, C, D, and H of the lower control arm are in the same plane. Therefore, a line through the kinematic points B and C crosses a line through the kinematic points D and H in the kinematic pole (see Fig. 13). The line through the roll pole and the kinematic pole is the instantaneous axis of rotation under suspension travel. The caster pole is the intersection point of the line through the roll pole and the kinematic pole and the XZ-plane through the wheel center (WC). Since the caster pole is located far in front of the wheel center, the bump caster of the integral bushing rear suspension system is relatively low. Since the kinematic pole is above the wheel center (WC), kinematic recession into jounce as well as anti-lift and anti-squat can be provided by the integral bushing rear suspension system.

Brake Pole. The intersection point of the line through the kinematic points C and D and the line though the kinematic points F and G is the brake pole (see Fig. 14). In case a braking force is acting at the tire contact patch in longitudinal direction, the wheel and tire assembly rotates around the brake pole.

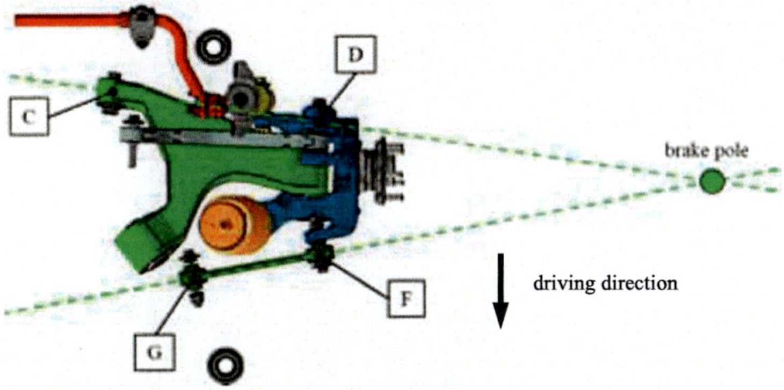

Fig. 14. Integral bushing rear suspension system (left-hand side suspension corner) – brake pole (top view)

3 Series Production Design

Fig. 15 shows the series production design of the integral bushing rear suspension system in top view.

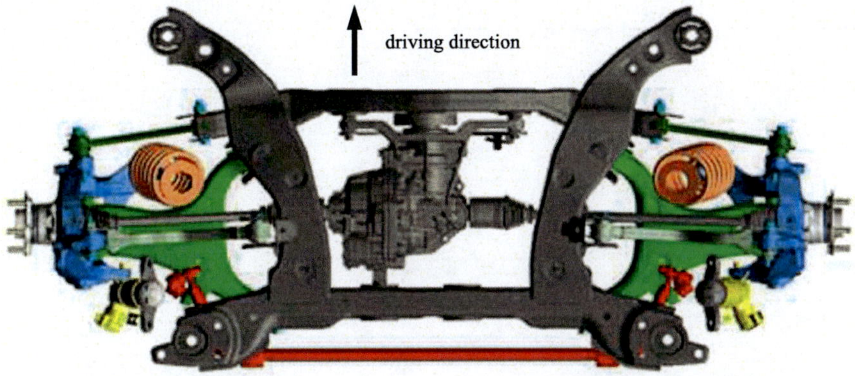

Fig. 15. Series production design of the integral bushing rear suspension system (top view)

Fig. 16 shows the series production design of the integral bushing rear suspension system in isometric view.

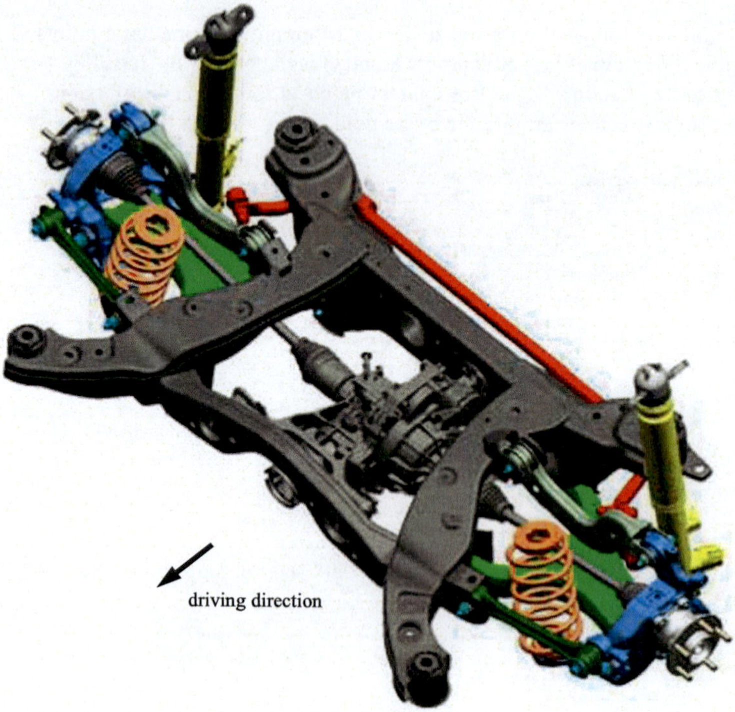

Fig. 16. Series production design of the integral bushing rear suspension system (isometric view)

Fig. 17 shows the series production design of the integral bushing rear suspension system in front view and Fig. 18 in rear view.

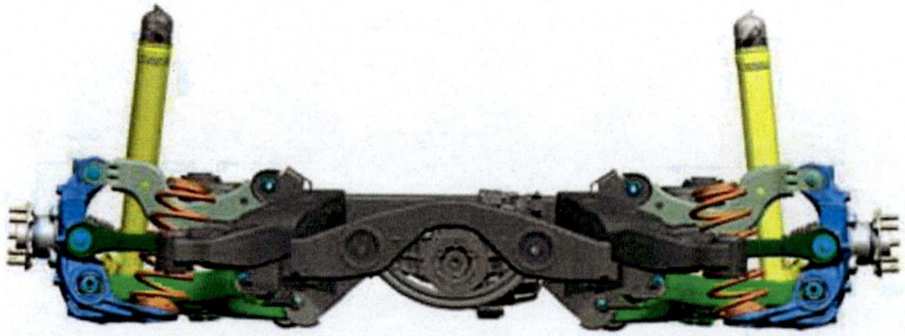

Fig. 17. Series production design of the integral bushing rear suspension system (front view)

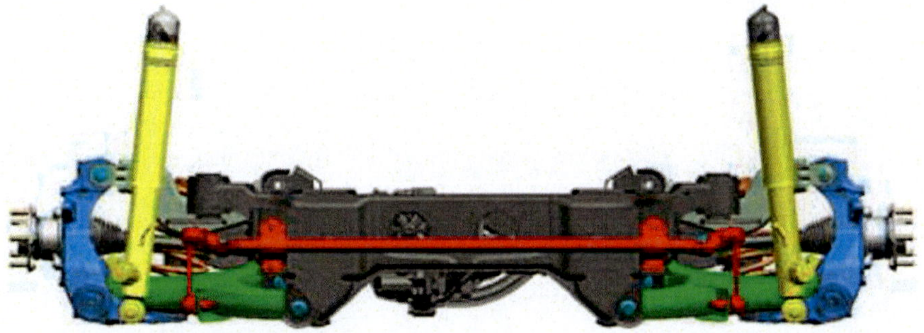

Fig. 18. Series production design of the integral bushing rear suspension system (rear view)

Fig. 19 shows the series production design of the integral bushing rear suspension system in bottom view.

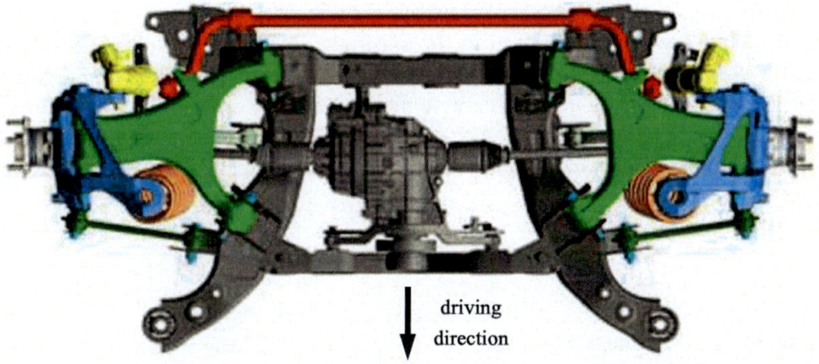

Fig. 19. Series production design of the integral bushing rear suspension system (bottom view)

Fig. 20. Series production design of the integral bushing rear suspension system (left: right view, right: left view)

4 Summary

The premium integral bushing rear suspension architecture of the new Lincoln Corsair is the evolutionary advancement of the premium integral link rear suspension architecture of Ford's global midsize class platform in terms of attribute performance and weight reduction. The prime development focus of the integral bushing rear suspension architecture for the new Lincoln Corsair was to significantly improve the performance in terms of impact harshness as well as noise, vibration and harshness (NVH) behavior compared to the Lincoln MKC featuring a control blade rear suspension architecture. Especially in the fields of interior noise the performance level could be significantly improved.

The architecture offers various sophisticated features that enabled the successful implementation in the new Lincoln Corsair. Both the springs as well as the shock absorbers are directly connected to the wheel carriers to optimize the motion ratios to maximize the vehicle dynamics performance and subsequently reduce cost and weight of the spring and damper system. The novel lower control arm design (Y-arm design) is driven by compactness and simplicity to enable a robust and cost-efficient sheet metal construction.

To optimize material utilization in the subframe design only a single longitudinal load path was implemented per side. The camber link is realized as a planar structure to minimize cost and weight. In contrast to the integral link rear suspension architecture the integral link is superseded by an advanced integral bushing integrated into the structure of the wheel carrier. From a functional perspective the integral bushing mimics the integral link.

References

1. Zandbergen, P., Girelli Consolaro, A.: Ford Motor Company's New Rear Suspension Architecture for the Global CD Platform. In: 34th FISITA World Automotive Congress 2012, F2012-G01-005. Fisita, London, UK (2012).
2. Zandbergen, P., David, W., Knoy, E.: The new rear suspension for Ford's global CD platform. In: 21st Aachen Colloquium Automobile and Engine Technology 2012, pp. 1503 – 1524, fka Forschungsgesellschaft Kraftfahrwesen, Aachen, Germany (2012).

An Efficient Evaluation Method for Longitudinal Driver Assistance Systems within a Consistent KPI based Development Process

Nesensohn, J.[1] and Lefèvre, S.[2] and Allgeier, D.[1] and Schick, B.[1] and Fuhr, F.[2]

[1] University of Applied Science Kempten Germany
[2] Dr.-Ing. h.c. F. Porsche AG, Germany

Abstract. One of the most commonly used advanced driver assistance system is adaptive cruise control. Although many cars are equipped with such a driver assistance system, the development is still based on subjective evaluation indices. The design of adaptive cruise control could be more powerful and reach more customer acceptance when a design process with specific values to ensure good driving characteristics is used. Such an objective based process is still state of the art of classic chassis development, but typically not used for advanced driver assistance systems.

To achieve this, use case-based scenarios are deduced from properties. A scenario simulation is used to generate information about important KPIs and collect data to develop an automated data analysis tool. To correlate subjective KPIs with objective measurement values an inertial navigation system in each car with differential GPS (RTK) and a wireless connection between both cars is installed. With the aid of this measurement system differential speed, acceleration and distances are calculated in a precise way. It also contains a fully controllable target vehicle in longitudinal and lateral direction to generate manoeuvres with high reproducibility. An analysis algorithm automatically calculates KPIs and displays important diagrams after each test run. This leads to a complete system analysis test in two days for each car. The comparison of target KPIs and subjective evaluation criteria shows whether the system reached its goals or if improvements are necessary. For this purpose the project partners Porsche, University of Applied Sciences Kempten and MdynamiX have joined their forces.

Keywords: Advanced Driver Assistance Systems, Adaptive Cruise Control, Efficient development process, subjective and objective evaluation method

P. E. Pfeffer (ed.), *11th International Munich Chassis Symposium 2020*, Proceedings, https://doi.org/10.1007/978-3-662-63193-5_7

1 MOTIVATION AND AIM OF WORK

Advanced Driver Assistance Systems (ADAS) are a major development objective of car manufacturers and are facilitators for highly automated driving functions. One of the most well-known advanced driver assistance systems is adaptive cruise control (ACC). The first ACC systems were offered in 2000. In 2007, TÜV SÜD presented an objective evaluation method for ACC enabling property and KPI based development [1]. Such systems could be implemented in the V-Model. However, this method requires a lot of effort because of the complexity of reproducible tests and their analysis. Instead of a KPI based development process with specified targets, driving characteristics with ACC are usually based on functional development, which are more dependent on factors like sensors, drivetrain or brake behaviour than on clear properties and objective KPIs. This leads to different ACC driving characteristics in almost all manufacturers' model series and less acceptance and trust by customers because their needs are not being fulfilled. One of the main missing customer requirements for higher acceptance is reproducible and predictable driving behaviour.
To reach a practical and cost-efficient continuous development process, subjective and objective evaluation parameters are necessary. This paper is the consistent continuation of the method shown in [19] and presents an approach how a property and KPI based development process for ACC can be implemented in an efficient manner.

2 STATE OF THE ART

2.1 Testing Method

For highly accurate and synchronous measurement of the vehicle movement and position of two vehicles, TÜV SÜD developed the following measurement setup in cooperation with Genesys / Dewetron in 2006: A reference measuring system is installed in each vehicle. In order to be able to collect data for each respective vehicle and drive corresponding parameters, the (1+n) measurement systems are linked to form a Car to multi Car relationship. This allows the relative motion parameters to be display and evaluate dynamically in all spatial directions [3].

2.2 Subjective Evaluation

The ACC function was already documented in 1984 [4]. Due to the long history of the ACC, many system tests have already been carried out. The best known are a large-scale study by TÜV Rheinland [5] or UMTRI (University of Michigan Transportation Research Institute) [6]. The main objective of their study is to examine the acceptance of the ACC system by customers as well as its operability and handling. Further tests carried out with customer drivers can be found in [7], [8], [9], [10] and [11]. A dissertation by Muriel Didier investigated driving comfort while using ACC. A total of 36 drivers are involved in this test, who subjectively evaluate two different series of ACCs. The acceleration and braking behavior are evaluated based on reaction time as well as the level of brake intervention when approaching a vehicle. The aim of this study is to find out whether the criteria met the expectations of the test subjects. The results showed that the test subjects were able to detect differences in driving comfort, although the two ACC systems have similar functionalities [12]. This result can also be found in [7]. In this study, different ACC maneuvers are evaluated on a test track with various relative speeds of 10 km/h up to 80 km/h. It can be concluded that average drivers are capable of recognizing different system behaviors and selecting one of these that fulfill or surpass their expectations.

2.3 Objective Evaluation

One possibility for objective description of ACC systems is presented in [13], where the experiment is performed on a Proving Ground (PG). Schick [14] shows tools and methods to facilitate the depiction of realistic scenarios for the use of driver assistance systems in simulation. The German automotive magazine 'Auto Motor Sport' published consumer tests of these methods in 2006 and 2010. In this objective test in 2006, first generation ACC systems are compared to identify subjectively perceived differences and their characteristic values. Driving manoeuvres are defined and evaluated by awarding points [15].

At the 13th Braunschweig Symposium in 2012 on the theme of "Automation Systems, Assistance Systems and Embedded Systems for Transportation Vehicles", Pawlek, Torben and Liesner present an analysis for the evaluation and design of ACC systems. Their research is based on a catalogue of manoeuvres in which various test parameters are examined [16]. The manoeuvre catalogue covers pure free and column driving as well as transitions between these phases. All relevant measurement signals such as vehicle movement variables and CAN data are recorded by an measurement system. The data exchange is realized via a WLAN interface. The analysis is based on three different characteristics in terms of comfort, safety and dynamics. In addition, this publication demonstrates the procedure for system objectification. Sensitive parameters are analysed using a correlation analysis between objective parameters and subjective grades. Subsequently, regression analysis is used to identify mathematical correlations of objective parameters to identify objective notes [16].

3 EFFICENT GENERIC EVALUTATION METHODOLOGY

3.1 Manoeuvre generation method

The efficient evaluation methodology is based on a reduction of testcases through use case-based manoeuvres and its property derived KPIs. A base of 33 manoeuvres is created in an expert workshop considering all dimensions of freedom for ego and traffic vehicle. Taking into account route profile, traffic law, vehicle specification, traffic participants, driving style and weather, manoeuvre-IDs are created which include ego- and traffic-trajectories like speed, deceleration or acceleration.

In order to reduce test efforts, these manoeuvres are assigned priorities in sequence of their occurrence in the traffic. According to Strasser and Frey the most relevant scenarios are stationary drive, approaching on column, overtaking scenarios, deceleration and acceleration in column [17][18].

For reducing ACC adjustment parameters studies about the time gap behaviour of drivers in comparison between manual driving and driving with ACC are considered. In the Abendroth [22] study, the subjects chose an average minimum time gap of 1.1 s for both manual and ACC-supported test drives. In Filzek [23], the test subjects chose an average ACC time gap of 1.4 s. As a result of the studies, the majority of drivers prefer small time gaps in the ACC settings.

The next step is to include controller behavior in the consideration. The main input variables of an ACC controller are delta distance, delta velocity, delta acceleration and its target variables like desired distance to the vehicle in front and desired speed. When a slower vehicle in front is detected by the sensors, the controller adjusts a desired distance. If no vehicle is detected, a requested speed is adjusted. This is achieved by the output value desired acceleration which is sent to the powertrain and brake coordinator. Based on this information a simple variation of the ID input values reduces the test efforts.

An intelligent design of experiment approach reduces testing IDs. With the knowledge of boundary areas of the system and knowledge of the defined use case range, intermediate results can be created through interpolation with a reduction of the ID variations.

3.2 KPI generation method

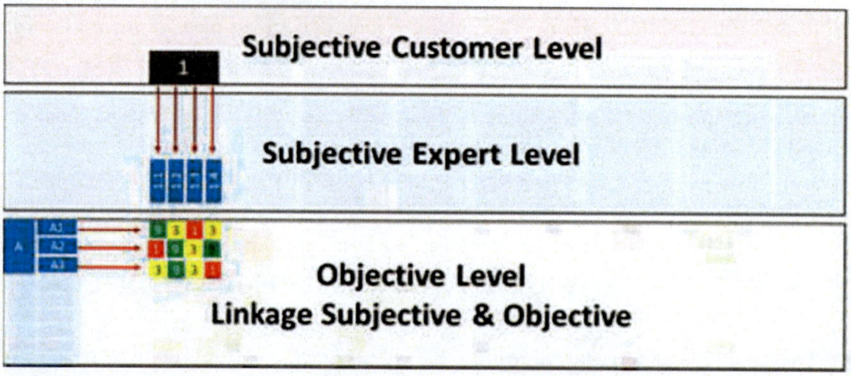

Fig. 1. Visualisation of objective and subjective correlation [19]

Property based KPIs are necessary to ensure high quality of data from the test results. Therefore, the presented method in [19, 21] is used to create a level model based on the Quality Function Deployment (QFD) approach. At first, a subjective evaluation sheet is created by a group of experts based on customers' expectations. In the second step all important measurement and indicators are defined by the same group of experts. For those subjective and objective KPIs a matrix with the values of 0 (no linkage), 1, 3 and 9 (large linkage) is created [Fig. 1], to show the strength of the linkage between objective and subjective KPIs. A bigger number suggests a stronger connection between subjective and objective KPIs. The strength of the connection is an important criterium for the rating system and results in higher acceptance by customers.

3.3 Data generation method

A consistent data generation process is necessary for a KPI based evaluation system. This could be performed in Simulation, HiL, SiL, ViL or on PG. The generation of objective KPIs requires a clear and ideal environment without obstacles, a steady state starting situation, repeatable scenarios with a controllable target vehicle. Furthermore, speed, acceleration and jerk data from the ego vehicle is necessary, as well as a high information density about traffic with ideal sensors to measure ground truth. These basic conditions are fulfilled in SiL, MiL, HiL as well as in VehicleHiL and ViL. PG tests are utilised to evaluate the entire system and to detect influences from different vehicle components like sensor or black box models.
A simulation concept provides a virtual test environment. In this test environment all manoeuvres and states can be pre-defined and tested precisely.

3.4 Data acquisition method

For a fast data acquisition method two data streams are saved. A Simulink model in the measurement setup saves all necessary measurement channels into a small Matlab file, which can be used directly for the evaluation without further conversion of measurement data. Matlab scripts are used for the automatic evaluation of objective KPIs. The evaluation scripts calculate all required objective KPI's and create plots which are necessary to evaluate the ACC functions. A second stream is saved as an mf4-file including all additional channels.
The evaluation with the scripts is based on the measured data determined in the simulation. A variation of a manoeuvre, for example by using different speeds, could be implemented and tested efficiently with these evaluation scripts. Therefore, it is possible to provide all the necessary scripts before the real test attempt. The evaluation scripts summarize the appropriate IDs based on the nomenclature of the measurement data. The exact allocation of the measurement data to a manoeuvre by the nomenclature makes it possible to show all data with the same ID in one plot. This allows a quick overview of all performed tests with the same manoeuvre settings.
All measurements are named automatically by an ACC-defined namespace to include meta information. This is needed to evaluate the test runs by their filename. Another advantage is the possibility to evaluate measurement data directly in the vehicle. The co-driver can therefore calculate and graphically evaluate the KPIs during the measurement runs. Possible errors in the measurement recording can be detected instantly and the test drive can be repeated if necessary.

4 IMPLEMENTATION OF THE EVALUATION METHODOLOGY

4.1 Simulation Implementation

The testing process is based on a simulation. The simulation implementation serves as a virtual pre-test to optimise the proving ground process. In this test environment all manoeuvres and test states can be pre-defined accurately and tested. Furthermore, the simulation determines manoeuvres and test conditions which are not practicable or safe to test. It creates the possibility to simulate different sensor signals and controller behaviour.

The simulation establishes a database and enables the creation and testing of data acquisition algorithms. Using a simulation results in an overall time benefit as the acquisition algorithms are ready to use before a single PG test has been done.

Fig. 2 Simulation approach with Hunter and Target vehicle

Because almost no models and data for the evaluation process is available, a vehicle model with a simple longitudinal powertrain, gearbox, brake application is used and parameterized with open public data like weight, gearbox ratios and engine performance from one of the test vehicles. The ACC-controller function is based on the input values delta speed, delta distance, desired speed and desired distance. The ACC-controller is described by formula number (1) and is limited to the acceleration limits in ISO 22179:2009 [20]. Ks and kv are dividing values for controller adjustment. The acceleration is provided through a PI-controller for gas and brake control.

$$a_x = \frac{(\Delta s - \Delta s_{desired})}{ks} + \frac{\Delta v}{kv} \quad (1)$$

An ideal radar sensor [Fig. 2] is simulated with typical sensor ranges, field of view, orientation and position based on the data sheet of the test vehicles radar. The simulation has got the same sensor setup as the proving ground tests. For this purpose, a sensor was attached to the centre of the forward most point of the vehicle. This sensor measures the distance to the rear-most point and the rear-most point in the centre of the Target vehicle to calculate the time to collision and simulate the PG measurement setup as accurately as possible. Virtual IMUs (Inertial Measurement Unit) are placed at all locations where measurement POIs (Point of Interest) are set in the proving ground setup. This provides data consistency throughout the whole process.

The test environment reflects a straight-line highway without disturbances and contains only vehicles that are directly related to the driving scenario.
The traffic vehicles do not have any parametrisation and depending on the manoeuvre either remain stationary or move according to a set acceleration value.
The namespace from the generated data is converted into a generic namespace. The same process applies to the PG data. Therefore, data acquisition algorithms can be developed and KPIs and manoeuvres can be evaluated.

4.2 Proving Ground Tests

Measurement

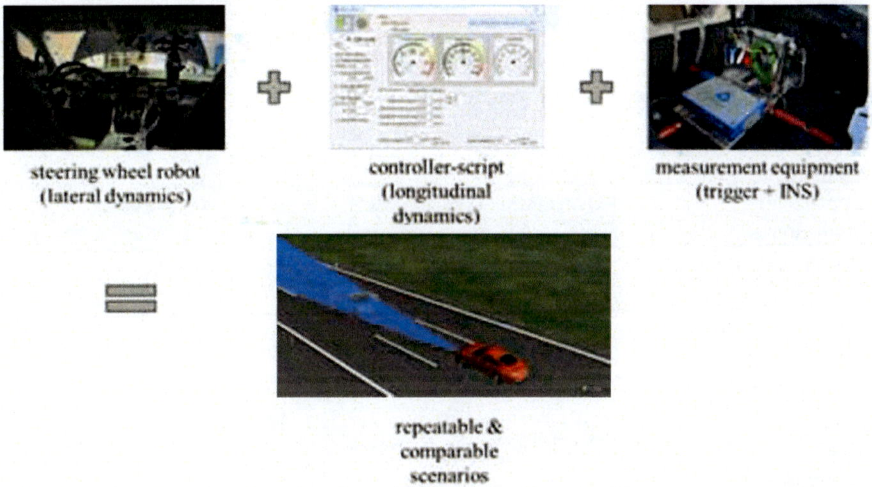

Fig. 3. Measurement method for repeatable and comparable manoeuvre

The high amount of measurement, car and proving ground costs and the complexity of the test equipment makes the measurement process one of the most expensive and time-consuming processes. Furthermore, the measurement equipment is often a limited resource. To test several cars in an efficient way, a high quantity of test equipment should be available. Alternatively, a self-aligning platform rig can be used [Fig.3]. It contains all measurement components. This platform is equipped with an Inertial Navigation System (INS) as well as a data acquisition system and some network and CAN/FlexRay-Interfaces. More cost-effective measurement sensors like analogue sensors and cameras are attached to the inside of the car.
To evaluate the distances between the two vehicles, a measuring system with a higher accuracy than the built-in sensor is necessary. Therefore, one INS in each car is used with a Real Time Kinematic (RTK) system to ensure high position accuracy. These systems are connected via a WIFI-Network. The front vehicle (Target Vehicle)

transmits its position to the back vehicle (Hunter Vehicle) including a GPS-Timestamp. The INS calculates the lateral and longitudinal distance between both cars in the coordination system of the vehicles. The calculated distances are also used to trigger acceleration or steering wheel actions.
The Target vehicle is equipped with a steering robot for exact lateral movement. The steering-robot is connected to the INS of the car and is triggered by a script for any lateral manoeuvre. For longitudinal control, the target car is equipped with an acceleration interface which has direct access to the car's acceleration controller. The acceleration is also triggered and controlled over the script per INS-Data.

Manoeuvres

The most common driving manoeuvres are presented in Table 1. These manoeuvres are Approaching Target, Target brakes, Target accelerates, Cut-Out Hunter, Cut-In Hunter and Cut-Out Target. All these tests are carried out on a test track to ensure a maximum of safety and no influence of other road users.

Table 1. Manoeuvre catalogue for objectification of the ACC.

Approaching Manoeuvres	Overtaking Manoeuvres	Other Manoeuvres
Approaching Target	Cut-Out Hunter (with/without indicators)	Stationary drive
Approaching Target (Target applies brakes)	Cut-In Hunter	Curve Drive
Approaching standing Target	Cut-Out Target with indicators	Reduce / Increase velocity using ACC
Target brakes	Cut-In Target	
Target accelerates		
Transient Response		
Stop and Go / Ramp drive		
Crawl		

A detailed description of the manoeuvre 'Approaching Target' can be found below.

Approaching Target

Within this manoeuvre, the Target drives at constant velocity and is outside of the sensor range of the Hunter. The Hunter, using an activated ACC, drives at a higher velocity than the Target and reaches the Target during the manoeuvre (see Fig. 4).

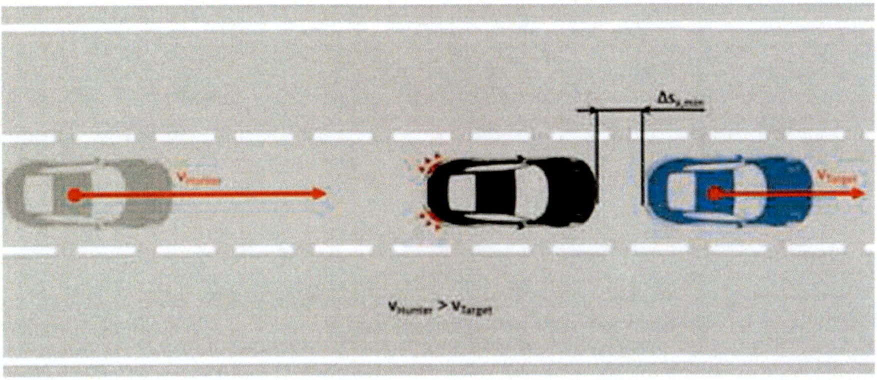

Fig. 4. Manoeuvre: Approaching Target

The reaction of the Hunter is measured. For example, in Fig. 5 some of the identified KPI´s are presented by plotting the distance between the Hunter and the Target over time. The dotted red line shows the ¼ speedometer distance in meters between the two vehicles. According to German road traffic regulations, this distance corresponds to the limit of a traffic offence. The first KPI is the time of reaction (t_React) and the response distance (sx_React). This time describes when the Hunter is braking higher than -0.3 m/s². The second KPI is the time of risk (t_Risk) which represents the time while the Hunter is under ¼-distance between both cars. The minimal distance (sx_min) is the minimum distance between both cars during the whole manoeuvre and the relative undershoot distance (sx_S_Us_rel) is the distance between the minimal distance and the ¼-distance (red dashed line). [1]

The maximum dynamic distance (max_dyn_sx) represents the distance between the Target and the Hunter after braking. The distance sx_10_static and sx_5_static are the distances during which the Hunter is within the 95 or 90 percentile of the adjusted distance after the manoeuvre. The time t_rebound_static_5 represents the time during which the Hunter is within the 95th percentile of the adjusted distance. The rebound time (t_rebound_min) represents the time the Hunter needs to reach the minimal distance and exceeding the ¼-distance line.

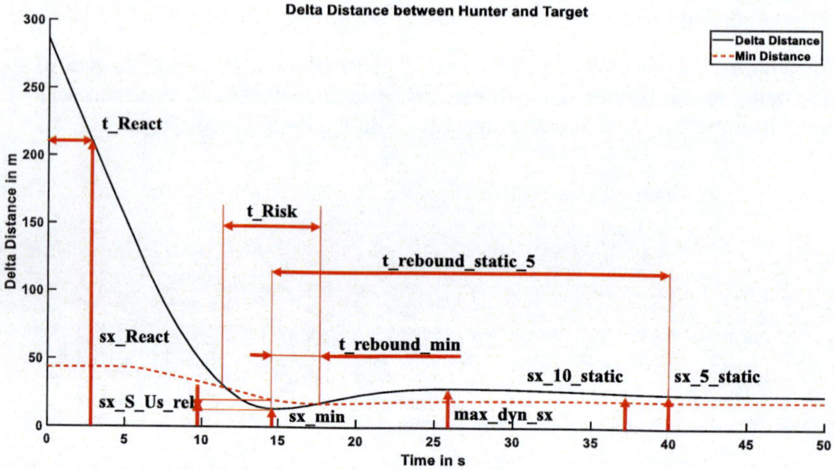

Fig. 5. KPI's for a repeatable and comparable manoeuvre

5 DATA EVALUATION

5.1 Test reproducibility and KPI estimation

To illustrate a high reproducibility, approaching to target is plotted in [Fig. 6&7] to show consistency in the test procedure. In this figure three different vehicles are tested and plotted in the same diagram. All driving tests were carried out under the same test conditions. The diagram shows the approaching with a differential speed from 180 to 80 km/h. The plot illustrates that all measurements start at a delta distance of 250 m and have nearly the same end distance. This scenario is a typical highway - scenario with a fast-moving vehicle approaching to a truck. 180 to 80 km/h is also the maximum manageable differential speed at PG. For higher differential speeds, no valid comparison could be accomplished. The bar plots in [Fig. 6] shows the KPI min distance and its standard deviation. It also shows that standard deviation rises with higher speed differences. In [Fig. 7] the hunter velocity is plotted with a starting speed of 180 km/h (speedometer) of the benchmark vehicles. It shows a slightly different starting speed. This problem accrues due to the restricted ACC-speed step option of some OEMs and poor speedometer precisions. This difference must be considered in the results.

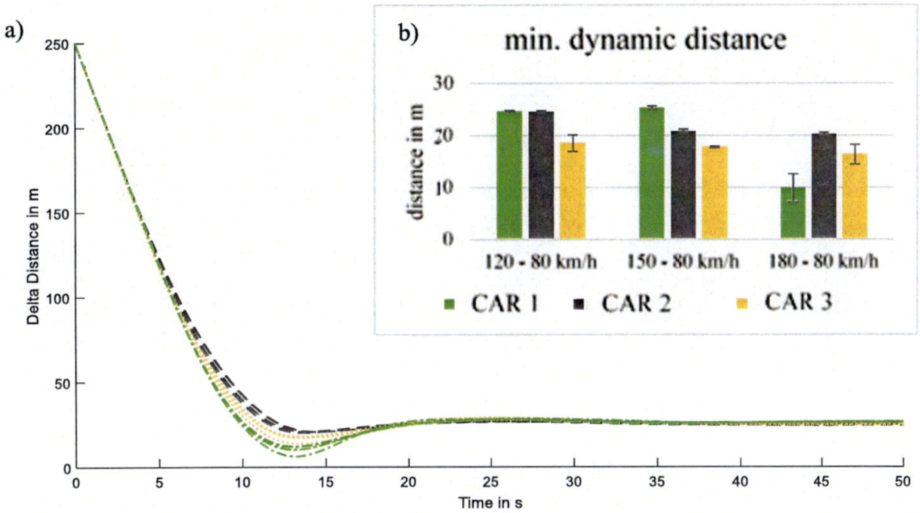

Fig. 6. a) Hunter Target delta distance at approaching to target 180 – 80 km/h. **b)** Comparison between three benchmarks

The plot also shows the reaction time of the cars. This is also an information about the detection quality. Car 1 reacts at a delta distance of 126.6+/-0.8 m. Car 2 and Car 3 react at almost the same distance of 166.98+/-9.31 m and 165.28+/-9.43 m. The standard deviation shows that Car 1 has a more consistent and a lower detection range. Because of the late reaction time, Car 1 is required to decelerate further. Despite the higher deceleration, Car 1 cannot compensate the late reaction time. This can be seen in [Fig. 7] with the min distance between Target and Hunter vehicle. Car 1 is approaching up to 9.87+/-2.7 m. Car 2 and Car 3 are approaching up to 20.28+/-0.3 m and 16.29+/-1.9 m. The average minimal distance kept by Car 1 at 150-80 km/h and 120-80 km/h is about 25 m. In all tested speeds Car 3 falls below the min distance, Car 2 does not have any positive undershoot values and Car 1 has one positive ID with 10.18+/-2.73 m at 180-80 km/h. The data shows the limit of system and sensor setup. Furthermore, [Fig. 7] shows the maximum deceleration of this manoeuvre. The maximum allowed deceleration of -3.5 m/s^2 for 2 sec [20] is achieved by Car 1 and 3. Due to the fact that Car 1 applies a harder braking strategy in the beginning, it is also theoretically possible to fall below the minimum distance for Car 3. Because of its faster brake application, Car 2 achieves the negative target acceleration sooner and does not fall below the minimum distance.
The measurement shows that Car 1 has an inappropriately selected sensor system for the area of operation. The ACC-system of Car 1 can be set up to 210 km/h. The tests show that the system cannot handle higher speed differences than 180-80 km/h in an ideal environment. In this test case the driver of Car 1 waits for 5 s after t_react to detect an issue with the management of a minimum distance.

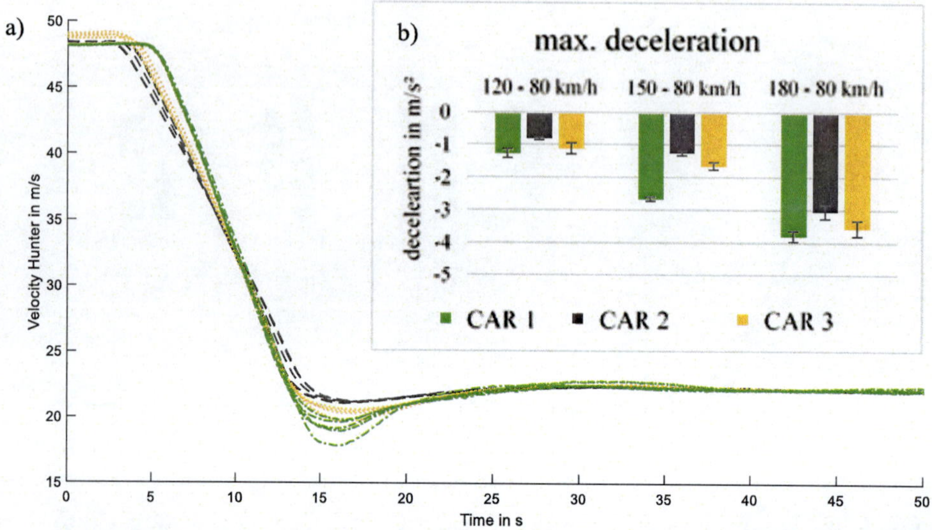

Fig. 7. a) Hunter speed at approaching Target at 180-80km/h. **b)** Comparison between three benchmarks

5.2 Comparison with subjective data

The next step is to compare the subjective and objective KPIs. As an example, two KPIs from two different cars of the manoeuvre approaching target are compared. The first KPI is the minimal distance. The left diagram represents the subjective assessment of the expert group. The experts rated Car 3 with around 7.1 and Car 1 with 8.2. A higher value inferring a better performance [Fig. 8]. As a result, Car 2 subjectively maintained a larger minimum distance than Car 3. The right diagram shows the measured minimal distance with two different initial velocities. The minimal distance of Car 1 is higher in both manoeuvres. This correlates with the experts' evaluation.

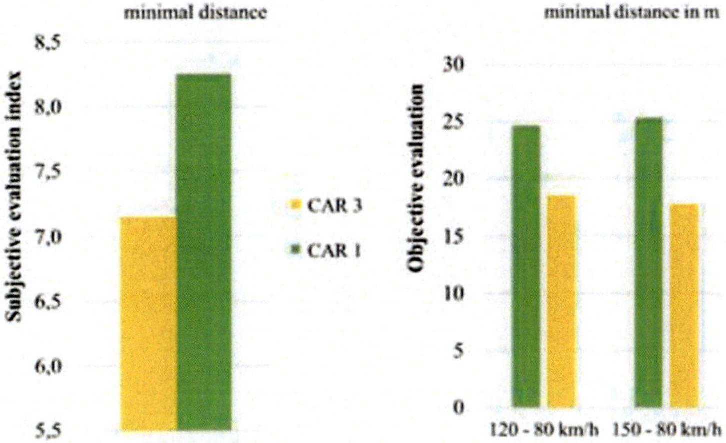

Fig. 8. Comparison of subjective and objective criteria (minimal distance) for manoeuvre approaching Target of two different cars

The next example represents the same two cars as before. The considered value is the response distance, which describes the response distance at the first noticeable deceleration. This value also provides information about the detection quality. Car 3 has a higher subjective evaluation index with around 8.0 than Car 1 with a value of 6.6. A higher value inferring a better performance [Fig. 9]

This means that from a subjective point of view, the first noticeable deceleration starts earlier with Car 1 than with Car 3. The feeling of safety is therefore perceived higher in Car 1. The measured values are presented in the right diagram. When Car 3 / Car 1 drives at a velocity of 120 km/h and reaches the Target which drives with a velocity of 80 km/h, the response distance is around 100 m (Car 3) and 90 m (Car 1). When Car 3 / Car 1 drives 150 km/h, the response distance is 170 m (Car 3) and 115 m (Car 1).

Car 3 shows the fastest reaction time with a significant difference to Car 1 with a start speed of 150 km/h. Therefore, the measurement results correspond with the subjective evaluation index. The experts also rated Car 3 higher than Car 1 from a subjective point of view.

In summary, the subjective assessment of the experts largely coincides with the objective measurements. It also shows that human subjective perception is directly demonstrable in measured data and a property-based development, in relation to the level model, is necessary.

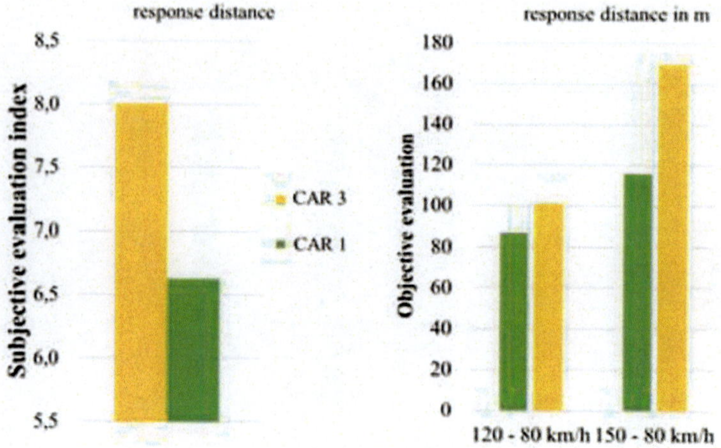

Fig. 9. Comparison of subjective and objective criteria (response distance) for manoeuvre approaching Target of two different cars

6 CONCLUSION AND OUTLOOK

The established method shows an efficient property-based evaluation for ACC. The current development of driver assistance systems leads to a function-based system selection instead of a property-based system selection. Consequently, sensors and actuators are not selected according to customer requirements. Customer requirements and properties must then be considered afterwards through parameterization with the existing sensor/actuator set.

Driving experts determine subjective evaluation parameters while objective evaluation parameters are developed based on these subjective evaluation parameters. The objective parameter development is supported by simulation data and linked to subjective KPIs using a step model in order to show important relationships. For a comparison of objective and subjective parameters, a fully controllable target vehicle was created. This vehicle is controlled by a network INS technology to generate results as efficiently and reproducible as possible. The networked INS technology allows for distances to be determined precisely. Simulation-based evaluation scripts enable an immediate data assessment inside the vehicle. A self-aligning platform generates high efficiency when installing the test vehicles. The comparison of subjective and objective evaluation parameters of two different cars coincide which leads to an efficient method to evaluate adaptive cruise control.

This method can be used as a basis for future development of adaptive cruise control and enabler for better customer acceptance of advanced driver assistance systems. The development of such a system cannot only be achieved with subjective evaluation indices. Additionally, it can be used for objective evaluation indices.

The methods are currently being integrated into an automated assessment tool for industrialization. In addition, the overall vehicle behaviour is to be broken down into the Sens-Plan-Act cause-effect-chain. Among other things, the sensor performance is to be evaluated by means of KPI's and their influence on the overall vehicle behaviour is to be analysable and specifiable. In the future, the method will be developed for other driver assistance systems, e.g. predictive adaptive cruise control.

References

1. SCHICK, Bernhard, et al. Bewertung der Funktion und Güte von Fahrerassistenzsystemen bei aktivem Bremseingriff. *ATZ-Automobiltechnische Zeitschrift*, 2007, 109. Jg., Nr. 5, S. 414-425. Author, F., Author, S.: Title of a proceedings paper. In: Editor, F., Editor, S. (eds.) CONFERENCE 2016, LNCS, vol. 9999, pp. 1–13. Springer, Heidelberg (2016).
2. SCHICK, B., et al. Attribute-based development of Advanced Driver Assistance Systems. Presented 19.03.2019 In: *19. Internationales Stuttgarter Symposium*.
3. HUBER, B.; SCHICK, B. Methoden zur Rekonstruktion von Versuchsszenarien und die Bewertung von Funktion und Güte von Fahrerassistenzsystemen. In: 2. Tagung Aktive Sicherheit durch Fahrerassistenz. 2006.
4. HAKULI, Stephan; KRUG, Markus. Virtuelle integration. In: Handbuch Fahrerassistenzsysteme. Springer Vieweg, Wiesbaden, 2015. S. 125-138.
5. BECKER, S.; SONNTAG, J.; KRAUSE, R. Prometheus CED 5 (AICC): Zur Auswirkung eines intelligenten Tempomaten auf die mentale Belastung des Fahrers, seine Sicherheitsüberzeugungen und (kompensatorischen) Verhaltensweisen. Bericht zum Testauftrag der Firma Daimler Benz AG, Stuttgart, im Rahmen des PROMETHEUS-Projektes" Method for Measuring Driver Capacity" des TÜV Rheinland (EU 45 Phase II, TV 8940 4), Köln: TÜV Rheinland, 1994.
6. FANCHER, P. S., et al. Intelligent cruise control: Performance studies based upon an operating prototype. In: Moving Toward Deployment. Proceedings of the IVHS America Annual Meeting. 2 VolumesIVHS America. 1994.
7. BULD, S., et al. Die Fahrsimulation als Methode bei der Entwicklung von Fahrerassistenzsystemen. VDI-Berichte, 2003, S. 153-175.
8. SACHER, Heike. Wie nutzen Fahrzeugführer das adaptive cruise control?. In: 2. Tagung Aktive Sicherheit durch Fahrerassistenz. 2006.
9. FREYER, Jörn. Vernetzung von Fahrerassistenzsystemen zur Verbesserung des Spurwechselverhaltens von ACC. Cuvillier Verlag, 2008.
10. RAJAONAH, Bako; ANCEAUX, Françoise; VIENNE, Fabrice. Trust and the use of adaptive cruise control: a study of a cut-in situation. Cognition, Technology & Work, 2006, 8. Jg., Nr. 2, S. 146-155.
11. HOEDEMAEKER, Diede Marika. Driving with intelligent vehicles: driving behaviour with adaptive cruise control and the acceptance by individual drivers. 1999. Doktorarbeit. TU Delft, Delft University of Technology.

12. IDIER, Muriel. Ein Verfahren zur Messung des Komforts von Abstandsregelsystemen (ACC-Systemen). Ergonomia-Verlag, 2006.

13. SCHICK, Bernhard, et al. Bewertung der Funktion und Güte von Fahrerassistenzsystemen bei aktivem Bremseingriff. ATZ-Automobiltechnische Zeitschrift, 2007, 109. Jg., Nr. 5, S. 414-425.

14. SCHICK, Bernhard, et al. Simulationsmethoden zur Evaluierung und Verifizierung von Funktion, Güte und Sicherheit von Fahrerassistenzsystemen im durchgängigen MIL-, SIL-und HIL-Prozess. In: 3. Tagung Aktive Sicherheit durch Fahrerassistenz. 2008.

15. THOMAS, J. Abstandsregeltempomat – Mit List und Lücke. Auto Motor Sport, Ausg: 2006 - 10, 2009.

16. TORBEN, Liesner, et al. Eine Methodik zur Analyse, Bewertung und Auslegung von ACC - Systemen. 13. Braunschweiger Symposium, Braunschweig, 2012.

17. Freyer, J.: Vernetzung von Fahrerassistenzsystemen zur Verbesserung des Spurwech-selverhaltens von ACC. 1. Aufl. Göttingen: Cuvillier Verlag (Audi Dissertationsreihe, 9), 2008.

18. STRASSER, Benedikt. Vernetzung von Test-und Simulationsmethoden für die Entwicklung von Fahrerassistenzsystemen. Cuvillier Verlag, 2012.

19. SCHICK, Bernhard, et al. Eigenschaftsbasierte Entwicklung von Fahrerassistenzsystemen. ATZ-Automobiltechnische Zeitschrift, 2019, 121. Jg., Nr. 4, S. 70-75.

20. NORM, I. S. O. 22179 2009. Intelligent transport systems–Full speed range adaptive cruise control (FSRA) systems–Performance requirements and test procedures.

21. SCHICK, Bernhard, et al. Optimization of steering behavior through systematic implementation of customer requirements in technical targets on the basis of quality function deployment. Fisbita 2006, 2006.

22. ABENDROTH, Bettina. Gestaltungspotentiale für ein PKW-Abstandsregelsystem unter Berücksichtigung verschiedener Fahrertypen. Ergon, 2001.

23. FANCHER, P. Intelligent cruise control field operational test. Final report. Volume I: Technical report. 1998.

The Power of Oil – Influence of Shock Absorber Oil on Vehicle Ride and Handling Performance

Dr. Shinji Kato [1], Dr. Navid Samadi [2]

1. KYB Corporation, Japan

2. KYB Technical Center Europe (Munich), Germany

Abstract.
When the suspension deforms, damping force is generated as the piston of the shock absorber moves, but a time difference occurs before the large hydraulic damping force rises. At the same time the frictional force in the shock absorber is playing an important role in generating effective damping force against deformation of the suspension, the generated force itself is small but its response is high. The frictional force in the shock absorber occurs in reciprocating motion in various speed ranges from amplitude of less than one millimeter to larger amplitudes, its behavior changes dynamically and is non-linear. Recent studies have revealed that controlling the friction is more important than reducing it for improvement of the performance. Because of this we have investigated the dynamic friction characteristics of shock absorber oil and sliding parts, by the means of an own developed device that can measure the speed dependence of friction force in reciprocating motion with high accuracy.
The result of our analysis shows that friction dissipates energy at a speed level of 0,002m/s where hydraulic dissipation is not working yet. By knowing this phenomenon, we have developed different kind of oil types in order to generate various dynamic friction characteristics.
The influence of our developed oil types on the ride and handling performance was tested and analyzed in different vehicles. Subjective assessments and objective measurement show the vehicle performance improvement, which can be generated by using more advanced oil types.

Keywords: shock absorber, damping force, dynamic friction, riding comfort, stability, vertical dynamics

P. E. Pfeffer (ed.), *11th International Munich Chassis Symposium 2020*, Proceedings, https://doi.org/10.1007/978-3-662-63193-5_8

1 Introduction

The damping force of the shock absorbers (SAs) is one of the elements making up the characteristics of the suspension system, and it mainly affects vehicle stability and ride comfort. In recent years, vehicles experience smaller fluctuations in the vertical direction even at high speed as road conditions have improved. Therefore, SAs are often operated at high frequencies with small amplitude, where the direction of SA motion is frequently reversed and the SAs are mainly operated at the low speed.
SA generates damping force by hydraulic oil passing through valves attached to a piston with its piston rod movement, SA responsiveness is affected by elastic deformation of oil (oil column rigidity) [1] especially at the point where the direction of SAs is reversed. There are new technologies that adjust hydraulic force, for example, a hydraulic valve structure [2] and an electronically controlled SA [3] aiming for generating appropriate damping force according to the road surface conditions by changing hydraulic force. However, these technologies are limited in terms of their ability to improve the damping force when the direction of motion is reversed because of de-formation of oil. The current situation calls for SAs that improve ride comfort and safety (stability, steering response) when driving on good roads.
Experience with on-vehicle subjective evaluation of SA performance indicates that ride feeling on good roads can vary greatly with hydraulic oil difference. In this paper, we focus on effect of hydraulic oil on vehicle ride and handling. Therefore, in order to clarify the factors that influence the change in feeling, we developed our own hydraulic oil to adjust the damping force in the very low speed range where direction of motion reverses.

2 Shock Absorber Design

2.1 Conventional damping force measurement method and its problems

A conventional indicator of the performance of SA is shown in Fig.1. The right figure is a magnified one of the surrouded area by the red square in the left figure. Here, the performance of the SA is evaluated by a plotting the peak damping force generated by the SA versus stroke speed, which is called “damping force - speed characteristic line diagram”. Generally, it does not show speeds of 10 mm/s or below. It is a quasi-static characteristics and not suitable for the evaluation of dynamic performance. Hence, in order to evaluate the SA performance on good roads, we believed that it was necessary to evaluate the dynamic characteristics from the time of reversing the moving direction to the very low speed region, which has not been noticed in the conventional performance evaluation.

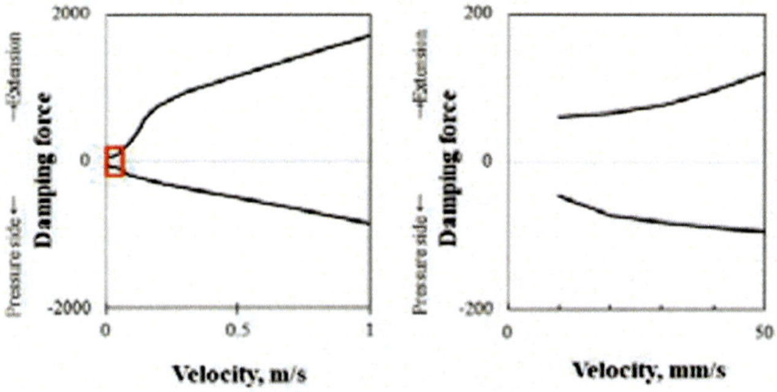

Fig. 1. Damping force - velocity characteristics line diagram

2.2 Fluid resistance and friction force in damping force

Fig. 2 shows the structure and damping force components of a twin-tube-type SA. As shown in the figure, hydraulic force is generated by the pressure change as hydraulic oil passes through the valves [4], and friction force is generated from the oil seal, the rod guide, and the piston band. The damping force of SAs is the sum of the hydraulic force and the frictional force. We believe that in the very low speed region, a region which has not yet been extensively studied, the hydraulic force would have a smaller impact on the damping force, while the friction force, which has nonlinear characteristics, would have a greater impact [Fig.3].

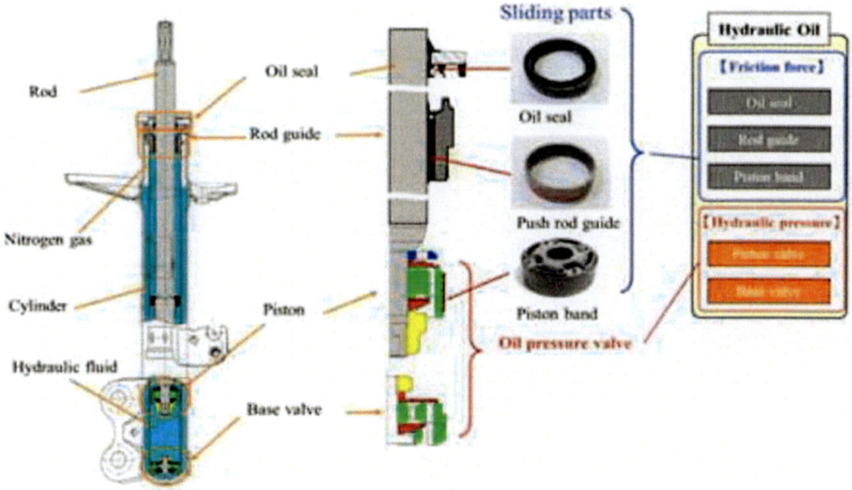

Fig. 2. Twin-tube-type shock absorber structure and damping force component

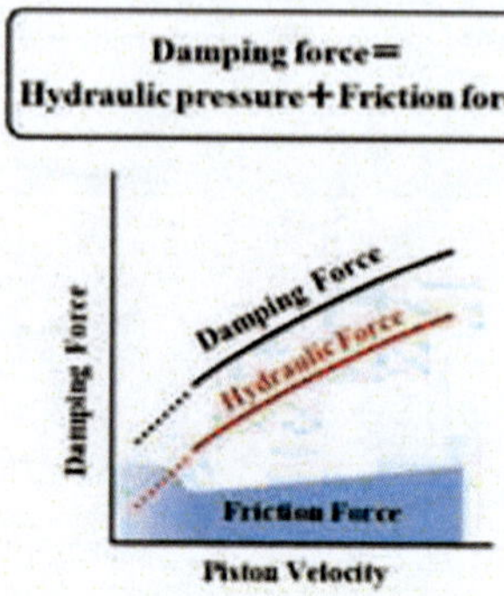

Fig. 3. Relationship between piston speed and damping force

In order to understand the relative contribution of hydraulic force and frictional force to the total damping force, we investigated the damping force-speed characteristics. The measurement experiments were conducted as follows.
First, we prepared a valve-less SA by removing valves in order to generate no hydraulic force, and we measured the damping force which now consisted only the friction force. This enables us to determine the speed characteristic of the hydraulic force from the difference between the damping force - speed characteristic of a normal SA and that of the valve-less SA. Fig. 4 shows the changes in the proportion of hydraulic force and the frictional force in the damping force as a function of the piston speed. The figure shows that the proportion of friction force is the highest in the very low speed region where the piston speed is 30 mm/s or less.

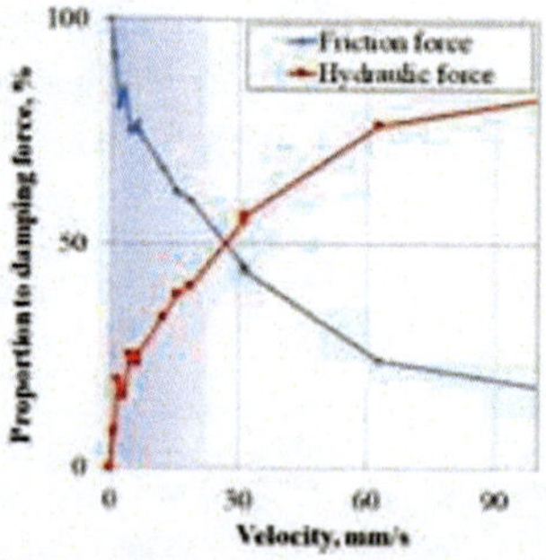

Fig. 4. Speed dependency of load ratio of hydraulic pressure and friction force in damping force

Next, we analyzed the behavior of the shock absorber on the test course in order to determine their functionality on a good road. In the experiment, we drove a passenger car with a displacement of 1.8 L at 60 km/h on a test course that reproduced the uneven road surface equivalent to a good road on a national highway. During the experiment, a displacement sensor was mounted in parallel to the SA, as shown in Fig.5, and the

displacement was measured at 2-kHz intervals, to calculate the frequency distribution of the stroke speed and the displacement. Fig. 6 shows the frequency distribution of the SA speed and Fig. 7 shows the displacement distribution from the neutral position of the SA. We found that for at least 50% of the time on a typical good road, the displacement of the SA would be less than or equal to ± 1.5 mm from the stroke center. We also found that for 50% or more of the time, the piston speed was less than or equal to ± 30 mm/s, and the most frequent speed was less than or equal to ± 10 mm/s.
In these low-speed and low-amplitude regions, little hydraulic force is generated, and the damping force consists mostly of the friction force. In other words, on good roads, the damping force of SAs is largely dominated by the friction force.

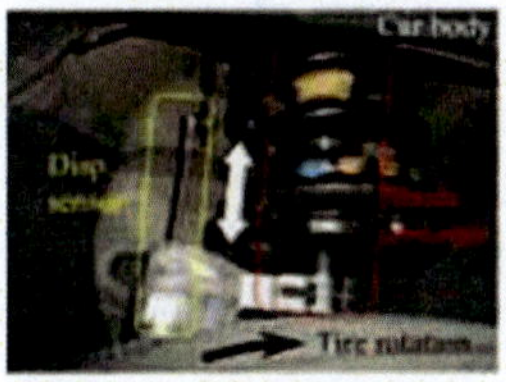

Fig. 5. External view of suspension displacement behavior measurement during actual vehicle running

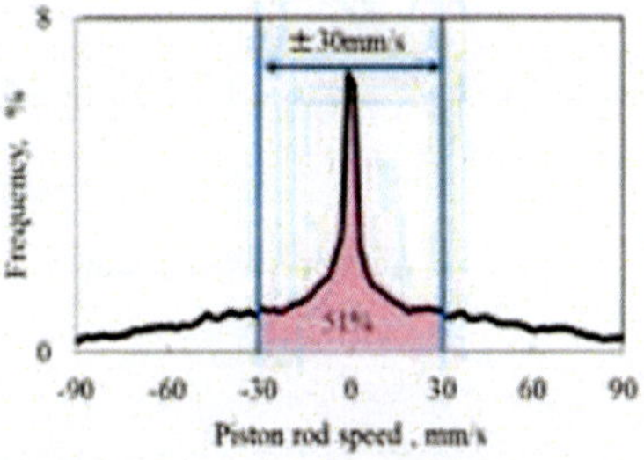

Fig. 6. Frequency distribution of piston speed

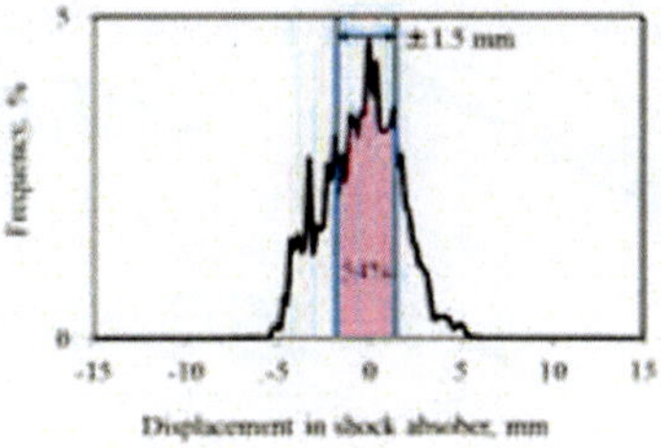

Fig. 7. Frequency distribution of displacement from neutral position

2.3 Evaluation methods and problems in the low-speed operation region

In order to evaluate the dynamic characteristics of the damping force, we use a plotting the force generated by SA versus stroke speed when the SA is operated at a constant frequency and amplitude, as shown in Fig.8. The low-speed region around the zero point (surrounded by the red square) in Fig.8 is magnified in Fig.9.
Damping force has hysteresis due to the compressibility of the hydraulic oil [1], that works in the form of oil column rigidity as shown in Fig.10, and in the very-low-speed zone starting at the zero line where the direction of motion reverses, the damping force of SAs is in the direction of acceleration. We believe that the SA response can be improved by reducing this zone.

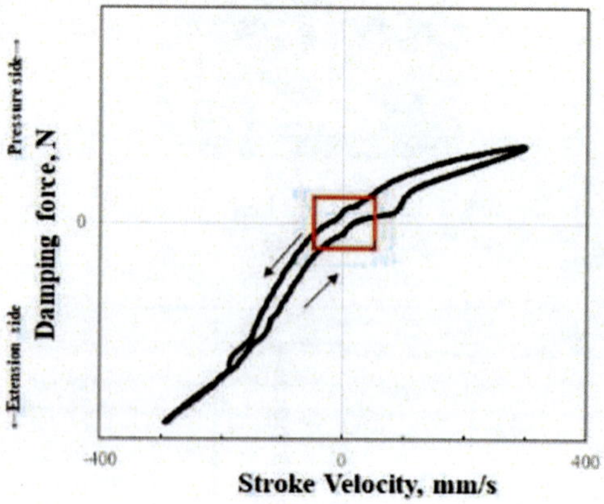

Fig. 8. F-V characteristics

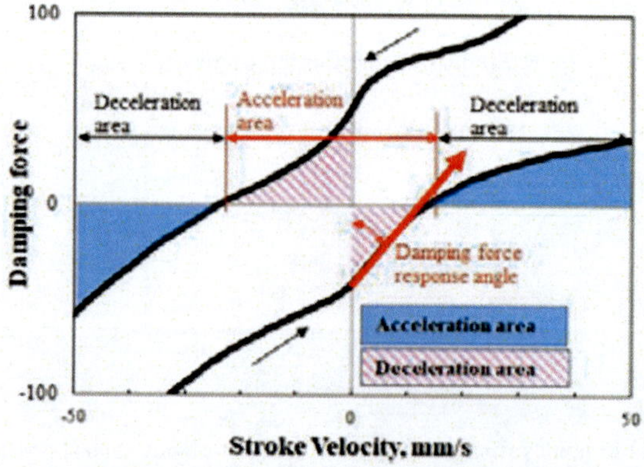

Fig. 9. F-V characteristic around the zero point

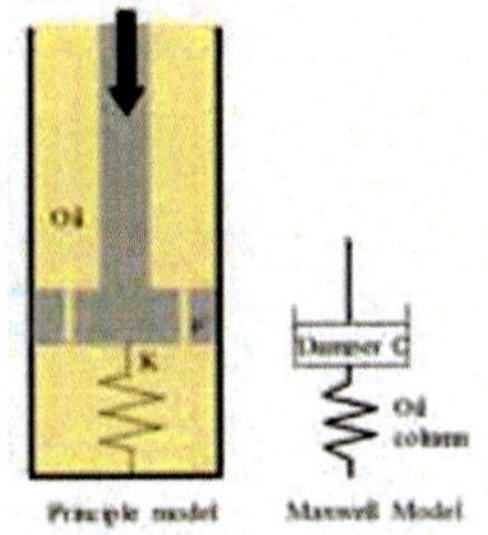

Fig. 10. Influence of oil column stiffness characteristics

Friction adjustment using hydraulic oil has been studied in order to improve SA responsiveness, but the response which was observed as the gradient around zero point and the magnitude of the friction force are simultaneously changed as shown in Fig.11 in order to avoid the occurrence of stick-slip. However, when this method is used, the responsiveness is improved, but at the same time, the ride comfort is deteriorated, and this trade-off characteristics have not been improved. Therefore, we thought that we can improve the responsiveness of SAs without deteriorating the ride comfort by adjusting only the gradient of the friction without changing its magnitude as shown in Fig.12.

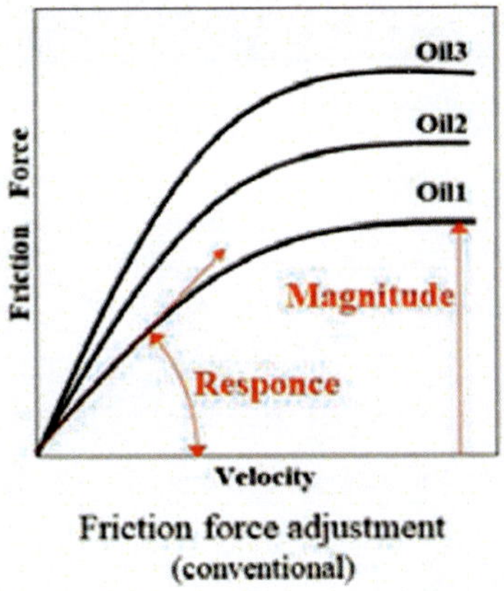

Fig. 11. Technology for changing the magnitude of friction

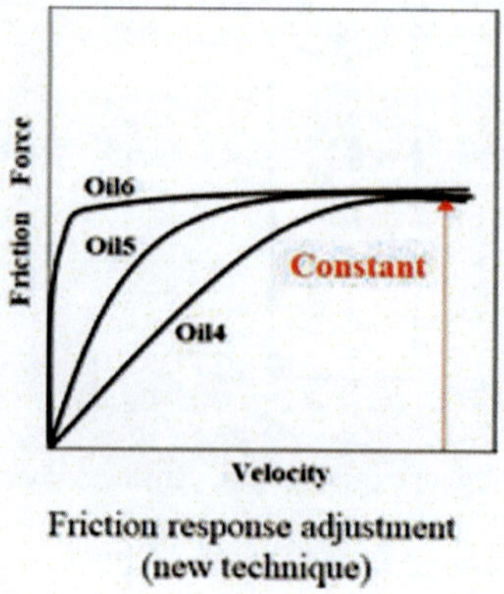

Fig. 12. Technology for adjusting responsiveness

2.4 Friction influence and related components

In order to evaluate friction characteristics while separating friction response and friction force and develop our own hydraulic fluid technology, we developed our own device as shown in Fig.13 and use it to examine the evaluation index [5].

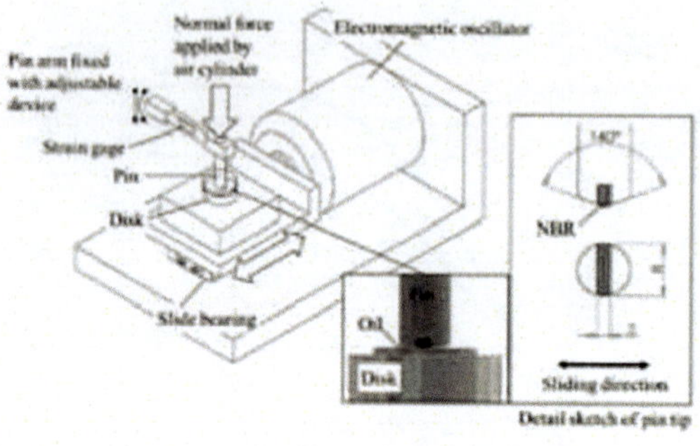

Fig. 13. Schematic of reciprocating force-tester

2.5 Response index

The Response index (RI) was defined as a quantitative index to evaluate the response of friction when the direction of movement was reversed. Fig.14 shows the range where we extracted the RI in the time waveforms of displacement and friction. *Fsa* is the maximum friction in the range from the start of motion up to the phase of $\pi/4$ radians (that is, top and bottom dead centers, $\pi/2$ to $3\pi/4$), and normalizes the difference by the average friction. We divided the difference by the average in order to obtain the acceleration-side RI using formula (1).

$$Response\ index(RI) = (F_{sa}-F_{ave})/\ F_{ave},\quad \pi/2 \leqq \theta \leqq 3\pi/4 \tag{1}$$

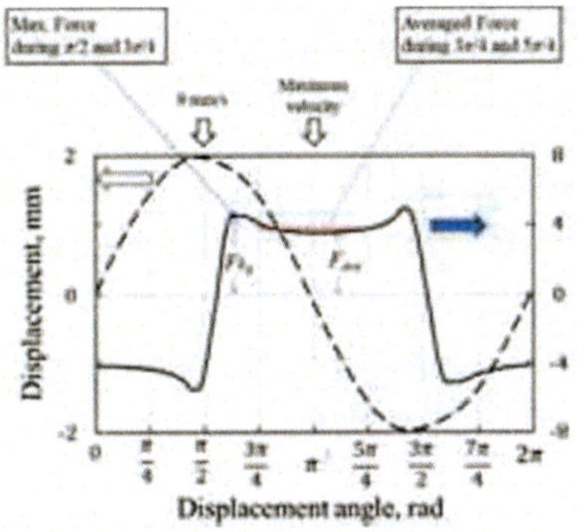

Fig. 14. Calculation concept of the Response index (*RI*)

3 New KYB OIL types and their characteristics

We used our specially made friction test device to develop our own hydraulic fluid technology which can adjust only the response when the direction of motion is reversed without changing the average friction force. Fig.15 compares the actual characteristics of the oils that we developed with the conventional hydraulic oil characteristics.

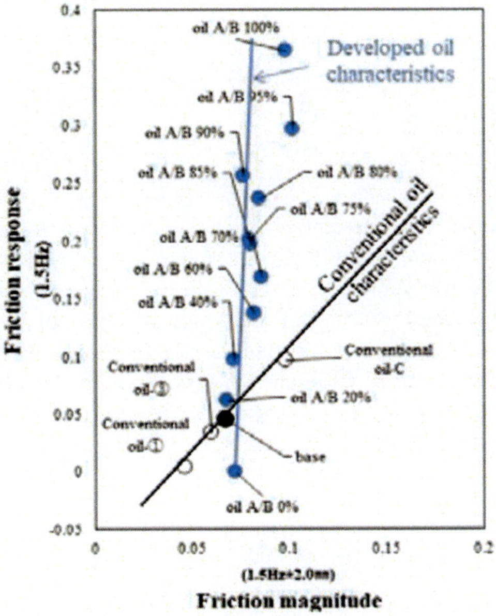

Fig. 15. Characteristics comparison between conventional oils and oils we developed

Hydraulic oil A is a high-response, and the hydraulic oil B is low-response, although the magnitude of friction is almost the same in each case. By changing the blending ratio of these hydraulic oils, it is possible to adjust only the response without changing the magnitude of the frictional force. This technique enables us to investigate the effect of the responsiveness on the basic performance in on-vehicle and bench tests.

4 Oil types Influence on Ride and Handling Performance (Vehicle Assessment Result)

The new KYB oil types and their performance variation on component level were analysed. In order to prove the impact of the oil types on vehicle ride and handling performance, an objective and subjective vehicle assessment is carried out by an engineering service company specialized in vehicle dynamics measurement and assessment.
For this analysis, three sets of SAs with three different oil types, Base (reference, not modified from the original equipment), KYB A/B 100% and KYB A/B 75% are set up. The viscosity of the two KYB oils were adjusted so that it would be the same as the Base specification and the influence of kinematic viscosity and the magnitude of friction would not appear [Table 1].

Table 1. Oil properties

		Base	A/B75	A/B100
Density (15°C)	g/cm ^{3}	0.83	0.85	0.86
Viscosity index		167	168	152
Kinematic viscosity (40°C)	mm2/s	11.7	12.9	12.9
Kinematic viscosity (100°C)	mm2/s	3.3	3.9	3.4

Apart from the oil, the SA hardware and tuning are not changed. It should be mentioned that by not changing the tuning of the SA while changing the oil, the performance might not be optimally balanced at vehicle level, since the tuning is made usually after choosing the oil. However, since the target of this analysis is to isolate the oil impact on vehicle performance, the tuning is not changed. Fig. 16 shows the damping force-velocity characteristics of the three different SA sets. It can be confirmed that there is almost no difference in damping force characteristics, and one would not expect that the vehicle ride and handling performance will be significantly differing.

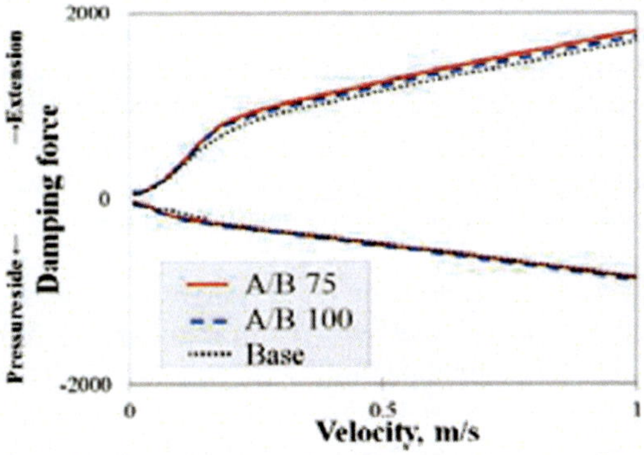

Fig. 16. Damping force-velocity characteristics

4.1 Objective assessment analysis conditions

The vehicle used for this analysis is a Mercedes-Benz A-Class (W177) which was equipped with various acceleration sensors on knuckle/wheel hubs, top mount at all 4 corners and the seat rail. The full sensor setup is shown in Fig.17.

The test track on which the measurements have been carried out consist of different road surface types: Nevada (smooth tarmac), South Africa (harsh tarmac) and Country-Road (bad local highway) and a special lane with 30 mm steps (3 positive and 3 negative). The vehicle speed was varied between 30-50 km/h in steps of 5 km/h and from 60-100 km/h in steps of 10 km/h. Measurements were conducted 3 times on every lane and at every speed step, and all of the measurements were averaged for the evaluation. All accelerations, body modes and movements were assessed for each lane, for low, high, and full speed range.

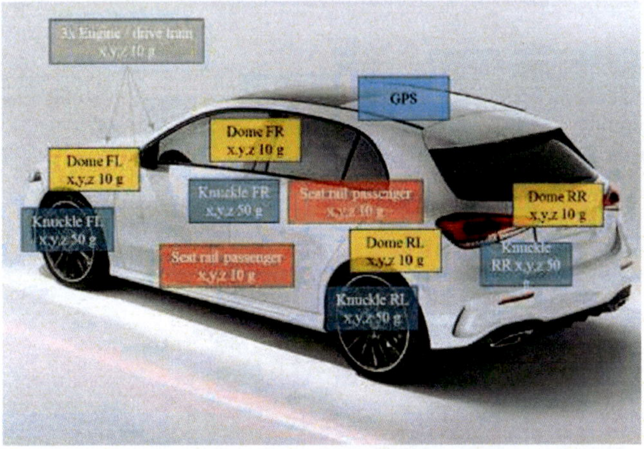

Fig. 17. Vehicle measurement setup

4.2 Objective assessment results

The results of the analysis are shown in Fig.18-19. Compared the base configuration shown on the dotted line, the A/B 100% oil shown on the dashed line exhibits a smaller body acceleration in the low-frequency region of 1-2 Hz on and larger body acceleration in medium-frequency region of 4-15 Hz. The A/B 75% oil shown on the solid line exhibits a smaller body acceleration than the base configuration in both the low- and medium-frequency region.

In general, there is a trade-off to be made when using valves to control the hydraulic force. If the body vibration near sprung mass resonance frequency of 1-2 Hz is reduced to emphasize safety, the body vibration of 4-15 Hz increases. To emphasize ride comfort, the opposite approach is required [6]. The characteristics are contradictory, making it difficult to achieve both goals. These results show that by simply adjusting the friction responsiveness while keeping the characteristics of the damping force unchanged, it is possible to produce a difference in the vibration levels to achieve both ride comfort and handling. The effect of friction response on the body vibration in the region can be summarized as follows.

Compared to the base configuration, our A/B 100%, which has the highest responsiveness, changes the vibration in a way that emphasizes safety. The A/B 75% oil, which has an intermediate responsiveness between the A/B 100% and the base con-figuration, does not change the vibration of the 7-15 Hz region, and yet decreases vibration in the 1-2 Hz region. This oil can be said to improve the stability of the vehicle body without impairing the riding comfort. This improvement, which is generally difficult to achieve with hydraulic valves, appears to be the result of the oil's improved responsiveness, which reduces the acceleration zone in the very low speed range.

We investigated the effect of a SA equipped with this more responsive hydraulic oil on an actual vehicle, and we confirmed that it has a significant impact on handling and ride comfort performance.

Traditionally, when the hydraulic force is adjusted to improve the riding comfort, it is at the expense of safety (manoeuvrability and stability), and vice-versa. We were able to enhance both the riding comfort and safety performance by changing the hydraulic oil to adjust the damping force after the reversal of the direction of motion in a region where it has been difficult to achieve both goals. The results verified that this is an important technique for meeting the requirements for ride comfort and safe-ty depending on the type of vehicle and the customer's preference.

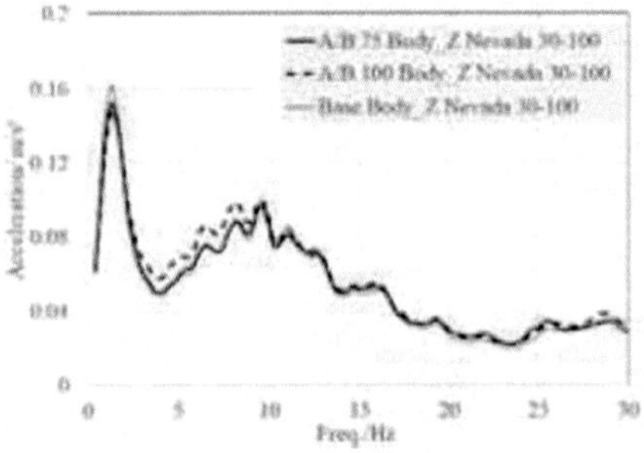

Fig. 18. Vertical Vehicle Body Acceleration for Nevada

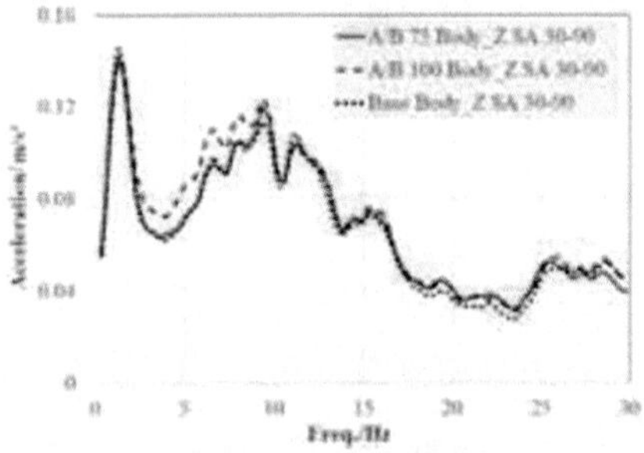

Fig. 19. Vertical Vehicle Body Acceleration for South Africa

4.3 Subjective assessment

The subjective evaluation of ride comfort and handling/safety was made on different roads and various conditions such as German motorway (with speed up to 180 km/h), well-used roads, potholes, asphalt patches and state roads with conditions between brand new and well-used. Further the assessment was made on the test course containing comfort tracks, simulated state road, single obstacles, handling track, long straight. As shown in Fig.20, varying the responsiveness of the oil resulted in significant changes in ride comfort and safety. In addition, we confirmed that it is possible to modify characteristics such as "safety" and "comfort" depending on the purpose and taste of the vehicle.

The subjective assessment of our A/B 100% and A/B 75% oils compared to the base configuration was also commented by the test driver as if a valve setting such as the bypass or bleed area of the SA were modified. Driving the vehicle with the new oils changes the vehicle dynamics like it has a different positioning e.g. a sport type derivate, or a different manufacturer tuning philosophy.

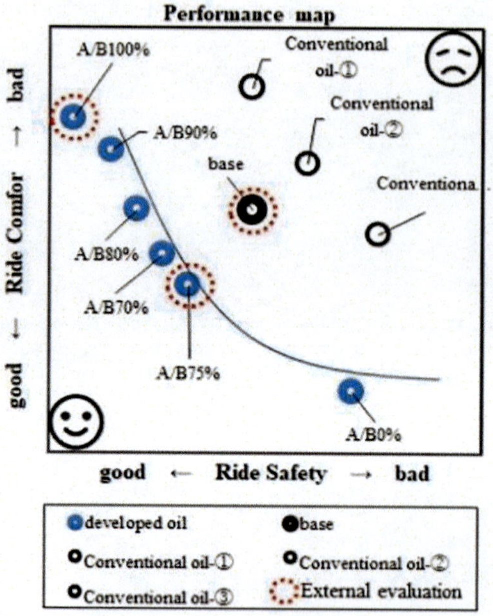

Fig. 20. Response adjustment hydraulic fluid performance

5 Conclusion

1. When driving on good roads, SAs typically operates in a very-low-speed region between zero (where the direction of motion is reversed) and 30 mm/s.
2. On good roads, the very-low-speed region of the SA has a greater effect on the friction force than the hydraulic force, which greatly affects the ride comfort and handling.
3. The conventional friction adjustment technique in a very-low-speed region of the SA, which involves varying the magnitude of friction, has drawback in that it negatively impacts the ride comfort.
4. We confirmed that the friction response is an important factor in meeting safety and ride comfort depending on the type of vehicle and customer preferences.
5. We believe that mechanism by which we were able to achieve this result is the improvement in performance produced by compensating for the damping force that is generated by compressibility of the hydraulic oil in the acceleration zone part of the very-low-speed region in which the direction of motion is reversed.
6. By using the KYB developed oils a new dimension of creating variations and optimizing the vehicle drive feel opens. As the result, we developed a technique that adjusts only the friction response and confirmed that it can improve safety without compromising ride comfort.

6 Outlook

In this paper, we proposed a technique for improving the performance using ordinary conventional SAs, but we anticipate that this technique will improve performance even further if applied to a SA with the latest sated-of-the-art valve structure. In the future, in addition to pursuing the practical application of these results, we plan to further study on the relationship between the friction response and vehicle performance, and will continue to develop the potential of this approach as a new performance enhancement technique.

References

1. Ernst, Walter. "Oil hydraulic power and its industrial applications." (1960).
2. KAMAKURA, Ryosuke, et al. "Development of Externally-Mounted Shock Absorber with Adjustable Solenoid Damping Force." (2017).
3. Klinger, Florian, Johannes Edelmann, and Manfred Plöchl. "Characterization and potential analysis of passive and (semi-) active suspension systems by means of equivalent suspension parameters." *10th International Munich Chassis Symposium 2019*. Springer Vieweg, Wiesbaden, (2020).
4. Dixon, John C. *The shock absorber handbook*. John Wiley & Sons, (2008).
5. Kato, Shinji, and Shinya Sasaki. "Effects of hydraulic oil and lubricant additives on dynamic friction properties under various reciprocating sliding conditions." *Friction* (2019).
6. Kaldas, Mina MS, et al. "Triple-control-mode for semi-active suspension system." SAE International Journal of Commercial Vehicles 8.2015-01-0621 (2015).

How to extract advantage from driving simulator technology in developing brand specific vehicle dynamics

Florian Goy[1], Antonino Pizzuto[1], Christian Ludwig[1], Alessandro Salgarello[1]

[1] Hyundai Motor Technical Center Europe GmbH, Germany

Abstract. The two major challenges in chassis development - both aiming at improved efficiency - the automotive industry is facing are:

- reducing development cost – e.g. less prototype vehicles, less physical testing
- increasing value for customers – e.g. more variants, higher complexity

In this challenge, virtual methods are efficiently serving the development process with data and key performance indicators in order to understand effects on an objective level. Nevertheless, shaping the overall vehicle dynamics characteristics to a brand-specific DNA and providing this DNA throughout the whole vehicle lineup often requires the need to enhance the understanding of objective knowledge by subjective measures.

Given multiple examples of new installations in the recent years, the rise of driving simulator technology in the industry is more than obvious. However, driving simulators itself are tools – and like other simulation tools – featured with several limits and boundary conditions in their application. As in physical testing, test procedures need to be designed in a way that specific performances of the vehicle can be evaluated. The same holds for the usage of driving simulators, but unlike in physical testing several additional points need to be considered. Therefore, the paper describes as a first step a driving simulator-oriented test and maneuver design for efficient vehicle evaluation and lists advantages which are only present in virtual environments.

Whereas many publications are developing methods which focus on studies with many participants (“normal driver”), Hyundai has developed a methodology aiming at the standard vehicle dynamics process which is executed with the help of professional subjective test drivers. The paper describes how the virtual tests are prepared and how test drivers are smoothly integrated in the process.

The application of this process is shown on the example of vehicle variant development, which is in the real world only possible for some selected variants of a given vehicle. Based on the shown process, a driving simulator environment enables the subjective evaluation and consistency check for different weight, trim line or tire configurations for a wider range of variants, which finally improve the quality of the brand specific vehicle dynamics DNA.

P. E. Pfeffer (ed.), *11th International Munich Chassis Symposium 2020*,
Proceedings, https://doi.org/10.1007/978-3-662-63193-5_9

Keywords: Vehicle Dynamics, Virtual Development, Driving Simulators, Development Process

1 Introduction

Hyundai Motor Europe Technical Center (HMETC) has as European development and research center of Hyundai Motor Company (HMC) the responsibility to develop vehicle driving performance in order to match the taste of European customers and - by that - fit the strong needs of competitiveness on the European market.

In this role the focus lies on the technical aspects of Vehicle Driving Performance and especially on the topics of Chassis Development including all relevant items like Suspension Hardware, Tire, Steering and Chassis Control Systems. Nevertheless - as overseas development center - the availability of prototype vehicles and proving grounds is limited, and project timing often does not allow detailed tuning and optimization activities on physical prototypes. For this reason, virtual development is one of the growing key areas of expertise in HMETC which is continuously enhanced by improved methodologies and tools.

One of the technologies currently under investigation are moving base driving simulators and their claimed capability to shift subjective evaluation of driving performance into the virtual environment, reducing the mentioned trade-off between project timing and availability of vehicles, tuning hardware/software and proving grounds.

This paper is based on practical experience from several vehicle projects and driving simulator activities in the past 3 years. The underlying theory and fundamentals are considered, but as the driving simulator is a tool intended for subjective evaluation, the definition of a generalized methodology for application is not straightforward. Nevertheless - a general finding is, that the success of driving simulator application is strongly dependent on the environment, the company structure and finally everyone, who is part in the process. To succeed, it seems necessary that vehicle test and simulation disciplines are merging into a common entity, which requires that the people involved have knowledge or are open-minded in both domains.

2 Vehicle Development Process for Dynamic Driving Performance

The general understanding of the term driving performance is everything which influences the customer driving experience by means of driver input or forced oscillation by road irregularities and resulting vehicle reaction to driver and passengers.

2.1 Driving Performance in Chassis Development

In Chassis development the full vehicle Ride Comfort & Handling driving characteristics are in focus. Ride Comfort is mainly a forced excitation due to road irregularities, and therefore highly dependent on the condition of the road. In Ride Comfort especially the vertical, longitudinal, roll and pitch movements of the body (see Fig.1.b) up to 30 Hz are investigated.

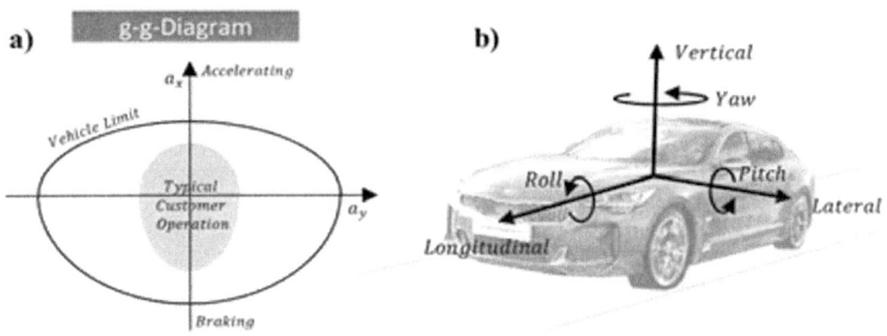

Fig. 1. a) g-g-Diagram with vehicle limit and customer operation b) full vehicle movements of interest for vehicle dynamics

In contrast handling describes the interaction between a driver input at the steering wheel and the resulting vehicle reaction, which happens especially in lateral, yaw and roll movements. Steering torque buildup behavior plays as well a significant role in how the driver can feel the vehicle and influences therefore the customer driving experience. Compared to Ride Comfort the frequency range is with up to 2-3 Hz lower [1] but lateral displacement and yaw angle are much higher for typical cornering maneuvers.

Compared to the forced and often unavoidable excitation due to road irregularities, in handling domain the driver can decide on his own how quickly or in which acceleration range he operates the vehicle. In developing driving characteristics for passenger vehicles it is still important to know the range of operation or at least in which performance envelope the majority of customers will operate for the predominant amount of time. The g-g-Diagram in **Fehler! Verweisquelle konnte nicht gefunden werden.**a) shows the limit performance envelope for lateral and longitudinal acceleration. Inside this envelope, the area of typical customer operation is sketched which covers mainly the linear range of driving up to approximately 0,4g [2]. Even if for some specific and sporty oriented vehicles the behavior in near-limit driving is of higher attention, the normal driving range needs to give the driver in all vehicles a high confidence and a pleasant feeling.

An additional differentiation in operational range for vehicle handling can be made between the before mentioned normal anticipatory and sudden unpredictable driving tasks like an obstacle avoidance. In this situation, the vehicle will need to be steered very abrupt which results as a consequence in a fast vehicle reaction up to the lateral limit. Also in this exceptional situation the vehicle controllability and stability needs to be maintained.

2.2 Virtual and physical development process

Ride Comfort & Handling characteristics are evaluated objectively and subjectively [3] during the development process with both pros and cons. Subjective evaluation is the closest way to future customers use, but comes with the necessity of an available physical vehicle prototype. If this is not given the objective evaluation by full vehicle simulation is the only possibility to evaluate driving characteristic prior to the availability of a prototype – with the challenge that resulting numbers and KPIs need to be interpreted in a customer oriented way. A driving simulator is a tool, which can be used to generate subjective evaluations from virtual vehicle prototypes – merging both worlds and reducing the before mentioned constraints.

During Concept and Platform development typically all design decisions have to rely on objective data coming for example from simulation models. At this time only Virtual Prototypes are available. As soon as physical prototypes are available, subjective evaluation is possible – still in the early phase physical prototypes often are only representative for the new platform and lack comparability to the driving characteristics of the desired final product. The highest impact of driving simulator integration in the development process can be expected in the early phase, when subjective evaluation can be made possible. But as well in the phase of serial development, the simulation technology does have advantages over physical testing to increase efficiency – e.g. if many setup iterations or variants have to be evaluated in a short amount of time.

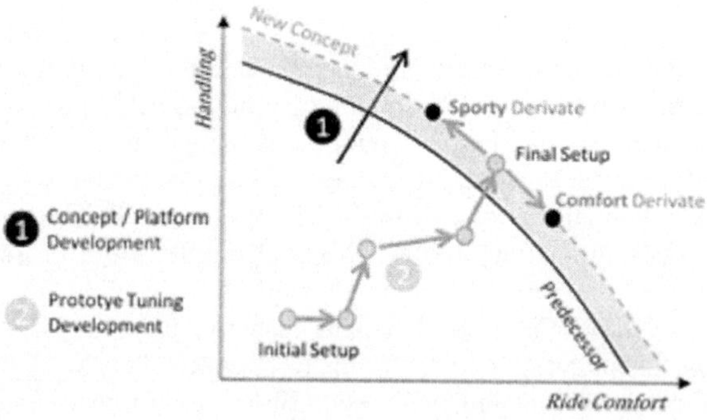

Fig. 2. Ride Comfort & Handling performance envelope & aim of development phase

Finally in a vehicle development process many parameters are defined straight from the beginning as they need to fit the target customer group and thereby the size, usability level and target driving character. In consequence vehicle properties with strong dependency on Ride Comfort and Handling like wheelbase & -track, weight & weight distribution, tire size and driveline layout are only slightly adjustable or even fixed. These parameters are not always aligned to achieve the expected driving character – e.g. shape sporty driving character to large SUV-styled vehicles.

Even if the before mentioned boundary conditions tend to diverge more and more - the goal in chassis development is to improve the performance from one generation to another, highlighted in **Fig. 2**. If the general design specifications of subsystems and components which influence driving dynamics have been defined in concept & platform phase, the performance envelope of the new concept should be larger in both Ride Comfort & Handling, reducing the need for trade-off solutions as much as possible. In the phase of serial development the target is to setup the tunable elements of the chassis like bushings, springs, dampers and control software in a way that the maximum of the new platform performance envelope can be utilized.

2.3 Importance and limitations of subjective evaluation

How effective subjective evaluation is, becomes obvious for everyone driving a new and unknown vehicle. Automatically a comparison to the known (daily used) vehicle starts. After some minutes of driving a – probably only subconscious - judgement has been made. The judgement detail could range from as simple as one might think of (like / don't like) to very complex and structured comparison of single performances - the latter one typical for professional test engineers.

The generation of an objective overall rating is much more difficult, even if enormous efforts have been made in research and industry by trying to objectivize the subjective evaluation [4]. There are a couple of reasons which explain this difficulty. Assuming that all key performance indicators (KPIs) are independent, unique and sensitive to changes of vehicle reaction, there are still following challenging questions to be answered:

- How to calculate an overall rating from multiple KPIs?
- How to weight individual KPIs to each other to match a desired vehicle character (e.g. sporty, comfortable)?
- How big is the acceptable variance in KPIs not affecting or deteriorating the overall rating?

Partly those questions can be answered by analyzing benchmark vehicles (if available), but if the definition of a unique brand DNA is in focus, benchmarking methods are limited. Nevertheless, subjective evaluation methods have as well limitations, which were mentioned before briefly:

- Physical vehicle prototypes need to be available
- Many variants (engine, transmission, trim line) need to be checked for consistency
- Setup changes on tunable elements like springs, dampers, bushings take time
- Early prototypes lack comparability to final product

Nearly all of the bullets can be supported by the integration of driving simulation methods in the process as they are scalable, time efficient and highly available.

3 Driving simulator application: how to make a valuable tool for test engineers

As every other tool in the development process driving simulators do require the understanding of the special boundary conditions and limitations coming along with the usage. Therefore HMETC has developed a testing methodology, which is specifically designed for the use on driving simulators. It consists of the testing process in general as well as the driver adaption and accordingly the maneuver design.

3.1 Driving Simulators for Driving Performance Evaluation

Nowadays several types of moving base driving simulator are known and used from research institutes and the automotive industry. The main differentiation between known systems is the size/weight, travel ranges and the concept of the moving base system [5, 6]

Fig. 3. a) VIgrade DiM250 @ IDIADA, Spain [7] b) ABDynamics aVDS [8]

However – the main focus of this work is put on commercially available solutions of mid-size driving simulators which allow due to their typical low moving mass higher frequencies to be excited and are therefore usable for work in the Ride Comfort domain, e.g. offered from companies like ABDynamics[1], Cruden[2], McLaren Applied

[1] https://www.abdynamics.com/en/products/lab-testing/driving-simulators
[2] https://www.cruden.com/automotive-driving-simulators/

Technology/MTS[3] or VIgrade[4]. The systems under spotlight are the aVDS driving simulator from ABDynamics and the DiM250 driving simulator from VIgrade both having a comparable lateral travel range of approx. 2.5m and both offering a linear frequency behavior of up to 50Hz. The systems are shown in **Fig. 3**.

Whereas the system of VIgrade is using a combined tripod/hexapod design, the system of ABDynamics uses 2 rails with eight linear motors and four wedges for vertical motions. Both simulators can carry the mock-up made of a real vehicle on top of the moving base. The visual system consists of multiple projectors (minimum 5) and a cylindrical / conical screen with diameter of up to 9 m. According to the supplier both systems are designed for low latency operation of the moving base and the visual system. There are typically several additional items which can be added to a driving simulator in order to improve the immersion, for example active seat and belt tensioners, high frequency shakers or active pedals. The simulators in spotlight were especially selected by HMETC as they are offering high dynamic motion and low latency visual feedback which is both expected to have a high importance for Ride Comfort & Handling application.

3.2 Strengths and limitations of "engineering" mid-sized driving simulators

In an ideal world a test driver would not be able to distinguish driving a simulated or real vehicle. However – this is not possible as today and an experienced test driver will feel clear differences driving a virtual vehicle in a driving simulator. A major part of the difference can be explained by the limited moving space. In order to explain this in more detail **Fig. 4** are comparing the lateral acceleration (a) and displacement (b) resulting from typical driving scenarios and maneuvers. On the x-axis the steering wheel input dynamics quantity is shown without unit. It can be interpreted as e.g. maximum steering wheel speed or the steering wheel angle frequency from a sinusoidal input – whatever is more appropriate for the driving task in focus.

As mentioned in section 2.1 the normal customer operation from field operational tests show lateral accelerations below $4\,m/s^2$. Steering wheel angle velocities are rather slow for smooth driving operation e.g. on country road sections. As soon as the maneuvers get more dynamic, the achieved accelerations can go up to the vehicle limit (e.g. $< 10\,m/s^2$). The obstacle avoidance as the most abrupt maneuver shows typically the highest steering wheel velocities especially if an oversteering reaction needs to get stabilized. The acceleration limits of the driving simulators are much higher than the mentioned $10m/s^2$ and are not considered as a limitation compared to the situation of lateral displacements shown in **Fig. 4** b). A simple calculation shows, that driving for only 1s with a lat. acc. of 4m/s^2, the vehicle will be laterally displaced by 8m. Even for simulators with much higher travel range than 2.5m the limi-

[3] https://www.mclaren.com/applied/case-study/mts-vehicle-dynamics-simulator/
[4] https://www.vi-grade.com/en/products/dim-dynamic-simulator/

tation in travel will be reached quickly afterwards. To overcome these (strong) limitations many motion cueing approaches have been developed from tilt coordination [9] to model predictive based moving space optimization [10] or track-based prepositioning [11] or simply scaling down the movements to a reduced level. All of the known motion cueing strategies are trying to "cue" a motion to the driver in order to "feel" the vehicle reaction in the most realistic way. Whereas the travel limitation could be a potential constraint for handling application, the situation for ride comfort is different. The vehicle vertical movements and pitch and roll angles are even for badly maintained roads in the travel range of the moving base and a limitation will be only the frequency limit until which the moving base can deliver a consistent excitation, see section 3.1.

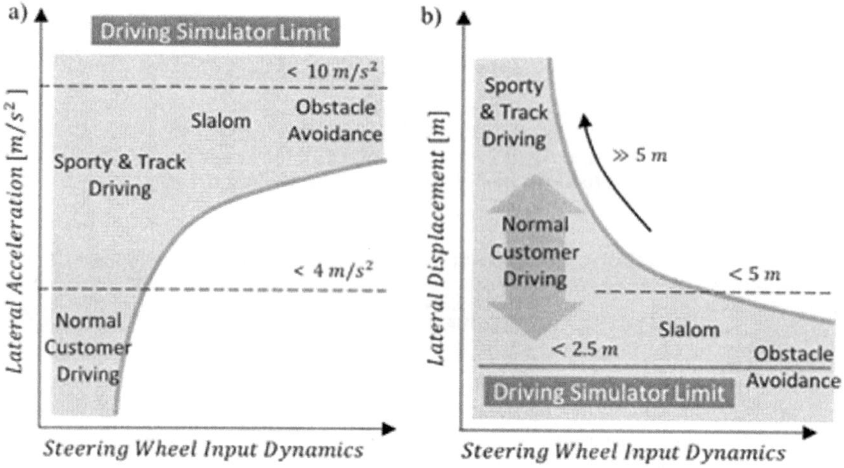

Fig. 4. Lateral acceleration (a) and displacement (b) for selected driving tasks

3.3 Requirements on test drivers and adaption to virtual environments

Subjective test drivers for Ride Comfort & Handling are trained to identify shortcomings in a vehicle e.g. compared to other vehicles or competitors and to report and eventually counteract these shortcomings. To do so, the feedback from vehicle motion, steering wheel torque as well as vibrations and noise are essential. The section before has shown that especially handling characteristics of a virtual vehicle rendered on a moving base driving simulator will in most cases feel different for the test driver than a real vehicle, e.g.:

- body motions will be scaled down, therefore not matching the exact behavior of the vehicle, especially for motions larger than the motion space
- body motions will wash out and not persist after a short amount of time (no steady state accelerations)

Apart from the differences perceivable in the mentioned motion feedback there are several other feedback systems, like the visual, audio and steering feedback system. State of the art visual systems work with a refresh rate of 120Hz, which gives in most automotive based situations a fluid and smooth perception of the picture. Additionally to the motion feedback the visual system is essential for the driver for localization and especially for low frequency – high amplitude body movements. Therefore it needs to be synced to the motion system – or in other words the latency needs to be low. One of the most common constraints of driving simulator usage is the occurrence of the so called motion sickness or more specific simulator sickness, which is dependent on the mentioned latency but as well on other not yet fully known effects. In consequence this leads to conclusion that a test driver who can work efficiently in virtual environments needs to have:

- mental capability to accept reduced or even wrong motion compared to the expected behavior of a real vehicle
- capability to transfer findings on a driving simulator into the real world, considering the missing or reduced feedback channels
- no strong tendency to get motion sick (often already known from travelling as passenger)

Upper requirements can be found out by giving a test driver several trial runs on a driving simulator and potentially switching between a real vehicle and a virtual one side-by-side. Therefore a special introduction and adaption strategy was elaborated, which consists of following steps:

1. Detailed introduction of the moving base driving simulator system (3.1)
2. Explanation of the behavior and limitations to be expected (3.2)
3. Demonstration of the moving-base behavior to the candidate as external viewer
4. Experience of only visual, steering feedback and sound without capturing the full field of view (potentially to be done on a static simulator)
5. Experience of full system operation on endless straight road (without traffic)
6. Increasing the complexity of driving tasks

It proved beneficial in several experiments to introduce the system more from a simulation perspective – approaching step-by-step to the full system operation. The adaption phase (5 & 6) can take several days each having only short times of exposure per day.

3.4 Maneuver Design for Handling & Ride Comfort

In order to overcome the specifics of driving simulators described in the earlier sections the highest improvement for handling evaluation is possible by designing the maneuver in accordance with the system in place. Therefore the question was which driving maneuvers require the least amount of cueing or which driving maneuvers can be even rendered 1:1:

- on-center – staying in one lane – driving on a straight (endless) road section

- short, high dynamic steering input & high lateral acceleration maneuvers, with low overall displacement e.g. lane change, double lane change, slalom

As a result a straight three lane road section was selected to serve as basic and first step in vehicle handling evaluation. In **Fig. 5** this road model is visualized. Looking from a first view very simplistic, this road model proved very powerful and offers additional setup possibilities like a double lane change or slalom setup. Another advantage using this kind of road model compared to real testing is the possibility to setup repeated maneuvers like lane changes on the road. In this way it is easy to run several repetitions of a maneuver in a very short amount of time. Having as well the possibility to evaluate different vehicle velocities up to high speed without limitations in road section length turns out to be an additional flexibility over other roads.

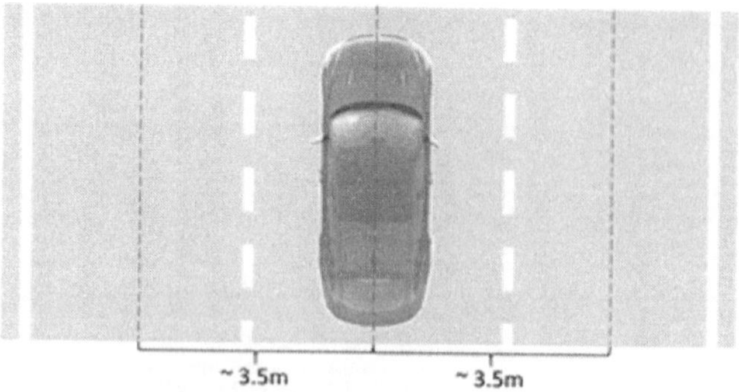

Fig. 5. 3 lane road section as basic for handling evaluation

A typical evaluation on the described road begins with a free-driving session followed by the placement of several double lane changes and slaloms by cones. Besides the evaluation on a straight road there are maneuvers and tracks which are affected more by a limitation of motion space:

- maneuvers with persistent and high steady-state accelerations
- long duration corners with small corner radius leading to high yaw angle difference between start and end of maneuver

These kind of road characteristics can be found e.g. on winding country road sections or on racetracks. Therefore an evaluation on country roads is done only on pre-selected sections where the before mentioned effects are not appearing. However – a major reduction in motion scaling will be unavoidable.

The maneuver and scenario design for the evaluation of ride comfort on driving simulators with the mentioned high frequency capabilities is more straightforward, as the typical excitations are happening in the range of mm rather than m. The experience still shows, that often high accurate road surface models from LIDAR scans are available (e.g. from offline simulation) but no according graphical model was made, which is required for a driving simulator experience. For this reason, HMETC created a synthetic scenario model, where short patches of high accurate road surface models can be included. The graphical scenario is based on the before mentioned 3 lane road setup. To integrate a patch of high accurate road surface into the existing straight & flat road scenario model, some adjustments to the measured road are performed:

- low frequency vertical content is removed, e.g. no altitude changes are existing anymore between start and end of the road surface patch
- low frequency lateral curvature content is removed – resulting road patch is straightened
- any high road banking will be transitioned smoothly into the rest of the flat road laterally and longitudinally

All operations are performed without altering the (higher) frequency character as well as amplitudes of the road. Another positive side effect from the described adjustments is that no large translational travel of the moving base are expected to happen, which effectively gives again the possibility of a 1:1 scaling.

4 Applying the guideline to setup brand specific vehicle driving performance

In this chapter two use-cases about the development of brand specific driving performance using driving simulator technology are shown. Both use cases were examined considering the process shown in chapter **Fehler! Verweisquelle konnte nicht gefunden werden.**.

4.1 Vehicle variant development

Over the past 15 years, the vehicle line-up complexity has dramatically increased. To define a consistent and unique brand-specific DNA, all variants in the vehicle line-up need to be considered.

The process starts with an objective analysis of the respective variants in the line-up. Typical setup-related KPIs are used to describe the vehicle, such as Ride Frequencies, Roll Moments and their balance between front & rear axle. These objectives, combined with a reliable estimation of the vehicle properties in terms of mass and inertia, gives the possibility of an initial comparison and adjustment if required. In addition, the brand differentiation is performed considering different KPI targets, depending on the brand character. An additional competitor analysis gives finally the boundary conditions for the vehicle development.

For the driving simulator case-study a vehicle with a defined baseline setup is the starting point: how much additional information can be delivered by a driving simulator in order to confirm – or readjust – the tuning direction?

In this scenario the driver is asked to drive two different variants of the baseline vehicle, where only springs and anti-roll bars have been modified. While the balance of front/rear vertical Eigen-frequencies has been left the same, the roll balance has been considerably changed. The objective values are shown in **Fig. 6**.

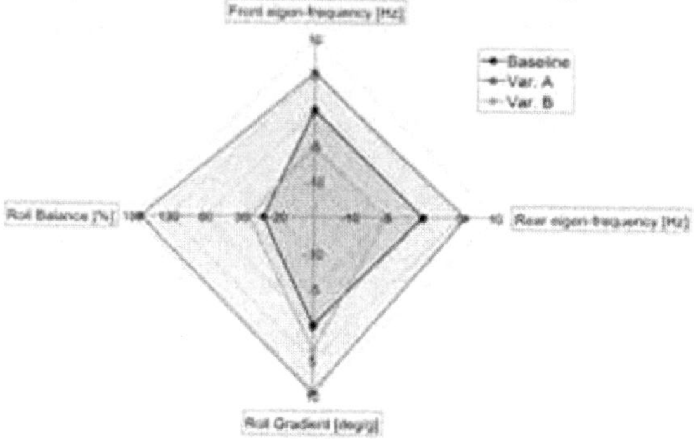

Fig. 6. Example of two vehicle variants: values represents delta (in percentage) vs. baseline

After running the maneuvers for handling evaluation, the driver has been asked to give his judgment and no variant was better judged than the baseline. Slightly different characters have been recognized with the variant **A** being judged better for cornering stability and response, with a degradation in terms of understeering level; the variant **B** has been instead judged very similar to the baseline, even though the softer spring setup.

To understand the subjective ratings better, several standard open-loop driving maneuvers (ramp steer and sweep steer) were simulated using the same vehicle model like used on the driving simulator. Looking at the objective results coming from them, the **understeer** gradient matches the subjective rating, with variant **A** judged more understeering than the others. The recognized **responsiveness** can be instead identified in the delay between steering angle (system input) and yaw rate (system output) at 0.5Hz (close to normal customer utilization range), which is lower than the baseline.

After running the maneuvers for ride comfort evaluation a clear difference could be recognized between the variants, with variant **B** judged overall better than baseline and variant **A**, with the only issue of a worse primary ride. Again an offline simulation with the vehicle model was performed on two different road surfaces (smooth and rough) and extracting the acceleration behavior in frequency domain. Looking at the objective results, it can be noticed that variant **B** has a lower acceleration than the others, with the exception of primary ride on a smooth road. Surprisingly variant **A** has been subjectively judged worse than the baseline, while objectively the accelerations are lower than the reference. After a detailed analysis, it has been recognized that variant **A** was having much higher accelerations on rough roads than the others, affecting the final judgment.

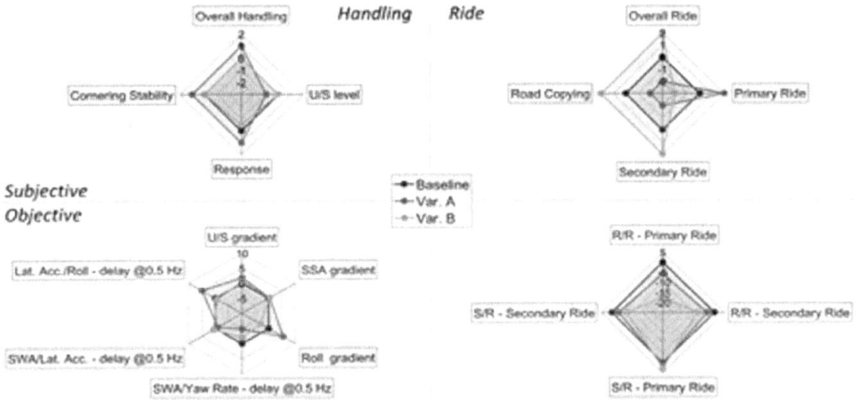

Fig. 7. Example of Subjective (+2 = Much better, -2 = Much Worse) vs. Objective (delta in percentage vs. baseline) of vehicle's Ride & Handling performance [U/S = Understeer, SWA = Steering Wheel Angle, SSA = Side Slip Angle S/R = Smooth Road, R/R = Rough Road]

At the moment variant **B** is considered the best variant from this study. It meets the handling targets and it offers better ride performance. How much further improvement could be reached by a dedicated damper tuning? Can the driving simulator as well deliver valuable information for this aspect?

Based on a spreadsheet calculation, a sporty and a comfort-oriented damper behavior have been designed, using now variant **B** as baseline setup. This time only Ride comfort behavior has been evaluated as the changes in damper curve were mainly targeted for this performance.

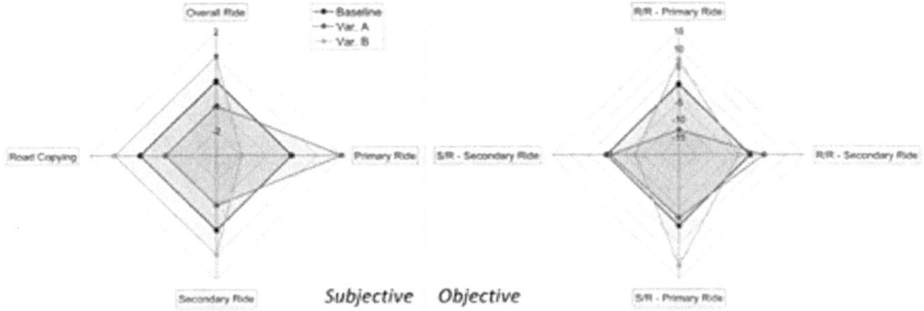

Fig. 8. Subjective (+2 = Much better, -2 = Much Worse) vs. Objective (delta in percentage vs. baseline)vehicle's Ride performance
[S/R = Smooth Road, R/R = Rough Road]

The driver recognized changes and judged the **comfort** damper as best variant. In primary ride the sporty-variant was rated better than the baseline. Nevertheless, the driver's opinion was that the difference was rather small. In the objective KPIs it is possible to find-back these judgments, as the comfort-oriented damper is making the vehicle moving more in primary ride but is as well filtering more in secondary ride. Difficult is instead to give a clear answer based only on objective parameters for what concerns the **sporty**-variant against the baseline. The sporty damper is giving a more controlled primary ride in both smooth and rough roads, but the secondary ride accelerations are comparable, especially for the smooth road.

As a result of the described study, it has been judged and objectively proved that the softer variant with the comfort-oriented damper curve is giving the best trade-off in the ride & handling performance of this vehicle. The findings will be spread over all variants in order to confirm consistency in between the vehicle line-up. To confirm the virtually optimized setup in the real world, the support of damper suppliers is required in order to translate the optimized hydraulic curve in a valve design. The possibility of direct valve tuning at the driving simulator, considering the effect of different valve constructions [12] is one of the future subjects to further improve the efficiency in the development process.

4.2 Tire mount variant development

The increase of variants due to line up diversification over the past years, has multiplied the released tire specifications in HMC. Each tire specification is determined by its market positioning in terms of:

- geography: EU, US, domestic
- segment
- brand DNA: Hyundai, Kia, Genesis

and design requirements such as wheel load, suspension design, maximum speed and available package space. Once the specification is defined the actual development process starts with the choice of a reference tire, against which the submitted tires in the development cycles are compared. Tire development targets are often expressed on the basis of this reference tire, these performances can be classified into two main groups:

- mostly subjectively evaluated performance: Handling, Ride, NVH
- mostly objectively evaluated performances: Rolling resistance, Wet & Dry Grip, NVH, Wear, Durability…

Some of the objective performances are further constrained by regulations regarding the tire itself or the combined vehicle-tire pairing (e.g. RRC, NVH, wet grip,...). For the driving simulator application, the main focus was therefore laying on the subjectively evaluated performances, i.e. handling and ride. Here the reference tire based development presents an almost perfect environment for driving simulator testing. The A-B comparison of two candidate tires with regular back to back confirmation drives against the reference tire is an easy task to achieve and can be executed with an unbeatable efficiency.

The tire force and moment (F&M) response is a central element in the model build up, to guarantee the validity of the overall vehicle response to driver inputs. Of course real time (RT) execution is a fundamental request to all tire models used for driving simulator tests.

For handling simulation a widely used empirical tire model is the Magic Formula (MF). The publicly available description of equations and the lean structure have contributed to the high use rate in vehicle dynamics simulation. The MF model parameters can be fitted to replicate the measured tires F&M response for quasi-static, transient and combined slip situations, with a satisfactory accuracy. Being based on a combination of mathematical equations, the MF model is RT capable. The MF model cannot represent thermal and speed effects on the F&M response, therefore one can expect differences in the limit behavior. This is as well true for situations where the tire is maintained under slipping conditions for longer durations.

In order to evaluate the difference in tire characteristics and not the difference in measurement protocol or fitting routines, the generation of the MF parameters should follow the same steps for each tire mount.

As discussed in chapter **Fehler! Verweisquelle konnte nicht gefunden werden.** the maneuver design was chosen to comply with the constrained moving space of the platform. Checking those maneuvers against the tire model induced limitations; we remark that the constraints are congruent.
We can apply therefore the before mentioned maneuvers, that are focused on the linear and the transient response of the tire. The two tire characteristics found to be governing the overall vehicle response for the above mentioned maneuvers are the cornering stiffness and relaxation length; refer as well to [13].

The above described maneuvers have been evaluated for a recent tire development of a SUV vehicle. The evaluation has been done in a two-step procedure, first on real track during the standard tire evaluation and second on a driving simulator. Both evaluations were based on the same set of tires or respectively their digital twin in form of a MF model.

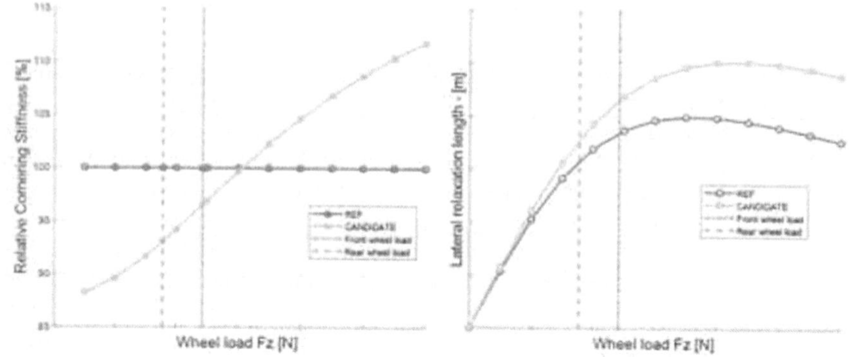

Fig. 9. Relative cornering stiffness & relaxation length

Fig. 9 shows the difference in cornering stiffness and their relaxation lengths. The driver ratings and description in **Table 1** showed a good level of correlation between the real track test and the driving simulator test.

Table 1: Driver rating (Real track, Driving Simulator)

	Real Track	**Driving Simulator**
REFERENCE	– Good stability – Slightly soft lateral response – Linearity is quite good and grip level is high	– Good Balance – Soft lateral response – Linear range responsive, grip level is good
CANDIDATE	– Overall balanced – Delay and linearity during the low-mid ay range – Needs more lateral support at high ay →more side slip than ref in the lane change	– Less US – Yaw gain increased, car is turning a lot – More side slip, Lane change: in 3rd gate is difficult to stabilize

The driving simulator has proven to serve as efficient tool for tire handling evaluation. A good matching of the driver's comments on variation of vehicle performance was observed. After promising first trials of the ride comfort evaluation on driving simulators, as well tire ride comfort evaluations are considered in a close future.

References

1. Mitschke, Manfred, and Henning Wallentowitz. "Dynamik der Kraftfahrzeuge. 5., überarb. u. erg. Aufl. 2014 Auflage.", Springer Vieweg, Wiesbaden
2. Ammon, D. "„Vehicle system dynamics challenges on the way to autonomous driving "." 4th International Munich Chassis Symposium. Hrsg. von ATZlive Springer Vieweg. Proceedings of the 4th International Munich Chassis Symposium. Wiesbaden: Springer Vieweg. 2013.
3. Heißing, Bernd, and Hans Jürgen Brandl. Subjektive Beurteilung des Fahrverhaltens. 2002.
4. Schimmel, Christian. Entwicklung eines fahrerbasierten Werkzeugs zur Objektivierung subjektiver Fahreindrücke. Diss. Technische Universität München, 2010.
5. Schöner HP., Morys B. (2015) Dynamische Fahrsimulatoren. In: Winner H., Hakuli S., Lotz F., Singer C. (eds) Handbuch Fahrerassistenzsysteme. ATZ/MTZ-Fachbuch. Springer Vieweg, Wiesbaden
6. Baumann, Gerd, et al. "How to build Europe's largest eight-axes motion simulator." Proceedings of the Driving Simulation Conference, Driving Simulation Association. 2012.
7. Applus+ IDIADA integrates virtual and physical testing with 2 new VI-grade driving simulators, Internet source: https://www.vi-grade.com/en/about/news/applus-idiada-integrates-virtual-and-physical-testing-with-2-new-vi-grade-driving-simulators_889/ (visited on 17.05.2020)
8. aVDS Advanced Vehicle Driving Simulator, internet source: https://www.abdynamics.com/en/products/lab-testing/driving-simulators/avds (visited on 17.05.2020)
9. Colombet, Florent, et al. "Tilt Thresholds for Acceleration Rendering in Driving Simulation." SIMULATION, vol. 93, no. 7, July 2017, pp. 595–603, doi:10.1177/0037549716675955.
10. Dagdelen, Mehmet, et al. "Model-based predictive motion cueing strategy for vehicle driving simulators." Control Engineering Practice 17.9 (2009): 995-1003.
11. Brems, Willibald. Querdynamische Eigenschaftsbewertung in einem Fahrsimulator. Springer-Verlag, 2018.
12. Salgarello A., Hahn D., Pizzuto A. (2020) Road to virtual tuning: support for damper tuning in Hyundai Motor Europe Technical Center with new lumped model and testing protocol. In: Pfeffer P. (eds) 10th International Munich Chassis Symposium 2019. Proceedings. Springer Vieweg, Wiesbaden
13. U.Peckelsen, Objective Tyre Development, KIT Scientific Publishing, Karlsruhe (2017), http://dx.doi.org/10.5445/KSP/1000073428

Designing and Assessing the Driving Experience of a Tilting Vehicle

Tom Lehmkuhl[1] and Prof. Dr. Lutz Eckstein[2]

[1] fka GmbH, Germany
[2] Head of Institute for Automotive Engineering (ika), RWTH Aachen University

Abstract: Small and narrow electric vehicles can represent a solution module for the impending traffic infrastructure and environmental problems in city centres. Along with this vehicle concept, either a deterioration in driving comfort due to a very hard chassis tuning has to be accepted or an active tilting system has to be integrated into the concept. On the one hand, this active tilting control requires an additional development effort for a robust and safe control system. On the other hand, however, it also offers new possibilities for a new, attractive driving experience. With this newly gained degree of freedom in the tilt motion it is possible e.g. to vary the perceived lateral acceleration in a wide range.

Keywords: Tilting, Experience, Perception.

1 Motivation

The continuing trend towards urbanisation presents cities with the great challenge of avoiding infrastructural collapse. The urban population in Asia and Africa is expected to double between 2000 and 2030 [1]. However, mobility must continue to be possible, as it is an important prerequisite for individual and social prosperity [2]. In addition to numerous mobility concepts such as car sharing, people mover etc., individual mobility will continue to play a very important role. Small and narrow vehicles for a maximum of two passengers can be a solution module. The existing traffic areas can be used more efficiently with such concepts.

Such vehicle concepts have a comparable center of gravity height as a conventional vehicle. However, the relationship between small track widths and the center of gravity height is unfavorable for driving stability. To avoid roll over, a hard vertical dynamic tuning is required. This, however, deteriorate driving comfort. Another possibility to ensure driving safety is the implementation of an active title control.

Several concept of tilting vehicles and concept for the tilting control have already been published. Some are only theoretical studies, others presented implementations of such concepts as prototypes or production vehicles (e.g. [3], [4], [5], [6], [7]). But these publications only aim at the feasibility and the assurance of driving stability.

P. E. Pfeffer (ed.), *11th International Munich Chassis Symposium 2020*, Proceedings, https://doi.org/10.1007/978-3-662-63193-5_10

In most cases, the tilting strategy simply aims at completely compensating lateral acceleration while cornering.

However, active tilt control can do more than just ensure safe mobility. The newly gained degree of freedom in the tilt movement can offer a completely new driving experience.

2 Technical Introduction

The considered vehicle class of this concept is the small vehicle class l7e. This vehicle class addresses the needs of individual and sustainable electro mobility. The most important limitations are shown in Table 1.

Table 1. Regulation of l7e vehicle class.

Vehicle curb weight	$m_{CW} < 400$ kg
Powertrain	Electro mobility
Electrical Power	P< 15 kW

As an exemplary vehicle layout, a two-seater tandem configuration (the passenger seat is behind the driver seat) with a track with below one meter (s < 1 m) is introduced. The initially formulated requirements for an efficient use of space can thus be achieved. The scope of this paper is focused on the driving experience for a normal driver in standard driving scenarios, thus excluding e.g. high accelerations and high frequency steering inputs. Table 2 summarizes the main limits on driving states and vehicle behavior.

Table 2. Basic Driving State Range

velocity	v <100 km / h
Lateral acceleration	$|a_y| < 4.5$ m/s²
Steering frequency	f < 0.5 Hz
Tilt behavior	monotonically increasing while increasing lateral accelerating
Driving behavior	Overall perceptible for normal driver

Especially in crowded cities, where the driver has to focus on the traffic situation, vehicle guidance should be simple and predictable. In the literature, there are various options for tilt control. Steering-tilt control, direct tilt control or a combination of both. The differences lie in the determination and control of the desired tilt angle. There are some concepts where the driver has to tilt the body by steering input. A motorcycle is an example of this type of steering tilt control. Direct tilt control uses an actuator that executes the tilt. In a control loop the desired target angle is calculated and the actuator implements the request. In order to allow the driver to concentrate only on his driving task, direct tilt control is preferable. This type of tilt control is considered in this concept

In this paper, the possible scope for design is shown virtually and evaluated only on the basis of objective simulation data. The simulation environment in this paper is a multi-body-simulation. The front and rear suspension are planar double wishbone configurations, with parallel upper and lower control arms. For this type of configuration, the effect of wheel alignment on driving behavior is almost neglectable. Therefore, driving safety and experience are determined by the setup of the tilting control system. One tilting actuator acts at each axle and is mounted at the body. A relative tilting torque between body and lower control arms realizes the tilting movement.

A steady state ramp steering and a slalom maneuver are used to develop and assess the driving experience of this tilting vehicle. With the first steady state maneuver, the steady tilting experience and especially the compensating strategy can be shown and assessed. The slalom maneuver focuses on the resultant tilting reaction of the vehicle with respect to the drivers input. The slalom is limited in velocity and pylon distances. A driver with no tilting vehicle experience can follow the paths and can lead the vehicle with the resultant tilting reaction through the maneuver. The slalom simulations are performed by an open-loop sinusoidal steering wheel input.

2.1 Tilting Controls

This paper determines a desired titling angle by using a simple enhanced single track (ESM) model. The conventional single track model considers tire force as the product of side slip stiffness $C_{S,alpha,FA/RA}$ and tire side slip angle (α_{FA}, α_{RA}) by equation

$$\alpha_{FA} = \delta + \beta - \frac{l_{FA} \cdot \dot{\psi}}{v} \tag{1}$$

and

$$\alpha_{RA} = \beta + \frac{l_{RA} \cdot \dot{\psi}}{v} \tag{2}$$

This approach is enhanced by considering wheel camber angle and the resulting additional lateral forces, that cannot be neglected considering the high camber angles of a tilting vehicle. The considered tires are also more like those of a motorcycle tire, where the camber forces are dominant. In this concept the camber angle correlates with the tilt angle of the body. For tilting vehicles, there is a relationship between the lateral acceleration and the gravity acceleration. Taking into account a compensation strategy $C(a_y)$, an equation can be determined to calculate a corresponding body tilt angle.

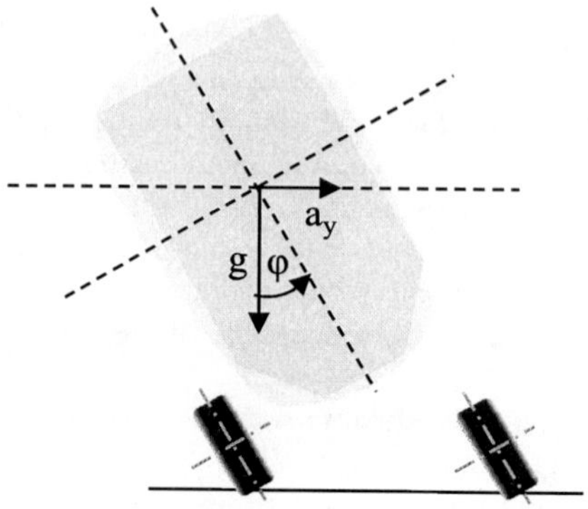

Fig. 1. Derivation of the tilting angle

With a perfect compensation (C = 100%), the vehicle body has to be aligned with the sum vector of gravity and lateral acceleration equation (1).

$$\varphi = \operatorname{atan}\left(\frac{C \cdot a_Y}{9{,}81\frac{m}{s^2}}\right) \tag{3}$$

The passengers perceive no lateral acceleration and there is no tilting torque, that has to be supported by a wheel load difference.

Linear camber stiffnesses $C_{S,\varphi,FA/RA}$ are considered to calculate the additional tire force $F_{S,\varphi}$ in lateral orientation.

$$F_{S,FA/RA} = F_{S,\alpha,FA/RA} + F_{S,\varphi,FA/RA} = c_{S,\alpha,FA/RA} \cdot \alpha_{FA/RA} + c_{S,\varphi,FA/RA} \cdot \varphi \tag{4}$$

With equation (1), (2), (3) and (4) a lateral acceleration (equation 5) and yaw rate (equation 6) can be numerically determined.

$$a_y = \frac{1}{m} \cdot (F_{S,FA} + F_{S,RA}) \tag{5}$$

$$\ddot{\psi} = \frac{1}{I_{zz}} \cdot (F_{S,FA} \cdot l_f - F_{S,RA} \cdot l_r) \tag{6}$$

With suitable parameterization of the enhances single track model it is possible to calculate the vehicle states lateral acceleration, yaw rate and tilt angle by taking the steering wheel angle and velocity into account.

With the help of the tilting actuators the desired tilting angle is realized. The actuators get a torque request to the component controller, to tilt the body. These torques are determined in a feedforward and feedback loop. The feedforward loop uses the desired tilt angle and a body tiling model (Fig 2.).

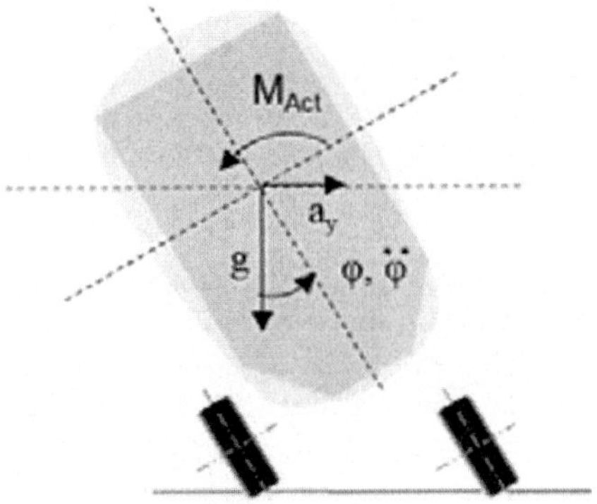

Fig. 2. Tilting torque model

The resulting torque are determined by Eq. 5

$$I_{xx} \cdot \ddot{\varphi} \quad = M_{Act} + h \cdot \cos(\varphi) \cdot \mathrm{m} \cdot 9{,}81\frac{m}{s^2} - h \cdot \sin(\varphi) \cdot \mathrm{m} \cdot a_y$$

$$\Leftrightarrow M_{Act} = I_{xx} \cdot \ddot{\varphi} \quad - h \cdot \cos(\varphi) \cdot m \cdot 9{,}81\frac{m}{s^2} + h \cdot \sin(\varphi) \cdot m \cdot \ a_y \tag{7}$$

The tilt acceleration is calculated by the double time derivative of the desired tilting angle.

This (ESM) model-based approach has the advantage that the desired tilt angle depends on the steering input and the vehicle behavior is more predictable for the driver. Other approaches in the literature use the measured lateral acceleration as a feedback loop to calculate the desired tilt angle. The vehicle reactions influence the desired tilt angle, resulting in a less predictable tilt angle. The disadvantage of this ESM approach is that the accuracy of the desired tilt angle depends on the accuracy of the model used. In non-linear driving situations, the model may not be accurate enough to calculate a tilt angle that exactly matches this behavior. The tilting model for calculating the tilting torque (Eq. 5) is a further simplification that may lead to further inaccuracy. Therefore, a feedback loop is also taken in to account in overall control strategy. This feedback loop compares the desired with the actual tilt angle. These differences are the input for a PD- controller and a feedback torque is the output. This is added to the feedforward torque.

Another advantage of this kind of control strategy is the calculation of different driving safety characteristics. With the appropriate parameterization, it is possible to monitoring the yaw reaction. The desired (EMS) yaw rate can be compared by a measured yaw rate. As for conventional vehicles, the deviation between desired and actual yaw rate can be used as a trigger for ESC interventions [8]. A second stability assessment can be made by monitoring the output of the PD-controller. In situations with high steering velocity, the tilt controller demands a high tilt acceleration, which can lead to loss of wheel contact on the inside of the curve. However, this loss is not critical. Firstly, this wheel load loss can be predicted on the basis of equations. Secondly, the lateral acceleration counteracts the rollover and causes the body to straighten up. If the Feedback-Torque decrease fast enough, the situation with the loss of tire contact can be assessed as uncritical and stable. A decreasing feedback part means that the body will soon have achieved the desired tilt angle

With this controller approach, there are two ways to generate different reactions of the tilt control with different resulting driving experience. The first one is the compensation factor C in the ESM approach (Eq. 3). It can be set to a constant value for the entire driving status range. It can also vary depending on the lateral acceleration C(ay). With this variable factor it is possible to realize different perceptible driving states. The result of the ESM is the desired tilt angle $\varphi_{des,ESM}$. The second way to design the driving experience is to change the dynamic build up of the desired tilt angle to vary the tilting and vehicle dynamics. The ESM output is modified by a PT1 filter to the desired dynamic tilt angle $\varphi_{des,DYN}$, which is considered in the tilting torque model approach (Eq. 7).

Six different controller setups, which are used in the further analyses, are presented in Table 3

Table 3. exemplary controller setups

Setup	**Compensation**	**PT1-Filter**
100 %	C = 100 % - constant over the entire a_y range	0.01
75 %	C = 75 % - constant over the entire a_y range	0.01
50 %	C = 50 % - constant over the entire a_y range	0.01
#1	$C(a_y) = \begin{pmatrix} 75\% & a_y < 2.5\, m/s^2 \\ 75\,\% - 50\,\% & 2.5\frac{m}{s^2} < a_y < 7\, m/s^2 \end{pmatrix}$	0,3
#2		0,1
#3	$C(a_y) = \begin{pmatrix} 100\% & a_y < 1.5\, m/s^2 \\ 100\,\% - 75\,\% & 1.5\frac{m}{s^2} < a_y < 6.5\, m/s^2 \end{pmatrix}$	0.1

2.2 Assessment Criterias

In the assessment of conventional vehicle, the vehicle reactions regarding lateral acceleration and yaw rate are usually taken into account. Rolling behavior, especially rolling dynamics, is only assessed by experts, but has a minor role in a development process. In conventional vehicles, the influence of rolling motion on vehicle dynamic or driving safety is more important than the influence on passenger comfort. For driving safety the rollover evaluation is crucial in the development process and some stability criteria have been published. (e.g.[9]). These criteria typically take lateral acceleration and roll acceleration into account with regard to vehicle height and the track width. Rolling acceleration and jerk are aspects for the evaluation of driving comfort. E.g. ISO 2631-1 [10] includes a frequency depended consideration of the rolling accelerations. Both, the driving safety and driving comfort evaluation with regard to rolling behavior only deal with roll accelerating and small roll angles. However, these limitations only apply to conventional vehicles. For tilting vehicle, the range of limitations may be wider. Especially for the tilting acceleration. High tilting accelerations are necessary to ensure driving safety and an adequate driving experience. However, high tilt acceleration would lead to a poor assessment of driving safety and comfort in the conventional development process for conventional vehicles.

Through the open-loop execution of the simulations, only the influences of the different setups on the vehicle reactions can be evaluated. Driver-related corrections cannot be covered in this objective study. For the evaluation of acceptance and drivability, these corrections can be derived as objective variables. The steering wheel angle to achieve comparable lateral acceleration is considered to assess the objective steering effort in the open-loop simulation.

When assessing the driving experience, the tilting dynamic and the vehicle reaction should be taken into account. The tilting dynamic can be measured as the tilt velocity or acceleration of the tilting body. The vehicle reaction is a combination of different factors. The tilt angle, yaw rate and lateral acceleration response can be evaluated in relation to the steering wheel input. In this paper, the slalom maneuver is used to present assessment criteria of the tilting dynamic, which would be suitable for customer studies. The phase of the vehicle response regarding the steering wheel angle can be determined as time difference of the zero crossings. High differences would lead to a sluggish vehicle evaluation. A fast reaction with small phases can be assessed dynamic. Nevertheless, it could also be assesses as to nervous. Especially in situations, where the driver would continuously correct the steering wheel to follow a desired path.

Another aspect is the perceptible lateral acceleration. This can be influenced by the newly gained degrees of freedom of the active tilt control. This perceptible acceleration can be calculated with equation (6) for a steady state

$$a_{Y,Per} = a_Y \cdot \cos(\varphi) - 9{,}81\frac{m}{s^2} \cdot \sin(\varphi) \tag{8}$$

In case of 100 % compensation, the perceptible lateral acceleration is zero. The driver would not feel any lateral acceleration

With regard to the previously formulated driving experience range, it must be taken into account that a driving condition-dependent compensation strategy always considers a monotonically increasing tilting angle with increasing lateral acceleration. The driver would notice a higher perceived lateral acceleration, but the driver would also notice a uprighting of the body. This could become an untrustworthy evaluation of the tilt control.

The overall assessment of the driving experience of a tilting vehicle considers the three parts tilting dynamic, response phases and compensation. In the slalom maneuver following criteria can be calculated and assessed

Table 4. Assessment of the driving experience of a tilting vehicle

Category	Criteria	Value
Tilting dynamics	Tilting Velocity	$\dot{\varphi}_{Max/Min}$
	Tilting Acceleration	$\ddot{\varphi}_{Max/Min}$
Vehicle Response	Phase Tilt Angle w.rt. Steering Wheel	$\Delta\varphi\delta = t_{\varphi=0°} - t_{\delta=0°}$
	Phase Yaw Rate w.rt. Tilt Angle	$\Delta\varphi\dot{\psi} = t_{\varphi=0°} - t_{\dot{\psi}=0°}$
Compensation	Actual Lateral acceleration	$a_{y,Actual,MAX}$
	Perceptible Lateral acceleration	$a_{y,Perc,MAX}$
Steering effort	Steering Angle	δ_{MAX}
	Steering Velocity	$\dot{\delta}_{MAX}$

The steering effort is used to compare and evaluate the effect of the different tilting strategy in relation to the driver's input.

3 Driving Experience

In a first analysis step, the effect the first three controller setups with a fixed compensation strategy are examined. The steady state ramp steer simulation is used to work out general limits. The tilting dynamic do not play any role in this analysis. In addition, the three other different tilting control setups are presented to illustrate the realization of three possible different driving experience (see table 4). The setups #1 & #2 do not differ in the compensation strategy, but in the tilting dynamic. The third considers higher compensation percentages.

In the second step the tilting dynamic is assessed. At the beginning, the focus is again on possible limits in the design of the driving tilting experience. The three setups are evaluated in a following analysis.

3.1 Compensation strategy

The following figure shows the perceptible lateral acceleration depending on the actual lateral acceleration and the compensation strategy. The drivers input are a constant velocity of 50 km/h and a ramp steering input, with a slow steering gradient. The difference in the first and second setup in the dynamic tilting behavior has no effect on the simulation results in the ramp steer simulation. Next to the three vehicle setups the resultant perceptible lateral acceleration for constant compensation values are shown. The constant compensation 100%, 75% and 50% are considered. With these three constant compensation factor the upper and lower limits can be discussed.

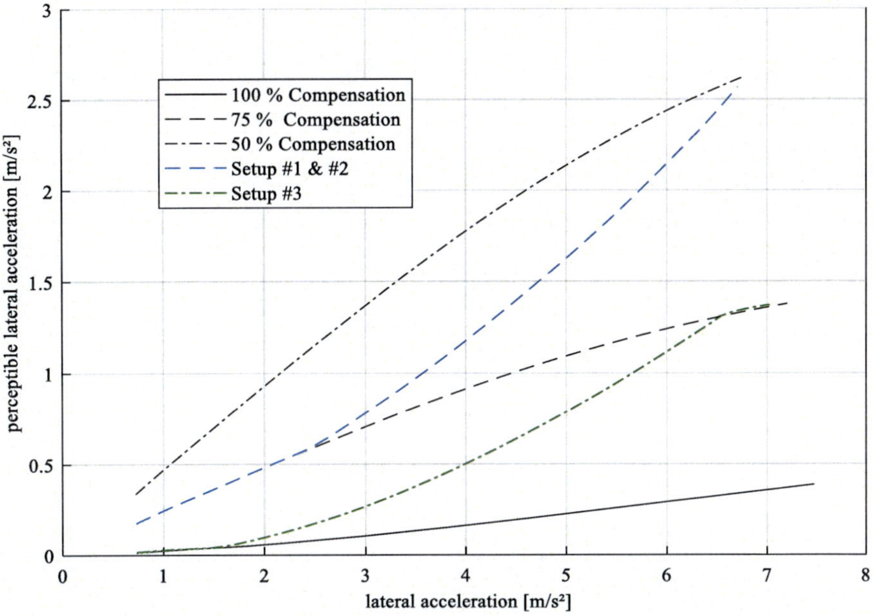

Fig. 3. Steady state perceptible lateral acceleration depending on Compensation strategy.

The upper limit is the 100% compensation. The multi-body simulation takes into account effects (e.g. kinematic chains from body tilt to suspension tilt) that can lead to non-linear body movements. The assumptions of the linear ESM approach have a further influence. The accuracy deteriorates with increasing lateral acceleration. For this reason, the 100% compensation does not show perfect simulation results in the vehicle reaction with a constantly perceptible lateral acceleration of zero. However, in relation to the actual lateral acceleration of the vehicle measured on the road surface, the perceptible lateral acceleration is very low (lower than 4 % at $a_y = 6\ ^{m}/_{s^2}$). Compensations above 100% would lead to a perceptible acceleration, which would point inwards into the curve. The driver gets the feeling of falling into the curve. This should be avoided and the upper limit is in this context 100%.

With the 100 % compensation the diver do not perceive any lateral. This motorcycle feeling is a driving experience that can be realized by a tilting vehicle, but has several drawbacks. First, the driver loses lateral acceleration as an imported indicator for the perception of driving limits. Furthermore, 100% compensation in curves with a very small radii feels untrustworthy. Lateral acceleration increases very quickly in very small radii at low vehicle speeds and high steering wheel angles. This leads to very high tilting angles. But all the driving feedback and the surrounding stimuli (e.g. parking zones) do not match this very high tilt angle. The driver might get the feeling of falling into the curve. Therefore, the driving experience tuning should consider compensations strategies, which take the perception of the driving states especially the physical limits and a trustworthy tilt angle in low velocity situations into account.

The lower limit of the compensation can be identified using the resulting wheel loads at constant radius cornering. The vehicle must always have enough wheel load reserves at the most relieved wheel during steady state driving conditions to ensure driving safety in sudden dynamic maneuvers.

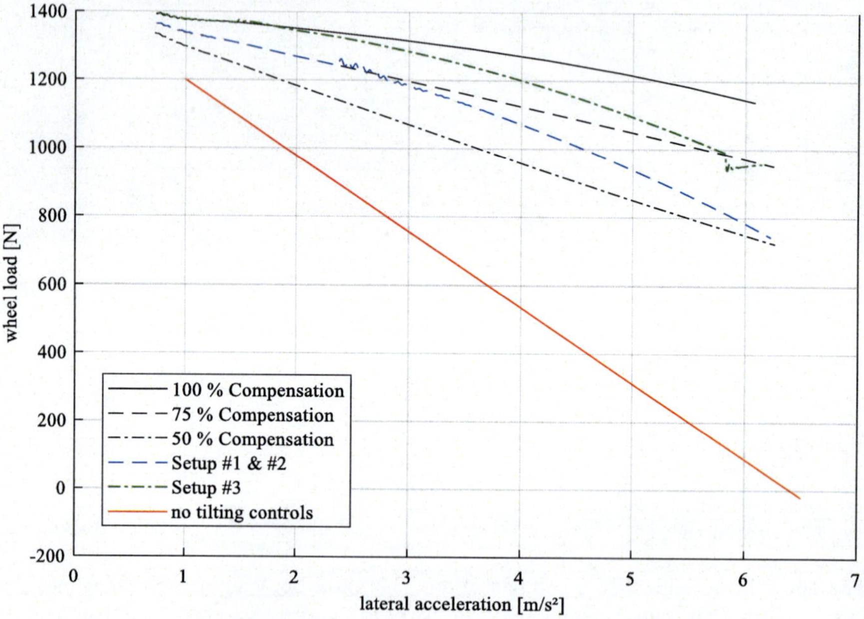

Fig. 4. Steady state wheel load depending on the compensation strategy.

If the compensation is limited to 50 %, the wheel loads are more than half of the static loads up to a lateral acceleration of 6 m/s². This should be sufficient to ensure a safety driving state. Dynamic influences that further reduce the wheel load or even lead to loss of road contact can be evaluated by the PD output already mentioned.

A non-constant strategy can be used to realize an increasing driving state feedback with increasing lateral acceleration for the driver. This feedback option is used in the three considered setups. Fig. 3 shows the perceptible accelerations for the three setups. Setup #1 and #2 starts with constant 75 % compensation at low lateral acceleration .With increasing lateral acceleration the drivers perceive higher lateral acceleration up to the 75% compensation level. The driver receives feedback about the current driving conditions by perceiving a lateral acceleration comparable to the feeling in a conventional vehicle. Setup #3 provides motorcycling feedback in low lateral acceleration ranges (ay < 1 m/s²). As the accelerations increase, the perceived accelerations also increase.

With the compensation percentage, the tilt angle increases too. A 100 % compensation has the highest tilt angles. A non-constant compensation strategy should ensure the initial requirements of a monotonically increasing tilt angle. The deviation from the highest to the lowest compensation is to be assessed using the built up of the tilt angle with respect to the lateral acceleration Fig 5.

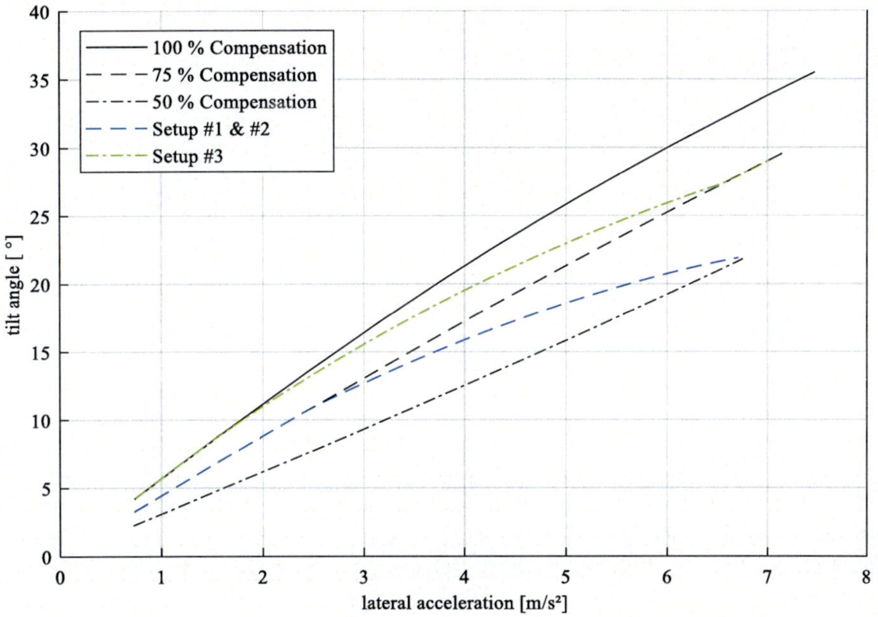

Fig. 5. Steady state tilt angle depending on the compensation angle

The three setups have a monotonously rising tilt angle and satisfy this requirement.

3.2 Assessment of tilting dynamics

The dynamic tilting behavior is assessed by evaluating the tilting reaction with respect to the steering wheel input during the slalom maneuver introduced above. The maneuver is defined in such a way that the steering frequency (f = 0.3 Hz) and amplitude (resulting peak lateral acceleration ay = 4 m/s²) correspond to an upper limit that fits the usual range in normal road traffic

The resulting tiling reaction depending on the PT1-Filter factor are shown in Fig 6. To assess this controller parameter simulations are done with a compensation strategy with a constant compensation factor of 100% and the steering wheel amplitude is fix for all simulation. The data shown in this figure can be used to define an upper limit for the PT1 filter parameter. The vehicle reaction (here tilting angle) has a PT1 depending delay. At a phase shift of 90°, drivers acceptance regarding delays could be lost at latest since the steering wheel angle would be zero when tilt angle its maximum.

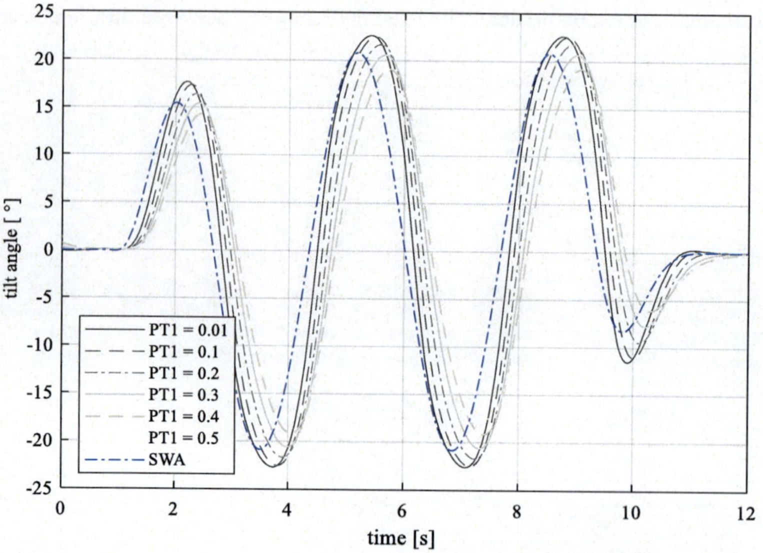

Fig. 6. Tilting reaction depending on PT1-Filter factor with 100 % compensation

In this slalom the steering frequency is 0.3 Hz. A 90° phase shift correlates with a time difference in the zero crossing of 0.83 s. However, the phases decrease with increasing control frequency, so that for the considered f = 0.3 Hz a sufficiently high phase reserve to the 90° limit should already be taken into account[1]. In this paper an exemplary 45° phase is selected as threshold for the evaluation of PT1 factors. This corresponds to an time difference of 0.42 s.

As an additional effect, a PT1 filter has damping effect on the resulting amplitude of the tilt angle. This leads to increasing perceptible lateral acceleration.

[1] [11] considers a steering frequency range up to 2 Hz in the development of a tilt controller. The 90° phase is used as a stability criterion for the entire frequency range.

The overall assessment by the criteria, mentioned above, of this first simulation study for the PT1 Variations and fixed 100 % Compensation is shown in Table 5

Table 5. Driving Experience Assessment – Simulation Fig. 6

Criteria	Steering effort		Tilting dynamics		Vehicle Response		Compensation	
PT1 Factor	δ_{MAX} [°]	$\dot{\delta}_{MAX}$ [°/s]	$\dot{\varphi}_{Max}$ [°/s]	$\ddot{\varphi}_{Max}$ [°/s²]	$\Delta\varphi\delta$ [s]	$\Delta\varphi\dot{\psi}$ [s]	$a_{y,Act}$ [m/s²]	$a_{y,Perc}$ [m/s²]
0,01	25,00	47,10	62,44	173,65	0,15	0,02	4,31	0,26
0,1	25,00	47,10	58,83	150,59	0,24	0,09	4,33	0,39
0,2	25,00	47,10	52,11	116,54	0,32	0,16	4,26	0,63
0,3	25,00	47,10	46,41	99,04	0,40	0,23	4,21	0,98
0,4	25,00	47,10	40,90	90,85	0,46	0,29	4,11	1,34
0,5	25,00	47,10	35,93	82,78	0,51	0,35	4,00	1,69

The damping effect of the PT1 filter can be derived in the tilting dynamics and vehicle response. The tilt velocity and acceleration decrease and the delays increase. The selected upper threshold with a phase of 45° is exceeded for PT1 greater than 0.3. The second delay (tilte angle w.r.t. yaw rate) does not exceed the Phase limits. Fig. 7 summarizes the influence of the PT1 filter regarding the tilt reaction with respect to the steering wheel input for the different constant compensation strategies 100 %, 75 % and 50 %. The perceptible lateral acceleration is shown on the abscissa axis. The ordinate axis represents the zero crossing differences. The different symbols represent the different compensation strategy.

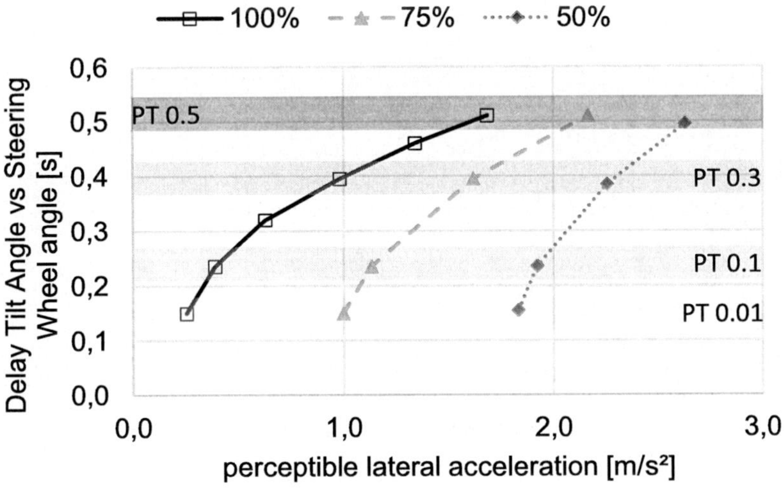

Fig. 7. PT1-Filter influence of phase tilt angle w.r.t steering wheel angle

With the help of the compensation factor and the PT1 filter, there is a wide range of design options to tune the vehicle's behavior and perceptible acceleration. The phase can be individually adjusted by the PT1 filter. With higher PT1 factors, however, the resulting perceptible lateral acceleration increases.

The effect of the compensation factor on tilting dynamics is shown in Fig. 9. Here, the compensation factors in the slalom maneuver are decreased and the PT1 factor is set to 0.01 in order to neglect the PT1 filter influence. With the highest 100 % compensation, the desired tilt angle is the highest. To achieve the desired value, a higher tilting velocity is required. The feed forward and feedback loop requests an appropriate torque to match the desired angle with the actual angle as quickly as possible. So the resultant tilt velocity can also be tuned by the compensation strategy. The compensation has only marginal effects on the phase between vehicle reaction and steering input.

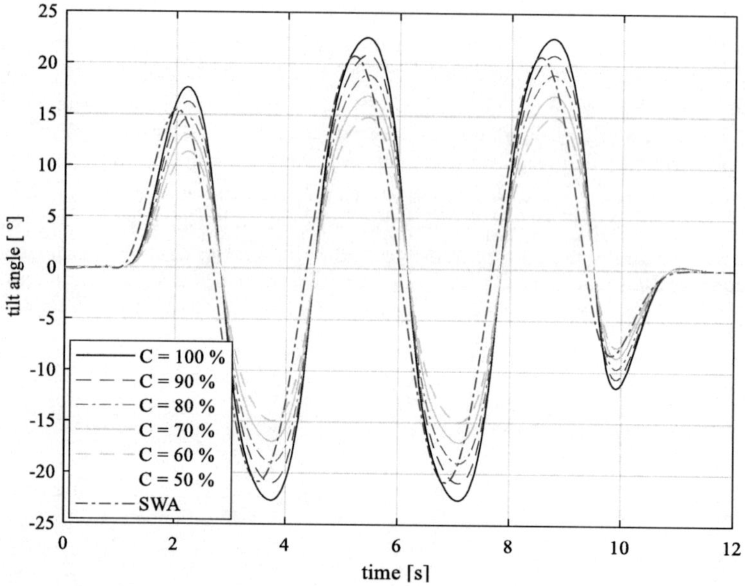

Fig. 8. Tilt angle depending on compensation and neglectable PT1

Finally, the results of the slalom simulation to evaluate the tilting experience for the setups defined in Table 4 are presented in Table 6. As already discussed, lateral acceleration during slalom maneuvers decreases with an increasing PT1 factor (see Table 5). Therefore, to achieve the same vehicle response, steering wheel angle has to be increased with increasing PT1 factors. This effect can be evaluated as an increasing steering effort or a virtual change in the steering ratio. The simulations for the different Setups are performed with corrected steering wheel angle, to achieve comparable actual lateral acceleration.

Table 6. Simulation Results of slalom maneuver

Criteria	Steering effort		Tilting dynamics		Vehicle Response		Compensation	
Setup	δ_{MAX} [°]	$\dot{\delta}_{MAX}$ [°/s]	$\dot{\varphi}_{Max}$ [°/s]	$\ddot{\varphi}_{Max}$ [°/s²]	$\Delta\varphi\delta$ [s]	$\Delta\varphi\dot{\psi}$ [s]	$a_{y,Act}$ [m/s²]	$a_{y,Perc}$ [m/s²]
100 %	25.00	47.10	62.44	173.65	0.15	0.02	4.31	0.26
75%	25.53	48.09	51.58	147.48	0.16	0.03	4.31	1.05
50 %	26.58	50.07	37.91	110.54	0.16	0.04	4.31	1.96
Setup #1	26.08	49.14	37.81	89.41	0.40	0.25	4.30	1.87
Setup #2	25.70	48.42	47.91	125.45	0.25	0.10	4.30	1.39
Setup #3	25.00	47.10	58.54	150.44	0.24	0.08	4.30	0.57

The steering effort is effected by the different setups. But the range of deviations in this concept design is small (approx. 6 % difference). Setup #1 has the lowest tilt dynamics and the highest phase delays. The delay is a result of the high PT1 filter value (0.3). With this and the compensation value of 50% the perceived lateral acceleration is highest for non-constant compensation strategy. Setup #2 has the second highest perceptible lateral acceleration with a small difference of 0.15 m/s^2. Thus, the first two setups could be evaluated as comparable with regard to the perceptible acceleration. But the tilting dynamic of Setup #2 is faster than Setup #1, even though the compensation strategy is the same. But the PT1-Factor causes the dynamics to higher tilting velocities. Setup #3 has comparable phases to Setup #2. But the tilt dynamic in the tilt velocity is higher. The difference between Setup #2 and Setup #3 is the compensation strategy. The controller allows higher tilt angle. These has to be achieved in a comparable time. So, the tilt velocity is higher and the perceptible acceleration is lower.

4 Summary

Tilt vehicles can be part of the solution to solve the traffic infrastructure problems in crowded cities and megacities. As small and narrow vehicles, they make more efficient use of existing traffic spaces and also offer a new kind of driving experience. However, the unfavorable relationship between the height of the center of gravity and the track width requires the active tilt control. Without an active system, the vertical dynamic setup of the vehicle must be very hard, which deteriorates the driving comfort. But an active tilt system allows new options to implement the novel driving experience. Several publications deal with the basic feasibility and the guarantee of stable vehicle behavior. But the new driving experience through active tilting is not presented in detail. With the active control, it is possible to realize different driving state feedbacks and experiences.

With the control strategy introduced, the desired tilt angle is calculated by an enhanced single track model (ESM) based on the drivers input (velocity and steering wheel angle). With this approach, a body tilt angle is determined in relation to the desired lateral acceleration and compensation strategy. A PT1 filter can be used to adjust the tilt dynamics. To evaluate the driving experience during tilt movements, there is the perceptible lateral acceleration, the tilting dynamics as tilting velocity and the phase between driver input and vehicle reaction.

With the compensation strategy and the PT1 filter, different driving experiences can be set. With the compensation strategy, a driving state feedback can be realized. On the one handy, it is possible to achieve a motorcycle like driving behavior without noticeable lateral acceleration. On the other hand, it is also possible to provide a setup with a moderate vehicle response to driver input, clearly noticeable feedback and perceptible vehicle dynamics limits

Active tilt control is necessary on the one hand to ensure driving safety with good driving comfort and on the other hand, to enable a driving experience that is exciting and innovative.

References

1. United Nations Population Fund: State of the world population 2007 – Unleashing the Potential of Urban Growth. United nation population fund, New Yorj, USA (2007)
2. United Nations Human Settlements Programme: State of the World's Cities 2012/2013. Properity of Cities. United nations human selttlements programme (UN-Habitat), Nairobi, Kenia (2014)
3. EGAN, P.Lean Machine Logic and substance from the Dreamer's Workshop, Road & Track (1983)
4. MERCEDES-BENZ, Das Kurvengenie – Mercedes-Benz F 300 Life Jet, Mercedes-Benz Frankfurt am Main, 1997
5. CARVER B.V., CarverAmsterdam 2018

6. KARNOPP, D., FANG, C.A Simple Model of Steering Controlled Banking Vehicles California 1992
7. KIDANE, S., ALEXANDER, L., RAJAMANI, R., STARR, P., DONATH, M.. A fundamental investigation of tilt control systems for narrow commuter vehicles. Vehicle System Dynamics, Heft 4, S. 295–322 (2008)
8. NEUKUM,A.,UFER,E..Kontrollierbarkeit potenzieller Fehleingriffe des Elektronischen Stabilitätsprogramms (ESP) Aachen, Aachen, 2010
9. BAUMANN, F. W.;Untersuchungen zur dynamischen Rollstabilität von Personenkraftwagen, Dissertation, Darmstadt 2003
10. ISO - INTERNATIONL ORGANISATION FOR STANDARDIZATION, Mechanical vibration and shock - Evaluation of human exposure the whole-body vibration - 2631 -1, 1997
11. [KÖH00] KÖHN, P. Zur Fahrmechanik von Zweispurfahrzeugen mit Kurvenneigung, Dissertation, Institut für Kraftfahrzeuge, RWTH Aachen, Aachen, 2000

Basic Study for Materials of Tire and Wheel for Dynamic Wireless Power Transfer System with In-Wheel Coil

Osamu Shimizu[1] Takashi Utsu Hiroshi Fujimoto Daisuke Gunji[2] Isao Kuwayama[3]

[1] The University of Tokyo, Japan

[2] NSK Ltd., Japan,

[3] Bridgestone Corporation, Japan

Abstract. Dynamic wireless power transfer (WPT) can solve the short-cruising range problem of electric vehicles. Energy transfer is one of the major problems of dynamic WPTs. Prevention of the entry of foreign objects physically between the transmitting and receiving coils is the best way to increase energy transfer for dynamic WPT. This research proposes a novel WPT system, named in-wheel coil, which can transfer through the tire and the wheel. The suitable materials of tires and wheels for in-wheel coil is also evaluated and proposed. Metal belt of tire is not efficient and rises temperature of tire. Instead of metal belt, fiber belt which has no conductivity is proposed and proved that it does not has any influence for WPT system at 85 kHz. This research also revealed that steel and aluminum and cross carbon fiber reinforced plastic (CFRP) has critical influence for WPT performance and uni-directional (UD) CFRP has little influence for WPT performance at 85 kHz. Combination of the wheel made of UD CFRP wheel and the tire whose belt is fiber belt is the best way to improve WPT efficiency.

Keywords: Dynamic Wireless Power Transfer, Tire, Wheel

1 Background

1.1 Dynamic Wireless Power Transfer

One of the obstacles in electric vehicles is the short cruising range. A simple solution for this problem is loading large amount of battery on electric vehicles. However, it makes electric vehicles heavy and not efficient. Moreover, there is risk of depletion of resources for battery. Therefore, to make sustainable electric vehicle society, conventional approach to extend cruising range with only heavy battery is not suitable.
To solve these problems of conventional approach, dynamic power transfer system, which is breakthrough technology, have been proposed. One is dynamic direct charging [1], the other is dynamic wireless power transfer (DWPT).
Strong point of direct charging is high efficiency between terminals, and proven system adopted for trolleybus. However, maintenances for worn direct charging terminals. Furthermore a risk of fire by spark when the terminal of vehicle side touch the

P. E. Pfeffer (ed.), *11th International Munich Chassis Symposium 2020*, Proceedings, https://doi.org/10.1007/978-3-662-63193-5_11

terminal of infrastructure are big problem. These problems are same as the problems of stationary auto charging system for auto guided vehicle, and there has been no effective solution yet. Overhead wires also destroy the beautiful scenery.

On the other hand, DWPT has no problems by charging terminals and the scenery is same as conventional road. The conventional electromagnetic induction WPT is not efficient with the large coil gap and misalignment between the transmitting coil and the receiving coil, therefore it is not suitable for dynamic application. However WPT with magnetic resonance coupling which is proposed in 2007 [2] can transmit high power and high efficiency with the large coil gap and misalignment between coils. Therefore some projects adopt magnetic resonance coupling for DWPT [3]-[5] system. DWPT system with electric field coupling [6] has been also proposed. Electric field coupling uses capacitance of the air between transmitting side and receiving side. When it is raining and water penetrates between transmitting side and receiving side, electric field coupling cannot be used easily. Therefore DWPT system with electric field coupling is not able to be used on public road.

2 Wireless In-wheel Motor

2.1 Concept of Wireless In-wheel Motor

We have proposed a novel system named wireless in-wheel motor (WIWM) [7] [8] which integrates driving system and dynamic charging system, motor, inverter, rectifier, control unit, cooling system, receiving coil, in the unsprung area. We have developed 3 types of WIWM. The latest one, 3rd generation of wireless in-wheel motor (WIWM-3), is displayed in Fig.1 and test vehicle with WIWM-3 is displayed in Fig.2

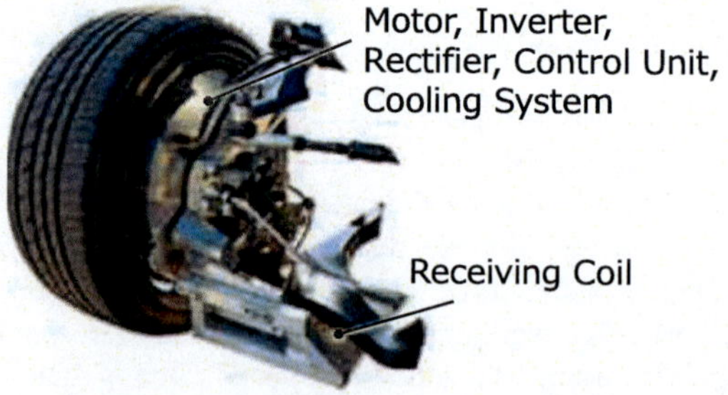

Fig. 1. 3rd Generation Wireless In-wheel Motor

Fig. 2. Test Vehicle with 3rd Generation Wireless In-wheel Motor

WIWM has 6 main strong point against the conventional drive system.
Strong point of in-wheel motor is below

- Light weight without drive shaft and differential gear
- Safe and comfort drive by high frequency control
- Wide cabin space by removing drive system from onboard to unsprung area

In-wheel motor is attached the wheel directly, therefore drive shaft and differential gear are not necessary. That makes vehicle lighter and more efficient. The transfer delay is also reduced. Then high frequency control can be realized.
Strong point of WPT system for WIWM is shown in below

- High efficiency can be achieved because the power is directly delivered to motors.
- The power receiving coil is arranged in the IWM, the air gap with the road surface coil is kept constant even when the suspension is displaced, and the air gap is minimized.
- Since power is supplied to each wheel, the output per transmitting coil can be reduced.

This system adopts SS (Series-Series) type magnetic resonance coupling circuit. Equivalent circuit of SS type is shown in Fig.1.

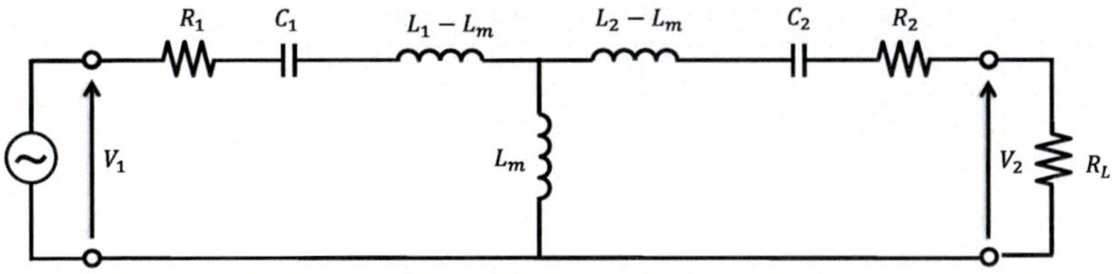

Fig. 3. Equivalent Circuit of SS Type Magnetic Resonance Coupling WPT System

L_m is mutual inductance between the transmitting coil and the receiving coil. R is AC resistance of coil which includes not only copper loss but also iron loss of magnetic circuit, L is inductance of coil, and C is capacitance. V is voltage and i is an identifying number, 1 is transmitting side and 2 is receiving side.
Theoretical maximum efficiency of SS circuit η_{max} is described as below:

$$k = \frac{L_m}{\sqrt{L_1 L_2}} \tag{1}$$

$$Q_i = \frac{\omega_o L_i}{R_i} \quad (i = 1 \text{ or } 2) \tag{2}$$

$$\eta_{max} = \frac{k^2 Q_1 Q_2}{(1+\sqrt{1+k^2 Q_1 Q_2})^2} \tag{3}$$

Where, k is coupling factor, Q is quality factor. ω_o is current frequency of transmitting coil.
Mutual inductance is important factor to increase efficiency, and the air gap between the transmitting coil and the receiving coil is the most effective factor of mutual inductance. Power of WPT P_{WPT} is described as below

$$P_{WPT} = \frac{(\omega_o L_m)^2}{(R_1(R_2+R_L)^2+(\omega_o L_m)^2)} V_1{}^2 \tag{4}$$

Here, R_L is equivalent load resistance. Batteries has power constraint due to safe charging. Thought parameters of WPT system should be constant to realize safe charging, distance from road is changed by vehicle weight or road condition.
The air gap with the road surface coil is kept constant even when the suspension is extended or retracted, and the air gap of WIWM is minimized. Even though minimum air gap of body side coil is 100mm given which is given SAE J2954 standard the air gap of unsprung coil which can avoid collision against steps of road by moving with tire can be a half of body coil's air gap. SAE J2954 is a standard of stationary WPT system for electric vehicle.

2.2 System Configuration

System configuration of WIWM-3 is displayed in Fig.3.
DC/DC converter is set between battery and inverter. Rectifier (AC/DC converter) is connected to output of DC/DC converter. DC voltage has relation to equivalent road. Then both power and efficiency of WPT system can be controlled by DC/DC converter. Frequency which is conforming to SAE J2954 is 85kHz.

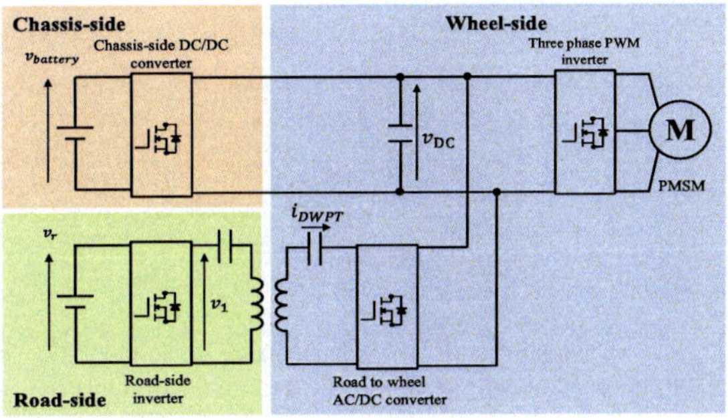

Fig. 4. System Configuration of WIWM-3

3 In-wheel Coil

3.1 Motivation of Developing

There are a lot of benefit to use DWPT system and WIWM. However there are also problems.

One of the biggest problem is foreign objects which are small animals or metal pieces between transmitting coil and receiving coil. If there are small animals or metal pieces, there is risk of damage animals or ignition by eddy current loss. When the magnetic field changes in metal materials, eddy current loss is occurred by induction change. Eddy current loss heats metal materials and it fires surrounding combustible materials, for example, plastic toys with batteries. It is dangerous situation and necessary to detect foreign objects and stop transmitting. WPT charging energy is decided by output, efficiency and charging time. Therefore, DWPT system has to detect the foreign objects much more quickly than stationary WPT system because of the short time from starting WPT to the end of WPT. Equivalent circuit model [9] for detection system is proposed to solve this problem, but the fundamental solution for foreign objects is to prevent them physically.

3.2 Structure of In-Wheel Coil

To prevent foreign matter objects between coils, it is necessary to fill the air gap with other objects. The receiving coil is set into the wheel, and air gap is filled with tire and wheel. This system is named In-wheel coil. The structure of In-wheel coil is displayed in Fig.5.

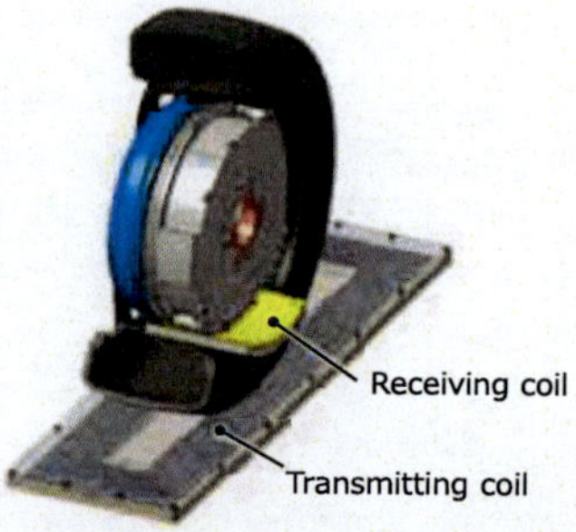

Fig. 5. Structure of In-wheel Coil

Receiving coil is set into the wheel and bent into arc-shaped due to minimizing air gap.

3.3 Problems of In-Wheel Coil

Vehicle tires consist of rubber and structural metal or organic materials. Structure of tire is shown in Fig.6.
Magnetic flux is through tread when WPT system is active. The metal belt wires are molded in tread and it is concerned eddy current loss in the belt. Then the effect by metal belt for WPT system is evaluated. Evaluation equipment is displayed in Fig.7 and the connection of WPT system is displayed in Fig.8.

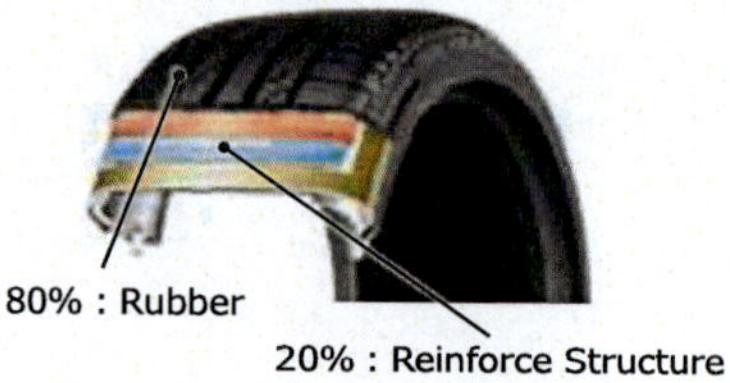

Fig. 6. Structure of Tire

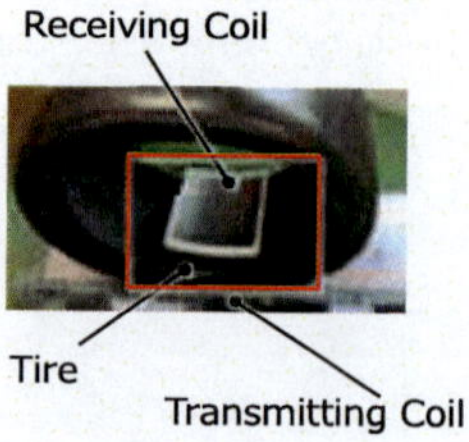

Fig. 7. Evaluation Equipment for WPT with Tire

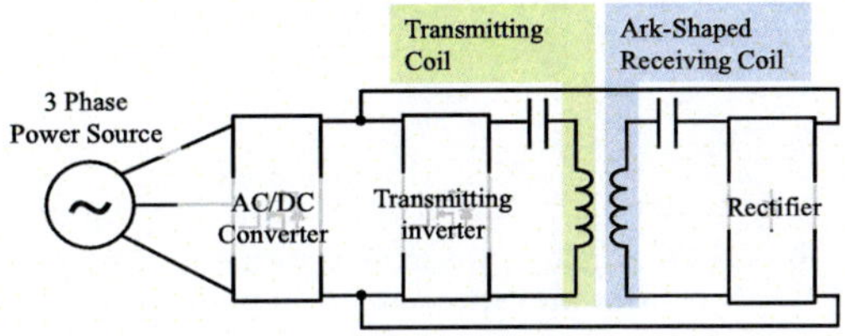

Fig. 8. Connection of Evaluation Equipment

The ark-shaped receiving coil and the tire are set on the center of the transmitting coil and the tire is put on the transmitting coil directly. The receiving coil is hook by resin wire into the tire. The transmitting side and the receiving sides are connected to same AC/DC converter output, thus the voltage of the transmitting side and receiving side is same.

Test condition is shown in Table.1 and the result is shown in Fig.9. The temperature of tire is measured by the thermographic camera. The test time is 18sec. The temperature of tire rises 20°C from the start to the end of test. High temperature and deviation of temperature affects bad the performance of the tire. Moreover the efficiency is also reduced. It shows in-wheel coil cannot use with conventional steel belt tires or metal wheels.

Table 1. WPT Test Condition with Steel Belt Tire

Items	Values
Airgap[mm]	100
Frequency[kHz]	85
Transmitting Voltage[V]	90

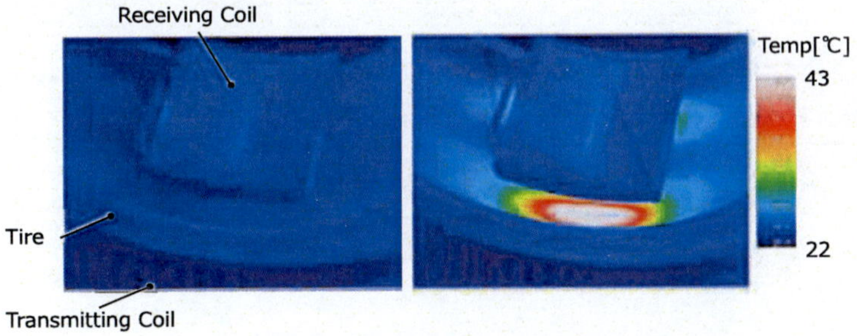

(a) Before Test (b) After Test (18sec later)

Fig. 9. WPT Test Result with Steel Belt Tire

4 Suitable Materials of Tire and Wheel for In-wheel Coil

4.1 Materials of Tire

The basic evaluation for materials of tire is done with the simple equipment shown in Fig.9. It can measure coil parameters, resistance, self-inductance, mutual inductance, with LCR meter. Then transmitting efficiency is calculated by equation (1)-(3) with coil parameters.

First, parameters change by rubber materials is measured. Types of test pieces are shown in Table.2. The calculated values by measured values are shown in Fig.11. It shows that there is no effect by rubbers and temperature rise with WPT system is only by the steel belt.

To prevent bad effect of steel belt, the fiber belt is proposed. The fiber belt is not conductive material, and it has enough toughness to use as tire. The fiber belt tire has already been adopted to air planes or racing cars. The effect by the fiber belt and the steel belt is evaluated with same equipment of rubber evaluation. Measured values are shown in Fig.12.

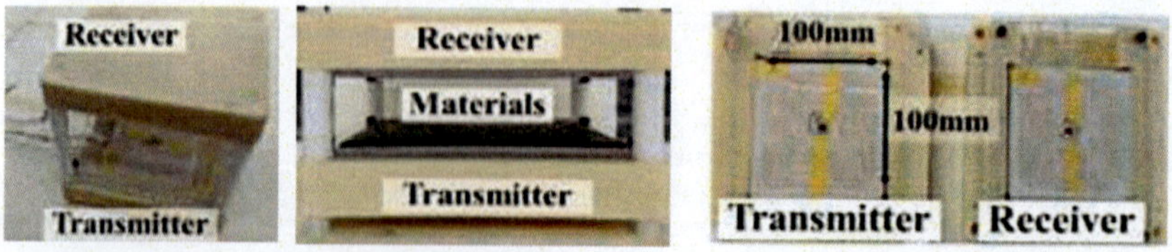

(a) Overview (b) Side view (c) Coil Dimension

Fig. 10. Evaluation Equipment for Materials

Table 2. Types of Test Pieces

Test Piece No.	Type of Tire	
	Vehicle Type	Use Type
1	Passenger Vehicle	Eco
2		Sports
3	Track	Eco
4		Sports

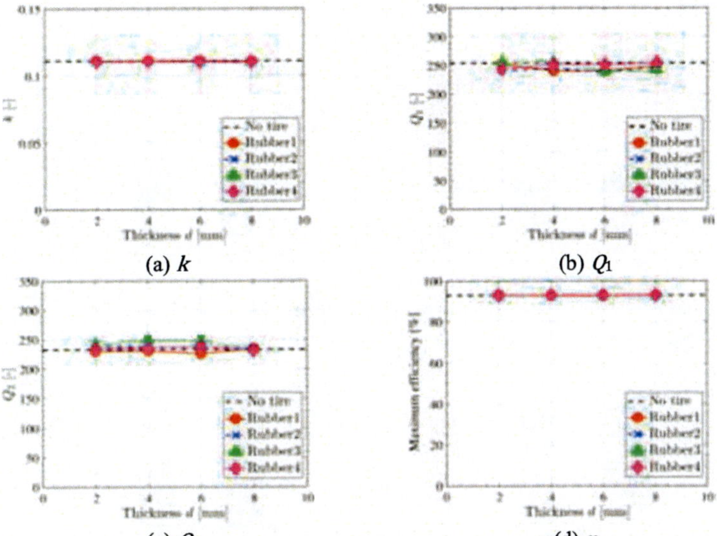

(a) k (b) Q_1 (c) Q_2 (d) η

Fig. 11. Evaluation Result of Rubber Materials (Frequency:85kHz)

Resistance rises by steel belt. The reason why the transmitting side has large influence is the distance from steel belt to the transmitting side is shorter than to the receiving side. Self-inductance of transmitting side rises with steel belt due to its high magnetic permeability. Therefore mutual inductance decreases by magnetic flux leakage.
Calculated values with these parameters are shown in Fig.13
The steel belt occurs about twice bigger WPT loss. As expected, it is confirmed that fiber belts does not effect for WPT efficiency. Then, the most suitable materials for tire of in-wheel coil is the fiber belt.

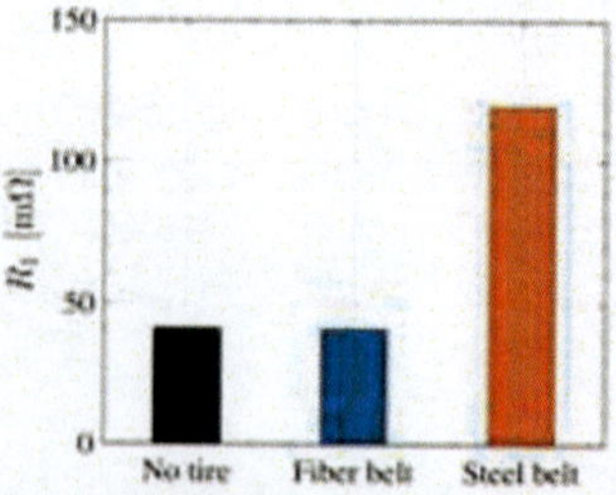

(a) Resistance of Transmitting Side

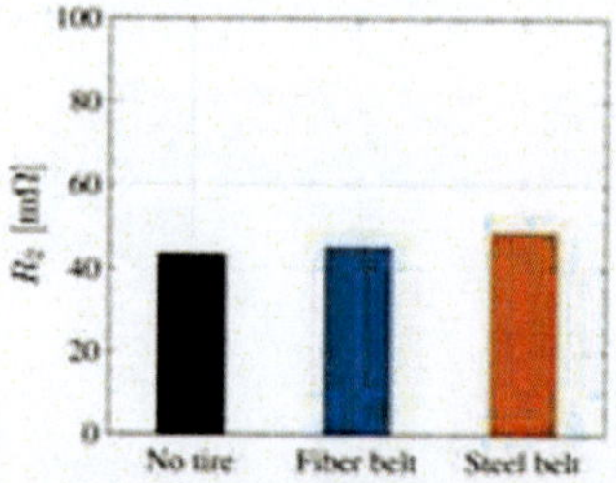

(b) Resistance of Receiving Side

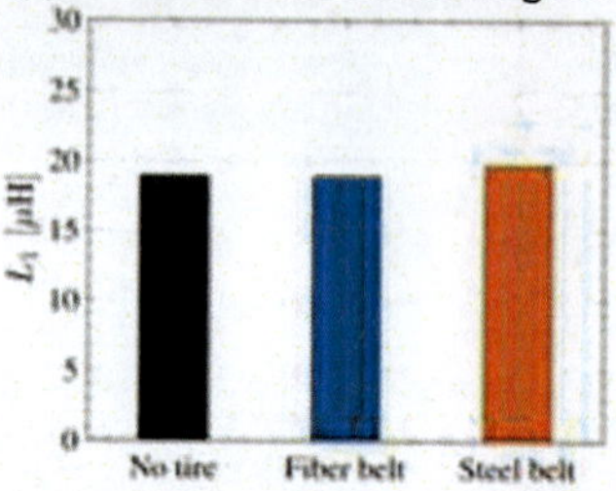

(c) Self-inductance of Transmitting Side

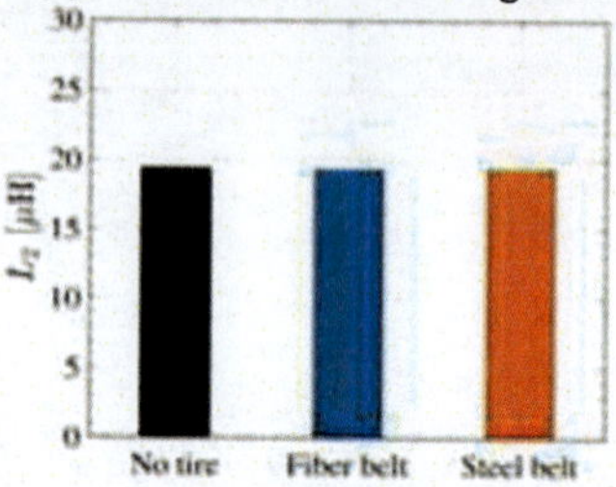

(d) Self-inductance of Receiving Side

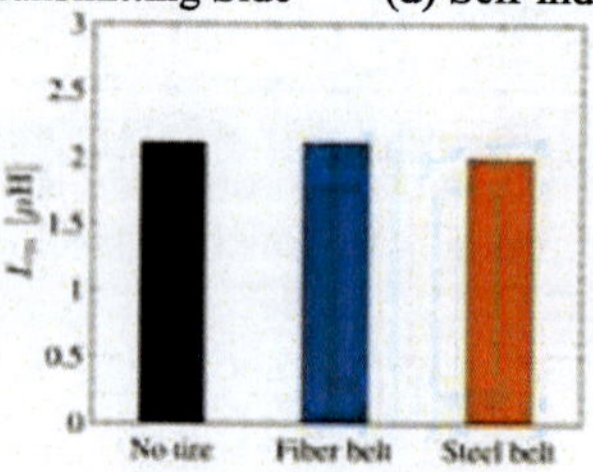

(e) Mutual Inductance

Fig. 12. Measured Values of Belt (Frequency:85kHz)

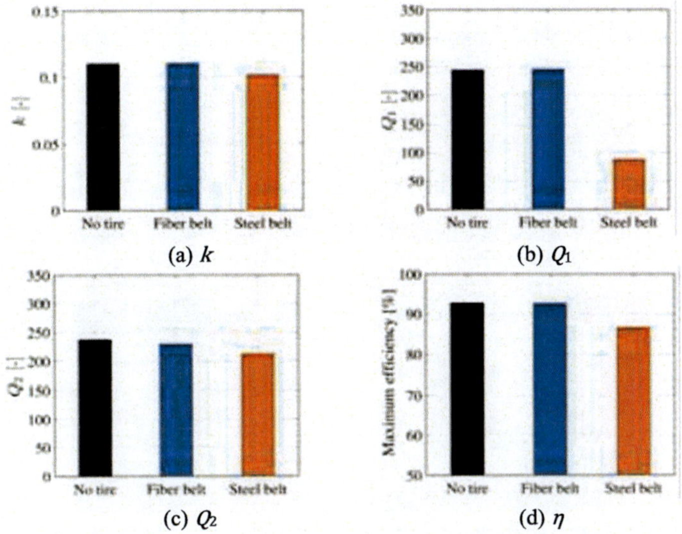

Fig. 13. Evaluation Result of Belt (Frequency:85kHz)

4.2 Materials of Wheel

The basic evaluation for materials of wheels is also done with the simple equipment. The evaluated materials are iron, aluminum, and 2 types of CFRP. The size of test pieces is 150*150*6. Test pieces are displayed in Fig.14.

These materials have been adopted for commercialized wheels. 2 types of CFRP are consisted of the same carbon fibers, however the methodology of knitting is different. One is cross CFRP and the other is unidirectional (UD) CFRP. Cross CFRP whose strength is isotropic is knitted like clothes, and UD CFRP whose strength is anisotropic is just bundled. UD CFRP has stronger direction than cross CFRP. Therefore UD CFRP is better than cross CFRP, when it is used for the products whose stress concentrates in one direction like wheel. Carbon fiber has a little conductivity [10], however it is impregnated with insulation resin. DC resistance of CFRP plate cannot be measured because of its resin courting. Then it is impossible to predict its eddy current loss by DC resistance tester.
Measured values are shown in Fig.14 and calculated values are shown in Fig.15.

(a) Iron Plate

(B) Aluminum Plate

(a) Cross Carbon Fiber Plate

(B) UD Carbon Fiber Plate

Fig. 14. Test Pieces of Wheel Materials

Both aluminum and iron effects WPT efficiency critically. The skin effects by eddy current is the biggest reason for efficiency reduction of metal materials, because mutual inductance is almost zero by metal materials. Cross CFRP also effects WPT efficiency critically. Mutual inductance is about the half of the test result without any materials. The skin effect is less than metal materials, however the eddy current loss increases because of its skin depth. UD CFRP effects WPT efficiency hardly. The carbon fibers of cross CFRP touch each other by knitting, on the other hands, the carbon fibers of UD CFRP does not touch each other. It is considered that little eddy current is occurred in UD CFRP, because its resin isolates each carbon fiber well. Then, the most suitable materials for wheel of in-wheel coil is UD CFRP.

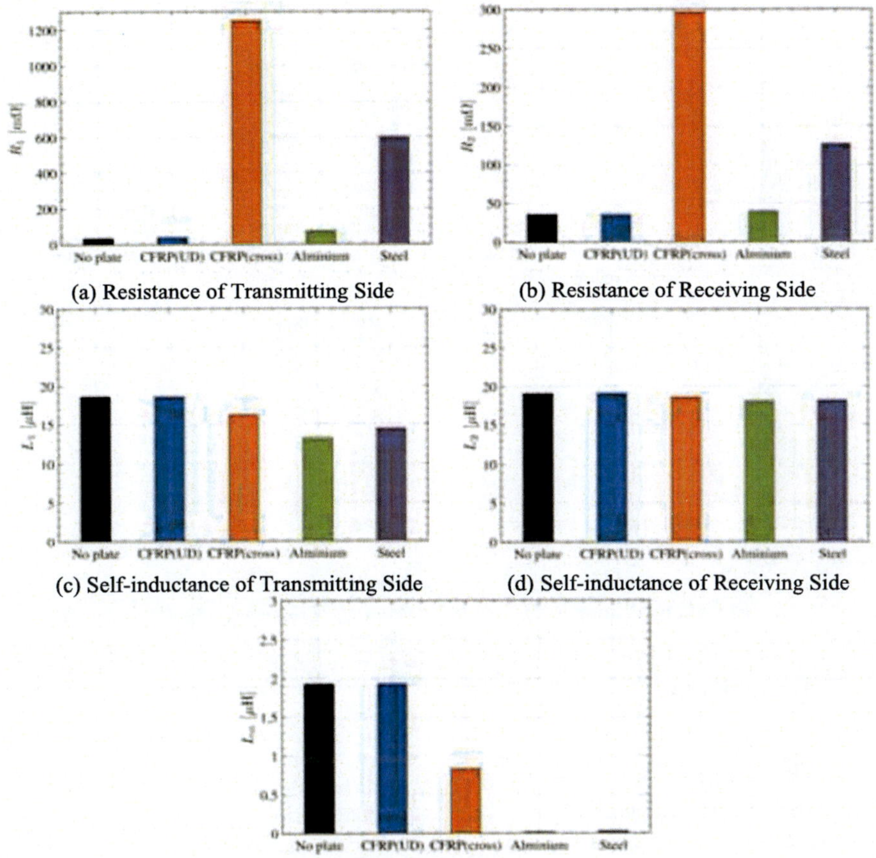

(a) Resistance of Transmitting Side

(b) Resistance of Receiving Side

(c) Self-inductance of Transmitting Side

(d) Self-inductance of Receiving Side

(e) Mutual Inductance

Fig. 15. Measured Values of Wheel Materials (Frequency:85kHz)

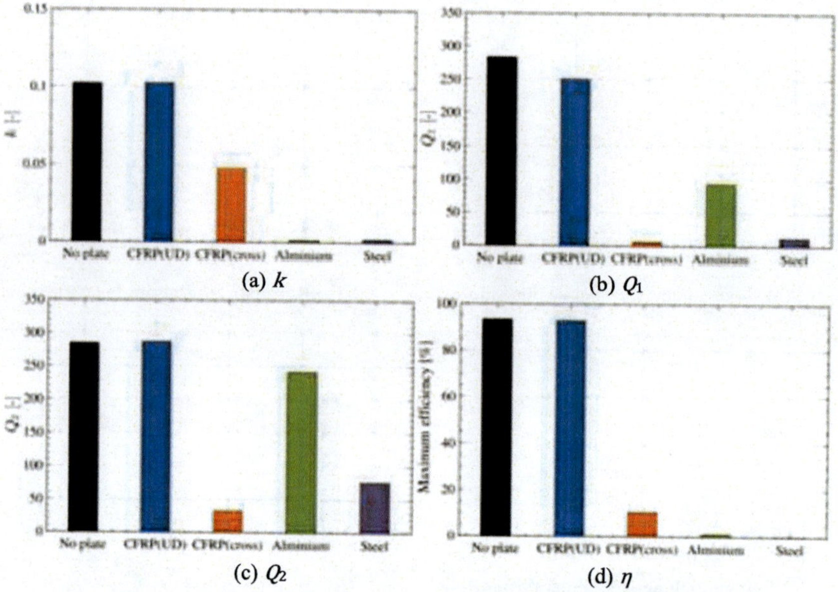

Fig. 16. Evaluation Result of Wheel Materials (Frequency:85kHz)

5 Conclusion

In this paper, the novel method for dynamic WPT system which can prevent foreign matter objects physically. Its suitable materials for wheel and tire are also proposed. The system can reduce possibility to foreign objects entering into between the transmitting coil and the receiving coil. The conclusion about materials of wheel and tire is as follows

1. The rubber does not affect WPT efficiency at 85kHz.
2. Fiber belt is the most suitable material for belt of tire because of its low loss.
3. UD CFRP is the most suitable material for wheel due to its low loss.

Combination of the wheel made of UD CFRP wheel and the tire whose belt is fiber belt is the best way to improve WPT efficiency.

6 Acknowledgement

This work was partly supported by JSPS KAKENHI Grant Number 18H03768 and JST-Mirai Program Grant Number JP-MJMI17EM, and the EIG CONCERT-Japan 4th Call under the Strategic international Research cooperative Program of JST, Japan.

References

1. Tajima, T.; Noguchi, W.; Aruga, T. : Study of a Dynamic Charging System for Achievement of Unlimited Cruising Range in EV: SAE Technical Paper 2015-01-1686, 2015
2. Kurs, A.; Moffatt, R. A.; Joannopoulos, J.D.; Fisher, P.; Soljačić, M.: Wireless Power Transfer via Strongly Coupled Magnetic Resonances, Science, Vol. 317, Issue 5834, pp. 83-86, 2007
3. Fujita, T.; Yasuda, T.; Akagi, H. :A Dynamic Wireless Power Transfer System Applicable to a Stationary System: IEEE Transactions on Industry Applications, Vol. 53, Issue 4, pp3748-3757, 2017
4. Covic, G. A.; Boys, J. T.; Kissin, M. L. G.; Lu , H. G.: A Three-Phase Inductive Power Transfer System for Roadway-Powered Vehicles, IEEE Transactions on Industrial Electronics ,Vol. 54, Issue 6, pp.3370-3378 ,2007
5. Mi, C. C.; Buja, G.; Choi, S. Y.; Rim , C. T.: Modern Advances in Wireless Power Transfer Systems for Roadway Powered Electric Vehicles, IEEE Transactions on Industrial Electronics, Vol. 63 ,Issue 10 , pp. 6533-6545 ,2016
6. Ohira, T.: Via-wheel power transfer to vehicles in motion: 2013 IEEE Wireless Power Transfer, 2013
7. Sato, M.; Yamamoto, G.; Gunji, D.; Imura, T.; Fujimoto, H.: Development of Wireless In-Wheel Motor using Magnetic Resonance Coupling: IEEE Transactions on Power Electronics, Vol. 31, Issue 7, pp. 5270-5278, (2016)
8. Fujimoto, H.; Takeuchi, T.; Hanajiri, K.; Hata, K.; Imura, T.; Sato, M.; Gunji, D.; Guidi ,G. : Development of Second Generation Wireless In-Wheel Motor with Dynamic Wireless Power Transfer, The 31st International Electric Vehicle Symposium & Exhibition and International Electric Vehicle Technology Conference 2018, 2018
9. Imura T. "Simple equivalent circuit model with foreign object on wireless power transfer via magnetic resonant coupling", 2017 IEEE Conference on Antenna Measurements & Applications, 2017
10. Smith, Y. R.; Bhattacharyyaa, D.; Willhard, T.; Misra M.: Adsorption of aqueous rare earth elements using carbon black derived from recycled tires, Chemical Engineering Journal, Volume 296,pp.102-111,2016

Analysis of tire wet grip mechanisms and their respective weightings in a wet braking test

Frédéric Biesse [1] and Violaine Todoroff [1]

[1] Michelin, France

Abstract.

Car drivers constantly demand a higher level of road safety, whether in terms of road infrastructure, vehicle and/or tires. They also ask for more information about the level of safety they have when driving. To meet this need, the tire label – which has existed since 2012 in Europe – gives them information about the tire's wet grip performance as well as its noise and fuel consumption impact. However, this information is worth for a new tire, and it is known that wet grip performance is one of the few tire performances that reduces as a tire wears.

In this study, the analysis of wet grip mechanisms is conducted through an innovative method, which is based on a detailed analysis of the regulatory wet braking test. This method enables a decomposition of the respective weighting of each mechanism, both for new and worn tires. It has been observed that there are two main mechanisms of wet grip: rubber friction and hydroplaning. While rubber friction enables the generation of a braking or steering force by the tire through the rubber-road contact, hydroplaning reduces its efficiency by diminishing the contact surface area.

This study demonstrates that the relative importance of the involved mechanisms depends on the tire state: while rubber grip is the predominant mechanism of a new tire, both hydroplaning and rubber friction are important for worn tires. This decomposition study also shows that the reduction of wet grip from a new to a worn tire is tire dependent as decrease may be linked to rubber friction, or to a reduced hydroplaning capability

Keywords: Wet braking, worn tire.

1 Introduction

In the today's world, enabling improved mobility is a key question for human development. In this scope, the first question that is raised by drivers relates to driving safety. As tires have a unique contribution to automotive mobility, this safety question is a main point of preoccupation for tire manufacturers. After safety, customers always speak about driving economy and environmental protection.

P. E. Pfeffer (ed.), *11th International Munich Chassis Symposium 2020*, Proceedings, https://doi.org/10.1007/978-3-662-63193-5_12

In many regions of the world and since 2012 in Europe, a tire labeling system gives improved information to the consumer about those items of safety, economy and environmental protection. For a tire manufacturer, it is an everyday challenge to improve those three aspects without conceding a tread-off on one of those items in order to improve another.

It is interesting to point out the evolution of the tire's performance (particularly wet and dry grip) during its whole life. Tires wear out due to driving mileage, which reduces their tread depth. Dry grip improves with wear, generally from 5% to 10%, while wet grip declines, as can be seen in **Fig. 1**. It can also be mentioned that tires are usually tested when they are new – this includes tire labeling testing as well as standard journalist testing. This implies that, for dry weather, testing is performed in the worst case (grip will improve through tire life), whereas for wet weather, only the best case is tested (grip will reduce with wear). This is not an optimal situation for customer information regarding driving safety.

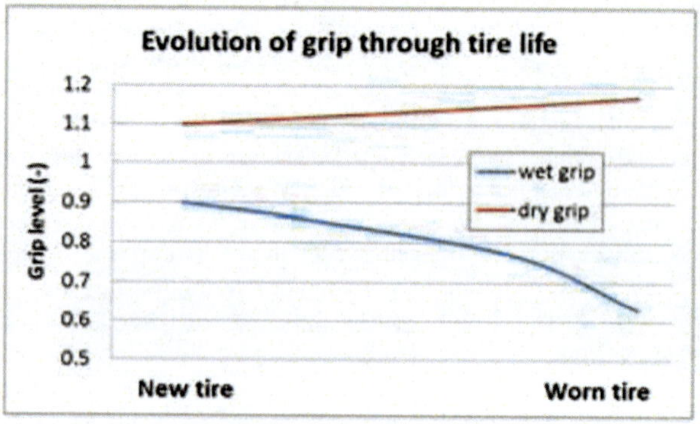

Fig. 1. Evolution of dry and wet grip through tire life (representative average value).

Two types of tests are used for labelling tires in the new state: on-vehicle braking and an analytical test (with a trailer or special evaluation vehicle). We will focus on the on-vehicle regulatory test which is a wet braking from 80 to 20km/h with water depth (WD) of 1±0.5mm on the road. Water depth of 1mm is defined such as there is 1mm of water above the top roughness of the road.

In this paper, we propose to evaluate the mechanisms involved in the regulatory wet braking test for new and also for worn tires, and then to understand the relative importance of the involved mechanisms depending on the tire tread depth. This will help also to better understand the interest of testing worn tires for wet grip, and will give indication to tire designer on the way to optimize tire design for a better worn grip.

2 Literature review

2.1 Grip mechanisms

It is well known that tire braking performance is related to the friction of rubber on a road. Numerous mechanisms are involved in friction such as adhesion, indentation and various kinds of viscous friction. [1] gives the following definitions for those mechanisms:

- Hysteretic friction is due to the deformation cycle imposed to the rubber by the surface roughness (from micrometers up to millimeters) which generates a dissipation of energy and thus a force opposed to the movement. It is one of the most important rubber friction mechanisms on rough surfaces.
- Adhesion friction can be generally described as 'a rise of physical bonding between two bodies when they are brought in a close contact'. Adhesion requires very clean contact, as water, mud, ... disturbs this phenomenon.
- Viscous friction is the friction of a rubber block on wet surface, when rubber slides over a water film, generating viscous friction. This mechanism, however, is believed to have minor importance at small sliding velocities less than 1 m/s.

We will group this set of mechanisms under the name of rubber friction. This effect exists whatever the vehicle speed and is influenced by the local sliding velocity, contact pressure, rubber temperature, and of course by the viscoelastic properties of the compound resulting of its design.

The second major mechanism that can be encountered while braking on wet and flooded roads is the hydroplaning mechanism. The hydroplaning mechanism is due to the water hydrodynamic pressure building up in front of the tire [2]. As hydrodynamic pressure is applied on the tire surface at the entrance of the contact area, a lift force is induced which carries part of the load initially carried by the tire. This lift force can become as big as the tire initial load. In that case, the tire loses all contact with the road and is no longer able to transfer any forces (which is defined as a full hydroplaning situation). Hydroplaning is linked to various parameters such as vehicle speed, water depth, road macrotexture (texture depth), contact area of the tire, and tread design. It will thus reduce the contact surface at high speed – and consequently the efficiency of rubber grip mechanisms – explaining the decline of the braking performance at the beginning (high speed part) of the braking test.

Fwa et al. [3] analysed the hydroplaning of a worn ASTM (standard ASTM E501 rib tire) at 1.6mm of remaining tread depth with a numerical approach. They also compared the hydroplaning speed of tires of various tread depths using their numerical approach. They demonstrated that tires with lower tread depths reach hydroplaning speed before new tires (on average, hydroplaning speed of worn tires is 15 km/h lower than those of the new tires) on all the tested water depths (from 1 to 10mm).

In synthesis, the literature review demonstrates that two major mechanisms are encountered during a wet braking test: rubber friction and hydroplaning. The literature also reveals that worn tires exhibit more sensitivity to the hydroplaning mechanism than new tires.

2.2 Relevancy of the testing conditions

The regulatory wet braking test defines an initial speed of 80 km/h for the test, as well as a 1mm water depth. Concerning the grip mechanisms, we have seen that both rubber friction and hydroplaning plays a role in the tire capability to generate braking and steering force on a wet road. But the question may be asked of the relevancy of those elements, and the frequency of their occurrence in the usual driving conditions.

In the scientific literature, several papers can be found about the driving speeds under rainy weather. For example in [4], Hartz observed that on German highways, the speed reduces from 10% to 24% versus dry road, depending of the rain intensity. In [5], the distribution of the initial speed of the accident described in the GIDAS database (German In-Depth Accident Study, 2.000 scientifically documented accidents per year since 1999) is given and compared with speeds in usage. As illustrated on **Fig. 2**, it is observed that initial speeds of accidents are much lower than the usage speed, and that more than 90% of the wet accident have their initial speed lower or equal to 80 kp/h.

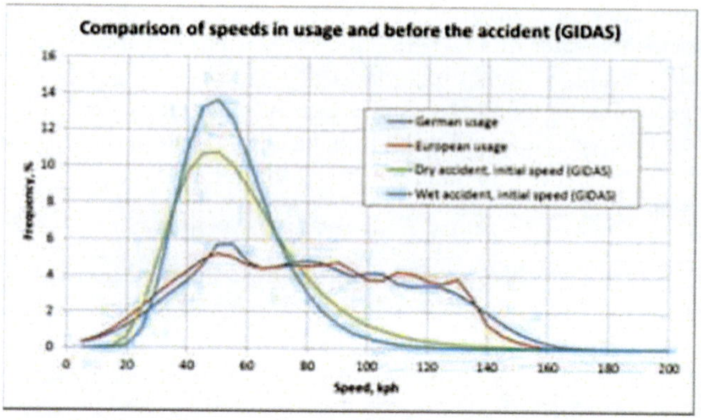

Fig. 2. Comparison of speeds in usage and initial speed before the accidents.

The distribution of water depth observed on the road is not well described in the literature. However, in [5], several water depth data were gathered and analyzed. A first set of data comes from road weather stations, these measurements are made by local weather stations that are spread all over the road. For Germany, the data are made freely available by the German weather forecast service ("Deutscher Wetterdienst", abbreviated DWD). Identical data were gathered also in France. In addition, water depth measurement on open roads were also conducted with a specially equipped vehicle. As illustrated on **Fig. 3**, the results show that in 99% of the wet road time, the water depth is lower or equal to 1mm. As a consequence, the relevancy of the reglementary wet grip test conditions is found to be in good agreement with the scientific publications. These reglementary test conditions can so be considered as reasonably severe.

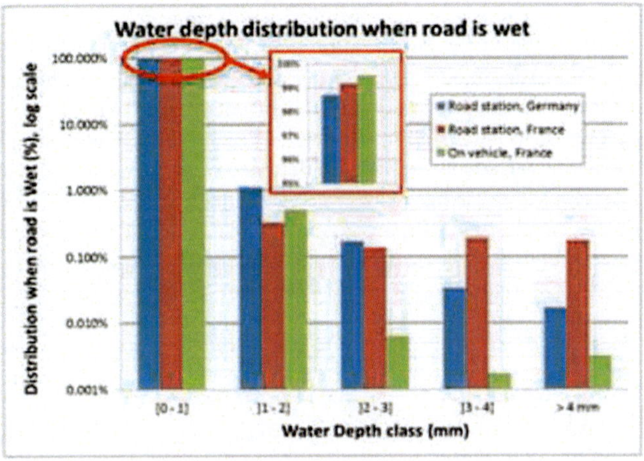

Fig. 3. Water depth distribution on wet roads, in Germany and France.

Concerning the importance of rubber friction and hydroplaning in the driving safety, the recently published paper [6] describe the analysis made on the GIDAS database to identify the share of full hydroplaning accident compared to the grip accidents. As shown on **Fig. 4**, it is observed in this study that on wet road, 62% of the accident are grip relevant, while 0.6% have full hydroplaning as main cause. We can then conclude that the two main wet grip mechanisms are observed in the accidents.

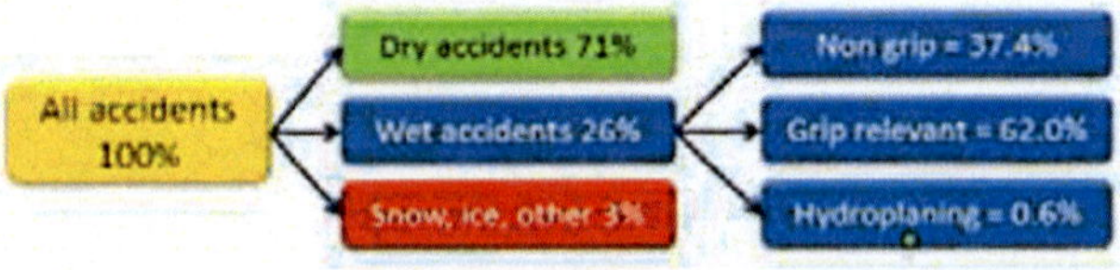

Fig. 4. Share of wet accident caused by hydroplaning, computed on GIDAS database.

3 Decomposition of the test result into the two mechanisms

The objective of this section is to describe the decomposition method on a specific example. An analysis over a big set of different tires will be conducted in section 4.

3.1 Influence of the two mechanisms during a wet braking test

In this paper, we propose to focus on the current regulatory wet braking test and to give a quantitative analysis of the share of the mechanisms described above for new and worn tires. To do so, we analyzed the deceleration data recorded during on-vehicle braking tests in regulatory conditions. Worn tires were obtained by the buffing process up to the tread wear indicator.

The testing method is given in detail in the European regulation N° 2009/1222. It is based on the calculation of the deceleration performance of the tested tires during wet braking, using an instrumented passenger car equipped with an Antilock Braking System (ABS). The wet condition is defined such as water depth is in the interval of 1±0.5 mm. The instrumentation required for this testing method is a sensor suitable for measuring speed on a wet surface as well as the distance covered between two speeds. Tire inflation pressure is set at 2.2 bars, the braking distance is computed between 80 km/h and 20 km/h.

Performance in regulatory condition is given in terms of μ (deceleration = μ.g, where g is the standard acceleration due to gravity) and not of braking distance. The relationship between those two is

$$\mu = \frac{V_1^2 - V_2^2}{2g(D_2 - D_1)} \tag{1}$$

where $V_1 = 80/3.6$ m/s and $V_2 = 20/3.6$ m/s and $D_2 - D_1$ the overall braking distance (m).

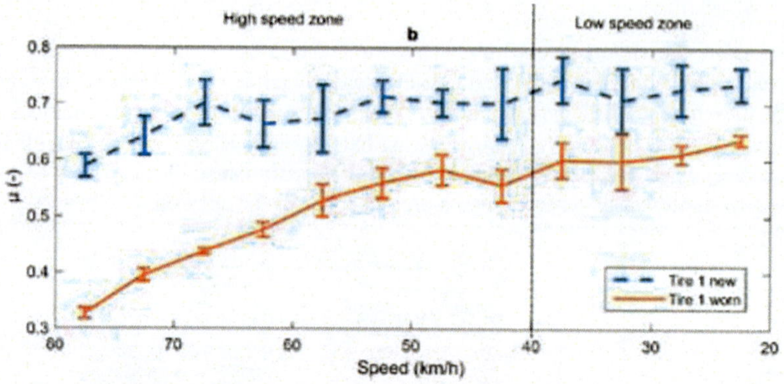

Fig. 5. μ as a function of speed for Tire 1 new and worn on regulatory wet braking conditions.

To illustrate our approach, we will consider the wet braking results of the Tire 1, in new and worn condition. As we could expect, the new tire performs better than the worn tire in wet braking. Yet, knowing the average braking distance or average μ level on a wet braking does not describe precisely the share of the mechanisms involved during the process. To do so, it is necessary to know the μ level for the different speeds encountered during the entire braking. In **Fig. 5** the grip coefficient μ for the various speeds encountered during the entire braking test is plotted. To do so, we use Formula 1 between two consecutive data points to compute the μ value in each interval. The μ value is averaged on 5 km/h speed intervals to reduce the variability of the raw signal. The following analysis can be done in **Fig. 5** :

- For the new tire, μ is slightly decreasing as a function of speed which confirms the fact that the deceleration appeared rather constant during the overall braking.
- For the worn tire, μ is rather constant at low speeds (up to 40 km/h) then it exhibits a significant decline at higher speeds.

We propose to separate the $\mu(v)$ curve into two zones to analyze the result:

- The High-speed zone shows a strong decline in μ level. This decline can be attributed to the partial hydroplaning mechanism. As seen earlier in the literature, this mechanism explains a reduction of the braking force at the higher speeds.
- The Low-speed zone shows a rather constant μ level. On this part, the hydroplaning mechanism is not visible and the μ value reflects only the friction mechanism of the rubber on the road.

On the new tire, the μ level is rather constant through the entire braking which shows that the hydroplaning mechanism does not impact its overall performance. The friction mechanism is explaining the overall wet braking performance of the new tire.

On this specific example, we can notice that the μ values at low speed for both tires are different. The tire at worn state displays a significantly lower value than its associated new tire. We will come back to that specific point in Section 4.

3.2 Weighting of the two mechanisms

In order to quantify the share of hydroplaning and rubber friction mechanisms on the overall wet braking performance, we propose the following decomposition of the performance:

- Rubber friction mechanism is well captured by the value of μ below 40 km/h
- Hydroplaning mechanism is responsible for the decline of the performance at higher speeds

This leads to the following decomposition of the global braking performance:

$$\mu_{80\text{-}20} = \mu_{WetGrip} \, . \, Eff_{Hydro}$$

where $\mu_{80\text{-}20}$ is the overall μ deduced from the braking distance, $\mu_{WetGrip}$ is the average value of μ in the 40–20 km/h speed interval and Eff_{Hydro} is defined as the hydro efficiency of the tire. The Eff_{Hydro} parameter quantifies the sensitivity of a tire to hydroplaning: when $Eff_{Hydro} = 1$ the tire is not sensitive to hydroplaning since it maintains its friction value above 40 km/h.

This conclusion can be visualized in **Fig. 6** for Tire 1 in the new and worn states. The dotted line displays the friction value at low speed: $\mu_{WetGrip}$. The shaded area illustrates the decline of performance due to hydroplaning, so that the larger the shaded area, the lower the hydro efficiency. For this specific case, we determine the wet grip friction value and the hydro efficiency of tire 1 in new and worn conditions in the **Table 1**.

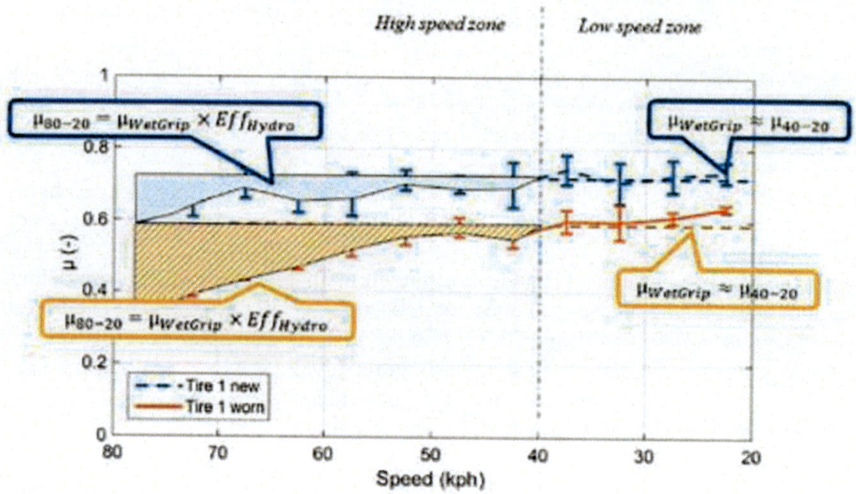

Fig. 6. Decomposition of the performance in two mechanisms.

Table 1. Decomposition of the test result for Tire 1.

Tire	$\mu_{80\text{-}20}$	$\mu_{WetGrip}$	Eff_{hydro}
Tire 1 new	0.67	0.73	0.92
Tire 1 worn	0.48	0.61	0.79

The hydro efficiency of the worn tire is below the one of the new tire which is in good agreement with its higher sensitivity to hydroplaning. As seen previously, the tire in the worn state has a significantly lower $\mu_{WetGrip}$ value; the decline of this parameter from new to worn is 16% in this specific case.

3.3 Focus on the hydro efficiency

In order to evaluate the limit of our approach, it is interesting to compare those same two tires tested under damp conditions. Damp condition is achieved after the track watering system is turned off. A limited time of waiting is given for any extra water to evacuate, so that the water depth is around 100-200 μm instead of 1mm. In this condition, hydroplaning is expected to be insignificant.

Table 2. Decomposition of the test result for Tire 1.

	Eff_{hydro} new tire	Eff_{hydro} new tire
Damp condition	0.93	0.92
Wet condition	0.92	0.79

Eff_{Hydro} is calculated for both conditions in **Table 2**. We see that the hydro efficiency of the new tire is almost constant for both conditions while the hydro efficiency of the worn tire is lower in wet condition. We also observe that hydro efficiency is nearly the same for the new and the worn tire in the damp condition, as expected.

The test on damp condition highlights another important point: hydro efficiency in damp condition does not equal one as would be expected, because on damp road no hydroplaning mechanism exists. Both new and worn tires have hydro efficiency close to 0.92 which is the same value that the one found for the new tire on 1mm water depth condition. Indeed, since hydro efficiency does not equal one on damp condition it means that μ is declining between low and high speeds and that the hydroplaning mechanism cannot be held responsible for the decrease. Three mechanisms can be responsible on this effect.

The first mechanism which could explain the decline of μ as the speed increases is the well-known sensitivity of μ to the sliding speed [7], which reduces the rubber frictions mechanism efficiency. The second point is the ABS system efficiency which might be less efficient at high speeds than at low speeds. The third point is correlated to the lubrication phenomenon. It is known that although appropriate road surface texture, thin water films (10–200 μm) can still be formed in the contact especially at higher speeds due to the water viscosity, which hides the smaller scales of the asperities to the rubber, resulting in a reduction of the hysteretic friction mechanism and consequently, a reduction of μ level [8].

In order to assess the validity of our hypotheses, we present in Figure 8 the hydro efficiency calculated for all tested tires (new and worn) in damp conditions. Over the last 2 years, multiple tests were conducted in Michelin and other test tracks for numerous new, intermediate tread depth and worn tires on damp condition. The histogram of **Fig. 7** integrates results from tires of various brands, dimensions, sculptures and materials.

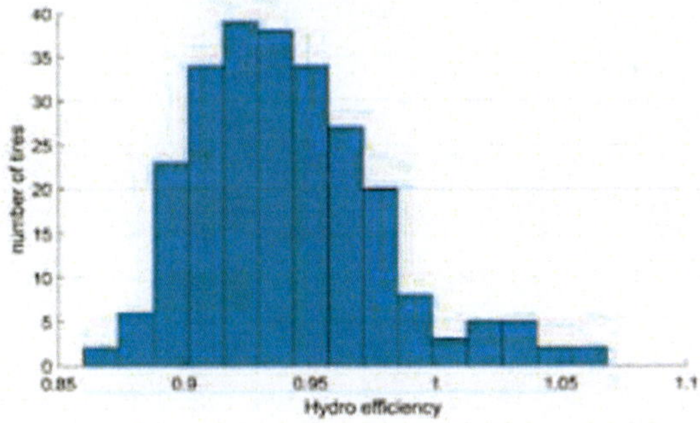

Fig. 7. Hydro efficiency of all tested tires (new + worn) in damp conditions.

As seen in this histogram, most tires have hydro efficiency included between 0.9 and 1 in damp conditions. The decline of μ at high speed, due to all the mechanisms previously cited, is rather limited. Nevertheless, based on this analysis, we can conclude that hydroplaning mechanism can properly be taken into account for hydro efficiencies below 0.9.

4 Statistical analysis of the mechanisms involved in a regulatory wet braking test

The previous section was based on a single example illustrating the share of both mechanisms on a specific new and worn tire. In order to draw more robust conclusions, we apply our analysis to all the tested tires in our database.

As a first result, we observed that the hydro efficiency of the new tires are included in the [0.86 ; 1.05] range, which indicates that for new tires, hydroplaning plays a very reduced role in the 80-20 braking test on 1mm water.

In **Fehler! Verweisquelle konnte nicht gefunden werden.**, we plot the overall performance ($\mu_{80\text{-}20}$) as a function of $\mu_{WetGrip}$ ($\mu_{40\text{-}20}$) as well as a linear fit and associated correlation coefficient on the data. This graph highlights the strong correlation between rubber friction and the overall wet braking performance and so, confirms the result seen in the first example. As a conclusion, we can state that current regulatory test applied on new tires ranks tires mainly on rubber friction mechanism and that hydroplaning mechanism has almost no impact on wet braking results in the tire labelling test in the new state.

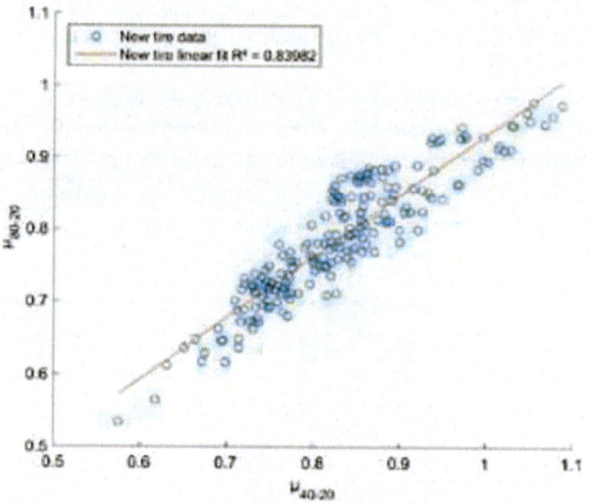

Fig. 8. $\mu_{80\text{-}20}$ as a function of $\mu_{40\text{-}20}$ for new tires

It is of more interest to study the results for worn tires. Hydro efficiencies in damp and wet conditions for worn tires, as well as normal fits, are presented in **Fig. 9**. Two major conclusions can be drawn from this plot:

- The wet and damp distributions are disjoined for the worn tires. It illustrates the fact that the hydroplaning mechanism is playing a key role on the wet braking test in a regulatory condition of worn tires. As expected for the damp condition, the Gaussian fit is centered around 0.95 while for the test in regulatory conditions it is centered around 0.74.

- The second point to highlight is related to the standard deviation of the hydro efficiency in the wet condition. This plot gives an overview of the variability in hydro efficiency that can be found on the numerous tires tested. 95% of the tested population has hydro efficiency included in the interval between 0.61 and 0.87. The broad range of this interval highlights the fact that the impact of the hydroplaning mechanism is not the same on all the tires of our database.

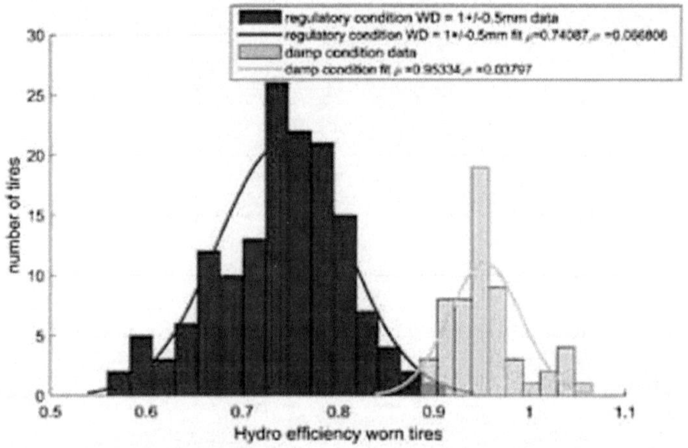

Fig. 9. Comparison of hydro efficiency on damp and regulatory conditions for worn tires.

Concerning the $\mu_{WetGrip}$, in order to get a relevant average value of decline, in **Fig. 10** we plot the decline of performance in grip potential from new to worn state in wet condition ($\mu_{WetGrip\ Worn}/\mu_{WetGrip\ New}$) on the entire database. We can then conclude that 95%of the tested population has a rubber friction coefficient ratio between 0.75 and 1. The broad range of this interval highlights the fact that the performance at worn state depends on the tire. Average decline is 12.2% for all tested tires.

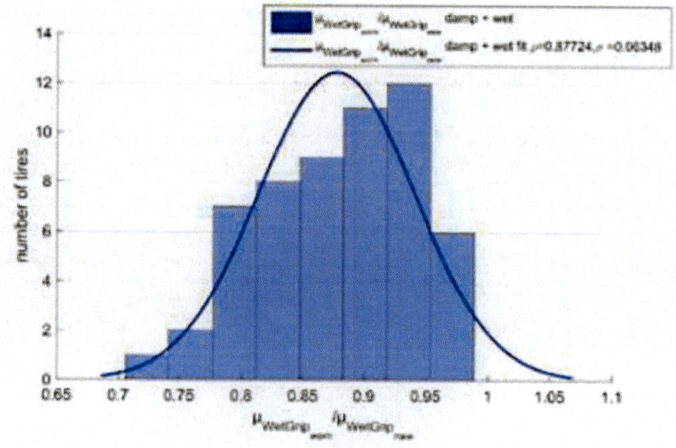

Fig. 10. Decline in grip potential from new to worn state for the entire database.

As the decline of performance varies significantly from one tire to another the performance in grip potential at worn state cannot be inferred from the performance of a tire in the new state. We illustrate this point in **Fig. 11**where we plotted the μ on damp condition of various pairs of new and worn tires obtained on a single track and sorted the tires according to their performance in the new state.

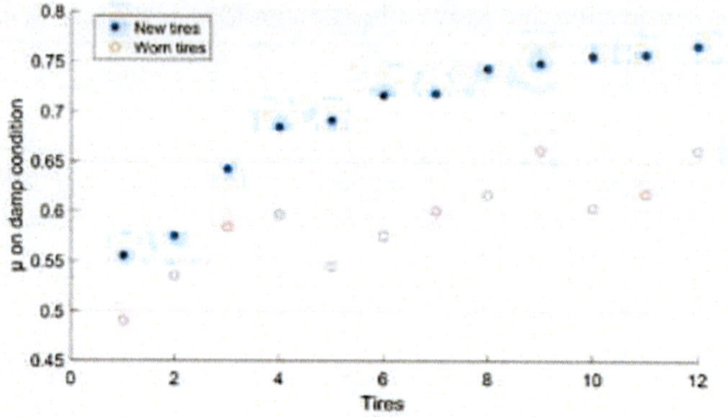

Fig. 11. μ on damp condition for some pairs of new and worn tires.

In this graph, we see clearly that ranking inversions can occur between the new tire ranking and the worn tire ranking. As an example, Tire 10 is the third best tire in the new state but it is only the fifth best in the worn state. Tire 9 which is really close to Tire 10 in the new state has a significantly better performance in the worn state. To conduct a detail analysis of the mechanisms to explain the decline of $\mu_{WetGrip}$ between new and worn tires is a subject of ongoing research that will not be developed in this paper.

5 Consequences of the mechanism weigthing on worn tires grip

5.1 Correlation between wet grip of new and worn tires

We have seen that worn tires undergoing a wet braking test in a regulatory condition are impacted by hydroplaning mechanisms as well as a loss in wet grip potential. We have demonstrated that the declines of performance from new to worn status due to those two mechanisms are not the same for all tested tires. This signifies that worn tires demonstrate different behaviors in terms of sensitivity to the hydroplaning mechanism as well as in grip potential due to their varied designs.

In **Fig. 12**, we plot those two parameters versus the overall wet braking performance. We see that no good correlation can be found by comparing either $\mu_{WetGrip}$ or the hydro efficiency to the overall wet braking performance. Overall wet braking performance of a worn tire is, therefore, a mix of the two mechanisms. **Fig. 11** and **Fig. 12** demonstrate that wet grip performance of a worn tire cannot be predicted from its new tire one's.

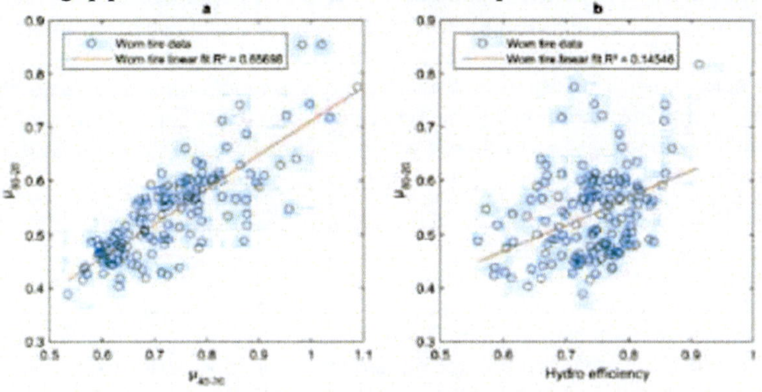

Fig. 12. Correlation between μ80-20 and μWetGrip (a)/hydro efficiency (b).

To finish the analysis, we plot on **Fig. 13** various pairs of new and worn tires and evaluate the influence of both cited mechanisms on the decline of performance from new to worn state. Seven tires were selected because their decline in performance cannot be attributed to the same share of rubber friction and hydroplaning mechanisms. This plot is showing the total decline of performance from the new tire to the worn tire Δμ and highlights the share of decline attributed to a loss of rubber friction (dashed blue) and to a loss in hydro efficiency (filled blue) from new to worn state.

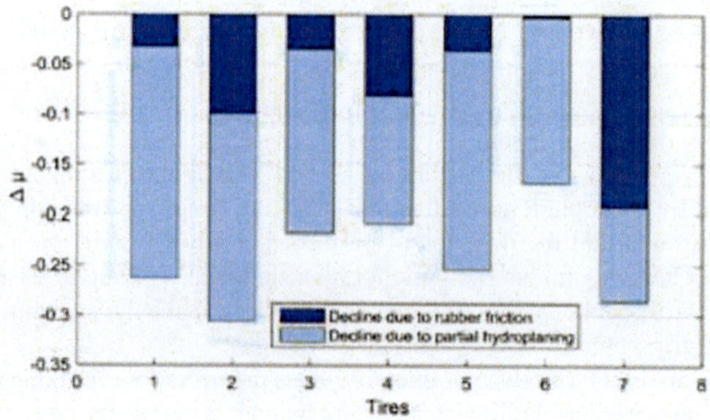

Fig. 13. Performance decline from new to worn: share of rubber friction and hydro.

We see that the loss of rubber friction in the decline of performance from new to worn tires varies from one tire to another from almost no decline (Tire 6) to almost 0.2 (Tire 7). The remaining decline from new to worn performance is attributed to the hydroplaning mechanism. The loss of μ attributed to this second mechanism varies also between tires, from a decline of 0.09 (Tire 7) to a decline of 0.23 (Tire 1). This analysis can be useful to help tire designers choose which design lever might give the best improvement of the overall wet braking performance at worn state.

5.2 Consequences on tire design

We illustrate this section with the tires presented in **Fig. 13**. In order to limit the decline from new to worn state on Tire 7, tire designers should focus on improving rubber friction to get the maximum improvement in wet braking performance. Hydro efficiency is already high and might not be better optimized. However, the comparison to other tires shows that the overall decline in rubber friction is significantly higher than the other tested tires.

On the contrary, Tire 1, Tire 3, Tire 5 and Tire 6 have limited loss of performance due to a loss of rubber friction between new and worn states. Tire designers should then focus on improving the hydro efficiency of those tires. Tire 2's loss of performance is attributed to both mechanisms. So working on either parameter should help designers improve the overall performance of the worn tire.

Finally, depending on the share of both mechanisms explaining the overall wet braking performance of their tire, tire designers can build design rules to improve hydro efficiency while also focusing on the grip performance of the rubber compound at worn state to ensure a better overall wet grip performance.

6 Conclusion

In this paper, we proposed an original analysis of the wet braking test signal to highlight the mechanisms taking place during wet braking under regulatory conditions with a focus on worn tires. The low-speed value of the μ(v) curve reflects the friction potential of the tested tire while the loss of performance at higher speed is attributed to the hydroplaning mechanism.

The testing conditions of the regulatory wet braking test are proved to be relevant by the scientific literature. We demonstrated that current regulatory wet braking test, when run with worn tires, exhibits both rubber friction and hydroplaning mechanisms. We also demonstrated that hydro efficiency and wet grip parameters have broad distributions for the worn tires tested in our database. This point implies that worn tire performance is specific and cannot be deduced from the new tire performance as:

- Grip potential of worn tires cannot be inferred from the grip potential of their associated new tire since the decline of performance may vary significantly from one tire to another.
- Testing tires in the new state does not give any information on the hydroplaning efficiency of worn tires.

This is the combination of both parameters that gives the overall performance of a worn tire. This original approach highlights the influence of each of the two parameters on the overall wet braking performance. This is then a good tool for tire designers as they can focus on the right design lever to ensure the best tire performance in the worn state.

Tire performance evolution with wear is among the less well known information by customers. Wet grip has been shown to be an important topic, in particular because of a greater tire role in wet safety compared to dry roads, and because this is a performance that reduces with wear. On the contrary, dry safety and customer fuel savings improves with tire wear.

Through this study, it has been shown that wet safety can be improved by the use of dedicated tests in the worn state. Such a test is necessary because the wet grip performance of a worn tire cannot be predicted from the new tire's test results. It has also been shown, through a detailed wet driving usage study, that a relevant test already exists: the R117 regulatory vehicle wet braking test, simply applied to worn tires.

References

1. Maycock, G.: Studies on the skidding resistance of passenger car tyres on wet surfaces. Proc Inst Mech Eng, 180:122–157 (1965–1966).
2. Horne, WB., Dreher RC.: Phenomena of pneumatic tire hydroplaning. Nasa Tech Note. (1963).
3. Fwa TF., Kumar S., Anupam TF., et al. : Effectiveness of tire-tread patterns in reducing the risk of hydroplaning. Transport Res Record J Transport Res Board, 2094:91–102 (2009).

4. Hartz, B.: Speed on German highways in heavy rain. 4th International Symposium on Highway Geometric Design (2010)
5. Biesse, F.: Analysis of wet road usage with a driving safety concern. VDA Technischer Kongress (2019).
6. Spitzhüttl, F., Goizet, F., et al.: The real impact of full hydroplaning on driving safety. Accident Analysis & Prevention (138), 105458 (2020).
7. Persson, BNJ.: Theory of rubber friction and contact mechanics. J Chem Phys., 115:3840–3861 (2001).
8. Mora, F., Sainsot, P., Le Chenadec, Y., et al : Lubrication of 2D soft elastohydrodynamic contacts: extension of the amplitude reduction theory. J Eng Tribol., 226(9):769–774 (2012).

Tire modeling for extreme handling maneuvers

Ioannis Konstantinou[1], Willem Versteden[1] and Konstantin Sedlan[2]

[1] Siemens Digital Industries, The Netherlands
[2] Volkswagen AG, Germany

Abstract. This paper describes the development of a methodology for representing tire behaviour during extreme vehicle handling simulations. Firstly, the tire operating conditions during a "fishhook" maneuver are analyzed. During this maneuver the tires experience significantly higher vertical loads, sideslip rates and camber angles than during nominal passenger car tire testing conditions. As these tire measurements are used for model parameterization, only testing the tire under nominal conditions is insufficient for accurately modeling the tire performance during extreme handling maneuvers. Three different tire testing protocols are used to parameterize the MF-Tyre/MF-Swift model. The three generated parameter sets are used in vehicle dynamic simulations of which the results are compared to vehicle measurements. The effect of the sideslip angle rate (steering rate) on the tire behaviour is found to be most influential. The effect of the steering rate is eminent when tire thermal effects are considered. Results using an extended version of MF-Tyre/MF-Swift that takes into account thermal effects show that the steering rate directly influences the peak friction that the tire reaches. It is concluded that if a tire model is used for extreme handling maneuvers, the tire operating conditions during the vehicle maneuver should be considered in the tire parameterization test protocol to increase accuracy.

Keywords: vehicle dynamics, tire/wheel, extreme handling, fishhook.

1 Introduction

Extreme handling maneuvers are part of the dynamic testing of newly developed vehicles, providing insight on the stability of a vehicle under emergency situations. Today, these tests are typically performed physically in the later stages of the development process when prototype vehicles are available. Vehicle simulations can add value as they provide possibilities for extended studies, can be performed in earlier development stages and are less cost intensive compared to physical testing.

The vehicle response during (extreme) handling maneuvers is up to a significant extent determined by the tire performance. Applying vehicle simulation in an accurate and reliable manner therefore requires predictive modeling of the tire behavior. The parameterization of today's industry standard tire models is based on tire testing, assuring a good match between measured and modelled forces and moments.

P. E. Pfeffer (ed.), *11th International Munich Chassis Symposium 2020*, Proceedings, https://doi.org/10.1007/978-3-662-63193-5_13

Due to time-budget limitations, the tire testing protocols provide a generic guideline on which the tire models can be generated. They typically assure a good quality model parameterization for soft and medium handling applications where the driving conditions are similar to the nominal tire testing conditions. However, extreme handling maneuvers bring the tire to its limit operating conditions which are typically not encompassed by default tire testing protocols. To this end, Siemens Digital Industries Software and Volkswagen cooperated with the goal to develop and validate a tire modeling methodology for extreme vehicle handling applications.

In the following sections the tire conditions during extreme handling are analyzed, then the tire modeling approach is illustrated and different test protocols for tire model parameterization are presented. Subsequently, the resulting model parameter sets are applied in full vehicle simulations in order to investigate their influence on the predicted vehicle response. Finally, the results of the study are being discussed and the final conclusions are given in the last section.

2 Tire conditions

In this section, the tire conditions reached during the extreme handling maneuvers are analyzed. The overall study focuses on predicting the response of a large SUV vehicle and includes multiple extreme handling maneuvers and multiple tire variants. The results of the study are presented in relative values and, for the sake of conciseness, this article focusses on the vehicle response throughout the well-known "fishhook" maneuver [1] with a 20 inch tire. This maneuver starts with the vehicle driving in a straight line at a velocity which is slightly higher than the desired entrance velocity. The throttle is then released and the steering wheel input as described in figure 1(a) is applied by means of a driving robot. The maneuver aims to introduce high dynamic load transfer in the vehicle in order to assess the vehicle's roll stability.

For standard passenger car vehicle handling applications, tire models are parameterized using measurements that are typically performed with vertical loads in the range of 40% to 120% of the tire's maximum load capacity and with tire inclination angles expected at normal handling conditions. Figure 1(b) compares the operating conditions of a single tire during a "fishhook" maneuver (simulated without ESC intervention) with the ones of standard applications. From these results, it is concluded that the operating conditions significantly exceed the default tire test conditions. Additionally, the tires also reach high slide slip angles during the maneuver.

During the maneuver, the vertical tire load reaches an approximately 40% higher value than in the default test protocol and the tire inclination angles reach 100% higher values. Furthermore, the tire force and moment measurements are usually performed with a moderate sideslip angle rate, such that the tire is in a quasi-steady state condition. Due to the high steering wheel angle rates throughout the "fishhook" maneuver, tire sideslip angle rates significantly increased.

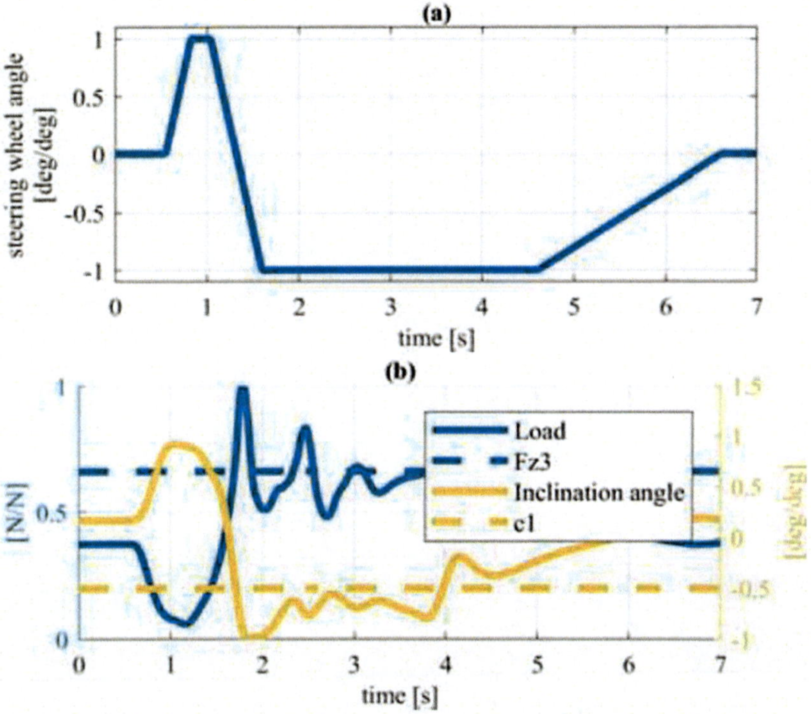

Fig. 1. Illustration of the normalized steering wheel input during the "fishhook" maneuver (a). Tire conditions of the rear left tire from normalized simulation results during the "fishhook" maneuver presented against nominal test conditions (b). Each axis is normalized with the maximum value in it.

3 Modeling tire properties

In this section, the tire modeling approach is presented. Subsequently, three different tire measurement protocols and the respective parameterization strategies are discussed.

3.1 Tire model

Vehicle dynamic simulations require an accurate and reliable representation of the tire behavior, for which the MF-Tyre/MF-Swift model of Siemens Digital Industries Software [2] is used throughout this study. For evaluating the vehicle performance under extreme handling conditions, the frequency bandwidth of interest is relatively limited; relevant dynamics are below 10 Hz. Furthermore, the road surface of interest does not contain short or mid-wavelength undulations and is hence modelled as a flat road.

To model the steady state force and moment response in the longitudinal and lateral directions, MF-Tyre/MF-Swift incorporates the well-known Magic Formula equations [3,4]. This approach consists of a set of semi-empirical expressions representing the tire forces and moments as a function of the tire slip with respect to the road.

In the "fishhook" maneuver the main interest lies in the lateral tire response. The magic formula for the lateral pure slip tire response is [3,4]:

$$Fy(x) = D\,sin(C\,atan(Bx - E(Bx - atan(Bx)))) + S_V \tag{1}$$

$$x = \alpha - S_H \tag{2}$$

Where, α is the side slip angle which depends on the relative tire orientation to forward velocity. The model parameters B, C, D, E further depend on the tire's load and inclination angle conditions. The parameters S_V and S_H represent the lateral force asymmetry and they are also depending on tire conditions.

Due to the finite stiffness of the tire construction, the tire response to fast changes in slip includes transient phenomena. The tire model provides multiple modeling approaches for simulating the tire transient behavior, the results presented in this article are obtained using the linear transient modeling approach [3,4]:

$$\sigma\dot{\alpha} = -V_x\alpha + V_{sy} \tag{3}$$

$$\sigma = \frac{\text{CFa}}{C_y} \tag{4}$$

Where, σ is the tire's relaxation length, V_x is the forward velocity, V_{sy} is the lateral slip speed, CFa is the tire's cornering stiffness and C_y is the overall lateral stiffness.

Finally, the tire vertical response can be modelled by means of a non-linear spring-damper combination, see figure 2, since the frequency bandwidth is relatively limited for the maneuvers under consideration [3].

$$F_z = \text{f}_{\text{corr}}\left(q_{Fz1}\frac{\rho_z}{R_0} + q_{Fz2}\left(\frac{\rho_z}{R_0}\right)^2\right)F_{z0} - k\frac{d\rho_z}{dt} \tag{5}$$

$$f_{corr} = 1 + q_{v2}\frac{R_0}{V_0}|\Omega| \tag{6}$$

Where, ρ_z is the vertical deflection, R_0 is the tire unloaded radius, Ω is the rolling speed, V_0 the nominal tire speed, F_{z0} the nominal tire load and k the vertical damping. The q_{Fz} parameters define the vertical stiffness of the tire and q_{v2} parameterize the rolling radius speed dependency. The tire model response further depends on the mass of the tire and the connection to the vehicle through the suspensions.

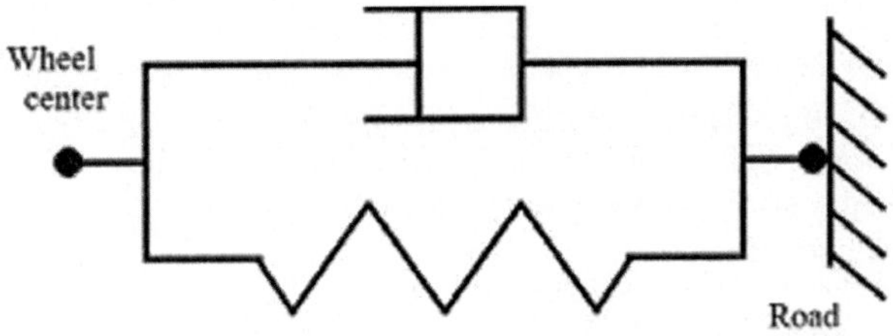

Fig. 2. Illustration of spring damper combination of vertical tire model excluding rigid ring dynamics [4].

3.2 Testing protocol and parameterization strategies

The parameterization of the tire model is based on tire measurements conducted on an MTS Flat-Trac [5] machine in defined operating conditions. There are two important requirements for tire tests that are used for model parameterization. Firstly, several tire conditions and states are not explicitly modelled and are hence inherited from the tire measurements. During tire testing, conditions such as the tire thermal state and tire wear need to be managed such that they are representative of the desired full vehicle operating conditions. Secondly, for the conditions that are explicitly modelled (e.g. sideslip angle, inclination angle, vertical load), one needs to make sure that the tire testing protocol covers the operating envelope of the full vehicle application in a representative manner.

Table 1. Summary of measurement protocols conditions (Fz1<Fz2...<Fz6 and c1<c2).

Measurements	Vertical load	Inclination angle	Steering rate
Standard	Fz1, Fz2, Fz3	-c1, 0, c1	Steady state
High load	Fz4, Fz5, Fz6	-c1, 0, c1	Steady state
High steering rate	Fz4, Fz5, Fz6	0, c2	High steering rate

Keeping these two requirements in mind, different tire test protocols have been defined and executed. The measurement conditions of the different protocols are summarized in table 1. The “standard” measurement protocol presents the testing conditions that are chosen for parameterizing a tire model for regular handling applications. In a next step this protocol is extended with additional “high load” measurements at higher vertical loads to better match the tire loading conditions during the maneuvers under consideration. Finally, measurements with significantly higher steering rates are added by means of the “high steering rate” protocol to investigate the impact of the rapidly changing slide slip angle conditions. In this protocol also the inclination angle was also adjusted to match the vehicle conditions.

Using these measurements, three respective parameter sets for the tire model are identified. During the “standard” and the “high load” tire measurements, the tire forces and moments are in steady state and hence the model parameters were identified using the default optimization procedure of MF-Tool [6]. The model parameterization based on the "high rate" measurements presents a higher complexity because in this case the tire response cannot be considered steady state anymore. The high steering rate induces transient tire behavior which imposes the need of a novel parameterization procedure. During this developed procedure, the tire model steady state response was parameterized evaluating the full model (steady state and transient model) on the measurement data (figure 3).

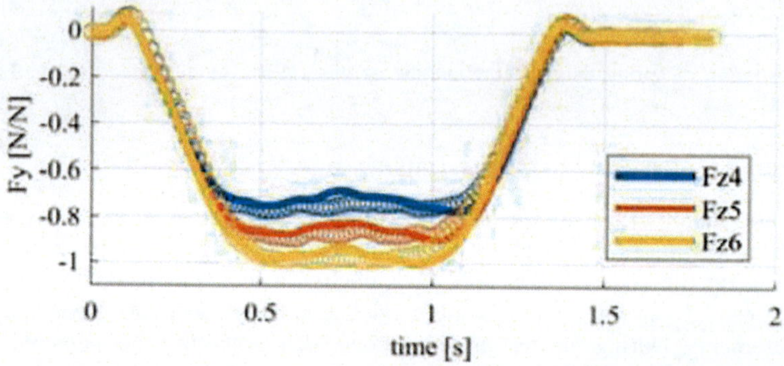

Fig. 3. Normalized simulation results and measurements with the high steering rate ("-": model, "o": measurement). The figure's axis is normalized with the maximum value in it.

Figure 4(a) illustrates the steady state lateral force response of the three generated models from the measurement protocols when being evaluated at the highest tested vertical load ("Fz6"). It is noticeable that the model response based on the "standard" and the "high load" parameter sets show a similar behavior, but the "standard" parameterization has a higher peak friction and more significant reduction in lateral force at large slide slip angles. Note that by means of the evaluation at "Fz6" load the "standard" model is effectively evaluated at a vertical load which is significantly outside of the tire measurement operating conditions. The "high rate" model shows a peak friction which is 7% and 8% higher than the "standard" model and "high load" model respectively for zero camber angle. Additionally, it is 3% and 8% higher than the "standard" model and "high load" model respectively for camber angle "c2".

The cornering stiffness of the three models is illustrated in figure 4(b). Cornering stiffness is evaluated at the different tested loads of the three measurement protocols. The maximum deviation between the "high rate" and the "standard" and the "high-load" models are 6% and 4% respectively. The main reason for the deviation of the "standard" and the "high rate" model is that both models have been parameterized in different load conditions, so the comparison is always at model extrapolated estimations.

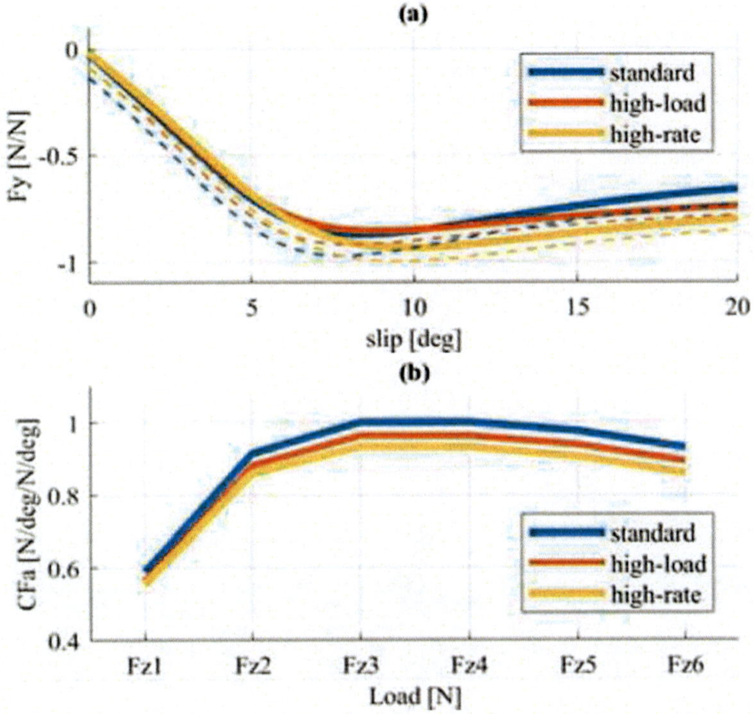

Fig. 4. Steady state lateral force of the three tire models normalized response at vertical load of "Fz6" ("-": at camber 0, "--":at camber "c2") (a). Cornering stiffness of the three tire models normalized response (b). Each axis is normalized with the maximum value in it.

4 Results and discussion

The three tire model parameter sets are applied in full vehicle simulation and the results are analyzed and discussed in this section. Furthermore, a qualitative study is presented to provide insight of the governing effects of the steering rate on the tire force response.

4.1 Vehicle handling simulation

Volkswagen has developed an accurate multibody model of the SUV vehicle in MSC Adams coupled with MF-Tyre/MF-Swift. The simulated lateral acceleration and yaw rate response during the fishhook maneuver are compared with vehicle measurement results in figure 5. The "MFTS" in the figure legend marks the MF-Tyre/MF-Swift model and the "STD", "HL" and "HR" refer to the parameterization strategies using the "standard", the "high load" and "high rate" measurements, respectively.

Two quality metrics are defined in order to evaluate the different modeling approaches: the deviation between simulated and measured lateral acceleration and the (absolute) maximum yaw rate reached between 1.5 to 2 seconds during the "fishhook" maneuver. The deviation was calculated according to equation (7). The lateral acceleration is used on the grounds that it is highly influenced by the tire behavior. On the other hand, a single point of comparison was used on the yaw rate results in order to specifically judge the vehicle response at the most critical point of the fishhook maneuver.

$$Deviation = 100\sqrt{\frac{\Sigma(y-\hat{y})^2}{\Sigma y^2}} \quad (7)$$

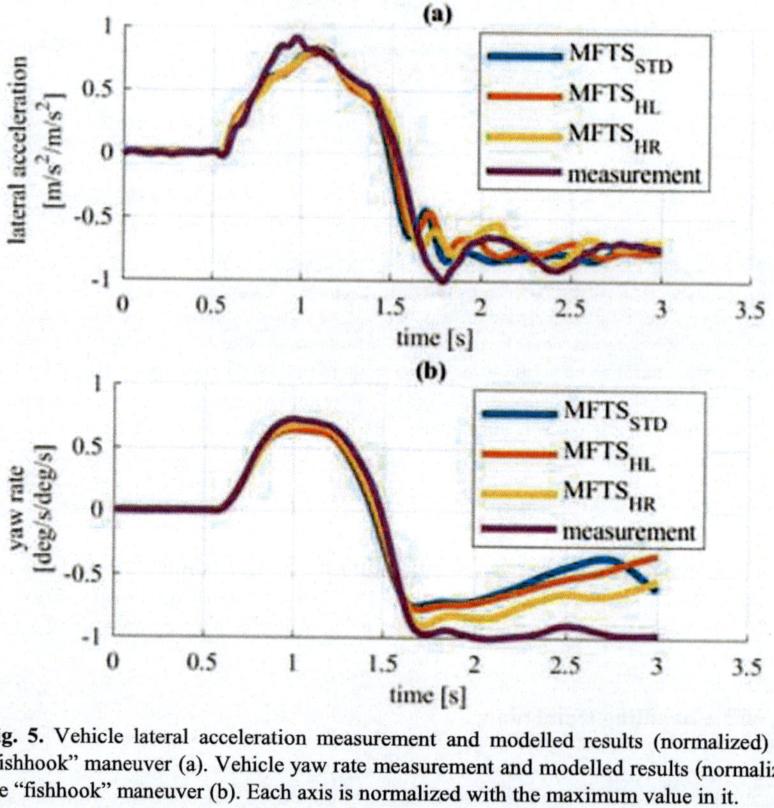

Fig. 5. Vehicle lateral acceleration measurement and modelled results (normalized) for the "fishhook" maneuver (a). Vehicle yaw rate measurement and modelled results (normalized) for the "fishhook" maneuver (b). Each axis is normalized with the maximum value in it.

Figure 6 presents a comparison of these quality metrics obtained for the different tire model parameterization strategies. Specifically, figure 6(a) presents the lateral acceleration quality metric which clearly indicates that the parameterization method with the higher steering rate measurements outperforms the other methods. The simulated vehicle response follows both the amplitude and the phase of the measured response well. Furthermore, the comparison of the yaw rate quality metric depicted in figure 6(b) confirms that the "high rate" parameterization provides an improved correlation. Quantitatively, an improvement of 5% is obtained for the lateral acceleration quality metric and 15% for the yaw rate quality metric.

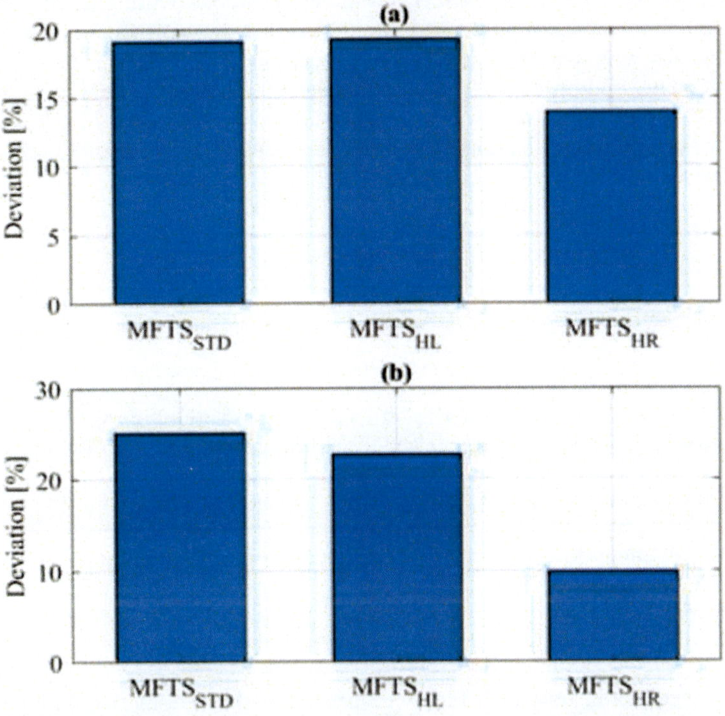

Fig. 6. Evaluation of generated models based on lateral acceleration deviation with measurement for the "fishhook" maneuver (a). Evaluation of generated models based on (absolute) maximum yaw rate deviation with measurement reached between 1.5 to 2 seconds during the "fishhook" maneuver (b).

From the obtained results it is concluded that significant improvements can be obtained by applying the "high rate" measurement protocol for model parameterization. Furthermore, during the study additional tire variants and vehicle maneuvers have been evaluated. The results show that by means of introducing the novel "high rate" parameterization method more reliable simulation results can be obtained for the studied maneuvers. Especially the fact that, by means of simulation, the sensitivity of the vehicle response to using different tire variants can be well predicted is proven to be valuable to Volkswagen.

4.2 Steering rate effect

From the previous results it is obvious that the tire parameterization with the "high rate" measurements provides the most accurate full vehicle simulation results. This can be explained by the fact that the tire operating conditions during these measurements more closely mimic the on-vehicle operating conditions. During the "high rate" measurements less energy is introduced into the tire and hence it reaches the higher slide slip angles with a lower temperature. Recent studies prove that the frictional behavior of general-purpose passenger car tires deteriorates with increasing temperatures [7,8].

The latter can be proven by considering the frictional power generated in the contact patch of the tire:

$$\dot{Q}_{\mathrm{fr}} = \mathrm{F_y} V_{sy} \tag{8}$$

To calculate the amount of the frictional thermal energy generated in a maneuver, equation (8) is integrated:

$$\mathrm{Q_{fr}} = \int F_y V_x \tan(a) dt \tag{9}$$

When reducing the scope to a tire ramp steer maneuver, the time variable can be expressed using a constant steering rate and the side slip angle:

$$\mathrm{t} = \frac{\alpha}{steering\ rate} \tag{10}$$

Inserting equation (10) to equation (9), the steering rate effect on the friction energy generation can be expressed:

$$\mathrm{Q_{fr}} = \frac{V_x}{steering\ rate} \int F_y \tan(\alpha)\, da \tag{11}$$

The produced thermal energy is highly influenced by the steering rate during a tire test procedure. The effect is however not directly inversely proportional since the lateral force is also affected by the temperature state of the tire.

In order to get a better understanding of this phenomena, the recently developed modular thermal extension of the MF-Tyre/MF-Swift model [9] is applied. The model extension includes the major thermal effects of the tire, i.e. convections and conduction with environment and road, frictional heat and rolling resistance. It is coupled with the Magic Formula based MF-Tyre/MF-Swift model and provides scaling and offset corrections on the B, C, D, E coefficients depending on the tire temperature state. A qualitative study is performed since the parameters of the thermal model extension are derived for a similar size SUV tire and not for the exact tire variant used in the current study.

Single tire ramp steer simulations are performed with the different model variants at a vertical load "Fz6" and a camber angle of 0 degrees. These ramp steer simulations are performed twice with steering rates corresponding to these of the "standard" and "high steering rate" tire measurement procedures.

Figure 7(a) illustrates the obtained lateral force with the "standard" steering rate. A comparison between the "standard" model, the "high rate" model and the "standard" model extended with thermal effects is provided. Note that the obtained results can be considered steady state tire slip characteristics due to the relatively low steering rate. It is shown that in these conditions the thermal model behavior closely corresponds with the "standard" model. Figure 7(b) presents the comparison with a steering rate that matches the "high steering rate" protocol. Due to the higher steering rate, the thermal model predicts a significantly lower tire surface temperature which leads to less deterioration of the tire performance. Therefore, the thermal model matches the peak friction of the "high rate" model. Note that due to the high steering rate the obtained characteristics include both the steady state slip characteristics as well as the transient response of the tire.

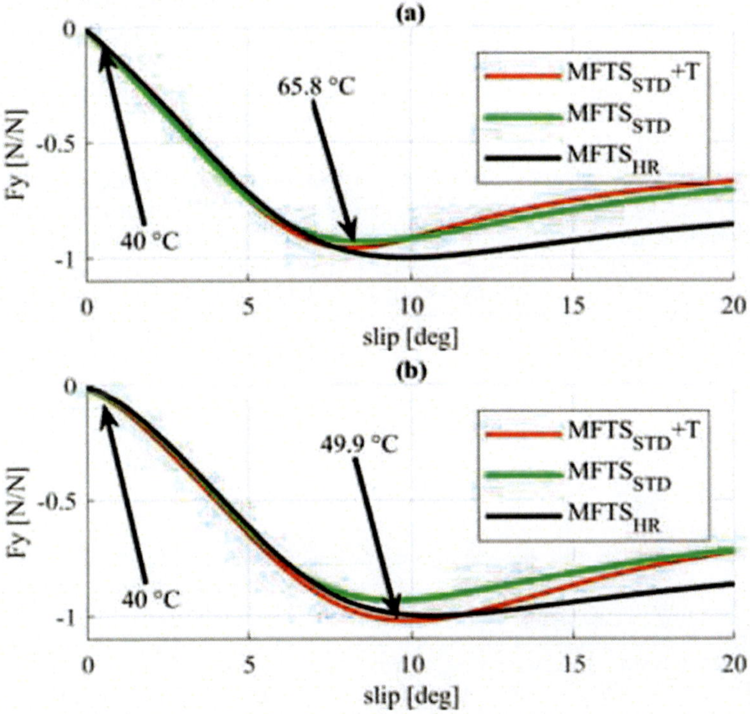

Fig. 7. Illustration of the lateral force of the "standard" (with and without thermal effect) and "high rate" models at load of "Fz6" at steering rates ensuring steady state conditions (a). Illustration of the lateral force of the "standard" (with and without thermal effect) and "high rate" models at load of "Fz6" at high steering rates (b). Both figures are normalized using the maximum "high-rate" response.

The major advantage of the thermal model extension lies in the fact that the model can capture the tire behavior over a range of tire temperatures and predicts the temperature propagation throughout the maneuver, potentially leading to more accurate simulation results. Further, this extension allows to use one single model for both soft handling as well as extreme handling maneuvers and therefore avoids dedicated tire testing and parameterization for these specific maneuvers. Finally, the model extension provides the possibility to initialize the simulated maneuver at different tire temperatures with which the seasonal effects and/or the thermal history of the tire on the vehicle response can be analyzed.

5 Conclusions

During extreme vehicle handling maneuvers the tires on a vehicle experience significantly different operating conditions than during normal driving. Siemens Digital Industries Software and Volkswagen have therefore developed a tire modeling methodology to model tire behavior for these specific maneuvers. Different tire testing and parameterization approaches of the MF-Tyre/MF-Swift model have been evaluated. From the obtained results it can be concluded that the key influencing factor is the level of peak friction reached during the quick maneuvering. This characteristic is highly influenced by the temperature state of the tire which therefore must be considered in the tire modeling methodology. The resolution presented in this article is to perform the tire model parameterization based on tire measurements under conditions that are as close as possible to the driving maneuver conditions.

Although important advancements are made, figure 5 also shows that there is further room for improving the correlation between vehicle simulations and measurements. Especially the prediction of the resulting yaw rate in the final stages of the fishhook maneuver can be improved as the measurement results in a stable yaw rate whereas in simulation the yaw rate steadily decreases. Volkswagen and Siemens Digital Industries Software are therefore continuing and extending the current study.

References

1. Rajamani, R.: Vehicle dynamics and control, 2nd edn. Springer, (2012)
2. Siemens Digital Industries Software MF-Tyre/MF-Swift Homepage, https://www.plm.automation.siemens.com/global/en/products/simcenter/simcenter-tire.html, , last accessed 2020/05/11.
3. MF-Tyre/MF-Swift 6.2 Equation manual, TNO, The Netherlands (2013).
4. Pacejka, H.: Tire and vehicle dynamics, 3rd edn. Butterworth-Heinemann, (2012)
5. MTS systems Homepage, https://www.mts.com/en/products/producttype/test-systems/simulation-systems/tire/flat-trac, last accessed 2020/05/11.
6. MF-Tool Help manual, TNO, The Netherlands (2014).
7. Angrick, C., van Putten, S., and Prokop, G.: Influence of tire core and surface temperature on lateral tire characteristics, SAE Int. J. Passeng. Cars - Mech. Syst. Journal 7(2), 468–481 (2014).
8. Lugaro, C., Alirezaei, M., Konstantinou, I., and Behera, A.: A study on the effect of tire temperature and rolling speed on the vehicle handling response. SAE Technical Paper 2020-01-1235, (2020).

9. Lugaro, C., Konstantinou, I., Mazzeo, M.: A tire model extension for predicting temperature and rolling speed influences on tire performance, JSAE (2020).

Transient Motion Analysis Based Objective Quantification of Cornering Feedback and Its Application to Vehicle Motion Control

Lee, Jun-Mo[1] and Adolfo Tec[1]

[1] Hyundai-Kia America Tech Center Inc., USA

Abstract. The Cornering feedback from the vehicle is important information to control the path and the yaw stability, especially in limit handling situation. To improve this cornering feedback, in-depth analysis of vehicle cornering motion should be advanced. But, due to the transient characteristics of vehicle cornering motion, effective analysis of cornering feedback has remained as a very challenging task. In this article, 1) important properties of transient cornering motion have been studied and 2) yaw de-composition based new transient motion analysis method is proposed. This new analysis method enables separate analysis of vehicle yaw motions by its sources, from path following and from yaw spinning, which is the key element in objectification of cornering feedback. Additionally, an in-depth analysis of yaw center migration and its effect on driver feeling has been investigated. Based on these analysis methods, subjective vs. objective correlation study has been conducted and major cornering feedbacks includes carving feel have been objectively analyzed and quantified. In this article, side slip motion compliance & resistance in path following also has been investigated to analyze vehicle turn-in response. Additionally, based on this newly proposed transient motion analysis, its application idea includes active control of vehicle motion in various situations is also proposed.

Keywords: Transient Motion, Yaw Rate Decomposition, Cornering Feedback.

1. Introduction

To correctly understand transient cornering behavior and its feedback to driver, it is important to use transient motion optimized analysis method. Traditional method is either based on steady state cornering analysis or frequency analysis of vehicle response[1][2][3]. In case of steady state analysis based method, as steady state doesn't change its state over time, it naturally lacks past history information of vehicle motion and difficult to see and predict the trend of motion change. In case of frequency domain analysis, due to the un-intuitiveness of open loop test and analysis, it is difficult to correlate to driver's feeling of specific events in cornering. Considering these limitations of traditional analysis methods, for enhanced quantification of driver's feeling

P. E. Pfeffer (ed.), *11th International Munich Chassis Symposium 2020*, Proceedings, https://doi.org/10.1007/978-3-662-63193-5_14

of cornering feedback, more effective analysis method is required. In this paper, below three items are introduced and proposed as new approach.

1) Yaw decomposition based analysis method for transient cornering motion
2) Subjective vs. Objective correlation study
3) Application idea of vehicle transient motion control

1.1 Definition of Transient Condition

Transient condition includes transient input and/or transient output. In cornering, input can be either intended or unintended. Output can be any type of vehicle response to given input. Most of cornering situation can be categorized as below. Case 1) is probably the most common transient condition of cornering as driver is continuously adjusting driver input based on the feedback during cornering.

1) Change in both of input and output
2) Change in input only (no vehicle response yet)
3) Change in output only (input is completed but vehicle is still responding)

1.2 Planar Motion Analysis

Analysis of planar motion of rigid body can be accomplished with velocity information of any two points on the rigid body at given moment [4]. Below Fig.1 shows various ways of describing vehicle planar motion at given moment. Even though all of these methods provide same velocity information at given moment, only method 3) can be used for expanded analysis of transient cornering 'motion' as this is the only method which involves time history information of vehicle motion which is critical for correct analysis of driver's accumulated feeling over period of time. Method 1) works for steady state cornering but doesn't work for transient cornering as it is not a pure rotation around fixed point.

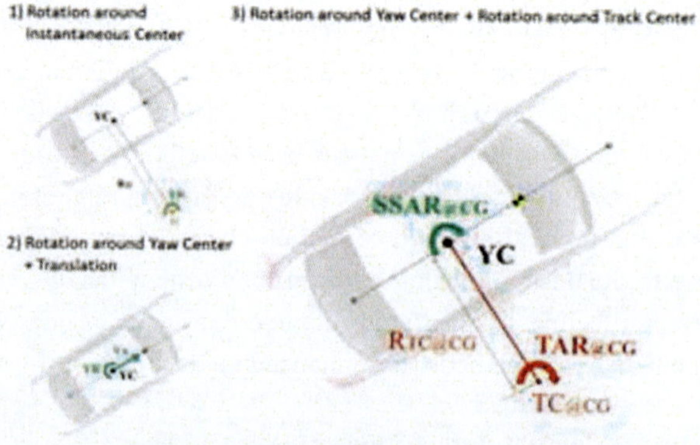

Fig. 1. Types of Planar Motion Analysis

Method 3) contains very important properties of planar and transient motion analysis as transient cornering is not pure rotation around one point but combined result of below two rotations – path following and yaw spinning.

1) Rotation around track center on the ground
2) Rotation around yaw center on the vehicle body

1.3 Cornering Feedback

As cornering feedback is very subjective, it comes with various terminologies subjectively describing it and wide discrepancy in preference. Below is the list of cornering feedbacks used for correlation study in this paper.

1) Carving Feeling
2) Path Precision
3) Turn In Response
4) Playfulness

2. Main Subject

2.1 Vehicle Transient Motion Analysis

In this paper, transient analysis is focused on its motion analysis based on [1] new analysis method by yaw rate decomposition and [2] subjective vs. objective correlation study.

Steady State Cornering Analysis

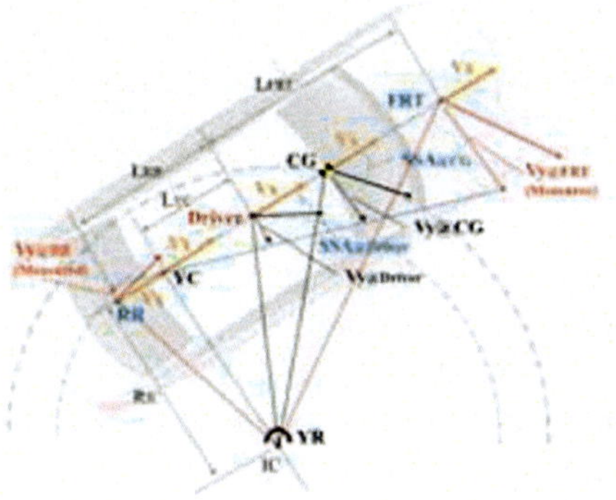

$$Vy_{@FRT} - Vy_{@RR} = (L_{FRT} + L_{RR}).YR$$
$$YR = (Vy_{@FRT} - Vy_{@RR})/(L_{FRT} + L_{RR})$$

$$Vx = R_{IC}.YR$$
$$R_{IC} = Vx.\ (L_{FRT} + L_{RR})/(Vy_{@FRT} - Vy_{@RR})$$

Fig. 2. Steady State Cornering

In steady state cornering, yaw rate can be representative information as vehicle is rotating around fixed point (IC – Instantaneous Center of Rotation) on the ground without change in side slip angle over time as described in Fig.2. Here the side slip angle of each point on the vehicle body maintains their values throughout the cornering as long as it is in steady state. So do yaw rate, turning radius and IC location.

Requirement for Transient Motion Analysis

As important part of cornering feedback is the change of vehicle motion sensed by the driver over certain time span, there are two important requirements for proper analysis of cornering feedback in transient cornering.

1) Time history information of vehicle motion: Vehicle motion information of just one given moment is not enough for transient motion analysis. Transient motion analysis requires time history information of motion to analyze the trend of motion which driver has perceived and to better predict following motion. Driver is using accumulated cornering feedback to predict next vehicle motion with any input change. As an example, in Fig.7, with just given information of given moment without the history of side slip angle change, it can be falsely analyzed as pure translation and it may lead to failure in predicting upcoming motion change.

2) Point of analysis: As all the velocity information used for motion analysis is dependent to location of point of interest, defining analysis point is important for proper analysis of driver feeling in cornering. In this article, driver seating position is used as an analysis point.

Sources of Yaw Motion in Transient Cornering

In transient cornering, vehicle yaw motion is the combined result of several yaw motions from different sources as below

1) Rotation from path following (cornering) : most dominant source which is a rotation around center of track (path) at given moment
2) Rotation from yaw spin around yaw center : next dominant source which is a rotation around yaw center and measured as change in side slip angle
3) Yaw from roll axis inclination : fairly small so not considered here

Yaw Rate Decomposition

Steady state cornering analysis method in Fig.2 can provide velocity information of each point on the vehicle body at given moment but it can't provide information about past history or future prediction of vehicle motion. As explained in 1.2, for proper transition analysis, we need to know how much of the rotation is from path following and how much is from yaw spinning. This information enables better prediction of upcoming event and better quantification of cornering feeling in transient condition. Fig.3 shows how yaw motion is decomposed into two components. First one is the yaw motion from path following which is rotation around TC (track center) of given moment at rotation rate of Track Angle Rate. This TC is on the line which vertically meets longitudinal axis of vehicle at yaw center as like IC but at different location (In steady state condition, due to the absence of Side Slip Angle Rate, Track Angle Rate equals to Yaw Rate and both of TC and IC are at the same location) The other yaw motion is the rotation around yaw center at rotation rate of Side Slip Angle Rate.

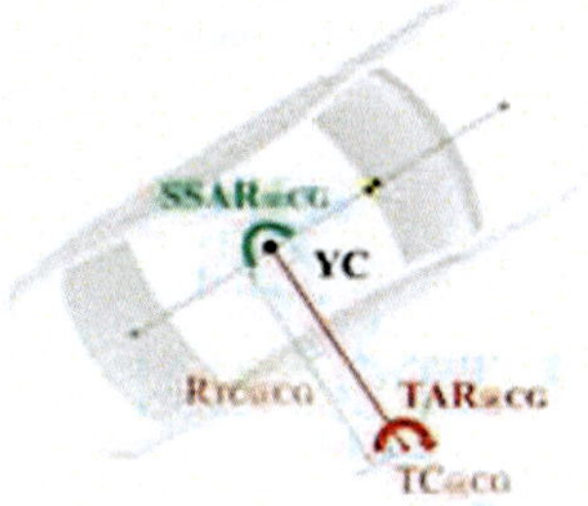

Fig. 3. Yaw Decomposition for Transient Cornering

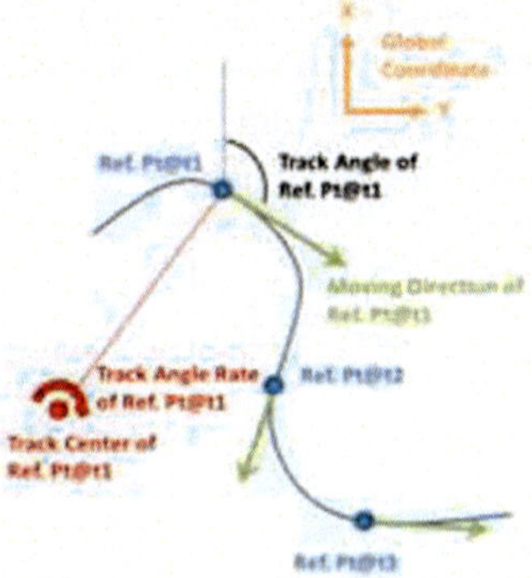

In Fig. 4. Track Angle and Track Center

In Fig.3, track angle, side slip angle and their rate of change are all presented with respect to CG point. If we change this analysis point, track angle and side slip angle related values will be changed even in the same motion at same moment. This characteristic is very important part of cornering feel analysis, More detail will be explained later. For driver feeling correlation, driver position will be used as an analysis point.

Track Angle and Its Rate of Change

In cornering, vehicle's yaw motion is mainly from path following and some from yaw spinning around yaw center. Fig.4 shows the trajectory of analysis point. The measured angle of moving direction with respect to global coordinate is the track angle. This track angle also can be expressed w.r.t body fixed local coordinate. As, in most of transient case, the center of this track (path) is not fixed and the shape of track (path) is hard to be complete circle, the track center is also migrating during cornering unless it is in steady state. Track angle rate is time derivative of track angle. Due to the difficulty of accurate measure of path and track angle, in this paper, new calculation method of track angle rate using lateral velocity and yaw rate is introduced.

Side Slip Angle and Its Rate of Change

Side slip angle is one of the most important but also least intuitive properties of vehicle motion especially in transient condition. It is an indicator of how much of sideway motion the vehicle is experiencing during cornering or even in some limited pure translating situation. And, like track angle, this side slip angle is also analysis point dependent. As shown in Fig.5, side slip angle is calculated from lateral velocity of each point which can be calculated from yaw rate and lateral velocity of any known point (See the planar motion analysis case 2 in Fig. 1.). Yaw center is the point on the longitudinal axis of vehicle where the lateral velocity is zero at given moment. This yaw center also migrates as vehicle motion changes. Time derivative of this side slip angle is side slip angle rate of change and this is important information of yaw spin trend. Below are two critical properties of side slip angle rate explained in Fig.6

1) Sign convention flip
2) Time history information

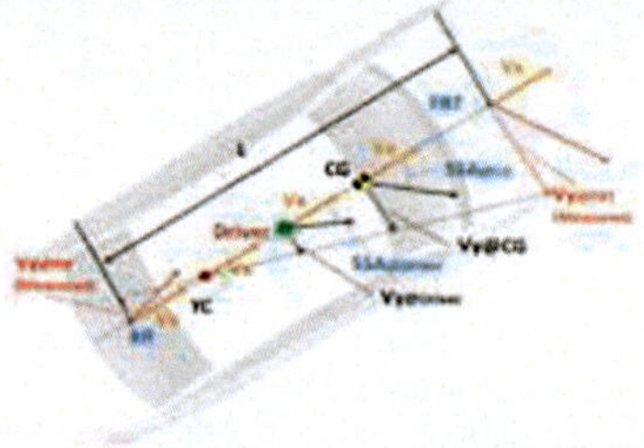

Fig. 5. Side slip calculation with lateral velocity

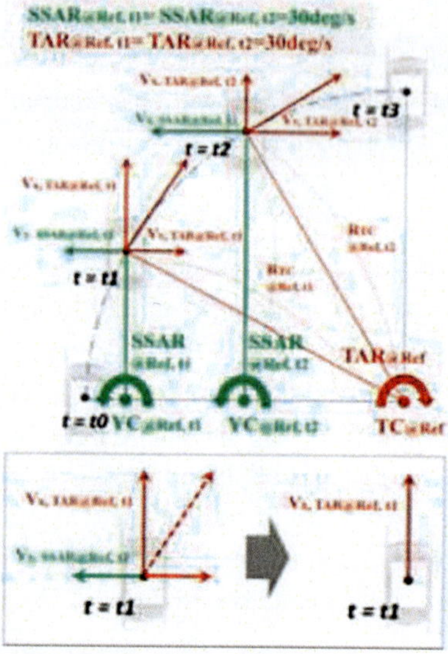

Fig. 6. Side Slip Angle Property: Sign Convention Flip and Time History Information

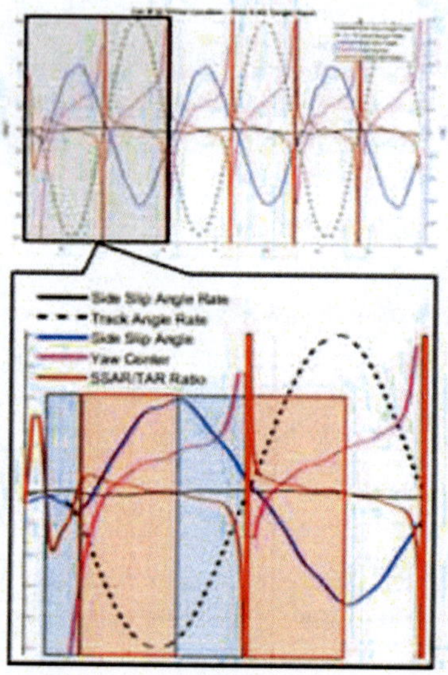

Fig. 7. Side Slip Resistance and Compliance

Fig.6 shows [1)] how side slip angle change is working together with track angle change and [2)] why the time history information from side slip angle rate is important by using extreme case which may not be physically possible in actual vehicle driving condition. Regarding sign convention issue, here the vehicle is rotating around fixed track center at the rate of 30deg/s while rotating around yaw center at the same rate of 30deg/s. If both of global coordinates and local coordinates follow the same sign convention, then both of track angle rate and side slip angle rate of reference point at time t1, as an example, will be calculated as positive 30deg/s. But, actual rotation around yaw center is in negative yaw direction w.r.t global coordinate. This confusion happens because side slip angle is calculated w.r.t body fixed local coordinates while all the other parameters are calculated w.r.t global coordinates. This shows the importance of sign convention flip of side slip angle rate in transient motion analysis.

Regarding importance of time history information, if we assume that we have vehicle status information of only one given moment at time t1, then the only information we can see is the longitudinal velocity. With this limited information, we can easily say that this is pure translation toward the front. But, if we have more information like side slip angle rate, track angle rate, track center location, then we can have better understanding of what has happened and we can also have better prediction of how the motion will develop in the next stage.

Additionally, side slip angle rate indicates whether the yaw spinning motion around yaw center is complying or resisting to path following. As an example, in Fig.7, in blue zone, side slip angle starts increasing until it reaches to its max. This is the point where side slip angle rate is crossing zero and also where side slip is starting to complying toward path turning. Until this point, side slip has been resisting against path turning which resulted in increase of yaw spin around yaw center in opposite direction of rotation around track center. In this situation, while actual path of vehicle is following the given target curve, vehicle yaw spin is not complying with this turning path. But, once after passing this side slip angle max point, as shown in red zone, side slip angle starts decreasing and complying with path turning. In this region, both of yaw motion from path following (@track angle rate) and yaw spin around yaw center (@ side slip angle rate) are working together toward the same direction of rotation. Again, for correct calculation, the sign of side slip angle rate needs to be flipped. This analysis of side slip compliance or resistance is important as it can be correlated to driver's feeling of turn in response (or resistance) of vehicle. As an example, in all front axle steer vehicles, driver always feels side slip resistance in the beginning of turn unless driver is seating right on or even behind the rear axle as yaw center always located at the rear axle in the beginning of turn. If there is too much side slip resistance in turn-in, driver may feel initial resistance in heading angle change but not in path following. Yaw rate itself can't explain this subtle feeling of turn-in response because this effect of side slip resistance is not separately shown in total yaw rate.

For effective analysis of driver feeling, yaw motions from path-following and yaw spinning need to be separately analyzed with driver position as the reference point.

Track Angle Rate and Side Slip Angle Rate Calculation

Due to the difficulty in measuring actual path of reference point with accuracy, in this paper, new method to calculate track angle is introduced. Fig.8 shows most complete form of transient motion analysis. For accurate measure of side slip (or lateral velocity) of vehicle, two of dual axis optical sensors are used in front and rear of test vehicles. Each sensor accurately measures both of longitudinal and lateral velocity without signal drifting. With these velocity information of two known points, yaw rate and lateral velocity of any reference points are calculated as shown in Fig.5. Once side slip angle and yaw rate are calculated, side slip angle rate can be calculated as time derivative. Then, track angle rate can be derived as below. (See Fig.8)

Yaw Rate = Track Angle Rate + Side Slip Angle Rate

※ Sign convention of Side Slip Angle Rate is flipped before this calculation to convert its value with respect to global coordinate

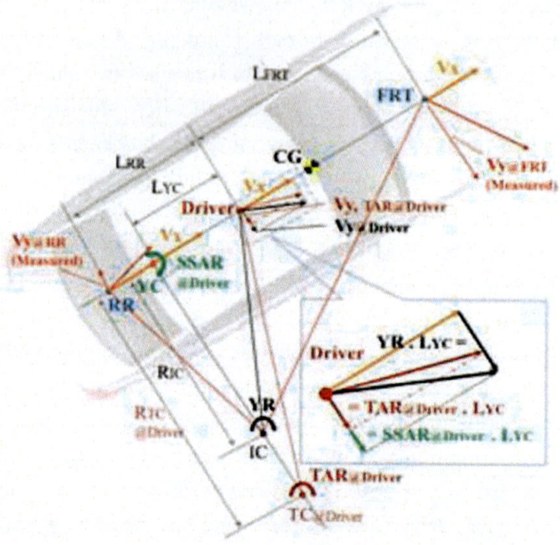

Fig. 8. Transient Cornering with Combined Rotation around Track Center and Yaw Center at Rotation Rate of Track Angle Rate and Side Slip Angle Rate

Side slip angle rate is the rotation rate around yaw center and this rotation does not create longitudinal velocity at any point on the longitudinal axis. It just contributes to small portion of lateral velocity. On the contrary, rotation around track center at track angle rate contributes to majority of lateral velocity and it is the sole source of longitudinal velocity. In Fig.8, lateral velocity components are explained in detail. By this method, track angle rate and track center location can be calculated without measuring actual path of reference point.

In Fig.8 example, both of the rotation around track center (TC) and the rotation around yaw center (YC) are working toward same direction. That means, track angle rate is slower than yaw rate by the side slip angle rate. This also means TC is located farther than instantaneous center of rotation (IC). As the longitudinal velocities from rotation around IC and rotation around TC are same, if TC is located farther, track angle rate is slower than yaw rate. Comparing relative location of TC and IC is another way to see how side slip angle is resisting or complying with path following. Here the TC and TAR is all calculated based on driver seating position as a reference point. If we change the reference point, the TC location and TAR will be changed accordingly due to the reference point dependency of track angle and side slip angle.

Yaw Center and Its Migration

Yaw center is the point on the vehicle body where lateral velocity is zero. Longitudinal velocity is the only velocity component on this point and it means TC and IC are on the same line which vertically meets longitudinal axis of vehicle at this point, the yaw center (Fig.8). As vehicle moves, this yaw center is also migrating unless vehicle is in steady state cornering. This yaw center migration is important information explaining how vehicle is developing its cornering motion and how it is affecting driver feeling. In case of rear axle passive steer vehicle, in initiation of turning, the yaw center starts migrating from rear axle and moving forward as turning develops.

Track Center and Instantaneous Center of Rotation

Once yaw center location, side slip angle rate, track angle rate and yaw rate are calculated, then TC and IC location can be calculated as in Fig.8.

- $R_{TC@Driver} = Vx/TAR_{@Driver}$
- $RIC = Vx/YR$

TC location is also reference point dependent. If we move the reference point, then side slip angle rate and track angle rate will be changed and it results in different TC location. But, YC location, IC location and YR are not dependent to reference point. The relative location between TC and IC is additional information about whether yaw motion around yaw center is complying or resisting to path turning. If the distance between TC and YC is shorter than distance between IC and YC, it indicates track angle rate is bigger than yaw rate and this means side slip angle rate is in opposite turning direction which is interpreted as side slip resistance since side slip motion is not helping cornering. It will be opposite situation if track center is located farther than instantaneous center. This case, both of yaw motions from rotation around yaw center and rotation from path turning are in same direction. This is considered as side slip compliance since side slip angle change is complying with path turning.

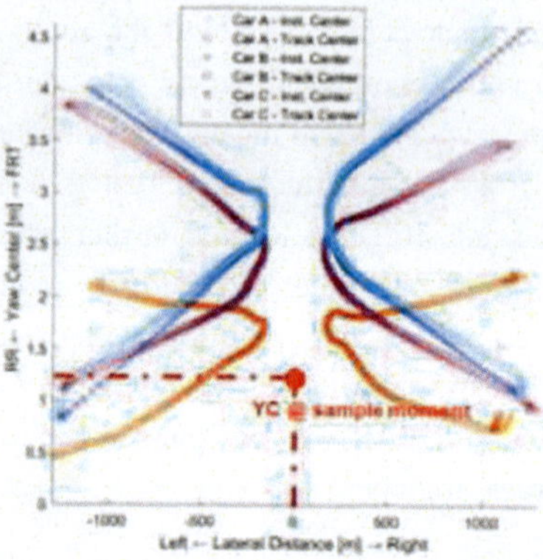

Fig. 9. Centrode - Track Center and Inst. Center (100kph, 02Hz, 0.4G sine weave)

Table. 1. Cornering Feel Subjective Ranking

Car	A	B	C
Grip	1st	2nd	3rd
Carving	1st	2nd	3rd
Path Precision (Mid Corner)	1st	2nd	3rd
Path Precision (Turn In)	3rd	1st	2nd
Playfulness (Throttle Off Response)	3rd	2nd	1st

Fig.9 shows the migration path (centrode) of TC and IC. Here x axis is lateral distance from the center and y axis is longitudinal distance from rear axle which is equivalent to YC distance from rear axle. Additionally, as seen in Fig.9, when TC centrode and IC centrode are crossing, this is the point where vehicle is in momentary steady state condition and this is the transition point between side slip compliance and side slip resistant. In Fig.9, Car A shows almost overlapped migration path of TC and IC. This means, , Car A's yaw motion is mostly from path following with very small contribution from side slip and this can be an important factor of carving feel.

2.2 Cornering Feedback Quantification

In this paper, three different vehicles are objectively and subjectively tested for correlation study. All vehicles are in the same segment of front drive hot hatch but with different driving characters. Limit handling cornering is controlling both of path and yaw stability by using intensive feedback like vehicle trajectory (path), side slip, lateral velocity, lateral acceleration, steer torque feedback, etc. Among them, separating yaw motion into two components (rotation around track center and rotation around yaw center) is one of the key driving skills for precise control of vehicle turning path and yaw spin motion. This might be an explanation for why the vision is always emphasized for good car control. Looking at farther where to go can provide driver more sense of the global coordinate and relative motion of body fixed local coordinate. This may enhance driver's separate sensibility of rotation around yaw center.

Carving Feeling

Carving feeling is one of the most preferred cornering feedbacks. When vehicle provides pure carving feeling, it enhances driver's confidence of path and spin control. Vehicle providing pure carving feeling normally has higher grip and more robustness against external disturbances. This makes it easy for driver to find quasi steady state condition in the middle of corner. Considering the importance of getting into maximum speed steady state condition in cornering, the pure carving feel can boost driver confidence and controllability of cornering. Not like ski carving where entire points on the edge of ski can be in pure carving situation, in vehicle cornering, only one point on the vehicle body can be in pure carving condition at given moment. So, it is important to find out which location of vehicle body is in (or close to) the pure carving condition with minimum sideway motion in given cornering. In theory, if this location does exist and is close to driver seating position, then driver can feel carving feeling. If time span of this event (condition) is extended through the cornering, it also enhances carving feeling. So, below would be requirement for carving feeling

1) Easy to find maximum speed steady state condition
2) Easy to maintain steady state condition against external disturbance
3) YC is located close to driver and stay longer

Car-A shows very distinctive pattern in side slip angle rate development (in Fig.10) and yaw center migration (in Fig.11). In Fig.10, throughout steer maneuver, side slip angle rate contribution in total yaw rate is relatively smaller than other vehicles. Side slip angle rate stays less than 5% of yaw rate. This means almost 95% of vehicle rotation is from path following while other vehicles show that path following induced rotation is reduced down to 85% of total yaw rate and also exhibit higher fluctuation in the ratio values which means side slip effect on total yaw rate keeps changing more in Car B and Car C. This means, throughout the maneuver, driver may not feel the quasi steady state condition when steer input is changing.

In Car A, ratio of side slip angle rate vs. yaw rate is not just relatively small but it also maintains its value in fairly stable manner even when steer input keeps changing. This may be interpreted as easiness of finding steady state condition in cornering. Additionally, in Car B and Car C, sometimes track angle rate is almost 15% bigger than total yaw rate. This can be interpreted as, from path following standpoint, more change in turning radius and more side slip resistance to path following which is supposed to degrade carving feeling.

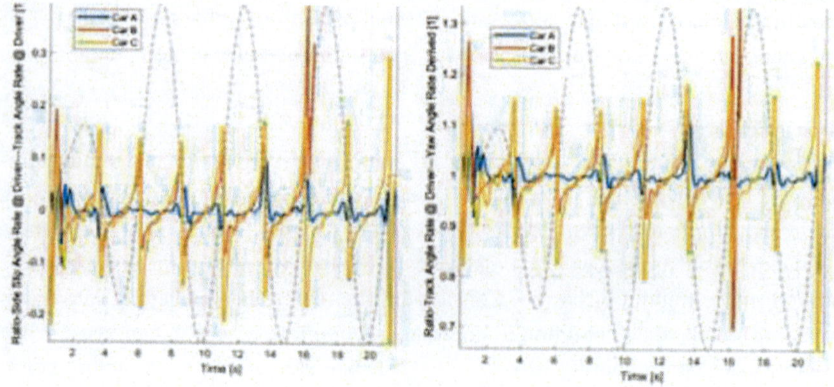

Fig. 10. Side Slip Angle Rate Contribution (100kph, 02Hz, 0.4G sine weave)

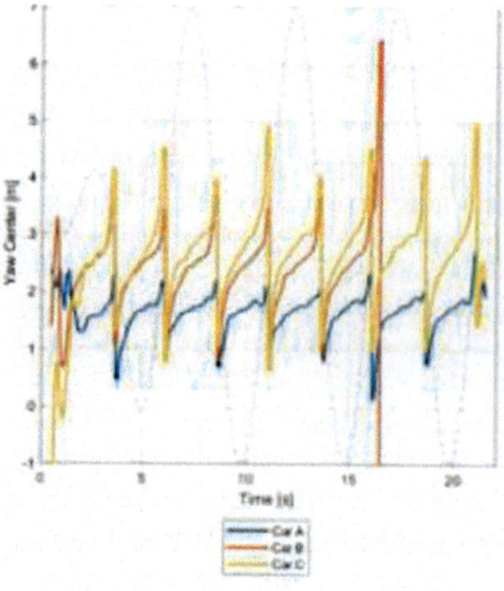

Fig. 11. Yaw Center Migration

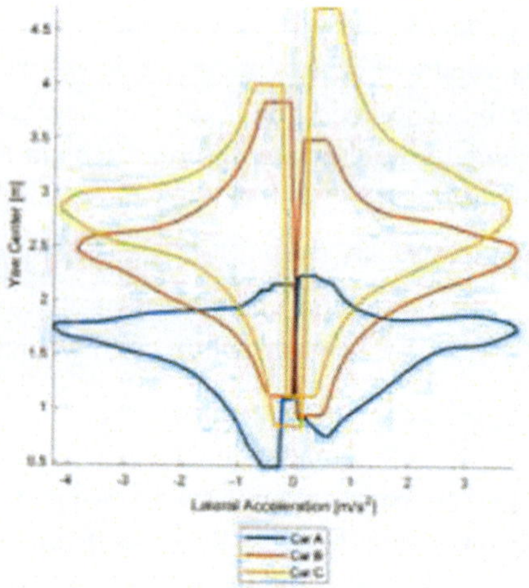

Fig. 12. Yaw Center Migration w.r.t. Latac

And, to define the effect of yaw center location with respect to driver seating position on carving feeling, yaw center migration also analyzed. In Fig.11, Car A shows significantly different yaw center migration pattern which can be directly correlated to subjective feeling and also to the theory of carving feel that side slip at driver position needs to be minimized. In Car A, yaw center starts migrating relatively slowly than other vehicles and shows very strong tendency of staging near driver seating position. Here the yaw center location data is with respect to rear axle. Considering the average wheel base of around 2.7m and driver position in near 4:6 ratio point between front and rear axle, Car A's yaw center staging near 1.7m from rear axle is very close to driver seating location. Initial slower migration of yaw center in Car A also can mean higher grip and cornering power in rear tire which requires smaller tire side slip and provides more pivoted turn in feeling at rear axle.

Fig.12 shows another important property of yaw center migration of Car A. By staging yaw center near driver location fairly longer than other vehicles, it achieves consistent yaw center location while lateral acceleration is continuously changing. As lateral acceleration is important feedback driver is using for vehicle control, this can be another important characteristic of carving feel. But, it also can be interpreted as disconnection between yaw center migration and lateral acceleration development. So, this character should be optimized by vehicle's overall dynamic concept.

In case of Car B and Car C, as side slip angle rate portion in total yaw rate keeps changing and as yaw center keeps migrating, driver may have to deal with continuous change in vehicle condition and it degrades pure carving feeling. But, this character of requiring continuous control may give feeling of driving on the edge of limit which can be enjoyed by driver looking for playfulness, rawness and joy of controlling.

Another aspect of carving feeling is sideway motion driver is experiencing by yaw center location. In Fig.13, when yaw center is behind driver, this sideway motion feels more of U/S as vehicle is looking at outside of the corner. If yaw center is located in front of driver, then this sideway motion feels more of O/S as vehicle is orienting into the corner in this case. But, as driver is not seating on the pure carving track in both case, neither case is preferred from carving feel standpoint.

Path Precision/Turn in Response

Path precision can be categorized as below.

1) Turn in path precision – ability to bring the vehicle to the corner entrance line
2) Mid corner path precision – responsive and linear path change ability

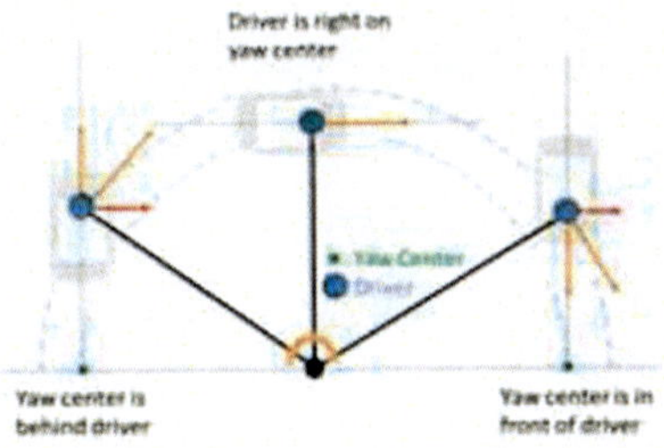

Fig. 13. Sideway Motion

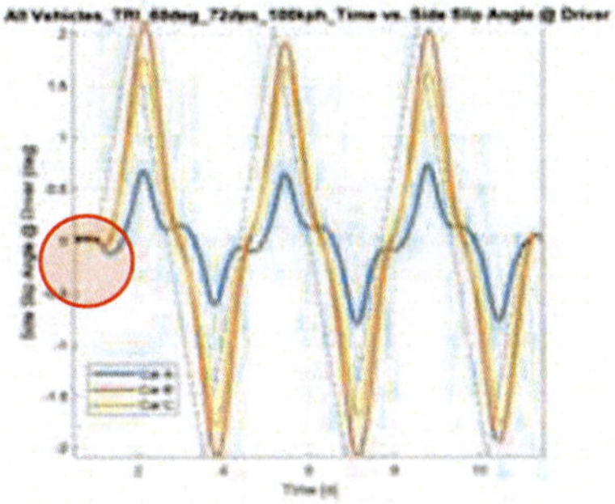

Fig. 14. Turn In Path Precision

Fig.14 shows that Car A's initial side slip resistance is bigger than other vehicles. This can be negatively affecting initial turn in precision which is required to bring the vehicle onto record line in corner entrance.

Playfulness

In front drive vehicle, power understeer is the common problem and one of the way to address it is 'balance throttle' or 'throttle off'. If the vehicle's yaw response gain to throttle control is high and linear, then driver feels it as 'playful'. From that playfulness standpoint, Car C increases its yaw motion very actively with throttle off maneuver. In case of Car C, yaw motion gain feels like combined result of front tuck in and rear slide out. In Fig.11 and Fig.12, it is shown that Car C's yaw center migration is most quick and linear than others (Fig.11) and it also shows highest level of side slip angle gain over steer input. This will correlate to playfulness of Car C. Even though rear axle of Car C lacks overall grip or planted feel, it helps throttle off response.

2.3 Vehicle Motion Control Application Idea

Emergency Handling Control

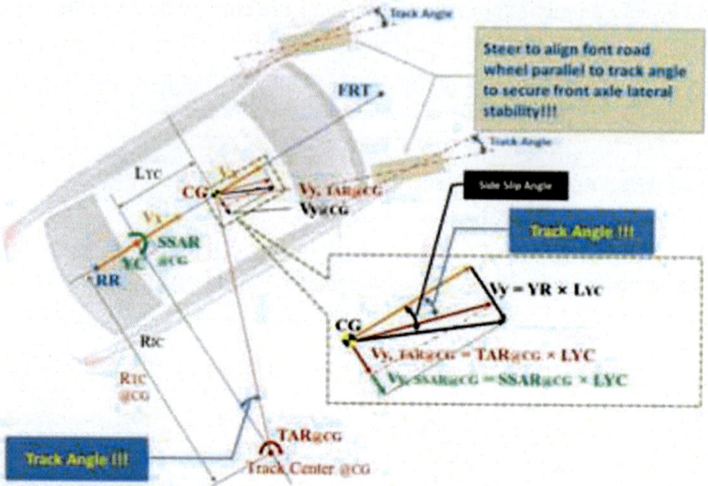

Fig. 15. Emergency handling control – Align front road wheel parallel to pure moving direction

In autonomous driving vehicle, proper steer control is critical for vehicle stability and path control. To stabilize vehicle in high side slip (or high yaw spin) situation, it is important to steer front road wheel to the direction of moving for lateral stability of front axle. In this situation, as lateral velocity component from side slip is not negligible anymore, using side slip angle only to calculate pure moving direction is not accurate as pure moving direction of particle is determined by track angle. Fig.15 shows this concept of aligning front road wheel to pure moving direction of vehicle. Here, lateral velocity is decomposed and pure moving direction of vehicle - the track angle is calculated. And, conveniently, as the track angle of CG is calculated w.r.t vehicle coordinate, actual heading angle of vehicle body w.r.t global coordinate is not necessary.

Driver Feeling Improvement

As driver seating location w.r.t YC is affecting the driver's feeling of all of side slip angle, track angle and their change rate, it is very important to define driver seating location which can be varied by vehicle type, especially in virtual tuning. As an example, if driver is close to rear axle, driver will be more exposed to sideway O/S with more lateral motion. This bigger lateral motion or side way may degrades carving feeling but it can enhance driver's sensibility of yaw spin around yaw center which may lead to enhanced yaw motion control in limit handling situation.

3. Conclusion

New concept of transient motion analysis is introduced with quantification of cornering feedback and application idea based on new transient motion analysis.

- New analysis method for vehicle transient motion– yaw rate decomposition
- Importance of time history information for proper analysis of transient motion
- Carving feeling and several other important cornering feedbacks
- Driver seating location effect in transient motion analysis
- Calculation method for pure moving direction of vehicle and steer control strat egy of emergency situation

In routine handling, yaw center migration should be considered as the most important factor for cornering feeling as side slip angle rate is relatively small. But in limit handling, side slip angle rate and track angle rate related information should be considered as additional factors in objective analysis of cornering feedback.

References

1. Gillespie, T. D., "Fundamentals of vehicle dynamics", SAE
2. Masato Abe, "Vehicle Handling Dynamics – Theory and Application", Elsevier
3. Joop P. Pauwelussen, "Essentials of Vehicle Dynamics", Elsevier
4. C. H. Suh, C. W. Radcliffe, "Kinematics and Mechanism Design", John Wiley & Sons
5. Lee, Jun-Mo, Yoo, Ju-Young Moses, "Effect of EPS response On Vehicle Steer response and Lateral Dynamics", FISITA 2016, F2016-VDCA-034

Enhanced Characterization of Suspension Systems for Virtual Chassis Development

Stefan Buechner[1], Norbert Deixler[1], Ralf Stroph[1] and Markus Lienkamp[2]

[1] BMW Group, Germany

[2] Institute of Automotive Technology, Technical University of Munich, Germany

Abstract. Simulation has established itself as an essential tool in the virtual chassis development process to reduce both development time and costs, especially for suspension design. In the literature, a variety of characteristic values are proposed to characterize and benchmark suspension systems. These characteristic values play a vital role for the design of suspension kinematics and compliance. This paper presents an approach for the analysis and characterization of suspension systems within multi-body simulation, which makes use of the suspension system's compliance matrix. A universal methodology for the calculation of the existing characteristic values, which characterize both steering feedback and wheel guidance, by means of the compliance matrix is outlined as well. The results demonstrate the application of the proposed approach to several exemplary use cases within suspension design. The newly gained insights with this method can be used to further enhance suspension design by improving the characterization and benchmarking of suspension systems.

Keywords: Vehicle Dynamics, Chassis Development, Suspension Analysis, Multi-body Simulation, Compliance Matrix, Virtual Product Development

1 Introduction

Simulation has established itself as an essential tool in suspension design to improve the virtual chassis development process. This paper deals with an approach for the enhanced characterization of suspension systems. It is based on [1] and summarizes the most important aspects. Further elaboration of motivation, state of the art, methodology and results can be found in [1].

1.1 Motivation

Despite the current trends in the automotive industry, such as electrification, digitalization and automated driving, the chassis of the vehicle is still of great importance in the future [2-3]. It defines vehicle dynamics, affects a lot of further vehicle characteristics and contributes to roughly half of the criteria which are relevant for the customer's purchase decision [3-4].

P. E. Pfeffer (ed.), *11th International Munich Chassis Symposium 2020*, Proceedings, https://doi.org/10.1007/978-3-662-63193-5_15

The ever-growing competition in the automotive industry has resulted in increasing cost pressure and shorter development cycles affecting the chassis as well [3-5]. More and more vehicle models and a simultaneous standardization of vehicle components increase the effort for validation and testing of the chassis. The result is a further increasing complexity of the chassis development process. To deal with these challenges, simulation has been established in the chassis development process and especially in suspension design to replace physical hardware tests and gain more detailed insights at an earlier stage [3, 6-7]. By means of simulation, development costs and time are reduced, complexity can be handled and chassis characteristics are enhanced.

In suspension design, suspension systems are evaluated by means of specific characteristic values [3, 8-9]. These characteristic values relate chassis parameters to vehicle characteristics. They make comparisons of different suspension systems possible and allow for the definition of respective target ranges. The characteristic values are also used for target cascading and are important for sensitivity analysis as well as optimization [3, 6, 8].

1.2 Goals

This paper deals with the enhanced characterization of suspension systems for the virtual chassis development process. The objective is to develop a universal approach to determine all characteristic values, which are relevant in suspension design, to improve the benchmarking of suspension systems. By analyzing these characteristic values, new insights into suspension design can be gained and recommendations for the chassis development process are given.

1.3 Structure of the Paper

The remainder of the paper is structured as follows. Section 2 gives an overview of the state of the art in the field of chassis development and suspension characteristics. A summary of relevant characteristic values in suspension design is given as well. In section 3, the approach for suspension analysis making use of the suspension system's compliance matrix is outlined. Section 4 presents exemplary results of the proposed approach. Section 5 summarizes the results of this work and gives an outlook on further research.

2 State of the Art

2.1 Virtual Chassis Development Process

Numerical simulation, especially multi-body simulation, plays a vital role in chassis development and suspension design [3, 6-7]. This subsection presents the different stages of suspension design in the virtual chassis development process (Fig. 1). Suspension design is an iterative process and involves different models, which become increasingly more complex during the development process [3, 10].

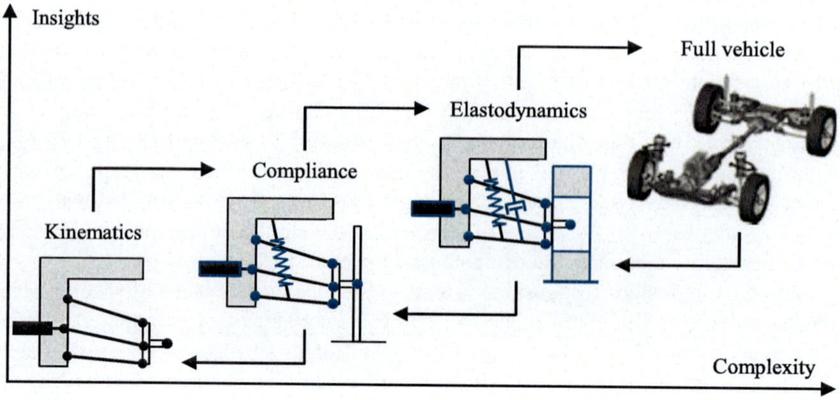

Fig. 1. Stages of suspension design in the virtual chassis development process based on [1]

Suspension design starts with the design of suspension kinematics [3, 6]. All components are considered as rigid and the wheel movement during wheel travel and steering is analyzed [3, 6, 11]. It is followed by the design of suspension compliance [3, 10]. The compliance of components, such as bushings, is considered and the wheel movement under external loads is analyzed [3, 6, 10]. Generally, the simulation is static, that means forces resulting from velocities and accelerations are neglected, and the forces do not result from a detailed tire model [12]. The objective is to adjust the stiffness characteristics of each component. Subsequently, dynamic forces and more detailed tire models are considered in a dynamic analysis of the suspension system, such as on a four-post rig or flat track [3, 13]. Most complex are full vehicle analyses. By contrast to the stages before, the suspension system is studied in the context of the full vehicle [3, 6]. The focus is on the vehicle behavior and the suspension system's influences.

2.2 Characteristic Values of Suspension Systems

In suspension design, specific characteristic values are essential for the comparison of different suspension systems, as they allow for the evaluation and benchmarking of suspension characteristics. This subsection presents an overview of the relevant suspension characteristics, which are predominantly used in the design of suspension kinematics and compliance, and describes their meaning in terms of vehicle dynamics. A complete and detailed summary of the literature on characteristic values in suspension design is found in [1].

Kinematics. A variety of characteristic values can already be determined from suspension kinematics [1]. They result from geometric relationships of the suspension system [3, 14]. These characteristic values can be divided into three groups referring to the wheel alignment, suspension geometry and steering geometry. The characteristic values belonging to suspension geometry include instant centers and component ratios. The ones referring to steering geometry include the steering axis and steering ratios.

Wheel Alignment. The wheel alignment includes the toe angle and the camber angle, which have an essential meaning for vehicle dynamics, as they determine the orientation of the tire relative to the road [3, 6, 12]. In addition to the static angles, the changes during wheel travel and steering are important. The toe angle and its changes influence directional stability as well as vehicle responsiveness and cause roll steer [3, 6, 10]. The camber angle affects lateral tire forces and therefore vehicle responsiveness as well as self-steering behavior [3, 6, 10].

Instant Centers and Angles. The instant centers of the suspension system can be determined in side view and front view [3, 6, 15]. The side view instant center results in the anti-lift/squat angles and the anti-dive/rise angles. The front view instant center results in the anti-roll angle. The instant centers determine the longitudinal and lateral wheel movement during wheel travel [10, 14]. They also affect the longitudinal and lateral load transfer, as they determine how the applied forces at the wheel are resolved into the components which are supported by linkages and springs [3, 6, 10, 14].

Component Ratios. The component ratios include spring ratio, damper ratio and anti-roll bar ratio. The location and orientation of spring, damper and anti-roll bar result in the respective ratios, which relate the movement of the wheel or the forces applied at the wheel to the corresponding movement or forces of the component [3, 10, 15-16].

Steering Axis. The location and orientation of the steering axis result in a number of characteristic values, which describe lever arms and angular relationships at the wheel. They characterize the influence of tire forces on the steering wheel torque and are important for steering feedback [3, 6, 10]. The characteristic values include scrub radius and caster trail, kingpin offset and caster offset, wheel load lever arm and anti-roll bar lever arm as well as kingpin angle and caster angle [3, 6, 10, 17].

Steering Ratios. The steering ratios include total steer ratio, steering linkage ratio and steering gear ratio. They are important, as they relate the motion of steering wheel, steering rack and wheels to each other [6, 10]. Furthermore, they affect the resulting steering rack force and steering wheel torque from tire forces [10, 18].

Compliance. Suspension compliance plays a vital role, as it affects the wheel alignment characteristics under external loads [14]. Another group of characteristic values includes stiffness characteristics. Furthermore, an elastokinematic steering axis results in several characteristic values. The characteristic values known from suspension kinematics can also be studied during the design of suspension compliance.

Wheel Alignment. When considering compliance of the suspension system, wheel alignment characteristics, such as toe angle, camber angle and caster angle, are affected by longitudinal, lateral and vertical forces [6, 14]. All of this influences vehicle dynamics and must be considered in suspension design. A focus is on the change of the toe angle due to lateral and longitudinal forces, as it contributes to vehicle stability [3, 6, 19]. Regarding the camber angle, the camber loss due to lateral forces is relevant [6, 14, 20].

Stiffness characteristics. Apart from wheel alignment characteristics, stiffness characteristics of the suspension system are in the center of interest during the design of suspension compliance, as they affect both comfort and handling [10, 14]. A low longitudinal stiffness is important for the comfort, as it reduces the impact of longitudinal shocks [10]. By contrast, a high lateral stiffness is requested for vehicle responsiveness and precise vehicle handling [21]. The vertical stiffness determines the vibration frequency and is important for the ride comfort [22]. Toe stiffness and camber stiffness are directly related to the wheel alignment characteristics [3, 23].

Elastokinematic Steering Axis. Apart from the steering axis studied in suspension kinematics, an elastokinematic (or elastic) steering axis is additionally described in the literature, which is determined by forces and refers to the wheel movement in form of the toe behavior instead of the steering wheel torque [18-20, 24]. Characteristic values similar to those in suspension kinematics can be derived from it, such as the longitudinal and lateral locations of the so-called toe neutral points [14, 24, 27]. Compared to the extensive research on the steering axis in suspension kinematics, the elastokinematic steering axis is not analyzed in detail so far.

Influence on Geometric Characteristics. The compliance of the suspension system affects the resulting movement of the suspension system and therefore might influence the characteristic values from suspension kinematics [3, 6]. Furthermore, these characteristic values are now dependent on the applied forces [3].

3 Approach

3.1 Compliance Matrix of the Suspension System

The suspension system can simply be regarded as an elastic system, which consists of linkages, bushings and springs, connecting the wheel to the chassis [24-25]. The elastic system, which is modeled by the system's compliance matrix (Fig. 2), comprehends the kinematic and elastokinematic properties of the suspension system and is equivalent to it [24]. Therefore, the elastic system can be used to analyze suspension characteristics [14, 24]. The compliance matrix C describes the resulting displacement vector $\vec{d}$ for a given force vector $\vec{F}$:

$$\vec{d} = C\vec{F}, \tag{1}$$

where

$$\vec{F} = (\vec{F}_t, \vec{F}_r)^T = (F_x, F_y, F_z, M_x, M_y, M_z)^T \text{ and} \tag{2}$$

$$\vec{d} = (\vec{d}_t, \vec{d}_r)^T = (d_{tx}, d_{ty}, d_{tz}, d_{rx}, d_{ry}, d_{rz})^T. \tag{3}$$

Each element $c_{i,j}$ of the compliance matrix describes the respective compliance for any degree-of-freedom i due to external loads referring to any degree-of-freedom j:

$$C = \begin{bmatrix} c_{1,1} & \cdots & c_{1,6} \\ \vdots & \ddots & \vdots \\ c_{6,1} & \cdots & c_{6,6} \end{bmatrix}. \tag{4}$$

The compliance matrix of the suspension system, which is determined by means of a subroutine in a multi-body simulation (Adams/Car) for each simulation step, is then used to calculate the suspension system's characteristic values.

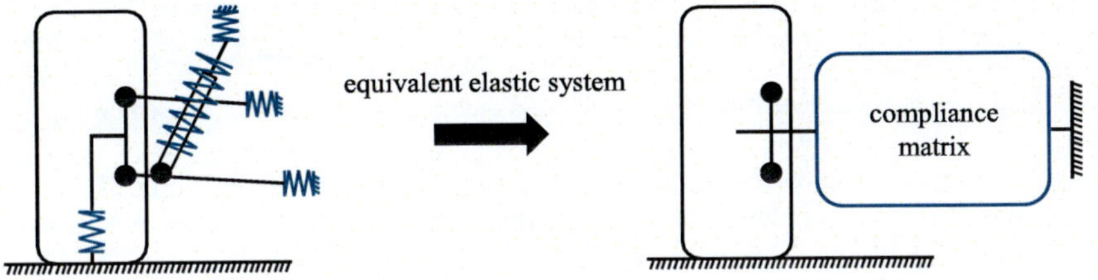

Fig. 2. Suspension system as an elastic system based on [1, 24]

Further elaboration of the compliance matrix and its properties is found in [1].

3.2 Calculation of Characteristic Values from the Compliance Matrix

This subsection deals with the calculation of the characteristic values described in the state of the art. The applied methodology is shortly outlined here. The derivation of the equations for the calculation of the characteristic values is done by using appropriate force-displacement-relations, which describe the meaning of the respective characteristic value. The characteristic values are divided into two groups characterizing either steering feedback or wheel guidance. Further elaboration of the methodology and the derived equations of the characteristic values is found in [1, 26-28].

Steering Feedback. The first group of characteristic values characterizes steering feedback in form of the resulting steering torque or rack force due to external loads. They include the characteristic values derived from the effective steering axis as well as steering ratios. The compliance matrix used in equation (1) is specified to describe the force-displacement-relations of the wheels and the steering rack or steering wheel.

Effective Steering Axis. The effective steering axis corresponds to the steering axis known from suspension kinematics, but differs from the kinematic one, as it considers suspension compliance. Scrub radius, caster trail, kingpin offset and caster offset are defined as the effective lever arms of longitudinal or lateral forces regarding the resulting steering feedback, which are derived from the steering torque neutral point (Fig. 3). Therefore, if the lever arm is zero, a corresponding force results in no steering rack force and no steering rack travel [1, 26]. This constraint is then used in equation (1):

$$d_{\text{rack,ty}} = 0. \tag{5}$$

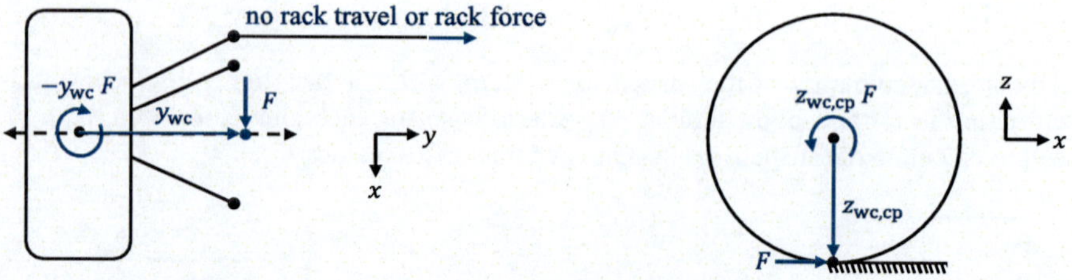

Fig. 3. Force-displacement-relation for the calculation of the scrub radius based on [1]

For each lever arm, the corresponding force vector (longitudinal/lateral force at contact patch height/wheel center height; resulting torques dependent on the lever arm) is used in equation (1) with the lever arm to be determined. As the compliance matrix is determined in the simulation, equation (1) can be solved for the lever arm. As an example, the scrub radius r_k is calculated as

$$r_k = \frac{c_{8,1} + c_{8,5}\, z_{wc,cp}}{c_{8,6}}, \tag{6}$$

where indices 1-6 refer to the forces/displacements of the left wheel and indices 7-12 refer to the ones of the steering rack.

A similar approach is used for the wheel load lever arm. Kingpin angle and caster angle are calculated geometrically from the corresponding lever arms.

Steering Ratios. For the calculation of the steering ratios, a steering rack force (for the steering linkage ratio) or a steering wheel torque (for the total steer ratio) is used in equation (1). With the resulting displacement vector due to the applied force vector, the steering ratios are calculated as the ratio of steering wheel angle/rack travel to the steer angle of the wheel [1]. As an example, the steering linkage ratio i_{link} is calculated as

$$i_{\text{link}} = \frac{c_{8,8}}{c_{6,8}}, \tag{7}$$

where indices 1-6 refer to the forces/displacements of the left wheel and indices 7-12 refer to the ones of the steering rack.

The steering gear ratio is calculated from the total steer ratio and steering linkage ratio.

Wheel Guidance. The second group of characteristic values describes the wheel guidance and characterizes wheel alignment as well as the transfer of applied forces. They include the characteristic values derived from the elastokinematic steering axis, specific stiffness characteristics as well as instant centers and component ratios. The compliance matrix used in equation (1) is specified to describe the force-displacement-relations of the wheels and – only for the component ratios – the respective component.

Elastokinematic Steering Axis. Unlike the effective steering axis, which refers to steering feedback, the elastokinematic steering axis describes the wheel movement in form of the toe angle properties. The elastokinematic steering axis yields the toe neutral points at contact patch height and wheel center height (Fig. 4). Longitudinal and lateral location of these toe neutral points are defined as the lever arms of lateral or longitudinal forces regarding the resulting toe angle change. Therefore, if the lever arm is zero, a corresponding force results in no toe angle change [1, 27].

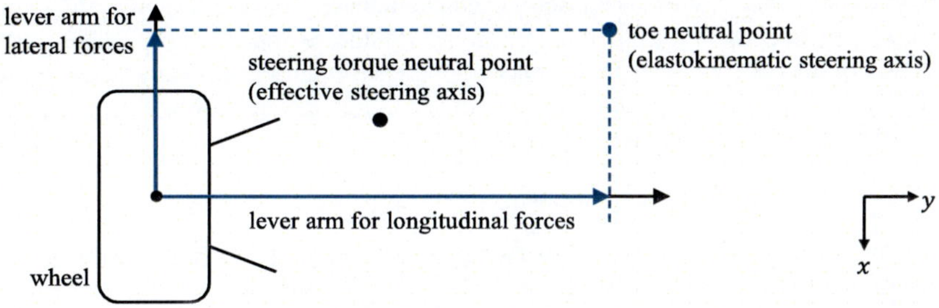

Fig. 4. Location of steering torque neutral point and toe neutral point based on [1]

This constraint is then used in equation (1):

$$d_{\text{wheel,rz}} = 0. \tag{8}$$

For each lever arm, the corresponding force vector (longitudinal/lateral force at contact patch height/wheel center height; resulting torques dependent on the lever arm) is used in equation (1) with the lever arm to be determined. As an example, the longitudinal location n_{e} of the toe neutral point at contact patch height (corresponds to an elastokinematic caster trail) is calculated as

$$n_{\text{e}} = \frac{c_{6,2} - c_{6,4}\, z_{\text{wc,cp}}}{c_{6,6}}. \tag{9}$$

A similar approach is used for the elastokinematic equivalent of the wheel load lever arm.

Stiffness and Compliance. The stiffness and compliance characteristics describe the translational and rotational movement of the wheel due to applied forces. For the translational degrees-of-freedom, the stiffness in each direction is calculated as the ratio of the applied force to the resulting translation [1, 26]. As an example, the longitudinal stiffness at wheel center height $K_{\text{x,wc}}$ is calculated as

$$K_{\text{x,wc}} = \frac{1}{c_{1,1}}, \tag{10}$$

where indices 1-6 refer to the forces/displacements of the left wheel.

For the rotational degrees-of-freedom, the compliance in each direction is calculated as the ratio of the respective rotation to the applied forces [1]. As an example, the toe angle compliance $C_{\mathrm{rx,Fy}}$ due to a lateral force at the contact patch (pneumatic trail not considered) is calculated as

$$C_{\mathrm{rx,Fy}} = c_{6,2} - c_{6,4} z_{\mathrm{wc,cp}}. \tag{11}$$

In general, stiffness and compliance characteristics can be calculated for each location in the respective planes, which can be visualized then.

Instant Centers and Angles. The instant centers and derived angles describe the movement of the wheel as well as the load transfer. Any forces applied at the wheels in the direction of the instant center in the respective plane do not result in any wheel travel, as they are completely supported by the suspension linkage and not by the springs [1]. This constraint is then used in equation (1):

$$d_{\mathrm{wheel,tz}} = 0. \tag{12}$$

By applying a vertical force as well as a longitudinal force (for side view instant center) or lateral force (for front view instant center), the longitudinal or lateral location as well as the vertical location of the instant centers are determined in the respective planes [1]. As an example, the lateral location $y_{\mathrm{IC,yz}}$ of the front view instant center is calculated as

$$y_{\mathrm{IC,yz}} = -\frac{c_{3,3}}{c_{3,4}}, \tag{13}$$

where indices 1-6 refer to the forces/displacements of the left wheel. The corresponding angles are derived from the geometric relationships.

Component Ratios. The component ratios for spring, damper and anti-roll bar relate the movement or forces of the wheel to the ones of the respective component. By applying a vertical force to the wheel, the resulting displacements of the wheel and the component are determined in equation (1) [1]. Spring ratio i_{spr} and damper ratio i_{dmp} are calculated as the ratio of the spring or damper displacement to the wheel travel:

$$i_{\mathrm{spr/dmp}} = \frac{\sqrt{c_{7,3}{}^2 + c_{8,3}{}^2 + c_{9,3}{}^2}}{c_{3,3}}, \tag{14}$$

where indices 1-6 refer to the forces/displacements of the left wheel and indices 7-12 refer to the ones of the spring or damper. The anti-roll bar ratio is determined in a similar way.

4 Results

After describing the methodology in the previous section, this section demonstrates the application of the proposed approach to several exemplary use cases. First, the influence of suspension compliance on the characteristic values is studied. It is followed by a comparison of the effective and elastokinematic steering axis. The next subsections deal with the influence of steering system properties and tire properties on suspension characteristics. The last subsection shows how suspension characteristics can be visualized. Further results with more detailed explanatory notes are found in [1, 26-28].

4.1 Characteristic Values Considering Suspension Compliance

This subsection deals with the influence of suspension compliance on the characteristic values. As an example, the characteristic values derived from the effective steering axis are studied for a purely kinematic, that is rigid, model (*kin*) and a model considering suspension compliance (*com*). As an example, the scrub radius is shown in Fig. 5. The kinematic scrub radius differs from the one calculated with the compliance matrix. The comparison of the steering torque calculated from the respective scrub radius with the actual steering torque resulting from the multi-body simulation (*mbs*) proves that neglecting suspension compliance results in an inaccurate scrub radius.

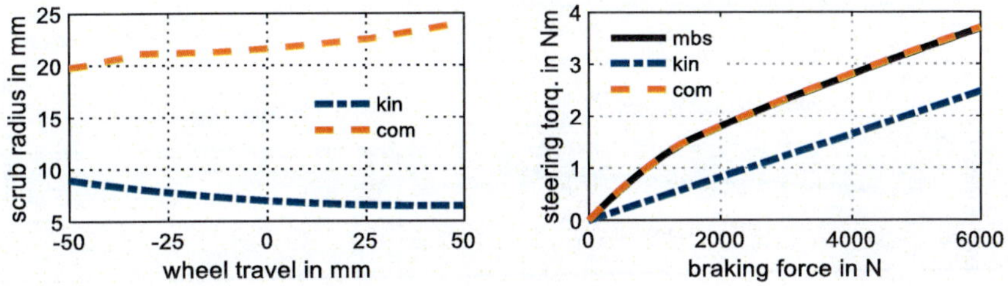

Fig. 5. Scrub radius during wheel travel and resulting steering torque due to braking forces

Furthermore, considering suspension compliance leads to a dependency of the characteristic values on the load case (Fig. 6), as the applied forces result in changes of suspension geometry and bushing stiffness characteristics.

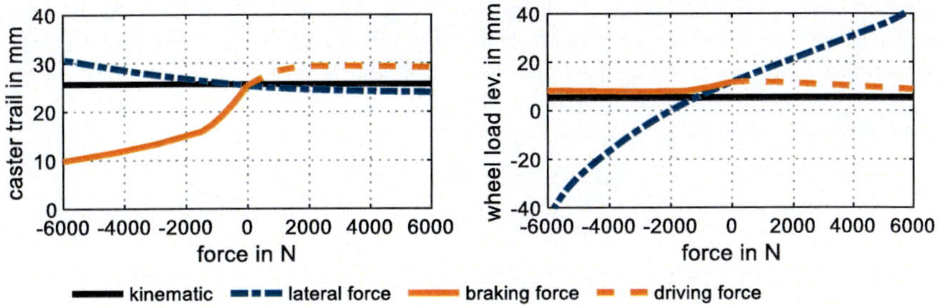

Fig. 6. Influence of different load cases on caster trail and wheel load lever arm

The results clearly demonstrate that a purely kinematic analysis of suspension characteristics is not sufficient and suspension compliance must be considered.

4.2 Effective and Elastokinematic Steering Axis

In the literature, two different kinds of steering axis are mentioned, but their differences are not analyzed in detail. In this subsection, the elastokinematic steering axis is compared to the effective steering axis. Fig. 7 shows the comparison of the respective lever arms derived from each steering axis during wheel travel. The lever arms for longitudinal forces (scrub radius and lateral location of toe neutral point), lateral forces (caster trail and longitudinal location of toe neutral point) and vertical forces (wheel load lever arm and elastokinematic equivalent) differ from each other significantly. The results demonstrate the importance to distinguish between these two steering axes.

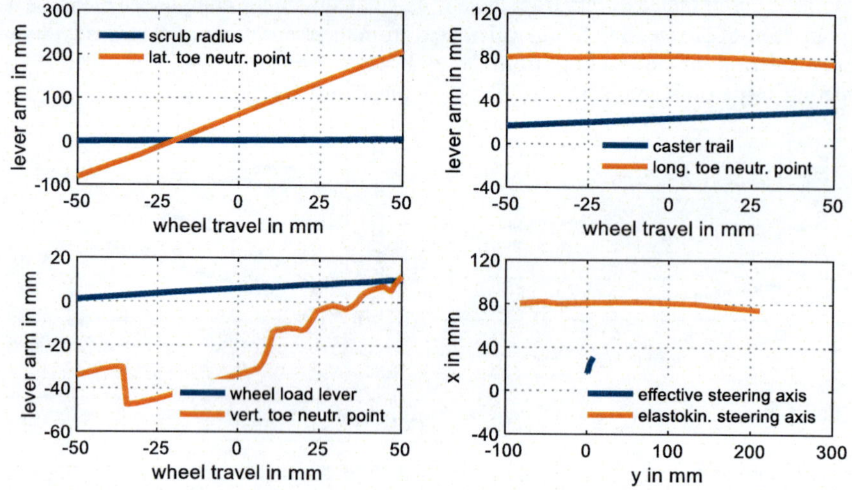

Fig. 7. Comparison of effective steering axis and elastokinematic steering axis

The location of the toe neutral point derived from the elastokinematic steering axis is basically dependent on the suspension concept (such as real or virtual steering axis) and the location of the steering gear as well as on the compliance characteristics (Fig. 8).

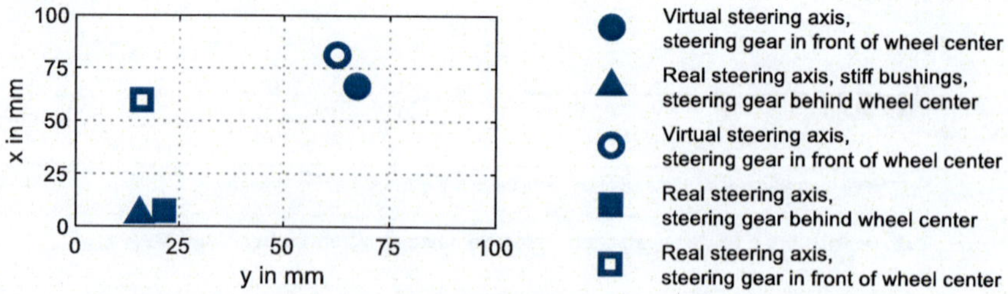

Fig. 8. Location of the toe neutral point for different suspension concepts

4.3 Influence of the Steering System Properties

This subsection deals with the influence of the steering system on suspension characteristics. The stiffness of the suspension system, which results from the torsion bar and the hardy disk (flexible couplings), contributes to the overall suspension compliance. It is affected by the speed-dependent steering assistance of the electric power steering. The steering assistance and therefore the resulting steering stiffness decrease with increasing vehicle speed. Fig. 9 shows the characteristic values derived from the elastokinematic steering axis for different levels of steering stiffness. They include no steering compliance at all (rigid steering system) as well as the steering stiffness for parking and driving by using appropriate steering assist characteristics. The characteristic values differ from each other significantly for the different levels of steering stiffness.

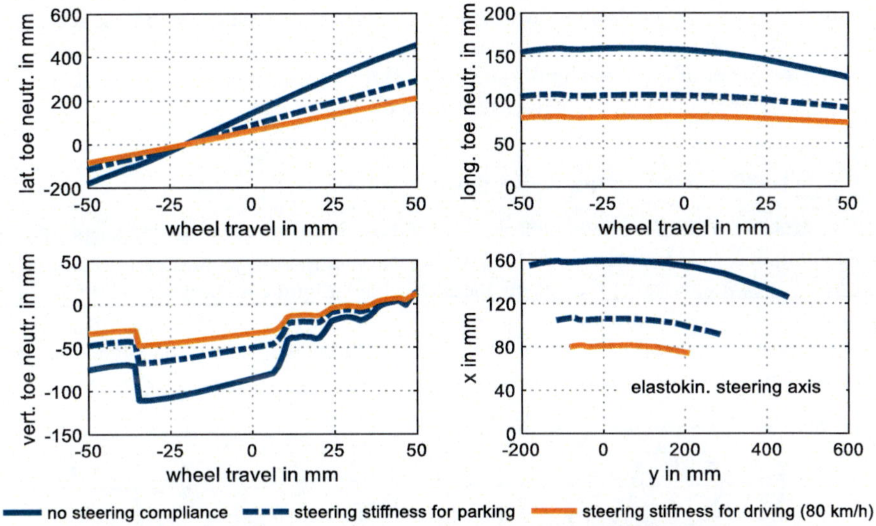

Fig. 9. Influence of the steering stiffness on the elastokinematic steering axis

The results show that it is essential to consider the steering system and its properties for suspension design in the chassis development process. The influence of steering assistance and steering stiffness on the toe angle is also shown on a K&C test rig [1].

4.4 Influence of the Tire Characteristics

The characteristics of the tire play a vital role for vehicle dynamics, as they define the forces and torques acting on the suspension system. Taking the pneumatic trail as an example, it is often considered as constant during suspension design, although it is dependent on the lateral force [10, 12, 18]. In Fig. 10, the influence of a constant and a force-dependent (based on the 'Magic Formula' [29]) pneumatic trail on the toe angle due to lateral forces is depicted.

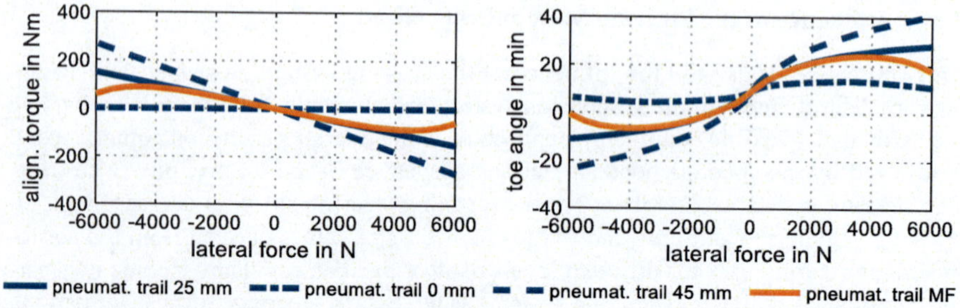

Fig. 10. Influence of the pneumatic trail on toe angle due to lateral forces

The results show that it is important to keep the properties of the pneumatic trail in mind during suspension design. Furthermore, a closer look should also be taken at the 'pneumatic scrub radius', which is the lateral equivalent of the pneumatic trail and describes the offset of a longitudinal force [10, 18]. It is mentioned in literature, but not analyzed in detail and mostly neglected in suspension design.

4.5 Visualization of Suspension Characteristics

An advantage of the described approach is the possibility to visualize the suspension characteristics, as it is proposed in [24]. Using the methodology outlined in section 3, suspension characteristics, such as stiffness or wheel alignment characteristics, can be calculated for each location. As an example, Fig. 11 shows longitudinal stiffness, toe neutral axis and windup neutral axis in the front view as well as the toe angle change and toe neutral axis in the side view.

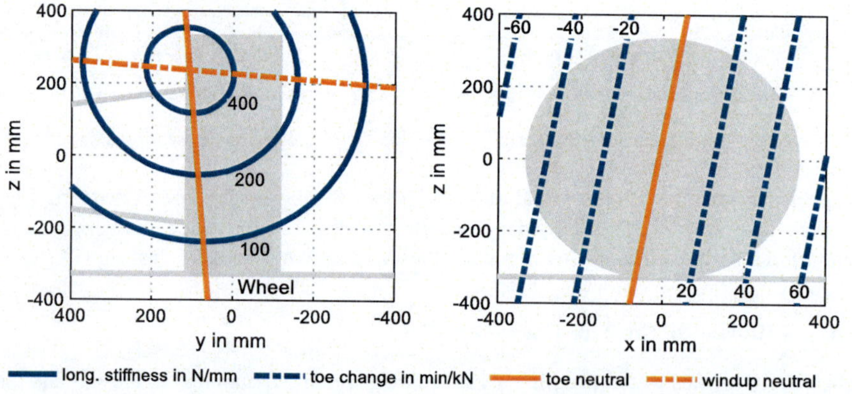

Fig. 11. Visualization of elastokinematic suspension characteristics

The visualization of suspension characteristics can provide new insights into suspension design and contribute to a clearer understanding of the complex dependencies in suspension design.

5 Conclusion and Outlook

This paper deals with an approach for the enhanced characterization of suspension systems in the virtual chassis development process.

In suspension design, a variety of characteristic values already exists for the design of suspension kinematics and compliance. Section 2 presents an overview of these characteristic values and their meaning for vehicle dynamics. The methodology proposed in section 3 for the analysis of suspension systems makes use of the suspension system's compliance matrix, which comprehends its kinematic and elastokinematic properties. The calculations of the characteristic values are derived from respective force-displacement relations referring to the definition of the characteristic values. In section 4, the application of the proposed approach to several use cases clearly demonstrates its suitability for the analysis and characterization of suspension systems. The results prove that compliance cannot be neglected in suspension design when studying characteristic values. The comparison of effective and elastokinematic steering axis shows that it is important to clearly distinguish between them, as they differ significantly from each other in their meaning. The influence of steering system properties and tire properties on the characteristic values is outlined as well. The visualization of suspension characteristics provides further insights into suspension characteristics.

Further work needs to be done on the implementation of the proposed approach in the virtual chassis development process. For each characteristic value, respective target ranges must be defined. Moreover, the characteristic values must be studied in the context of full vehicle analyses to analyze their influence on vehicle dynamics. This can be done by means of sensitivity analyses. The methodology can also be used to calculate further characteristic values. Additionally, it is suggested to develop an approach to determine the suspension system's compliance matrix analytically from the location of suspension hard points and bushing characteristics. This provides the possibility to calculate the characteristic values analytically without multi-body simulation.

Contributions

Stefan Buechner (as the first author) initiated and implemented this paper. He drew up the overall concept of the proposed methodology. This paper summarizes the essential aspects of his PhD thesis [1] submitted for publication.

Norbert Deixler contributed to a critical discussion of the proposed methodology. Ralf Stroph contributed to the conception of the research project and revised the manuscript critically for important intellectual content.

Markus Lienkamp made an essential contribution to the conception of the research project and contributed to a critical discussion of all aspects as well. He revised the PhD thesis [1] critically for important intellectual content.

Acknowledgement

This research was funded by the BMW Group. The author would like to thank them for their support.

References

1. Buechner, S.: Einfluss elastokinematischer Achskennwerte auf die Fahrdynamik. Dissertation, Institute of Automotive Technology, Technical University of Munich (in review).
2. KPMG Automotive Institute: KPMG's 20th consecutive Global Automotive Executive Survey 2019, https://automotive-institute.kpmg.de, last accessed 2019/11/21.
3. Ersoy, M., Gies, S.: Fahrwerkhandbuch. 5th edn. Springer Vieweg, Wiesbaden (2017).
4. Vivian, B., Held, V.: Goals and challenges of future chassis developments – solving a multidimensional optimization problem. In: 5th International Munich Chassis Symposium 2014, pp. 3–5. Springer Fachmedien, Wiesbaden (2014).
5. Schömann, S. O.: Produktentwicklung in der Automobilindustrie: Managementkonzepte vor dem Hintergrund gewandelter Herausforderungen. 1st edn. Gabler/Springer Fachmedien, Wiesbaden (2012).
6. Pischinger, S., Seiffert, U.: Handbuch Kraftfahrzeugtechnik. 8th edn. Springer Vieweg, Wiesbaden (2016).
7. Meywerk, M.: CAE-Methoden in der Fahrzeugtechnik. Springer-Verlag, Berlin Heidelberg (2007).
8. Wimmler, J., Schramm, D., Wahle, M., Zimmermann, M.: Concurrent design of vehicle tires and axles. In: 6th International Munich Chassis Symposium 2015, pp. 839–851. Springer Fachmedien, Wiesbaden (2016).
9. Hazare, M.: An Integrated Systems Engineering Methodology for Design of Vehicle Handling Dynamics. Dissertation, Department of Automotive Engineering, Clemson University (2014).
10. Matschinsky, W.: Radführungen der Straßenfahrzeuge: Kinematik, Elasto-Kinematik und Konstruktion. 3rd edn. Springer, Berlin, Heidelberg, New York (2007).
11. Adamski, D.: Simulation in der Fahrwerktechnik: Einführung in die Erstellung von Komponenten- und Gesamtfahrzeugmodellen. Springer Vieweg, Wiesbaden (2014).
12. Blundell, M., Harty, D.: The multibody systems approach to vehicle dynamics. 2nd edn. Butterworth-Heinemann, Oxford, Waltham (2015).
13. Bindauf, A., Angrick, C., Prokop, G.: Fahrwerkscharakterisierung an einem hochdynamischen Achsprüfstand. ATZ - Automobiltechnische Zeitschrift 116(12), pp. 76–81 (2014).
14. Dixon, J. C.: Suspension Geometry and Computation. Wiley, Chichester (2009).
15. Trzesniowski, M.: Rennwagentechnik: Grundlagen, Konstruktion, Komponenten, Systeme. 4th edn. Springer Vieweg, Wiesbaden (2014).
16. Milliken, W. F., Milliken, D. L.: Race Car Vehicle Dynamics. 14th edn. SAE International, Warrendale (1995).
17. Haberzettl, S., Zschocke, A. K., Gauterin, F.: Reduktion der niederfrequenten Lenkraddrehschwingungen bei wechselseitiger Fahrbahnanregung ohne Beeinflussung der fahrdynamischen Achskennwerte. In: 14. Internationale VDI-Tagung Reifen, Fahrwerk, Fahrbahn, pp. 257–272. VDI-Verlag, Düsseldorf (2013).
18. Henker, E.: Fahrwerktechnik: Grundlagen, Bauelemente, Auslegung. Vieweg+Teubner Verlag, Wiesbaden (1993).
19. Zomotor, A.: Fahrwerktechnik: Fahrverhalten. 2nd edn. Vogel, Würzburg (1991).

20. Schultz, E.: Analyse und Synthese elastokinematisch optimierter Verbundlenkerachsen: Entwicklung einer Verbundlenkerachse mit entkoppeltem Radträger. Dissertation. Shaker, Aachen (2013).
21. Zandbergen, P., David, W., Knoy, E.: The New Rear Suspension for Ford's Global CD Platform. In: 21st Aachen Colloquium Automobile and Engine Technology 2012, pp. 1503–1524. Aachen (2012).
22. Röski, K.: Eine Methode zur simulationsbasierten Grundauslegung von PKW-Fahrwerken mit Vertiefung der Betrachtungen zum Fahrkomfort. Dissertation, Chair of Automatic Control, Technical University of Munich (2012).
23. DIN ISO 8855. Straßenfahrzeuge - Fahrzeugdynamik und Fahrverhalten - Begriffe (2013).
24. Gerrard, M. B.: The Equivalent Elastic Mechanism: A Tool for the Analysis and the Design of Compliant Suspension Linkages. SAE Technical Paper Series (2005).
25. Yao, G.-f., Hou, J., Zhao, P.: A new methodology to calculate the equivalent stiffness matrix of the suspension structure with flexible linkages. Advances in Mechanical Engineering 9(7), pp. 1–8 (2017).
26. Büchner, S., et al.: Potential of elastodynamic analysis for robust suspension design in the early development stage. In: 9th International Munich Chassis Symposium 2018, pp. 367–386. Springer Fachmedien Wiesbaden, Wiesbaden (2019).
27. Buechner, S., Lienkamp, M.: Compliance Matrix Based Design and Analysis of Suspension Toe Behavior for Virtual Chassis Development. In: 2019 IEEE 10th International Conference on Mechanical and Intelligent Manufacturing Technologies (2019).
28. Buechner, S., Lienkamp, M.: Compliance Matrix Based Analysis and Design of Suspension Systems for Chassis Development. International Journal of Mechanical Engineering and Robotics Research 8(6), pp. 873–879 (2019).
29. Pacejka, H. B., Besselink, I.: Tire and Vehicle Dynamics. 3rd edn. Butterworth-Heinemann, Oxford, Waltham (2012).

The efficient use of HiL simulation and vehicle tests for in-house software development at Porsche

Benedikt Jooß[1], Thomas Rotter[2] and Prof. Dr.-Ing. Dr. h.c. Dieter Schramm[3]

[1] Dr. Ing. h.c. F. Porsche AG, Germany
[2] Hochschule Esslingen, Germany
[3] Universität Duisburg-Essen, Germany

Abstract. The challenges of modern vehicle software development are mainly due to the increasing complexity and growing number of functions that lead to high costs and time expenditure for software tests. To meet these challenges, in this paper a method is presented that assigns software test cases to the test instances Hardware-in-the-loop simulation and vehicle. The main aim is an efficient and effective use of both instances. As vehicle prototypes are tremendously expensive, the focus is on the reduction of time for vehicle tests. A four-stage method assigns individual test situations to the correct instance by means of a utility analysis. For this purpose, the identification of technical aspects of testing is the first step. Afterwards, filtering and weighting of these aspects is necessary for specific use cases. Subsequently, the identification of test situations is the next step before assigning them to the correct test instance with the utility analysis. To prove the validity of the developed method the paper contains an application example, which evaluates 151 test manoeuvres with the method. The result is a drastic reduction of the time required for vehicle tests.

Keywords: Software Testing, Hardware in the Loop, Simulation, Vehicle Testing.

1 Introduction

In-house software development of carmakers requires a high number of test activities, especially due to the increasing complexity and number of software and functions. Among the different methods for software development, the model-based software development approach requires particularly tailored test stages and test instances. Hereby the test objective differs depending on the instance. In addition, different development steps need different test objectives - starting with Model-in-the-Loop (MiL) tests up to the vehicle tests itself. Some instances are interchangeable, others are not. For instance, Hardware-in-the-Loop (HiL) tests can have the same test objective as vehicle tests. However, some of the tests need a specific test instance due to reasons like safety, environmental influences or driving emotion [1].

Recent advances in the automotive industry are leading to a change in testing strategies. The main reasons are high costs in research and development (R&D) and the change of technology leading to lower profit per car. To compensate for the lower

P. E. Pfeffer (ed.), *11th International Munich Chassis Symposium 2020*, Proceedings, https://doi.org/10.1007/978-3-662-63193-5_16

profit, new cost saving strategies dominate the automotive industry. Especially the extremely expensive vehicle prototypes or test vehicles exhibit high saving potential. The highest cost saving is achievable by cutting the number of cars by quite high amounts. Traditionally, the vehicle prototype has been the most important test instrument, especially for driver-focused cars like a Porsche. This reduction of test vehicles results in a test gap, which requires closing. Therefore, simulations come into focus. The vehicle is, for some tests, replaceable with virtual test instances such as HiL test benches.

To maintain or increase the quality of software, a method is necessary to decide which test instance is best suited for various test situations. In this paper, such a method is introduced. It concentrates on HiL test benches and vehicle prototypes used for tests. The method addresses the following questions:

1. Which aspects are necessary to decide for a test instance?
2. Which of the aspects are more important?
3. What test situations are relevant?
4. How to decide for a test instance?

This paper answers these questions and demonstrates the results using a concrete example.

Consequently, the paper is divided into three parts. The first part provides an overview of the development of the general method. The method is transferable to any other organisation with minor adjustments. Afterwards, an example of application will be explained together with the verification of the method. The described cases are software tests of the department chassis mechatronics of the Dr. Ing. h.c. F. Porsche AG. In the last part of the paper the results are summarized.

2 Development of the method

To test software on the ECU, it is possible to use either HiL test benches or vehicles. As already mentioned, it is necessary to use or prefer one or the other instance for certain test cases. The reasons for this are diverse. For example, it is always advisable to simulate high-risk manoeuvres with the help of HiL simulation. For other functionalities, the test needs to validate a concept and it is necessary to have a feeling for the function in a vehicle. To reduce costs, these kind of tests should be the exception. As test benches are cheaper over time compared to vehicle prototypes, as many tests as possible should be simulated on test benches. This leads to the first question: which aspects are important to decide for the correct test instance? This is answered with the use of two different approaches. The first was a literature search. Independently of this search, expert interviews helped to find the appropriate aspects.

In literature, many technical aspects are described which are necessary for testing. Each test case uses one or more of these aspects. A basic example will make it easier to understand: The test object is a function to detect wetness on the road surface. The aim of the test is to prove that the function works correctly. This means, that the function detects the wetness if the surface is wet and vice versa. This is only possible if the road is actually wet. If there is no wet environment available, it is not possible to test with a vehicle. However, it is possible to simulate the wet environment. The aspect in this case would be "environmental influences". That shows that regardless of the test object, the test instances HiL and vehicle have important technical properties. These properties are required for some test cases. Thus, the properties can be an aspect to decide for the correct instance within the method.

For the method, the difference between simulations and driving tests is important. For example, the reproducibility by simulation is easier than by driving because all the parameters will be set beforehand. So that all the parameters are the same for each test run. For vehicle tests, the driver of a manoeuvre would need to hit exactly the same speed and steering angle at the same point of the test for instance [2], [3]. Yet, model building is complicated especially if highly realistic physical conditions will be needed [4]. In addition, Langner et al. [2] explain that the complete target environment, which is available in the vehicle, shows different test results than simulated results due to the influences of actors and other functionalities. Moreover, environmental influences sometimes have also an effect on the results. Käppler [5] explains in this context that the vehicle shows more flexibility because unpredictable events can happen. In addition, it is only possible to gain subjective assessments in a vehicle [4]. However, more important than subjective assessments is the safety for drivers and passengers. Therefore, a simulation via HiL testing is the better choice if the manoeuvre is risky [6]. For this kind of manoeuvres, it is often necessary to inject faults in the system. Langner et al. [2] explains that HiL test benches make it easy to inject faults in a given system. On the other hand, they explain that the simulated environment leads to more sources of error and higher effort in troubleshooting.

In many projects, the availability of testing equipment is critical. Mostly simulation is available before the first vehicle prototypes drive on the roads [5], [7]. In addition, simulation is mostly less time consuming due to automation. Automation helps also for reproducibility. Nevertheless, as HiL uses also some hardware components the expected results need a tolerance range. This is because temperature, bus traffic load, etc. might have an impact on the ECU or the test bench [6], [8].

In the first step, we focused on the following (technical) aspects found in literature:

- Reproducibility
- Model building
- Target environment
- Environmental influences
- Subjective driving experience
- Safety
- Fault injection or manipulation
- Source of error
- Troubleshooting
- Availability in project
- Time required for testing

Since the aspects found in the literature are kept very general, literature research is not sufficient for a specific application.

For an application of the method in a certain environment (e.g. a company or a part of it), it is necessary to take a closer look to the company under investigation. Above all, an understanding of the software the company develops (test object) is required. Furthermore, the test strategy of the company must be understood. The best knowledge of these two things is held by the people who work in the company or part of the company under consideration. Consequently, the method uses expert interviews to confirm and to complete the important aspects, necessary to decide for a test instance obtained in the literature search. Experts are software developers and testers working with it every day. Interviewer is an independent person, who has enough knowledge to understand all the technical aspects. While conducting the interviews it is vital to get a particularly good understanding of the software itself and its application area. In addition, it is useful to understand the test activities already performed. This knowledge forms the basis for identifying the aspects relevant to the company. These aspects are mostly crucial technical aspects for the correct test execution such as safety or a needed test environment.

The next step to decide for the correct test instance is the weighting of these aspects to answer the second question: Which of the aspects are more important? As mentioned before some of the aspects lead directly to a decision, others do not. The following section explains how to weight the single aspects.

First, it is necessary to figure out which aspects lead directly to a decision. It must be possible to implement the aspects at both instances. It is not possible to simulate subjective impressions, in a way a driver can judge the vehicles behaviour, using HiL simulation. This means if a test case includes subjective judgement, the test instance of choice is the vehicle.

Afterwards, the method uses a pairwise comparison to weight all remaining aspects. The pairwise comparison uses a cross table to compare all aspects with each other and to gain a weighting factor for each aspect. The factor is necessary in the final phase of the method. Chapter 3 gives a deeper insight in this step with the help of a concrete use case.

The aspects build the basis but are not sufficient to conclude on a test instance. The third step is to identify different test situations and thus to develop categories of test situations. The combination of different situations results in a driving manoeuvre. Driving manoeuvre catalogues and HiL test specifications document all manoeuvres respectively test cases. A manual examination of a large number of manoeuvres has revealed commonalities. The result is a list with all situations. It is also apparent, that each manoeuvre represents an overall system consisting of driver, vehicle and environment. Each of these categories can influence a test manoeuvre. For example, a driver is responsible for the correct speed or steering angle, while the environment determines the friction coefficient or weather influences. Each test situation is assigned to a test category. This builds the basic framework for the last step and is the answer to the fourth question.

In the last step, the method uses a utility analysis to decide for a test instance. The evaluation shows to which extend the two test instances fulfil the aspects of the first step in the different test situations. First, points are given for degree of fulfilment of each aspect for the two instances. In the best case, this is evaluated by a group of experts. The points are on a scale form zero (no fulfilment) to four (high fulfilment). Afterwards, a multiplication of the points and the weighting factor from the second step gives a final score for each test instance. As seen in Table 1 for each test situation, the score indications the more suitable test instance. In this example, the HiL simulation is the better choice for test situation 1 but the vehicle for test situation 2. The following chapter a use case for the whole method is shown.

Table 1. Principle structure of the utility analysis

		Aspect 1	Aspect 2	Aspect 3	Aspect 4	Score
Weighting factor		40 %	35 %	15 %	10 %	
Test situation 1	Simulation	2	4	3	3	73.8 %
	Vehicle	1	3	4	2	56.3 %
Test situation 2	Simulation	2	3	0	4	56.3 %
	Vehicle	3	4	4	3	87.5 %

3 Application and verification of the method

3.1 Application Example

The easiest way to prove the validity of the method is to apply it to a relevant test case. As described before we will use the four steps. The given use case is the mechatronic chassis development of the Dr. Ing. h.c. F. Porsche AG. The paper concentrates on the software development of mechatronic chassis systems. A focus of interest will be on vertical and lateral dynamic systems. Therefore, the requirements e.g. for the simulation of tire models are particularly high. With the help of expert interviews, such special features are figured out and are part of the technical aspects for the given and any other example.

With the help of the experts, each aspect found in literature is evaluated for its relevance. The experts are mostly developers of the software functions as well as experienced SW-testers. In addition to the aspects from the literature, the experts have introduced new or adapted aspects such as the tire model. The following six aspects are relevant for the given use case:

- Model accuracy
  There is a direct link between the accuracy of vehicle, driver and environment models and the significance of the test results. It is also necessary that the model can react to events such as road bumps or potholes.
- Environmental influences
  This aspect shows the relevance for different geographical or weather influences, for instance friction coefficient, wetness, or gradient.

- Subjective driving experience
  To test some requirements, it is necessary to gain a subjective driving expierence. Otherwise, it might not be possible to know if a function is to harsh or soft.
- Safety
  This aspect considers the risk potential of a test manoeuvre. The safety for test drivers is most important while testing.
- Manipulability of signals
  This describes how easy it is to manipulate bus systems or functions. For example, the possibility to specify signal characteristics or the possibility to inject hardware and software faults.
- Interdependencies
  This sums up the relevance of the target environment for the tests. For instance, the interdependencies between the ECU, the software, and other mechanical components.

After the identification of the relevant aspects for the given use case, the next step is the weighting. As mentioned before the most important aspect is the safety. In this case, safety means the probability that a driver or other participant of the test will be injured or even killed. One aim of software testing is to find faults and thereby make the function safer. Simulation gives the possibility to test highly risky manoeuvres without the danger to the driver's health. Therefore, the virtual HiL environment is the right test instance for highly risky manoeuvres.

A second important aspect is the subjective driving experience. This is an especially important aspect for Porsche developers, as each Porsche should feel like a Porsche. Yet, it is not possible to simulate this feeling. Due to this reason, a vehicle test is essential for these kinds of tests.

For all other aspects, there is no direct link to a test instance. For the method, it is necessary to weight the aspects, though. A pairwise comparison guarantees the objectiveness during the weighting process. The avoidance of conscious or unconscious weighting by interests (as described, inter alia, in [9]) is important for the correct usage of this method. Close cooperation between testers and other key players in chassis software development eliminated a weighting by interests. Table 2 shows the weighting factors rounded to five percent in the right column.

Table 2. Results of pairwise comparison

Aspect	Comparison				Weighting
	M. a.	E. i.	S. m.	I.	
Model accuracy (M. a.)	-	2	1	2	40 %
Environmental influences (E. i.)	0	-	0	1	10 %
Signal manipulability (S. m.)	1	2	-	0	25 %
Interdependencies (I.)	0	1	2	-	25 %

In the next step, multiple manoeuvres were broken down to single pieces to identify test situations. Table 3 gives an overview of the situations relevant for the use case.

Table 3. Categorisation of test situations

Driver	Vehicle	Environment
Standstill	Implausible state	Wetness
Slow speed	Fault injection	Friction coefficient
Straight ahead	Initialisation	Gradient
Cornering	Display related	
Steady state		
Unsteady state		
Under-/Oversteering		

As described above, there are three categories of situations. The first one is driver. It includes all driving situations the driver can influence. The driver is, among other things, responsible for whether a vehicle is cornering, accelerating, or standing still. The second category includes all situations that depend on the vehicle dynamics. This includes, among other things, implausible states in which physical laws are deliberately ignored. For example, positive wheel speeds on one and negative wheel speeds on another wheel at the same time. Initialisation includes all test situations while the system starts. Some necessary values need training after the vehicle start before the system can process them. Display related tests include all situations that serve to ensure the display of chassis functions in the instrument cluster. The third category includes all situations depending on the environment such as wetness or inclination.

The individual test situations are then assigned to the two test instances by means of a utility analysis. For each of the four aspects (model accuracy, environmental influences, signal manipulability and interdependencies), the simplicity of implementation on a HiL test bench and in the vehicle are rated. The result is a degree of fulfilment for each aspect and both instances. Fig. 1 shows the assignment of test situations to test instances. The bisector splits the figure in simulation and vehicle test. For every situation that is above the bisector the better test instance is the HiL test bench. The situations below the bisector are more effectively to be tested with the vehicle. However, the figure shows that only the initialisation is below the bisector. The degree of fulfilment for the vehicle is 90 %, for the HiL test bench only 44 %. The reason is that many vehicle components are needed for the initialisation, thus there are high interdependencies between different ECUs. In addition, a high model accuracy of the complete vehicle is necessary for these tests. On the other hand, the implausible condition has a degree of fulfilment for HiL test benches of 88%, for the vehicle of 50%. This is mainly because the signal manipulability is easier and physical interdependencies are simulated.

Some test situations are close to the bisector. For these situations, HiL test benches have slight advantages. Yet a consideration of vehicle tests is possible, depending on availability of both instances. In this group, the standstill attracts the most attention. For both, the HiL simulation and the vehicle, the use value is around 90 %. The reasons are the low requirements concerning dynamic simulation and little interdependencies with the environment. For cornering, the use value is a bit less than 70 %, for the simulation a bit less than 60 %. This shows that complex situations have no ideal test instance. Both instances have advantages and disadvantages. Yet it is possible to perform these test situations with both instances. Somewhat more efficient and effective is the HiL test bench.

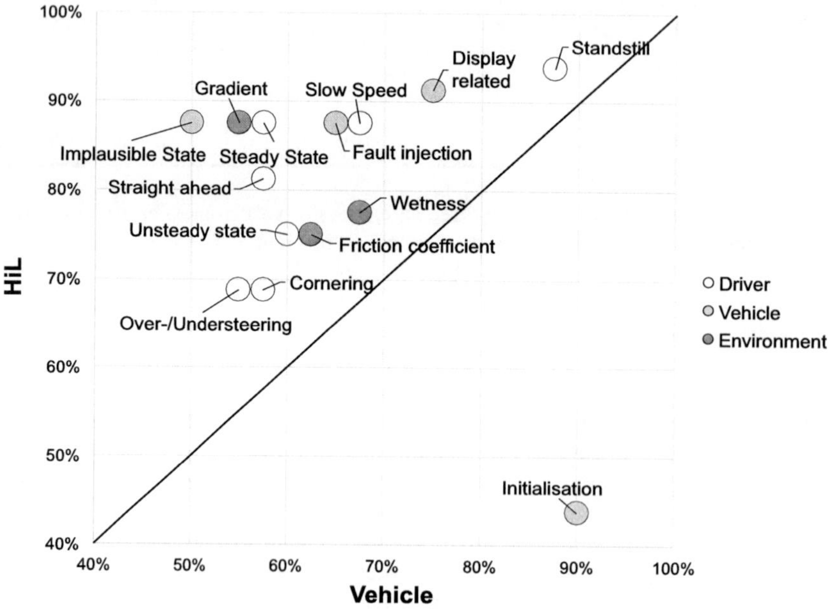

Fig. 1. Mapping of test situations to test instances

3.2 Verification

The following section explains the meaning of the methodology using basic test case examples:

The first is a test case in which the diagonally opposite wheels have the same wheel speed, but the other two wheels have different wheel speeds. As this is implausible, the instance of choice is the HiL test bench.

A further test case design estimates the lateral inclination. The estimation should work correctly while the vehicle is driving and while it is stationary. The test case contains the situations gradient and standstill. In principle, the test case can be carried out using a vehicle and a banked corner. However, the methodology recommends a simulation by HiL. A great advantage is more flexibility with a simulated gradient angle.

Many test cases are rather complex. Therefore, there are not always clear assignments to a situation. For instance, one test manoeuvre is a constant circular drive. On the one hand, the lateral acceleration and yaw rate, on the other hand, the tire temperatures are calculated. A steady state circular test is assigned to the test situations cornering and steady state. According to the methodology, a HiL simulation is the correct way to test these situations. To calculate the tire temperatures, an initialization process is necessary in advance. This situation on the other hand must definitely be tested in the vehicle. To be able to use the correct test instance, a splitting of the test case is possible. Otherwise, the initialization process is predominating, so that the test instance is the vehicle.

However, there are also test cases, which are not evaluable with the method that easily. These include, among others, a test case for estimating the friction coefficient during longitudinal dynamic driving. Although the methodology assigns the friction coefficient to a HiL test bench as a better test instance, this is not feasible in reality. The reason for this are the high demands on the tire model. In the current status the tire model has not the necessary accuracy.

These examples show that an evaluation with the methodology is very extensive. It is necessary to evaluate all test cases in individual steps. In the use case of mechatronic chassis development, this means an individual evaluation of more than a thousand test cases. A first check picks 200 test manoeuvres randomly for the verification. As a result, there was a new assignment of most test cases. Especially previous vehicle tests change to HiL tests.

Despite the initial high expenditure of time, an evaluation of all test cases is beneficial. HiL measurements in the above example take on average 3.12 minutes. In the sample, the times of individual measurements vary from about one minute to a few test cases with an execution time of 45 minutes or more. Since this is a real-time simulation, the duration of vehicle tests is comparable in theory. However, it is not considered that vehicle tests have to be repeated due to inaccuracies while driving. In addition, vehicle tests have a high manual effort for preparation and post-processing. Table 4 shows measured times for one vehicle test manoeuvre. On the one hand, the strong temporal fluctuations of the actual manoeuvre can be seen, but also the comparatively time-consuming preparation and post-processing. Reason for the fluctuation is a necessary parameter combination for the manoeuvre. It is necessary to hit the right combination of float angle, friction coefficient and speed.

Table 4. Time measurements of one vehicle test

Preparation [s]	Manoeuvre [s]	Post-processing [s]	**Total time [s]**
41.96	19.88	19.51	**81.35**
17.07	11.69	31.61	**60.37**
36.23	21.14	45.90	**103.27**
35.02	12.92	35.15	**83.09**
34.84	39.42	37.78	**112.04**
78.26	82.48	308.50	**469.24**

Bringing the vehicle into position and adjusting and restarting the measuring systems are preparation tasks among other things. The average time needed are 35 s. Since these are manual tasks, the fluctuation is expected. Tasks of post-processing are among others a brief analysis of the measurements, manual storing of the measurements or resetting of the measuring systems. This takes an average of 62 s.

The simulation with HiL test benches also needs time for preparation and post-processing. However, these are either one-time manual or automated and standardized tasks. Tedious tasks such as positioning a vehicle are not necessary.

As already mentioned, a simulated test case on the HiL test bench takes on average 3.12 minutes (192 seconds). This is a capacity of 150 test cases within 8 hours. Under ideal circumstances, vehicle test cases require the same time. Yet, on top, comes an average of 35 seconds for the preparation and 63 seconds for the post-processing. With an average total time of 290 seconds per test case, it is possible to execute a maximum of 99 test cases in the same period.

An example helps to illustrate the meaning. To test a single controller, 151 driving manoeuvres or test cases must be examined. Eighty of these are test cases for a HiL simulation and 71 are test manoeuvres for vehicle tests. This means that the HiL simulation will take a total of 4.27 hours and the vehicle tests will take a total of 5.72 hours.

Using the method, all 151 test cases are re-evaluated. The result is a distribution of 9 % vehicle test cases to 91 % HiL test cases. This reallocation of test cases results in a total time reduction of 1.54 hours for a complete run through of all test cases.

However, this is not the expected success. The critical path in testing are vehicle tests, as they require more manual effort and more person-hours. The applying of the method rises the time for simulation up to 7.31 hours and at the same time cuts down the time for vehicle tests from 4.27 hours to 1.13 hours. In addition, travel times to test sites are not considered. This reduces working hours even more. Also not considered is the advantage that the simulation can run 24 hours per day automatically (times for maintenance and commissioning are disregarded). This verification case shows the validity of the method used at the mechatronic chassis development of the Dr. Ing. h.c. F. Porsche AG.

4 Conclusion

The main challenge in today's vehicle development is the increasing complexity. Due to this, the development costs are exploding. Therefore, car manufactures put a lot effort into cost reduction. The increasing amount of software in vehicles lead to a high demand for testing. In the past, the test car was the only available test instance, but nowadays simulation supports testers and developers. For driver-focused cars, such as a Porsche, the vehicle test is still the most valuable. This is necessary to make a Porsche feel like a Porsche. However, prototypes for vehicle tests are awfully expensive, so that carmakers build as few vehicle prototypes as possible. Due to this, the utilisation of them needs verification and reduction where possible. For the software of mechatronic systems, the testing via HiL simulation is often more efficient.

Yet, it is not possible to test everything via simulation, as at the end the resulting car must feel like a Porsche. Therefore, this paper introduces a four-step method to decide for the correct test instance and answering the main question. The first step includes the identification of technical aspects, which are relevant for software testing. Examples are safety, model accuracy or needed time. In the next step, all identified aspects need weighting for importance. This is always depending on the use case and not possible to give a general answer. For aspects that do not clearly lead to one or another instance, the method uses a pairwise comparison to gain a weighting factor, which the fourth step (utility analysis) uses. Beforehand, the third step gains test situations by analysing existing test manoeuvres. There are three categories for situations. The situations of each category are combinable with the situations of the other category to create a test manoeuvre. The first are situations depending on the driver, the second depending on the vehicle and the third depending on the environment. The final step assigns each situation to a test instance. Therefore, it uses a utility analysis. The analysis evaluates each technical aspect for both instances. In combination with the weighting factor, a percentage figure is obtained, which indicates the utility of both instances for a test situation.

A test case shows the validity of the method. The evaluation for one controller means the evaluation of 151 vehicle test manoeuvres and HiL test cases. Before the evaluation, 47 % are vehicle tests manoeuvres. After the evaluation, just 9 % are left. The time for vehicle testing shrank from a theoretical 3.77 hours down to 1.13 hours.

This shows that the method answers the key question. Reducing the vehicle test manoeuvres down to a necessary minimum, lead to a more efficient use of test vehicles. With the more efficient use, it is possible to reduce the number of expensive vehicle prototypes. At the same time, the quality of the software stays the same and the complexity of modern vehicles is controllable with the effective use of HiL testing.

References

1. Wood, M., et al.: Safety First for automated Driving [white paper], 2019, https://www.aptiv.com/docs/default-source/white-papers/safety-first-for-automated-driving-aptiv-white-paper.pdf, last accessed 2019/03/20
2. Langner, F., Alsmann, U., Meyer, J.: Fahrversuch versus HiL-Test [white paper], 2009, https://docplayer.org/15484999-Fahrversuch-versus-hil-test.html, last accessed 2019/03/23
3. Schindler, E.: Fahrdynamik, Grundlagen des Lenkverhaltens, und ihre Anwendung für Fahrzeugregelsysteme. 3rd edn. expert-Verlag, Renningen (2019)
4. Onesti, P.: Verbindung von Human- und Hardware-in-the-Loop-Testing verkürzt Entwicklungszeiten. ATZ 111 (10/2009), 764-769 (2009)
5. Käppler, W. D.: Smart Vehicle Handling: Test und Evaluation in der Fahrzeugtechnik. 1st edn. Springer-Vieweg, Berline Heidelberg (2015)
6. Isermann, R.: Fahrdynamik-Regelung: Modellbildung, Fahrerassistenzsysteme, Mechatronik. 1st edn. Friedr. Vieweg & Sohn Verlag, Wiesbaden (2006)
7. Krimmel, H., Machmann, O., Diedt, S., et. al.: Chancen und Grenzen von HiL-Tests. ATZ Elektronik 1 (12/2006), 18-25 (2006)
8. Adamski, D.: Simulation in der Fahrwerktechnik: Einführung in die Erstellung von Komponenten- und Gesamtfahrzeugmodellen. Springer Fachmedien, Wiesbaden (2014)
9. Kühnapfel, J. B.: Nutzwertanalysen in Marketing und Vertrieb. 2nd edn. Springer Fachmedien Wiesbaden GmbH, Wiesbaden (2019)

Analysis of Suspension Dynamic Characteristics for Ride Comfort Development Using Suspension Rig Test

Seungmin Kwon, Yong-Sub Yi, Jieun Song, Jongmin Jung and Ho Joon Park

Hyundai Motor Company, Korea

Abstract. In the era of future mobility ahead of us, the importance of ride comfort performance will be highlighted further by the increasing demand from customers for comfort to lead daily lives such as reading and resting in cars while on the move. If it is possible to predict the ride comfort performance of the vehicle using only the system without the vehicle tests, ride comfort performance development at the early stage is very advantageous through target cascading and system level performance enhancement. In this paper, newly installed test rigs (suspension module test rig, dynamic comfort roadway, chassis dyno) were used to identify the effect of the vibration characteristics of suspension on the ride comfort performance, and the system's vibration characteristics and ride tests were carried out, and correlation between the ride comfort performance and system vibration characteristics was analyzed. Ride comfort performance enhancement process at the system level based on analysis of system vibration characteristics was proposed. The target of vibration characteristics of the suspension system to improve ride comfort performance and a process of component design was proposed.

Keywords: Suspension Module Test, Rig Test Facility, Ride Comfort, Impact Harshness, Vibration Characteristics of Suspension, Bushing Stiffness Design, Target Cascading, System Level Development

1 Introduction

Ride comfort is one of the main factors determining the performance of the vehicle, along with handling, crashworthiness, durability, and NVH. In the era of future mobility, consumers' demands for ride comfort for daily life in a vehicle on the road such as reading, sleeping, relax, and so on, further highlighting the importance of ride comfort performance. Though the ride comfort performance is determined by the vibration transmission characteristics of various subsystems, such as tires, suspension, powertrain, and so on, it traditionally has been developed by so-called "tuning" process which focuses on controlling specific vibration phenomena by changing characteristics of anti-vibration parts such as rubber bushings and mountings after subjective and objective evaluation of vehicle by driving test. However, this development procedure takes enormous time and cost, so vehicle manufacturers have had tried to find a

P. E. Pfeffer (ed.), *11th International Munich Chassis Symposium 2020*, Proceedings, https://doi.org/10.1007/978-3-662-63193-5_17

new efficient way of development to secure profitability and quickly respond to the market.

One of the most well-known approaches to overcome this situation is the V-cycle process which consists of two main branches of "target cascading" and "verification". Till now, many researches have been carried out to devise optimized and efficient methodology for this process, and showed good results in many industrial applications. However, for ride comfort performance development, the absence of the adequate equipment for subsystem level test and analysis method limits this task.

For target cascading process, it is essential to understand the contribution of each subsystem to the total system and correlate all the subsystems properly by analyzing interaction among them. Likewise, once the target for each subsystem is set and real hardware built, its functionality should be verified in the conditions which can represent practical operating conditions. Since it is really difficult to test a subsystem in those conditions, the multi-body simulation (MBS) has been used widely to correlate subsystems and whole system and analyze the underlying physics. Considering the weak points of modeling for simulation such as simplification, linearization and uncertainties for modeling parameters, simulation is not sufficient enough to fulfill this purpose.

Some of vehicle manufacturers built a dedicated test facility for suspension module and used it for ride comfort development [1, 2]. However, since test conditions were somewhat different from real ones of vehicle test, measured suspension dynamic characteristics such as natural frequencies were not realistic, and consequently the correlation between suspension characteristics and vehicle vibration behavior was imperfect and showed discrepancies.

In this research, a novel process for target cascading using state-of-the-art suspension module rig test facilities is suggested. Firstly, suspension module is tested in the very same conditions as a real vehicle driving test on the proving ground, and it is verified that test only with suspension module can capture the vehicle test evaluation results. Secondly, vibration transmissibility of suspension is investigated by dynamic characterization, Putting these two investigation results together, underlying physics of correlation between bushing properties, suspension system characteristics and vehicle performance is figured out. Finally, based on these findings, improved bushing characteristics suggested and its effect is verified by suspension module test again.

2 Rig Test Facilities

Since the development ride comfort performance by the objective and subjective evaluation of a vehicle is limited in analyzing the root cause at the system and part level, the next generation R&H performance development process using the test rigs was newly proposed as a solution [3, 4]. If the vehicle performance can be evlautated using only the system even before building a test vehicle, it can accelerate the development process greatly, in parallel with saving cost from the reduction in prototype vehicle and design change in late development stage. To fulfill this purpose, many vehicle manufacturers have conducted several studies on the suspension module tests,

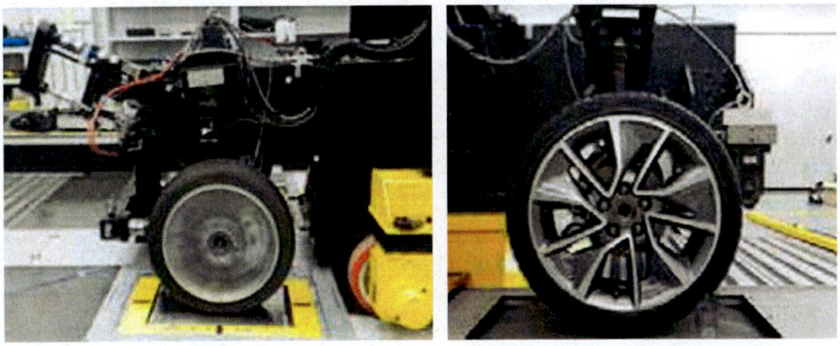

(a) (b)

Fig. 1. Suspension module test rig mounted on (a) Comfort Roadway and (b) Chassis Dyno.

but have had many difficulties simulating actual driving conditions at the system level [1, 5].

In this research, these problem is overcame by using state-of-the-art rig test facilities which are specially devised to simulate real driving conditions on the proving ground only with suspension module. The suspension module test rig used in this study is shown in Fig. 1. This test rig can be operated on several kinds of actuators, which simulate the interaction between road and tire contact patch. For ride comfort evaluation, two test machines of Comfort Roadway and Chassis Dyno are used. Comfort Roadway can mimic the road surface input to the tires by exciting each tire patch in vertical and longitudinal direction simultaneously from 0.5 to 40 Hz. Since both real vehicle and suspension module can be mounted on this machine, reaction of vehicle and suspension module to the very same road input fed to tire patch can be correlated directly. Virtual road input profiles used for this has been established so that they represent existing roads in the proving ground, and verified by comparing vehicle reaction data from the real road driving and rig test before this research. Chassis Dyno, also known as Roller Bench, is a typical test rig devised to drive the vehicle on drums with large diameter to measure driving power of a vehicle. In this research, two types of cleat, which have the identical shape and dimension to ones in the proving ground, are installed on the drum to evaluate impact harshness performance of a suspension module.

3 Evaluation Results

3.1 Vehicle Driving Test

A vehicle equipped with a torsion beam axle (TBA) type suspension on the rear axle is investigated in this study. As the TBA type suspension has a simple structure and there is only one pair of suspension bushings, it is easy to understand how it works and correlate parts, subsystems and vehicle characteristics. On the other hand, since there are not many design parameters to tweak, optimization of bushing characteristics is critical.

Ride comfort of a vehicle is evaluated by two maneuvers. Secondary ride is evaluated by root mean square (RMS) value of longitudinal acceleration level at front floor on the frequency range from 5 to 25 Hz while driving at 80 kph on a rough road. Impact harshness is evaluated by peak-to-peak change of longitudinal acceleration at the rear floor while passing two kinds of round-shape cleat at 40 kph. Two specs of bushings are used to analyze the effect of bushing characteristics. Bushing A and B have the identical shape and size, and only the shore hardness of elastomer in Bushing A is smaller than in Bushing B.

Evaluation results are shown in Table 1 and 2 in normalized value. While Bushing B (stiffer one) shows better Secondary Ride in rough road, contradictory results are observed in impact harshness. In case of small cleat passing test, Bushing A (softer one) shows lower peak to peak acceleration level, but Bushing B shows lower change in case of large cleat passing test.

3.2 Suspension Module Test

Suspension module of rear axle is tested in the same driving condition as a vehicle. Suspension module is attached on the fixture rigid enough to remove all the effect of inertial force and flexibility of a vehicle. As aforementioned, excitations which simulate rough road and cleats are given on the rolling tire patch by Comfort Roadway and Chassis Dyno. Since there's no body movement, forces transmitted through all the mounting points, which are measured by 3-axis load cells, are used to evaluate ride comfort. As main interests in this research are longitudinal vibrations of a vehicle, transmitted force in longitudinal directions through a bushing is investigated and the results from cleat passing and rough road driving maneuver are shown in Fig. 2 and 3 respectively.

Regardless of driving speed, less force is transmitted through softer Bushing A while passing a small cleat and large force through Bushing B on a large cleat. On rough road test, stiffer Bushing B transmitted less force up to 20 Hz, but Bushing A showed better results over 20 Hz frequency range. These results correlate well with

Table 1. Normalized RMS value of longitudinal acceleration at front and rear floor while driving on rough road.

Location	Bushing A	Bushing B
Front	1.02	1.01
Rear	1.00	0.99

Table 2. Normalized peak-to-peak value of longitudinal acceleration at rear floor while passing two types of cleats.

Obstacle	Bushing A	Bushing B
Small Cleat	1.00	1.03
Large Cleat	2.45	2.21

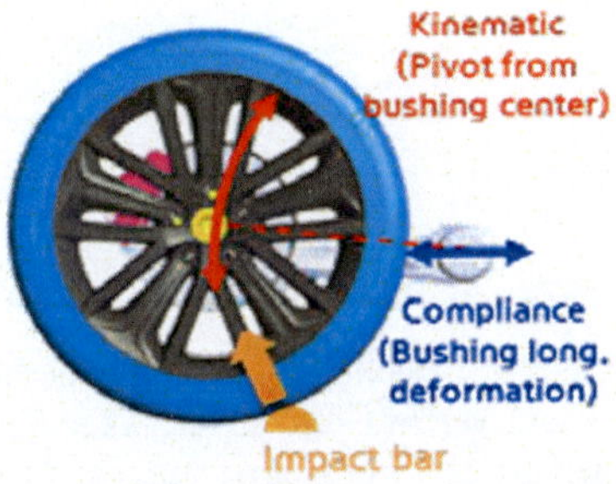

Fig. 4. Wheel motion mechanism in TBA suspension while passing a cleat.

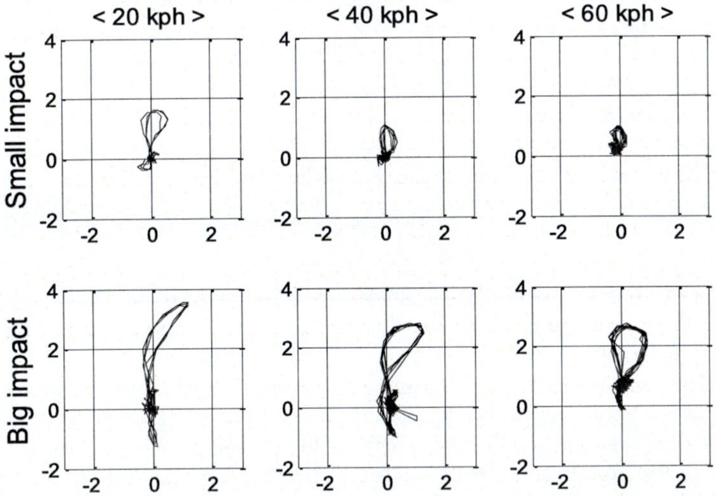

Fig. 5. Normalized wheel center displacement according to cleat size and driving speed

real vehicle driving test on the proving ground and show that suspension module test can be used to approximately predict the ride comfort performances of vehicle..

4 Characterization and Root Cause Analysis

Two kinds of further analysis are carried out to understand the mechanism how suspension transmits vibration to vehicle body and to find solutions which can improve ride comfort performance of a vehicle.

4.1 Wheel Motion Analysis while Passing a Cleat

Motion of a road wheel while passing a cleat is analyzed. A marker is attached on the wheel center, and behavior of suspension is recorded by high-speed camera in 200

Table 3. R value according to cleat size and driving speed.

Manuever	20 kph	40 kph	60 kph
Small cleat	2.45	2.08	1.71
Large cleat	3.13	2.73	2.33

frames per second (fps). Image processing software tracks the marker and gives its displacement in longitudinal and vertical directions. As shown in Fig. 4, motion of TBA type suspension comes mainly from two sources. One is kinematic motion which occurs by pivoting of whole axle structure from mounting point. Wheel center point follows the arc denoted by a red arrow. Since the initial location of wheel center is on the similar height to the pivoting point, the main component of its motions is vertical. The other is compliant motion which occurs by deformation of a bushing. In this case, most of the longitudinal force is transmitted through mounting bushing, and therefore the main component is longitudinal.

Wheel motions change according to driving speed and cleat size is shown in Fig. 5. To figure out which kind of motion is dominant for each test condition, a new index "R-value" is introduced as in eq. (1) and their values are shown in Table 3.

$$R = \frac{Vertical\ Displacement}{Longitudinal\ Displacement} \quad (1)$$

The R value is the ratio of vertical to longitudinal displacement, which means that the higher the R value, the greater vertical displacement. As the speed decreases and the cleat size increases, the R value increases, and it decreases as the speed increases and the cleat size decreases. This means that the vertical behavior due to kinematic motion is dominant in case of low speed and large cleat and the longitudinal behavior due to compliant motion is dominant in case of high speed and small cleat. It can also be seen that the impact time applied to the system increases in low-speed and large cleat conditions, resulting in relatively low-frequency excitation and vice versa. That is, as the excitation frequency to the system decreases, the R value increases, and as the excitation frequency increases, the R value decreases.

4.2 Transmissibility Analysis According to Bushing Stiffness

In principle, if the input from the road surface and the vibration transfer characteristics of all subsystems which consist of a vehicle are identified, it's possible to estimate its ride comfort performance under certain driving conditions. In the past, the biggest obstacle for this task was that input from the road surface and transmitted forces were hard to be measured. But the rig test facilities used in this research enables to overcome this limitation. Simulated road input of Comfort Roadway is described with vertical displacement and longitudinal speed of tire patch in time domain. And, load cells of suspension module test rig, which are installed at all the connection point to the vehicle body, makes it possible to measure transmitted force from suspension.

As a first step to build a new estimation process, vibration transfer characteristics of a suspension subsystem are analyzed and the effect of bushing properties on those are mainly focused on. Fig.6 is a conceptual diagram which shows how the vibration transmissibility is measured using suspension module test rig mounted on Comfort Raodway. The accelerations of the wheel center for vertical and longitudinal directions and the transmitted forces through the mounting bushing are measured while both tire patches, which are driven in 40 kph speed, are excited in vertical direction with swept sine (5-35 Hz) wave form.

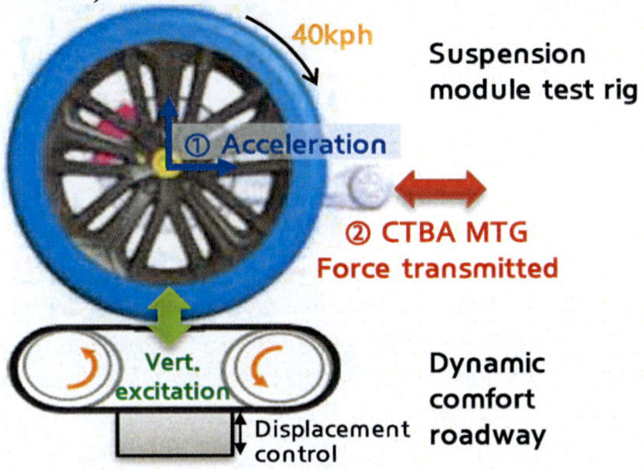

Fig. 6. Vibration transmissibility of a suspension measurement method using Comfort Roadway.

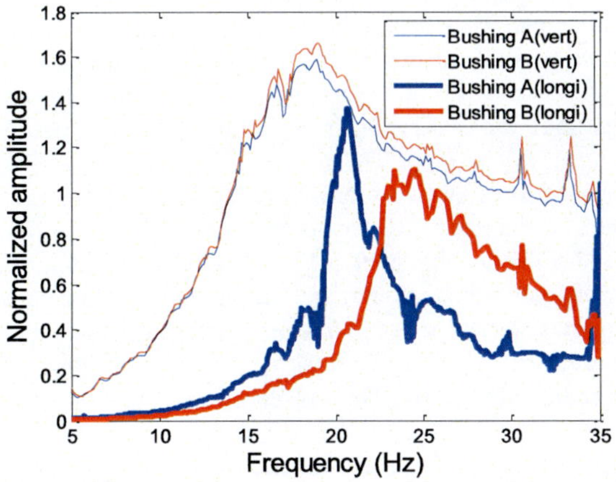

Fig. 7. Frequency Response of longitudinal and vertical acceleration of wheel carrier

Fig.7 shows the frequency response function (FRF) of the acceleration for vertical and longitudinal directions of the wheel center against vertical displacement of tire

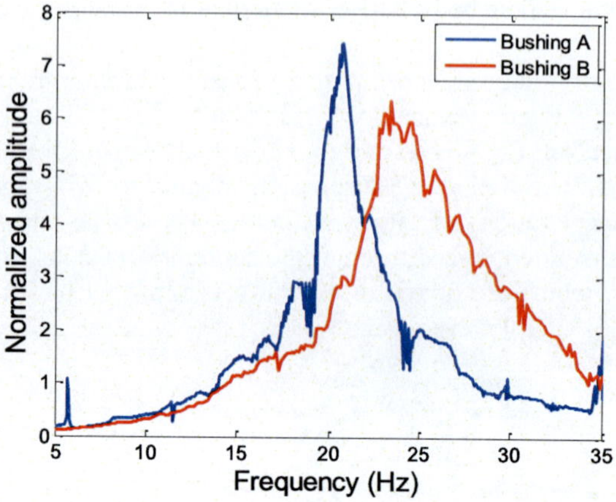

Fig. 8. Frequency Response of longitudinal forces transmitted through a mounting bushing.

patch. The vertical natural mode appears around 19 Hz regardless of bushing stiffness, it's because most of vertical load is dealt by spring and damper system and the bushing acts just as a pivot for kinematic motion, as explained in the previous section.

On the other hand, it can be seen that the characteristics of longitudinal direction vary depending on the bushing stiffness. The longitudinal mode of Bushing A is observed around 21 Hz, but Busing B around 24 Hz. This means that as the bushing stiffness for longitudinal direction increases, the longitudinal stiffness of the total system also increases, resulting in a higher natural frequency of vibration mode. Another interesting finding here is that even if the system is excited for only vertical direction without any longitudinal component, still quite large amount of longitudinal response, which is about 70 – 80% of vertical one, is seen. It's because longitudinal and vertical motion of TBA suspension is strongly coupled due to its kinematic constraint. In authors' other research, much smaller coupling between the vertical and longitudinal directions are observed for MacPherson Strut type and Multilink type suspensions.

FRF of transmitted force in longitudinal direction against tire patch displacement is shown in Fig. 8. The shape of FRF for force is quite similar to that for acceleration shown in Fig. 7. That tells us that inertial force from the vibration of unsprung mass is the main source of longitudinal excitation, and therefore it is essential to control the level of unsprung mass behavior in the frequency range of concern to improve vehicle vibration. Compared with Bushing B, Bushing A shows a bigger peak and transmits relatively higher load under 22 Hz. This correlates well with rough road test shown in Fig. 3.

It is the coupling effect of longitudinal and vertical motion that causes this phenomenon. In case of Bushing A, vertical mode (19 Hz) and longitudinal mode (21 Hz) is close. These two modes are superposed and makes bigger response around 20 Hz. On the other hand, Bushing B has relatively larger gap between vertical and longitudinal modes, which makes the response smaller under 20 Hz, and larger over 23 Hz.

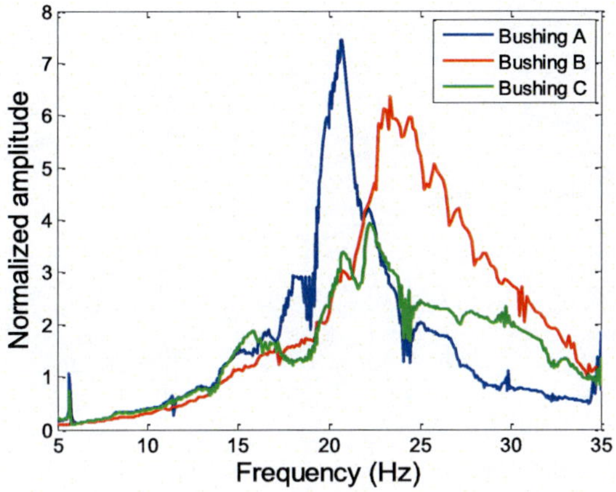

Fig. 9 Frequency Response of longitudinal forces transmitted through a mounting bushing. (Including the dual compound bushing result)

So we can conclude that stiffer bushing will show good performance in lower frequency excitation range and softer bushing in higher range.

This finding also applies for cleat test as well. As the driving speed is lower, and the cleat size is bigger, the excitation frequency gets higher. Considering this condition, contradictory results for impact harshness can be understood. In case of small cleat test shown in Fig. 2(b), Bushing A showed better performance regardless of driving speed. It seems because the main excitation frequency for small cleat is higer than 20 Hz. For large cleat test, on the other hand, Bushing B shows better results up to 50 kph and then the tendency is reversed. In this case, the main excitation frequency is under 20 Hz for low- and mid-range speed, and as speed gets higher, it goes over 20 Hz.

5 Solution and Verification

Using the findings and understandings from the previous section, new bushing properties to improve ride comfort are proposed. The target is so simple – "minimization of vibration transmissibility in whole frequency range of concern." To achieve this target, following two strategies are set.

Ⅰ. Separate vertical and longitudinal mode properly

Ⅱ. Increase Damping in longitudinal direction

With strategy I, mode superposition around 20 Hz can be prevented. However, as seen in Bushing B results, this causes bigger response in higher frequency range. To suppress the vibration in high frequency range, damping of whole system should be increased. Since the only vibration absorbing part TBA suspension in TBA suspension is mounting bushing, elastomer material needs to be changed.

It should be noted that this design change in mounting bushings can affect vehicle performance other than ride comfort. Stiff bushing with high damping has high dy-

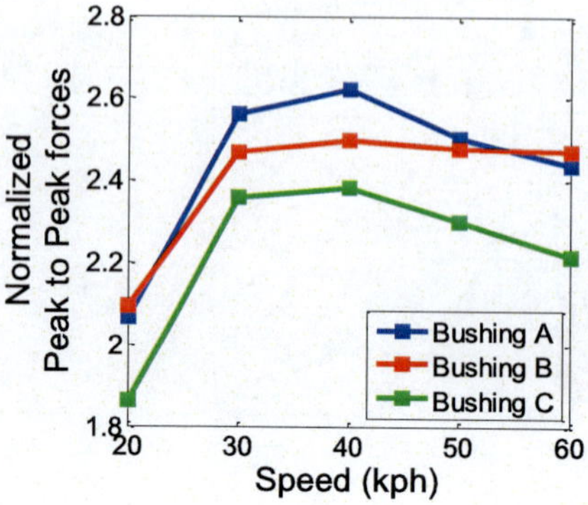

(a) big impact bar

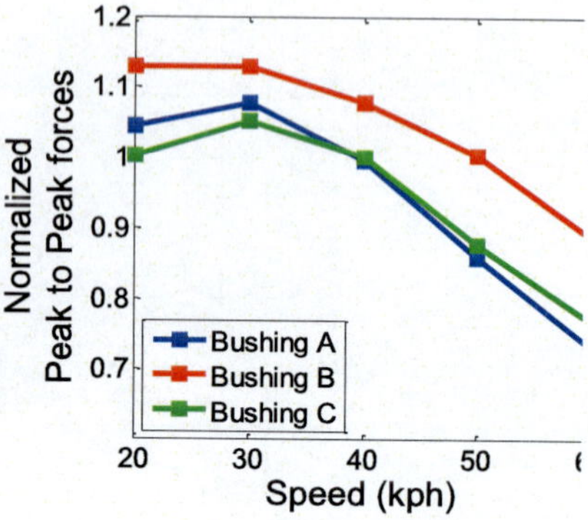

(b) small impact bar

Fig. 10. Impact results of longitudinal peak to peak forces at TBA bushing considering dual compound bushing.

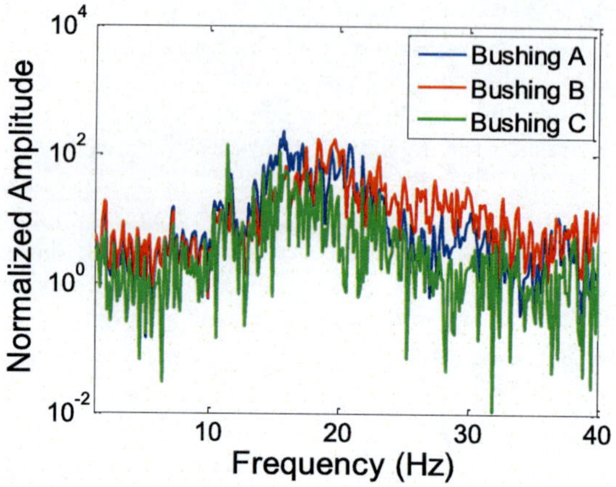

Fig. 11. TBA longitudinal transmitted forces on rough road at 80 kph considering dual compound bushing.

namic stiffness, and it can deteriorate road noise performance greatly. To solve this problem, dual compound bushing is newly designed.

Since it is well known that the vertical property dominates road noise performance, and what we want modify is the longitudinal property, two kinds of material are used for elastomer. For vertical direction, material with low hardness and low dynamic ratio, and for longitudinal direction, material with higher hardness and high damping are chosen for manufacturing.

The results from this new dual compound Bushing C is shown in Fig. 9, 10 and 11. As expected it shows best transmissibility performance in overall frequency range of concern, and in consequence, transmits the least force to body in all kinds of cleat test and rough road driving test. When it is tested on real car driving test, the same results are obtained, and it shows no disadvantage in NVH performance either.

6 Conclusions

In this study, ride comfort performance of TBA suspension subsystem is analyzed using vibration transmission characteristics and ride comfort tests, which are obtained by state-of-the-art rig test facilities. It is shown that ride comfort of a vehicle can be predicted only with suspension module test, and moreover, the further comprehension of the underlying physics of how suspension works, and how vehicle, suspension and bushings correlate. New bushing design and properties are proposed, based on the findings from suspension module test, and the improvement for ride comfort without deterioration of other performance is verified. This procedure can be used as base technology required for "Target Cascading" process. In the future, vibration characteristics of other subsystems such as vehicle body, powertrain and so on are measured, it is expected that the contribution of each system and interaction between them can be estimated accurately from the very early stage of development.

References

1. Sebastian Haberzettl, Alexander K Zschocke, and Frank Gauterin, "A new method for studying the longitudinal dynamic behaviour of a suspension on a test rig", Journal of Automobile Engineering, Vol. 230(8), pp. 1027-1039 (2016).
2. Mike DeLeeuw, Gerald Jung, and Werner Treusch, "Dynamic Suspension Test Rig Using Virtual Modeling", SAE Technical paper series, 2008 World Congress, Detroit, Michigan, April 14-17 (2008).
3. Yi, Y. –S., Jo, J., and Park, J., "Next Generation Vehicle Ride and Handling Performance Development Process Using Rig Test Facilities – Part 1) Infrastructure," Spring Conference of Korean Soc. Mech. Eng., Dyn. & Control Div., Seoul, Korea (2019).
4. Yi, Y. –S., Kwon, S. ,Choi, M., and Jung, J. "Next Generation Vehicle Ride and Handling Performance Development Process Using Rig Test Facilities – Part 2) Applications," Spring Conference of Korean Soc. Mech. Eng., Dyn. & Control Div., Seoul, Korea (2019).
5. Sell H., "Improvement of ride comfort – new ideas and confirmation by measurements," Chassis.tech plus, 2nd international Munich chassis symposium, Vol 2, pp. 1–16, Munich, Germany (2011).

Simulating and optimizing the dynamic chassis forces of the Audi e-tron

Stefan Uhlar[1]

[1] AUDI AG, Germany

Abstract. With battery electric vehicles (BEV), due to the absence of the combustion process, the rolling noise comes even more into play. The BEV technology also leads to different concepts of how to mount the electric engine in the car. Commonly, also applied with the Audi e-tron, the rear engine is mounted on a subframe, which again is connected to the body structure. This concept leads to a better insulation in the high frequency range, yet it bears some problems in designing the mounts for ride comfort (up to 20Hz) or body boom (up to 70Hz).

Commonly engine mounts are laid-out based on driving dynamics and driving comfort (up to 20Hz). The presentation proposes a new method to find an optimal mount design (concerning the stiffness) in order to reduce the dynamic chassis forces which are transferred to the body (20Hz – 70Hz). This directly comes along with a reduction of the sound pressure level for the 'body boom' phenomena.

We use multibody simulation along with a sophisticated tire model in the time domain in order to capture the correct excitation and structural dynamics behavior. Along with an optimizer, we rely on design of experiments (DOE) in order to find the optimal setup for engine-mount stiffness and therefore directly reduce the sound pressure level.

Keywords: Multibody dynamics, rear axle, bushings, NVH, CDTire, DoE.

1 Introduction

During the development of the Audi e-tron, the first fully electric vehicle developed by Audi AG, a lot of chassis issues were new and had no direct point of comparison with past experience to be drawn upon. Since the Audi e-tron shares its chassis platform with other MLBevo vehicles, the main question was how to introduce an electric drivetrain into the existing environment. One of the main issues in terms of vibration mechanics was the mounting of the electric drives. The concept of the Audi e-tron was to be an electric Quattro, meaning to install an electric engine on the front and the rear. Being

P. E. Pfeffer (ed.), *11th International Munich Chassis Symposium 2020*, Proceedings, https://doi.org/10.1007/978-3-662-63193-5_18

aware of the high-frequency excitation of an electric engine, in contrast to a conventional combustion engine, the rear drive-unit is mounted in the rear subframe. This leads to a better insulation in the high frequency range but it also comes along with a more sophisticated setup of the engine bushes, due to a strongly coupled mechanical system.

While dealing with this task during e-tron development, we relied on our existing design philosophy. This involves static analysis using defined load cases, and driving comfort simulation of the whole vehicle up to approx. 20Hz. First prototypes of the vehicle were equipped with the bushing setup, which was defined using this design philosophy. Later, the analysis of the prototypes showed a massive problem in terms of interior booming noise [1]. This revealed that our current virtual approach lacks of a description of this phenomenon and needs to be improved.

In order to alleviate this, we needed to enhance the design philosophy by a more sophisticated numerical model. The paper will give a detailed overview of the multibody model which was established along with a proper excitation. Employing design of experiments enabled us to understand the mechanical system more properly and to find the most promising setup by modifying the relevant chassis components (the final layout is always a tradeoff to all other disciplines such as driving dynamics, comfort and structure borne noise above 300Hz).

Booming noise can be excited by either the road or by the engine. Some references such as Schulze-Fehrenbach [3] focuses on the low-speed boom noise being excited by the combustion engine. Also concerning engine excitation Lee et. al [4] present a hybrid approach to the body boom phenomena. Focusing on the body structure and its influence on booming noise, Sung et. al [5] present a panel participation method.

In contrast to the approaches cited above, we focus specifically on road excitation and the structural dynamics of the chassis. We will presented a methodology of how to calculate the dynamic chassis forces and how to lower their amplitude level in order to reduce the interior booming noise due to road excitation.

The paper is structured as followed: Section 1 gives a more detailed overview of our current design philosophy. It will then be followed by showing some measurements on the first prototypes revealing the body boom problem. We will then present a new numerical method of how to accurately describe the effects of the body boom phenomena, induced by the chassis. A detailed description of the performed DOEs will follow and effectiveness of the new method will be confirmed by real measurements (Section 5). Finally, we will conclude and summarize our new approach in the final section.

2 Previous design philosophy of the chassis

Our previous philosophy considers a static calculation and the evaluation of the driving comfort of the whole vehicle.
For the static design, the engine mount geometry is of interest. This is directly derived from the available installation space and a specific mount travel. The mount characteristic (force-displacement curve) has a linear operating range and a progression range. The static load cases are divided into 'operating loads' and 'special loads'. All operating loads must lie within the linear range; special loads may run into progression. In addition to the design of the engine mounts, all other bushes of the axle (subframe, comfort mount of front axle) are also taken into account with the statics, as these are subject to additional load due to high drive torques (driving and recuperating).
Once the static stiffnesses have been determined, a whole vehicle model is set up using multi-body simulation (MBS). The MBS tool from MSC.ADAMS is used for this purpose. The numerical model for driving comfort is based directly on the real test bench. Figure 1 shows the acoustics test bench at AUDI (left) and the

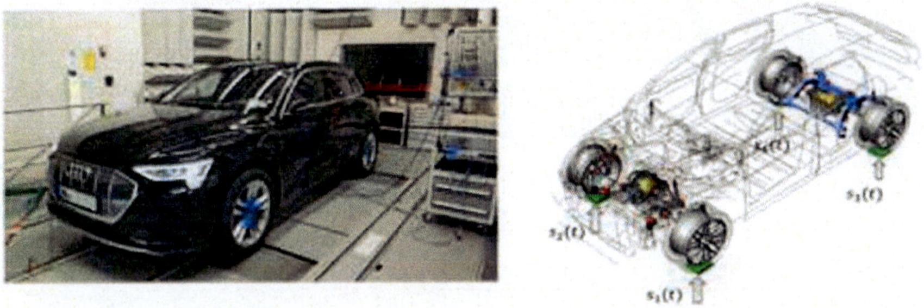

Fig. 1. Vibration comfort test bench, real (left) and in the multi-body simulation (right)

corresponding MBS model (the level of detail of the MBS model is directly linked to the frequency range of interest. For driving comfort simulation, a flexible body and a flexible rear subframe are mandatory). The vehicle stands on four posts and is excited by a synthetic signal at each post ($s_i(t)$); accelerations are measured at the seat rail and compared against a defined target value. This and other load cases, not described in detail, are used to evaluate the vibration comfort and again provide the opportunity to tune the mounting arrangement accordingly (in this case, the dynamic mount stiffnesses) to comply with the corresponding target values.
This design philosophy is currently a standard in every vehicle project, and applies before real vehicles are assembled. During development of the Audi e-tron, mounts were installed in the first real prototypes, which were defined using this design philosophy.

3 Analysis of the first prototypes

In addition to the vibration comfort test bench shown in Figure 1, the rough road test bench is crucial. On this test bench, the vehicle stands on four drums coated with a rough surface, and is accelerated based upon a velocity-time-function. Microphones in the passenger compartment at all four occupant positions measure the sound pressure level. Figure 2 shows the sound pressure level in dB(B) on the ordinate and the frequency on the abscissa. For this load case with a 'rough road acceleration', a target curve is defined for each vehicle segment, which must not be exceeded. The 30-50Hz range of the first e-tron prototypes clearly exceeds the target curve.

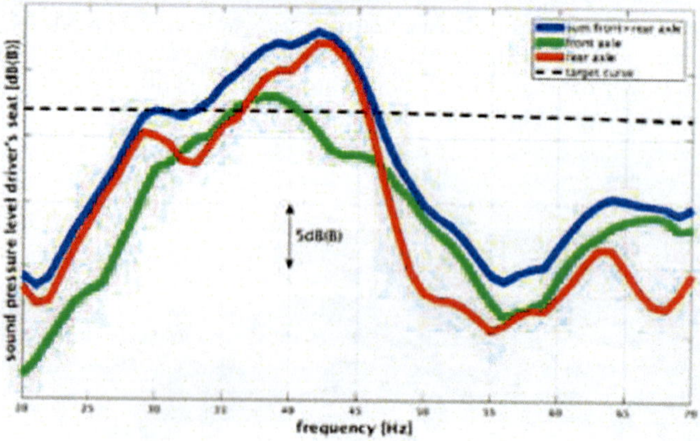

Fig. 2. Initial measurement of the sound pressure level at the driver's seat in first prototype.

As mentioned above, there are always two components and their interaction, involved in this phenomenon: The chassis and the body. Given the separation of function, the question is always which component is making what contribution to the final result. From the measurement in Figure 2, some initial suggestion is clear from looking at the contribution of the individual axles to the sound pressure level. It is obvious that the rear axle contributes the most in the frequency range of interest (Figure 2, red curve). A further measurement - this time the vehicle was excited by cleats mounted on the drum - confirms the measurement from the rough road test bench. The Audi e-tron shows a tendency for body boom. This again confirms that our current design philosophy lacks of an accurate description of this phenomenon and therefore has to be improved.

4 New design philosophy: reduction of the dynamic chassis forces

Multi-body simulation (MSC.ADAMS) is used in the acoustic department for chassis and drivetrain topics. The aim is to develop a numerical method that allows the existing problem to be simulated virtually for the first time, and to optimize the mechanical system employing design of experiments (DOE). The new methodology is to be used in the future, to develop a robust (towards axle parameters which change according to chassis adjustments; such as spring, dampers, stabilizer bar, drive shafts, position of brake caliper, tires and air pressure) axle system that minimizes the dynamic axle forces in a wide frequency range.

5 Numerical model

The test clearly showed that the body boom is also triggered by cleat impact excitation. As this is a deterministic excitation (in contrast to a rough road excitation), a MBS model is built up accordingly (Figure 3). For reasons of computing time and because the rear axle is mounted decoupled, in contrast to the front axle, the model consists only of this sub-level (this assumption is valid since the rear axle is decoupled. Simulations for the front axle must involve the whole vehicle model, since the front subframe is directly connected to the body structure). The rear axle is connected to a rigid body not displayed in Figure 3, which has only a vertical degree of freedom. The interaction of the tire with the axle plays an important role in terms of structural dynamics. The tire, as the link between road surface and axle, must not only adequately reproduce the transient excitation of the cleat impact, but also interacts strongly with the suspension due to its own vibration behavior. In an internal benchmark on tire models for structural dynamics (see [2]), the CDTire commercial tire model from Fraunhofer ITWM has proven to be very accurate, therefore we rely on this model. We received the corresponding data record for the 255/50 R20 tire from the tire supplier. All axle components are embedded as flexible bodies. The stiffnesses of the rubber mounts are added with their dynamic characteristic curves, and the hydraulic mounts of the subframe (Figure 4, HRV) are simulated non-linearly and dependent on amplitude using rheological models. The electric drive is a rigid-body model with kinematic connections (representing gear and differential).

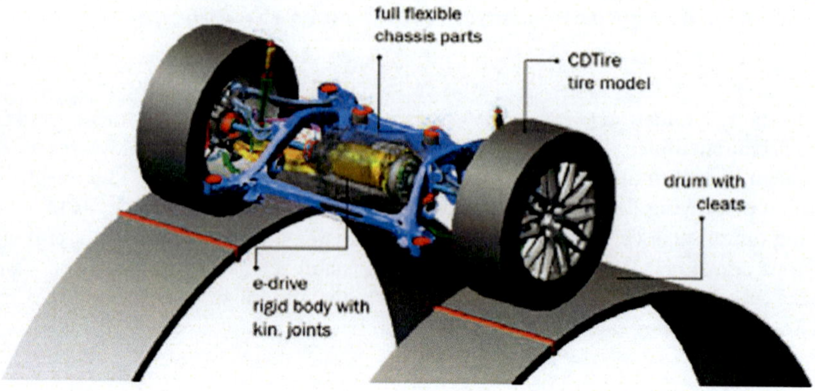

Fig. 3. MBS model of the Audi e-tron rear axle for calculating the dynamic axle forces.

The corresponding tire model also includes a numerical road. In this case, this is implemented as a drum with a circumference of 10m and a cleat of 8x20mm (height x width, Figure 3). The forces induced to the body are analyzed in all spatial directions (see Figure 4); i.e., the subframe connection points (HRV, HRH), as well as the force introduced by spring (FO) and damper (DO). The time signals of the forces (24 channels in total) are transformed into the frequency domain and visualized using a Python script (Figure 4 upper right).
The drum speed at which the greatest forces occur is determined in an initial simulation. The following DOE is then performed at this constant drum speed. Commonly the highest forces occur from 20 – 30km/h, above 30km/h the forces decrease.

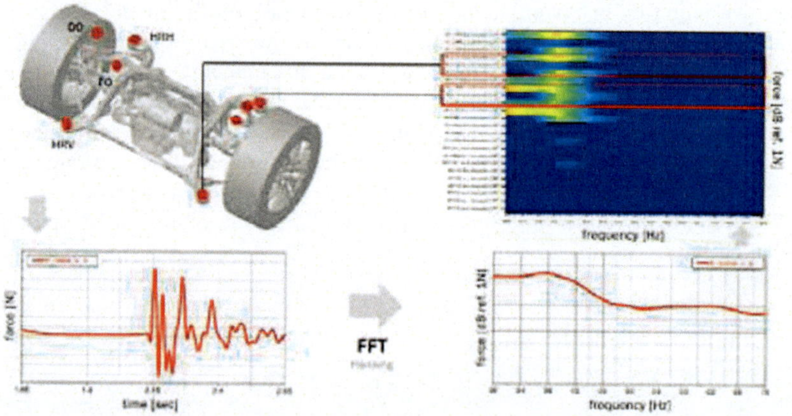

Fig. 4. Evaluation of the results, representation of all channels of the dynamic forces.

6 Design of experiments (DOE), optimization

As a total of 24 channels would have to be evaluated, it is very cumbersome and unnecessarily complicated for the DOE or optimization. Hence, as shown in Figure 5 (top left), an envelope is calculated across all forces (in all spatial directions) throughout the frequency range. This envelope force is then divided into frequency segments (30-40Hz, 40-50Hz and 50-70Hz) and the maximum in each case is monitored per segment. Three evaluation parameters are therefore available to the optimizer. The following input values are being modified in the DOE: stiffnesses of the subframe bushes and the engine mounts as well as the torsional stiffness of the drive shaft.
The DOE and optimization are performed using the OPTIMUS software from Noesis. Figure 5 (upper right) shows the corresponding workflow in OPTIMUS.
Despite the high level of model complexity, the calculation for a cleat impact is very straightforward and can be done in a few minutes (depending on the computing power). Hence, DOEs with thousands of experiments can be performed on a computer (multiple cores) in parallel within one day.

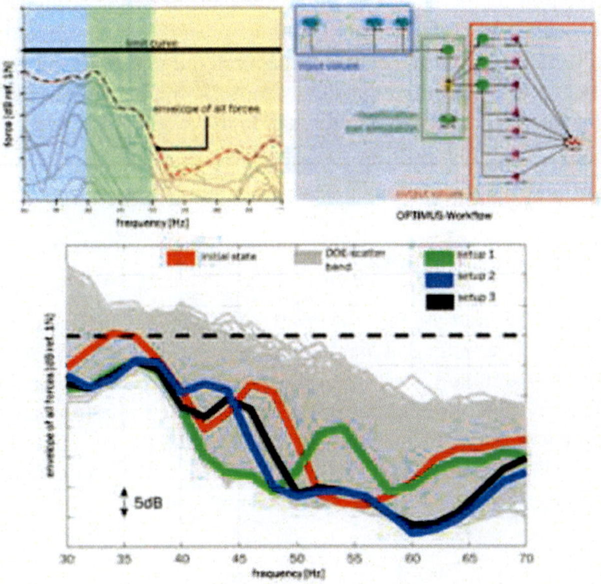

Fig. 5. Envelope of all dynamic axle forces (top left), Optimus workflow for DOE calculation (top right), result of DOE with best variants (bottom).

7 Results

One of the most important results of the DOE is the correlation of the input variables to the output variables. This reveals the parameters, which have the greatest influence on the result. Of course, these values are linked in a non-linear way, so it is not enough to modify just one value later on to get the best result. In the case of the Audi e-tron, thanks to the DOE, it is clear that the two front bushes of the electric drive have the greatest impact.

OPTIMUS can interpolate across all experiments of the DOE, and then can be used to generate the best initial value for the DOE before the actual optimization. In addition to the gray scatter band of all DOEs, the results of the best variants, which can be achieved with different mount setups, are shown in Figure 5 (bottom). You can see immediately, the dynamic forces can be reduced. However, one can also see that these are partly above the initial state in higher-frequency ranges. Therefore it is vital to know the sensitivities (defined in [Pa/N], at each connection point of the chassis to the body structure an impact hammer determines the corresponding value in each spatial direction. The sound pressure is measured in Pa at e.g. the driver's seat) of the body-structure, to be able to judge whether this bears a problem.

The most promising, virtually derived, variant was then installed in the vehicle as hardware, and another rough road acceleration test was carried out. The results of this test together with the initial state are shown in Figure 6. The reduction of forces directly leads to a reduction of the sound pressure level in the whole vehicle (Figure 6, red curve). The effect was also confirmed in all other prototypes, therefore these bushes are now installed in all series models. The optimization did not affect the driving dynamics nor the comfort in a negative way. As shown in Figure 6, the optimization was done using cleat impacts, but the effect is reflected on the rough road excitation as well, this was confirmed in multiple test runs.

8 Summary and outlook

The accurate description of the rear axle in the MBS model and the corresponding tire model reproduce the structural dynamics of the real component very well. Correct presentation of the excitation is essential (here, in the time domain on the cleat).

Given the relatively short computing times, it is possible to perform a DOE of all relevant axle parameters. This applies not only to the rubber mounts, but to all masses and stiffnesses, as well as entire components. The influence of all axle components and of the tire, including inflation pressure with different tire dimensions, can therefore be considered.

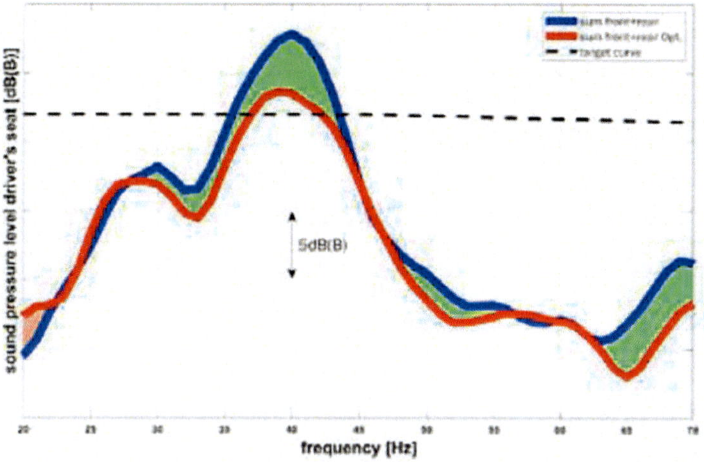

Fig. 6. Sound pressure level at the driver's seat before (blue) and after optimization (red).

The objective is to reduce the dynamic axle forces. Applying this new methodology, it is possible for the first time to evaluate the structural dynamics in the early phase, as soon as an axle design is available; and therefore to enhance the existing design philosophy by this important component.
The separation of function between axle and body also clearly shows the limits of each subsystem. The enhancement of the methodology to the frequency range up to 250Hz to evaluate noise transmission of the axle is currently in progress.

References

1. Brandstätter M., "Tieffrequente Geräusche in einem Kraftfahrzeug bei Unebenheitsanregung", PhD, TU-Berlin, 2013.
2. Uhlar S., Heyder F., König T., "Assessment of two physical tire models in relation to their NVH performance up to 300 Hz", 2019, Vehicle System Dynamics, doi:10.1080/00423114.2019.1681475.
3. Schulze-Fehrenbach D., Rumpel T., Spengler R., "Low-speed Boom Noise - Escalating Relevance According to CO2- Targets and High Torque Engines", 2012, ISNVH, doi:10.4271/2012-01-1547.
4. Lee D., Hwang W., Kim M., „Booming Noise Analysis in a Passenger Car Using a Hybrid-Integrated Approach", 2000, SAE World Congress Detroit, Michigan.
5. Shung S., Sung C., Hari L., Lee M., Structural-Acoustic Analysis of Vehicle Body Panel Participation to Interior Acoustic Boom Noise, 2011, SAE World Congress, doi:10.4271/2011-01-0496.

Lightweight Composite Spring for the new Transit Camper

Alberto Girelli Consolaro[1], Francesco Italiano[1], Paul Zandbergen[1], Nick Watson[2], Ian MacLachlan[3]

[1] Ford Research and Innovation Center Aachen, Germany,
[2] Ford Dunton Technical Centre, England,
[3] Ford Motor Company, Fairlane North 3, USA

Abstract. One challenge for the Ford chassis research team in the recent past was to investigate lightweight solutions for the upcoming Transit platform, for the end customer benefits of higher payload, reduced emission and better fuel consumption. As first application within the platform, the Camper / Motorhome was chosen. One of the promising items which was looked at were the rear longitudinal leaf springs. Replacing the standard steel springs with equivalent parts in composite material, generated a weight saving of about fifty percent, or about fifteen kilograms per vehicle. The spring had to be entirely redesigned, to obtain best in class performance, considering the new material properties and limitations. The project covered all the necessary steps, from the first research state to series implementation. Three important aspects which are discussed here are the design, the testing and finally the manufacturing process.

Keywords: Composite Spring, Leaf Spring

1 Introduction

To reduce the fuel consumption of future road vehicles, weight reduction is a key area of focus within the automotive industry. Therefore, chassis components made from fiber-reinforced plastics (composites) are becoming more and more popular relative to state-of-the-art components made from steel or aluminum.

Aeronautics and sporting goods industries have been following the path of composite technology for several years now, but today composites are increasingly used in the automotive industry, too. Interior parts, body panels and structures are currently most common applications. In the chassis area, the use of composites is still limited, at least in relation to structural parts. Components such as suspension links, wheel carriers or subframes have very challenging requirements in terms of strength, stiffness and durability. This limits the design freedom and makes a material substitution very difficult. In addition, the challenging requirements for structural chassis components often require carbon fiber reinforced plastics to be used, which increases part costs due to high material prices.

P. E. Pfeffer (ed.), *11th International Munich Chassis Symposium 2020*, Proceedings, https://doi.org/10.1007/978-3-662-63193-5_19

However, components where the implementation of composite materials can generate much better business cases are springs. The Ford Research and Innovation Center Aachen has been working on composite springs (longitudinal and lateral leaf springs as well as coil springs) for the last 15 years, focusing on design principles, materials, manufacturing processes, and testing as well as failure mode avoidance strategies.

For springs, glass fiber reinforced plastics are suitable to meet all functional requirements such as strain capabilities, stiffness and durability. The lower material price of glass fiber compared to carbon fiber supports the business case for various applications.

This paper focuses on the longitudinal rear spring which has been developed for part of the new Transit platform; the springs will be first applied to the Camper version; aside from fuel consumption benefits, the main reason for the choice of the application was to regain competitiveness against competition in terms of weight and payload.

There will be three main aspects discussed here: the design developed, the prototype parts with the tests performed and the manufacturing processes considered.

2 Design

2.1 Requirements

Longitudinal, wheel guiding springs are commonly used on the rear suspension of trucks and commercial vehicles, mainly because of their simplicity, robustness and cost effectiveness (See Fig. 1. Ford Transit Rear Suspension.

In most cases a leaf spring is connected to the vehicle body through a soft mount on the front end and by a shackle on the rear end. This is known as a Hotchkiss suspension. The function of the rear shackle is to allow for a length compensation under vertical spring travel; the leaf springs are then connected to a transverse solid axle by a center clamping system; the axle could be driven (for a Rear Wheel Drive or All Wheel Drive vehicle) or not (for a Front Wheel Drive vehicle); also, there are differences depending on how many leaves are used: single leaf or multi-leaves; in the first case, the vertical force generated by the spring will be linear to the wheel vertical travel; in case of a multi-leaves design instead, the spring reaction force will be progressive, meaning that the force increases non-linearly as the wheels move vertically upwards.

The vertical force generated over the vertical wheel travel (so called vertical stiffness) is one of the most important design parameters for the spring; every application has indeed its desired characteristic, mainly depending on the loading that is expected.

Next to vertical stiffness is the lateral stiffness, as it directly influences the steering and handling performance of the vehicle. It is important to notice here, that the system lateral stiffness, is a result of the different stiffnesses of the components (leaf spring, rubber bushings, metal brackets, etc.) and that is at the end the important parameter.

Safety has also an important role: as the springs control the wheel motion, a sudden failure of the spring could mean reduced vehicle control, which is certainly to be avoided. And safety is also guaranteed by the strength and durability requirements: the spring needs to maintain its function for the complete life of the vehicle, therefore it must be capable of withstanding repeated load cycles; which imply the robustness of the spring and its interfaces.

Last but not least, package and clearance requirements, can be sometimes difficult to meet because of the proximity of many other components (dampers, brackets, mounts, etc.) and because of the large movements of the suspension.

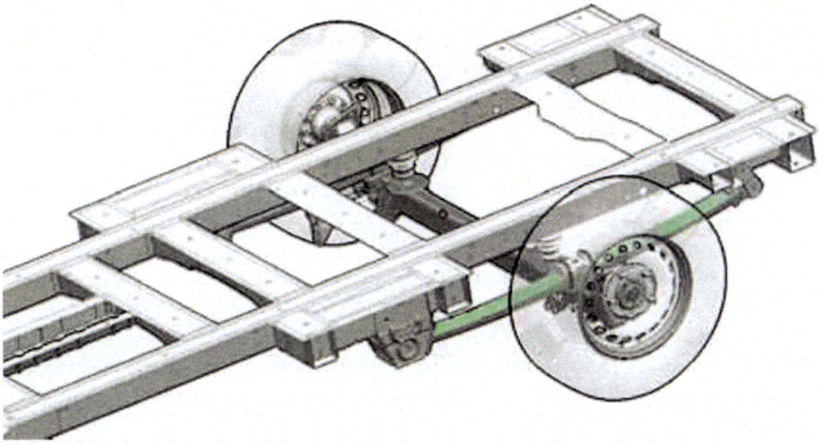

Fig. 1. Ford Transit Rear Suspension

2.2 Material Selection

The target set at the beginning of the project was to reduce the component weight; the density of the material was then one of the primary properties to dictate the material choice; but not the only one.

The material had to allow for large deformation, to allow the spring to bend with the suspension movements; it has to deliver the desired vertical and lateral stiffness, as well as being capable to withstand strength and fatigue loads; this can be achieved with a combination of material and optimum design of the part; the manufacturing process also plays an important role in the material selection as the quality of the material itself, and therefore its characteristics need to be guaranteed even when a large number of parts are produced at a high rate.

Cost is indeed another important factor and it includes raw material costs as well as process costs; if the material is more expensive than the reference material (steel), a standard metric to evaluate the proposed solution is the amount of money to be spent for each kilogram of mass reduction in the vehicle; the target for this metric varies depending on the application, production volumes, etc.

Regarding safety, a material with a safe and progressive "failure mode" is preferred to one where the failure occurs suddenly (e.g. snapping).

Other factors for the material choice to be considered include: the resistance to corrosion, heat, moisture and chemical agents (e.g. automotive fluids), resistance to creep (sag under constant load) and recyclability.

Composite materials (fiber reinforced plastics) are a good choice, as they can meet all the requirements above; for the Transit Camper project. E-glass fibers in Epoxy resin has been the preferred selection. The glass fibers are aligned to the spring main dimension; this means that the material cannot be treated as homogeneous, but the characteristics depends on the direction considered; the two directions, perpendicular to the fiber direction, will have equal properties, while the direction parallel to the fiber direction will have different properties; generally, the mechanical propertied in the fiber direction are higher than in the perpendicular directions.

The glass fiber content (in volume) can vary between 50% to 60%; with a lower fiber volume, the material would not have enough stiffness and strength; whereas with a higher fiber content the fatigue properties might be compromised. The density of the material is approximately 2 g/cm3, or about 25% of the density of steel. In terms of stiffness, the elastic modulus, in the fiber direction, is about 40 to 45 GPa, so about 20% of the elastic modulus of steel; which means that, compared to a steel spring, the geometry, and the cross section of the spring, has to change in order to deliver the required stiffness.

As springs are subjected to high deformations, an additional important parameter is the material allowable strain; the ultimate strain for glass fiber reinforced plastic can easily be above 2%; that is one of the differences for which glass fibers are preferred, for example, to carbon fibers.

In terms of cost, the absolute values for the materials depends on several factors (e.g. volume, manufacturing process, etc.); the composite material is generally more expensive than steel, but that can be, at least partially, offset by the reduced amount of material (mass) needed

2.3 Challenges and Solutions

This chapter will deal with some of the design solutions which guarantee, in combination with the material chosen, the fulfillment of all the targets and requirements described previously.

As already mentioned, to compensate for the lower elastic modulus of the glass fiber compared to steel, the cross section of a composite leaf spring will have to change; in particular, to obtain the same vertical (or bending) stiffness (*EI*), the height of the section (*h*) will need to increase; as the second moment of inertia is proportional to the height of the section on the third power, a small increase will be enough in most cases; also, in vertical direction, normally there are less package constrains and the height increase can be accommodated.

$$EI = E \frac{1}{12} h^3 w \tag{1}$$

In the perpendicular direction, there are similar considerations for lateral stiffness; in this case, increasing the width (*w*) of the spring, which can of course be done, but it brings more challenges, mainly related to package. Near the spring, there are typically other components such as the damper, stabilizer bar, brakes, and body rails; considering the total space occupied by the spring during its movement (mainly vertical) they quickly limit the maximum width of the spring that may be designed.

As stated before, the spring is connected to the vehicle body (on the front and, through the shackle on the rear) and to the suspension axle, through a center clamp (See Fig. 2.). In case of the Transit Camper, the center clamp is composed of a lower plate and an upper plate which hold the spring in position; the lower plate is welded on the axle, while two rounded U-bolts wrap around the upper plate and are tightened under the lower plate; a change in height of the spring can be relatively easily compensated for by using longer U-bolts; the other components do not need to change. on the other hand, if the width of the spring were to change, both plates and U-bolts would need to change also. That would result in additional weight (not only for the leaf spring but for the center clamp too) as well as additional costs: that is driven by the additional material and by the increased complexity in part numbers.
The leaf spring is only one of the several components in the system that determine the laterals stiffness; the rubber mounts and the body bracket have their contribution as well; which means, that the final degradation of lateral stiffness at a system level, will be less than the degradation measured or calculated at the spring.

Because of these considerations, with the agreement of the entire team involved in the project, in the case of the Transit Camper, it has been decided to keep the same width as the original steel spring, that is about 76mm.

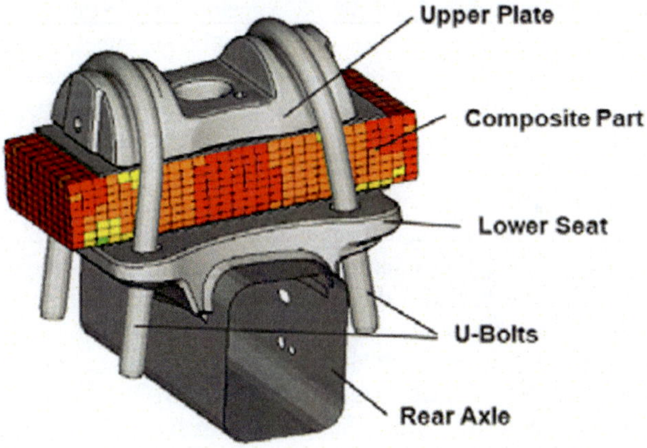

Fig. 2. Center Clamp group

Next, stress and strain must be verified and the design of the spring changed, if necessary; the center part of the spring is the most important for this regard; the system can be evaluated as a three-point-bending beam, and so the highest bending stresses are indeed at the load introduction point (axle), in the longitudinal (or fiber) direction, and they are linearly proportional to the spring height; which means that the highest tension and compression stresses are at the surface of the spring, respectively, top and bottom;

once the maximum allowable wheel travel is known, the maximums stress can be derived and compared to the existing targets; it should be noted that the choice of the target value needs to consider different factors; using the maximum material strength (or fatigue) at ambient condition, for example, would be too optimistic, as the vehicle will likely be used in more severe environmental conditions. Therefore, the target is adapted to the estimated worst-case operating scenario (e.g. high temperature and high humidity conditions); and finally, additional safety factors must be considered, resulting in decreased targets for maximum allowable stress, for example driven by the tolerances of material quality from the production line.

To estimate the main bending strain and stresses caused by the spring vertical movements a relatively simple model could be enough. On the other hand, in some areas, like the center clamp, much more refined modelling is necessary, because of the complex stress state; the torque applied to the U-bolts to secure the leaf spring to the axle, results in high compressive stress, perpendicular to the fiber direction; as this stress state is mainly counteracted by the matrix, one potential issue is material creep, a situation for which the material slowly deforms over time; because this phenomenon is mainly related to the matrix system, there is a relatively strong temperature dependency; a second effect of the compressive stresses is to weaken the fiber to matrix interfaces; this may result in fatigue issues, when in combination with cycling load on the fibers; in particular, on the lower side of the spring, where the fibers are subjected to compressive longitudinal stress, the so-called fiber buckling might occur.

As previously explained, the compressive stresses (or, the pressure) on the spring due to the clamp joint needs to be minimized; on the other hand, a minimum clamp force must be guaranteed; the best design principle is to have an homogeneous distribution of the pressure in the clamp area; this would be accomplished if the interface parts (upper and lower plate) were infinitely stiff and if the contact surfaces between the spring and clamping were perfect. In reality, this is not the case, but some improvement actions have been taken: the stiffness of the lower plate has been greatly improved through a better design and manufacturing process, while on both sides (top and bottom) of the spring, thin metal plates have been added, and attached to the spring through vulcanized rubber; with the help of the spring supplier, many iterations have been performed to find the optimum values for the layer thickness and rubber shore hardness, for the best trade-off between pressure distribution and minimalized creep behaviour.

Another aspect regarding the spring design and solution to be considered are the end fittings; on a conventional steel spring, the ends are generally forged; rolled to take the shape of rubber bushings; on a composite spring, this is not possible and a simpler solution with extruded aluminum brakes have been preferred. The connection between bracket and spring is obtained through two bolts and the fact that the spring has a slight increase in section near the ends (See Fig. 3. Composite Leaf Spring assembly.

3 Prototyping and Testing

3.1 Prototype parts

After a design for the spring and interfaces parts was prepared and checked in CAD and CAE, the next step in the development process was to produce and test prototype parts. The target for this phase is of course to verify that the requirements are met, but also to correlate the model used in the design phase. If there were any mismatch, there would then be the possibility to iterate and improve the model to increase correlation; this was hardly the case, as a good correlation was observed.

One important item is the accuracy of the prototype parts: accuracy in this case indicates how representative the prototypes are as compared to the final product; it is of course a trade-off between the speed and cost of producing parts and included design changes if necessary.

For the spring, for example, the manufacturing process and the material used was representative of what was planned for series production (pre-preg compression molding) but not exactly the same as production; more details of the manufacturing process will be discussed in the next chapter. Regarding the interfacing parts, most of them could be taken from existing (Transit) programs, with no or little adaptation.

Finally, for the end fittings, a machining process has been used instead of extrusion (followed by machining) as planned for series production.

Fig. 3. Composite Leaf Spring assembly

3.2 Testing

The development tests are divided into material, component and vehicle level; each of these groups gave important information which was then used to improve or confirm the design.

Material Tests. These tests focus on the composite material only; they are meant to check the quality and some mechanical or thermal characteristics.

Void content and fiber content are checked as they influence the material strength and durability, as well as stiffness characteristics. These are also verified through some mechanical tests at a coupon level, as the compressive or tensile tests; the Glass Transition Temperature and the Temperature Operational Envelope reveal how well the material, and in particular the resin system, function with different and extreme temperatures. Finally, aging, where the material is subjected to different agents before being mechanically tested, show the resistance to materials which are potentially harmful factors (UV light, engine or brake fluids, etc.)

Component Tests. These tests focus on the leaf spring itself; as the interfaces are also very important because they often represent the weakest link in the system, representative interfaces have always been used for these tests.

The leaf springs have first been tested for geometry, stiffness and load characteristics, as they are also the main design input.

Fatigue is tested at the spring level, with a vertical test; the input to the spring, provided by an actuator to the center clamp, using cyclical loads, with different amplitude and duration. this is a simplification compared to what happens in the vehicle, but it represents the main stresses and it offers significant advantages, being a fast and rather simple test; thanks to this characteristic, this test can be more easily repeated several times, for statistical evidence. Moreover, as the test set up is quite compact, it is possible to include the complete system and actuators in the climate chamber and test at different ambient conditions (high temperature, low temperature, high humidity, etc.); in the case of the Transit Camper spring, additional tests with springs damaged by stone impacts have been conducted also.

To bridge the gap between the simplification of the vertical testing and the rear vehicle testing, a system level (half vehicle) test was also performed, using measured road load data as input.

Creep was also tested at component level; in this case, the spring is loaded with a constant load and the change in reaction force was measured; this is to test and prevent that the vehicle sags over time. For the creep test, different climate conditions were used also (high temperature and humidity).

One of the challenges during the development of the Transit Camper spring was to define acceptance criteria for these tests. This is because most of the tests were specifically done for the composite material, and therefore there was no existing target which could be used; then, the relation between component and vehicle testing is not straightforward; to solve this problem, most of the tests have continued until failure, to understand the safety margin compared to the assumed target and the actual failure.

To evaluate the test results, two criteria have been used, functional and appearance; the functional acceptance criteria are based on stiffness and load degradation while the appearance criteria were based on the appearance of the part during and at completion of the tests.

Vehicle Tests. The last series of tests were performed in the actual vehicle; these tests are usually completed near the end of the development process because they are generally more time consuming than the component tests. If on a test rig, a mix of new and old parts can be assembled relatively quickly, it takes longer to properly assemble them on a vehicle, also considering the safety standard that needs to be guaranteed.

The quasi static characteristics can be checked on a special rig (Kinematic and Compliance rig) where the vehicle is positioned, and different loads and moments are applied to the wheels; this is to verify that the stiffnesses are on target and that all the movements are evaluated (no package conflicts).

The vehicle durability test is a long and severe test, with several event repeated multiple times to stress the components on the vehicle. The test definition and acceptance criteria are fixed by the Transit program; this test is used as final confirmation of the robustness of the system.

Finally, a lot of vehicle attribute tests have been performed also; they include steering, ride and handling tests; in most cases, they have been performed as back to back comparison with an equivalent steel spring vehicle, in order to better judge performances.

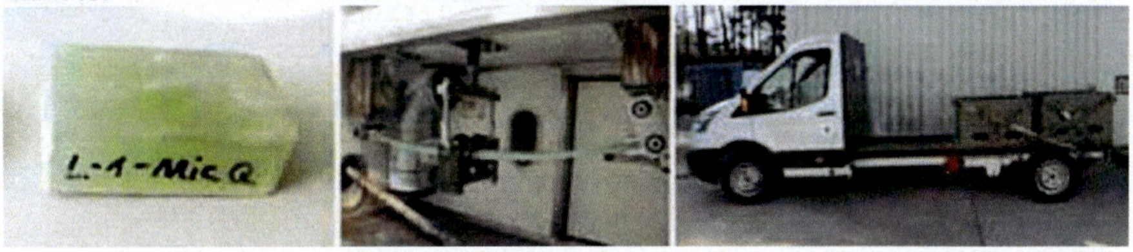

Fig. 4. Tests at Material, Component and Vehicle level

4 Manufacturing Process

Due to its rather simple design, there are several possible manufacturing processes which can be used to produce a composite leaf spring; Resin Transfer Molding (RTM), Pre-impregnated compression molding (Pre-Preg), Filament winding, Tow-preg winding, Pultrusion are some of them.

Each of them offers some benefits and some disadvantages compared to the others: they can be rated according to different metrics; the most important being quality and robustness, part performance, cost (both variable and fixed) and speed; some of them are related to each other, and the balance between them can be tuned and optimized within the same process.

In the recent years, as composite springs became more and more popular, the first two listed processes, namely RTM and Pre-Preg, have become the most widely used.

The production of the spring is done by a Ford supplier, who is then completely responsible for it. For the Transit program, the choice of the process to use, jointly done, is Pre-Preg compression molding. Without listing all the differences, advantages and disadvantages, the main factor for which the Pre-preg process was chosen lies in the cost, especially in the split between part (and material) and tooling cost; generally speaking, tools for an RTM line are more expensive than the ones needed for a Pre-Preg line. On the other hand, the piece price is normally lower for RTM. Therefore, RTM is the preference for very high-volume production, and Pre-Preg for rather low volume production.

In the Pre-Preg process, the material is already "pre-impregnated"; this means that the glass fibers are already within layer(s) of partially cured epoxy resin; the material, is placed in a sandwich of protective layers and it is generally delivered in rolls, which must be kept in controlled temperature to preserve material integrity.

Each layer of the material is cut from the roll and placed on top of the previous layer until the spring pre-form is complete. The shape, number and sequence of layers is determined before-hand based on the design and computer program of the supplier; once the pre-form is complete, it is placed in the mold which goes into the press, where pressure and temperature profiles are applied. The tool may have multiple cavities, depending on desired volume output, press characteristics, etc.

After the press operation, the springs are machined to their final shape; afterwards, the springs are placed in an oven for the post-curing step, to guarantee that the material reaches the desired degree of cure. After the spring is out of the oven and cooled, the other parts of the assembly (end fittings, bushings, central clamp plates) are mounted to the composite part. The last step of the process are the checks on the part, for the important mechanical properties (e.g. stiffness and load).

For the prototype phase, an extra requirement to the manufacturing process and material is that they should be as close as possible to what will be used for series production; that is to guarantee the trustworthiness of the test results and to avoid the need to repeat development tests during the late phases of the project; the changes and improvements made in the process during the project, should mainly be focused on scaling up the production from a few prototypes to large volume.

5 Conclusions

This paper describes, at a high level, the approach followed to develop a composite leaf spring for the new Transit Camper application; it describes the main considerations used for the design of the parts, as well as for the testing program and the manufacturing process.

With the use of composite material, a significant weight saved can be achieved on each vehicle; that directly improves fuel consumption, reduces CO2 emissions and results in higher payload for the user; this last characteristic is what made this technology especially interesting for the Transit Camper application.

The benefits of the composite springs are not limited to weight saving though, but include better durability performances, corrosion resistance and noise damping.

The importance of the project is not only related to this single implementation however but lies in the fact that it opened the way for more applications: even with sometimes different and unique requirements, leaf spring suspensions are used within Ford Motor Company in several vehicles, mainly light commercial vehicles and trucks.

A larger implementation volume will also improve what, at the moment, is the main disadvantage of the composite materials, namely the higher cost as compared to an equivalent steel part.

Design of an innovative car wheel suspension concept with a fiber-reinforced plastic spring/control arm

Dipl.-Ing. Ludwig Seethaler

BMW AG, Germany

Abstract:
Batteries and drive train components in future concepts for electrified vehicles require a significant amount of space. Because of their additional cost and weight penalty it is necessary for the chassis components to claim as little space as possible and contribute to the overall cost and weight savings of the whole vehicle. To achieve these goals a suspension concept with a reduced number of parts is introduced. By functional integration only a single part made of fiber-reinforced plastic is needed to support the wheel load and provide wheel guidance. This spring/control arm and the shock absorber are the sole means to control wheel travel instead of several control arms and bushings. The required space for the suspension system is minimized without affecting vehicle dynamics and ride comfort.
Test rig and misuse testing results of a prototype front axle are discussed, and simulation and road test results of a rear axle concept in a prototype vehicle are presented. Finally, possibilities of further cost and weight reduction are suggested.

Keywords: rear axle concept, FRP spring/control arm, functional integration of wheel support/wheel guidance.

1 Motivation, basic suspension concept

Wheel suspensions in passenger cars have been improved for decades from one generation to the next. Two main development directions can be discerned: increasing functionality resulting in complex double wishbone front axles and multi-link rear axles in upper class vehicles and on the other hand cost-optimized suspensions like strut front axles combined with H-arm rear axles for usually front wheel driven compact cars.

Now electric mobility comes into play: the most expensive part of a battery electric vehicle today is the high voltage storage (HVS) and probably will be for a considerable time, especially in compact cars. Cost of high voltage storage is closely related to its geometry: simple boxy designs are less expensive than complex geometries. Thus in order to optimize the available space for the HVS and achieve sufficient electric range it is crucial for other vehicle components like suspensions to require as little space as

P. E. Pfeffer (ed.), *11th International Munich Chassis Symposium 2020*,
Proceedings, https://doi.org/10.1007/978-3-662-63193-5_20

possible. Additionally, the usual pressure to cut costs and save weight is further increased by the considerable cost and weight penalties of a large high voltage storage.

To achieve these goals coming from well proven concepts like strut or H-frame suspensions the only feasible way seems to be a reduction of parts number by function integration. A similar approach has been described before [1]. The here proposed axle concept combines the wheel load support and the wheel guiding function into a single part made of fiber-reinforced plastic (FRP) referred to as spring/control arm. The FRP part together with a wheel guiding shock absorber comprises the rear suspension, there are no other links involved. This fact constitutes the difference to the state of the art. On the front axle there is necessarily a conventional track rod and a steering gear.

During preliminary considerations two conflicting targets were identified as a main issue: on the one hand the single FRP part has to be flexible enough to allow sufficient vertical wheel travel but on the other hand it has to provide adequate rigidity to bear horizontal loads and external moments. A third requirement for the compliance behavior of the FRP part are the usual elastokinematic goals represented by toe in and wheel camber characteristics under longitudinal, lateral and vertical loads during vehicle operation. The typical design process, starting with kinematic considerations, is a different one here: due to the deformable design of the spring/control arm there is no reasonable rigid body kinematic analysis [2] any more, forces and deformations have to be taken into account from the beginning.

A two-step approach was chosen:

- Testing a front axle concept on the test rig and in a vehicle misuse test to find out if the existing target conflict could be solved (the reason for choosing a front axle being the higher misuse loads on this axle)
- Road test of a rear axle concept in a prototype vehicle evaluating driving dynamics, ride comfort and NVH (the reason for choosing a rear axle being the more demanding compliance behavior of this axle)

2 Front axle concept

With respect to front wheel driven cars the front axle concept was designed to accommodate drive shafts and a final drive gearbox. Because vertical space is very limited in order to keep the bonnet line and the center of gravity low, no carry-through concept like a transverse leaf spring was possible. For this reason two separate spring/control arms connect the wheel carriers with the car body leaving space for the final drive gear box between the wheels near the center of the vehicle. The inner ends of the spring/control arms are rigidly clamped to the body while the outer ends are pivotably attached to the wheel carrier with rubber bushings to provide sufficient free length for deflection.

A split wheel carrier was chosen (revo knuckle principle [3]) to separate wheel travel (vertical movement) and steering angle (rotational movement). The spring/control arm participates in this way on the vertical movement alone. An additional benefit of the split wheel carrier is a reduced traction force radius for minimal interference of a traction force with the steering. The shock absorber is rigidly clamped with the non-steerable part of the wheel carrier and has a wheel guiding function like in a traditional strut front axle. Due to the fact that the top mount is not related to the kingpin axis a low position of shock absorber and top mount is possible. A conventional track rod connects the steerable part of the wheel carrier to the rack and pinion steering gear (Fig. 1).

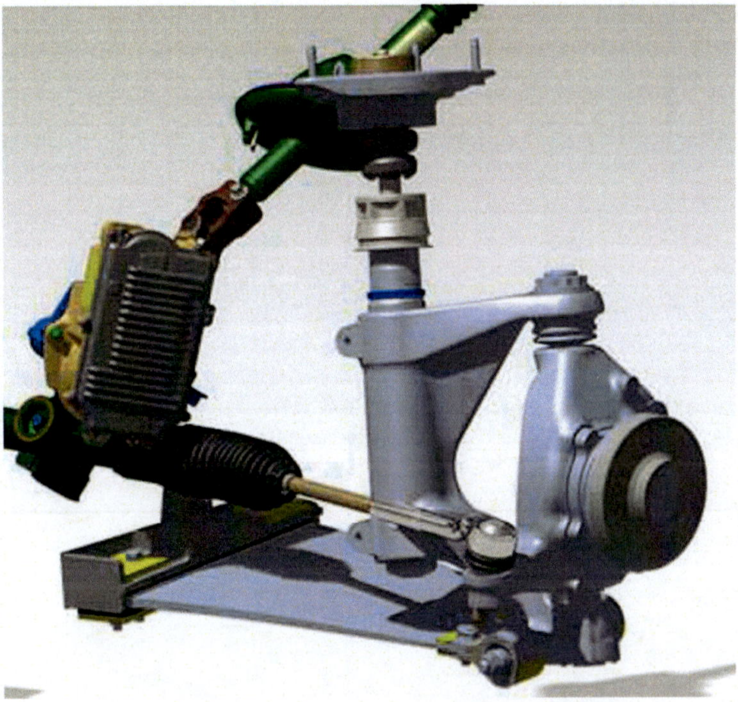

Fig. 1. Front axle concept with spring/control arm and split wheel carrier

2.1 Test rig run

The front axle was tested on a 4-component test rig with 3 translational forces and a braking torque, simulated with hydraulic cylinders (Fig. 2). The applied load collective was the original one from series development. A first test rig run was conducted at room temperature, a second one with changing temperatures (20°…60°C) and intermittent irrigation with a corrosive medium (5% NaCl-solution).

The result was a minor delamination of the bottom layer of the spring/control arm with no effect on the spring rate at the beginning of the test run. There was no apparent failure propagation over the next test rig operating hours. After approximately 33% of the running time required for series release a clash occurred between the outer bushing and the wheel carrier with a resulting failure of the clamping part and subsequently of the spring/control arm. Main reason for the clash was insufficient clearance of the conflicting parts with regard to the substantial axial travel of the rubber bushings. The second test run with temperature and corrosion exposure led to similar results.

The test results can be interpreted as not suitable for a series release (which was not intended) but appropriate to answer the question if the target conflict between rigidity and flexibility can be solved: obviously this seems to be possible.

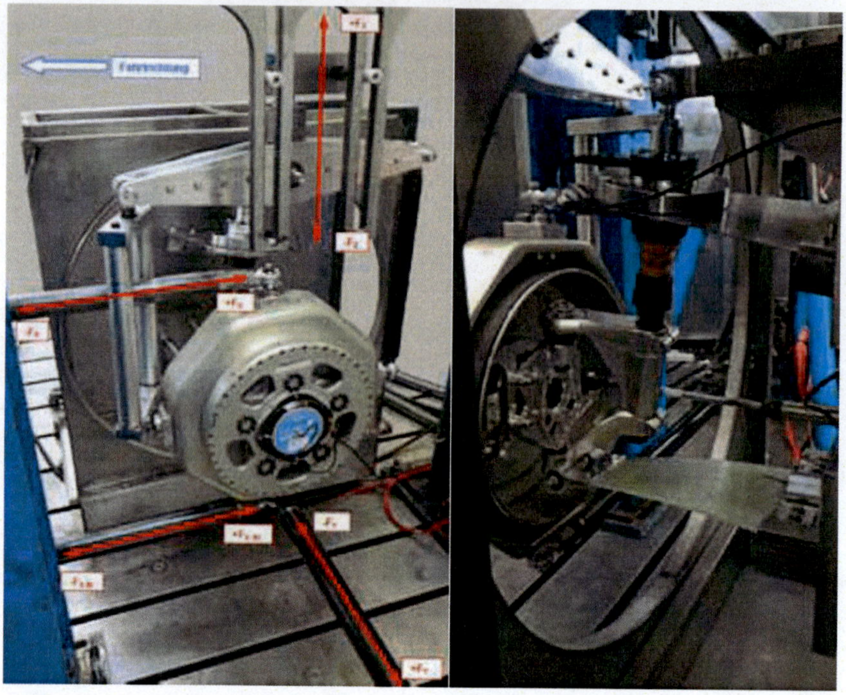

Fig. 2. 4-component front axle test rig

2.2 Misuse testing

An experimental vehicle was equipped with a spring/control arm front axle to evaluate misuse behavior. Because the drive train was not operational, the misuse test was conducted by towing the vehicle over specified obstacles as defined in a standardized OEM test catalog, a carry-over from the series release process. A positive and a negative obstacle (bump and pothole) has to be passed with different speeds (40 and 60 km/h) representing a special event and a misuse case. At the low speed pass no damage is acceptable, the high speed pass may lead to minor damage but the vehicle must remain under control. The pothole high speed pass represents the load case with the highest vertical wheel loads superimposed by longitudinal loads which is the reason why this test was conducted (Fig. 3).

Fig. 3. Misuse test of the front axle with an experimental vehicle

The main question was whether the two test stages could be passed and how the FRP part's behavior would look like – sudden complete failure or minor crack with no loss of functionality. After the first pass with 40 km/h no damage occurred at all. The second pass with 60 km/h caused a failure of a rear axle link, a prototype part which was not a subject of the test. The front axle components were closely inspected after the test; some minor cracks on welding seams of the prototype front axle carrier were detected, but no damage on the FRP spring/control arm. The test has been passed, but we still have no detailed knowledge about the failure behavior in misuse cases.

3 Rear axle concept

To evaluate driving dynamics, ride comfort and NVH, a prototype vehicle was equipped with a rear axle comprising the FRP spring/control arm. Starting point was a stock BMW i3, a rear wheel driven battery electric vehicle. Based on similar considerations as for the front axle no transverse leaf spring concept was selected, but with respect to the final gear package, ground clearance and center of gravity height two separate spring/control arms were implemented.

3.1 Prototype vehicle

The standard rear axle, a 5-link-suspension, was removed and transverse beams made of milled aluminum were installed to connect the inner clamping parts to the vehicle body. A wheel guiding shock absorber was attached to the wheel carrier and the top mount with an adapter part to bridge the distance from the original to the new top mount position. Wheel carrier and adapter are also made from milled aluminum. Possibilities to adjust toe and camber angles are provided at the wheel carrier and the top mount adapter. The elimination of 5 links per side with bushings and bolts as well as the helical steel spring provides new available space in the rear end package (Fig. 4).

Fig. 4. Rear axle concept and prototype vehicle

Compared with the series rear axle concept of a compact car the spring/control arm rear axle allows a different solution for the high voltage storage package: instead of 2 separate volumes a one-box design of the HVS is possible with considerable effects on cost, available volume and electrical range (Fig. 5).

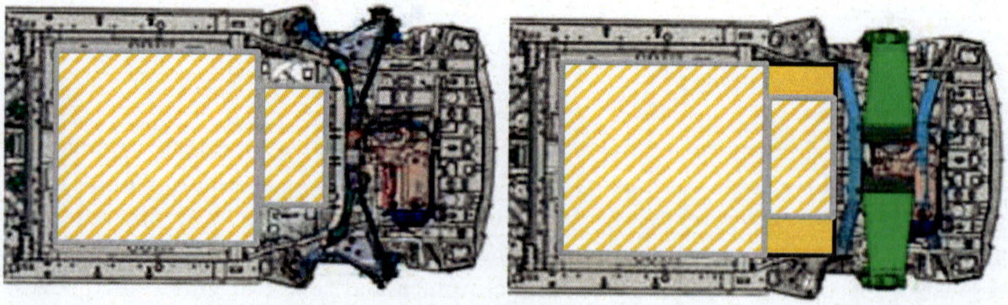

Fig. 5. Comparison of available space for high voltage storage in an electrified compact car (left: series rear axle, right: spring/control arm rear axle)

A direct cost and weight comparison to the standard axle is not very useful due to the different stages of development: a prototype solution with milled aluminum parts on the one hand and a production rear axle refined over several model generations on the other hand. A major influence on cost and weight are the interface parts between the spring/control arm and the vehicle body because of the support of both forces and moments. The realized inner clamping parts are split in two halves and avoid any through-holes in the FRP part, the connecting bolts and nuts are arranged beside and behind the spring/control arm. This solution provides an undisturbed cross section of the FRP part but leads to rather massive clamping parts.

A projection of cost and weight for a possible series solution of the spring/control arm concept leads to numbers roughly comparable with the stock rear axle: slight but more reliable advantages in weight combined with a considerable but fuzzy potential in cost.

3.2 KnC testing

The prototype vehicle equipped with the spring/control arm rear axle was tested on the kinematics and compliance (KnC) test bench, where wheel position (toe in, camber) is recorded during wheel travel and under longitudinal as well as lateral loads.

The results of vertical test (Fig. 6, 7), longitudinal (Fig. 8, 9) and lateral force test (Fig. 10, 11) show the toe in and camber changes in relation to the static values (red: right side, green: left side).

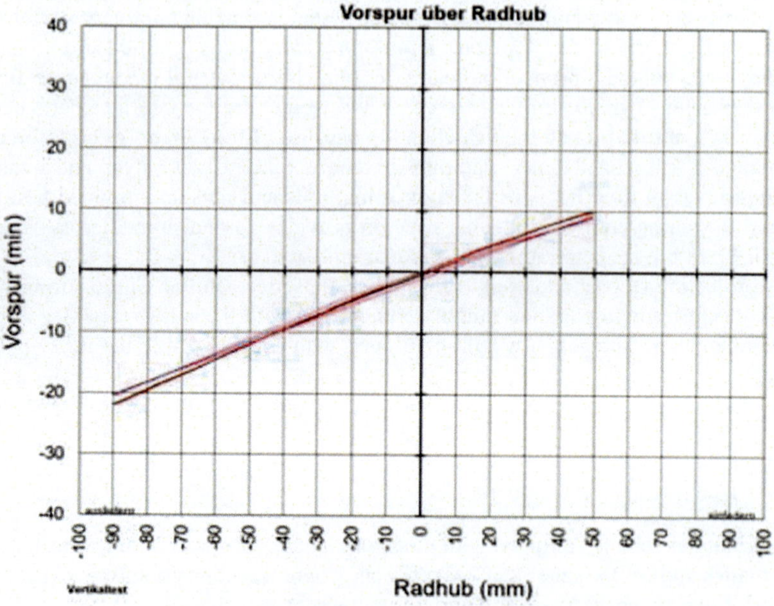

Fig. 6. Toe change over vertical wheel travel

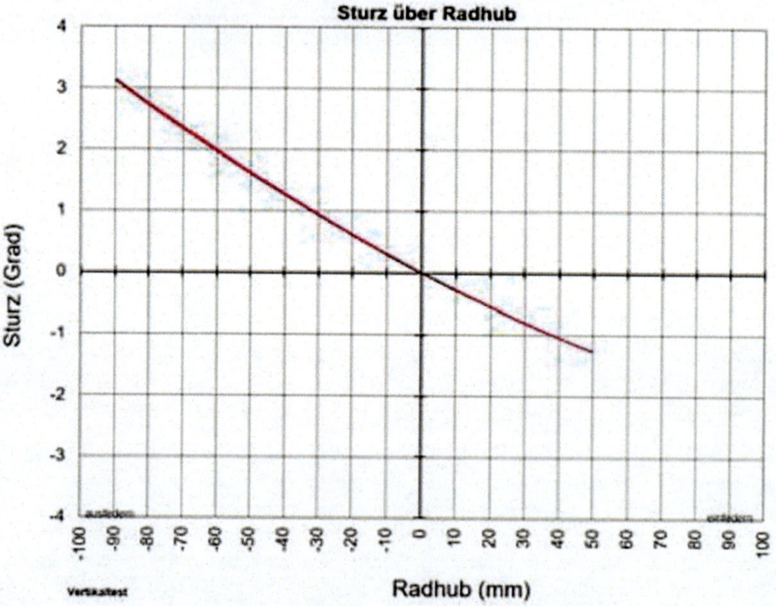

Fig. 7. Camber change over vertical wheel travel

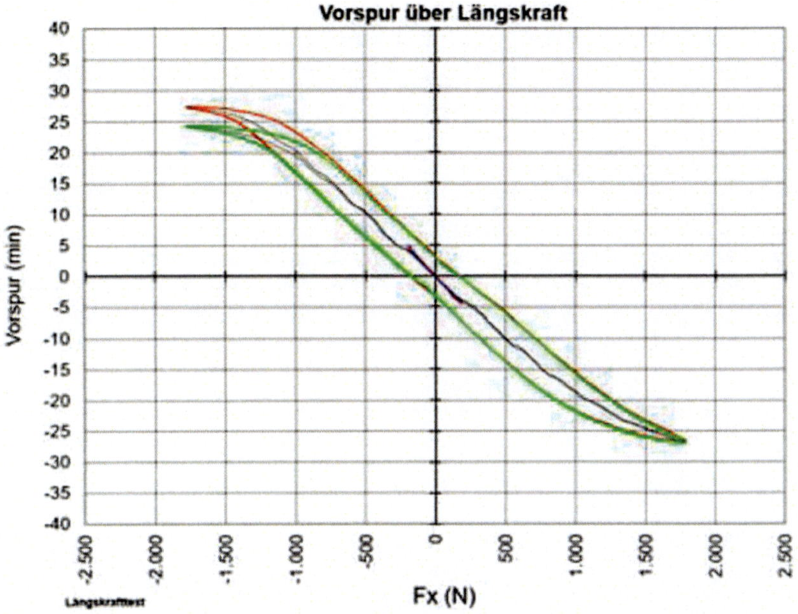

Fig. 8. Toe change over traction force

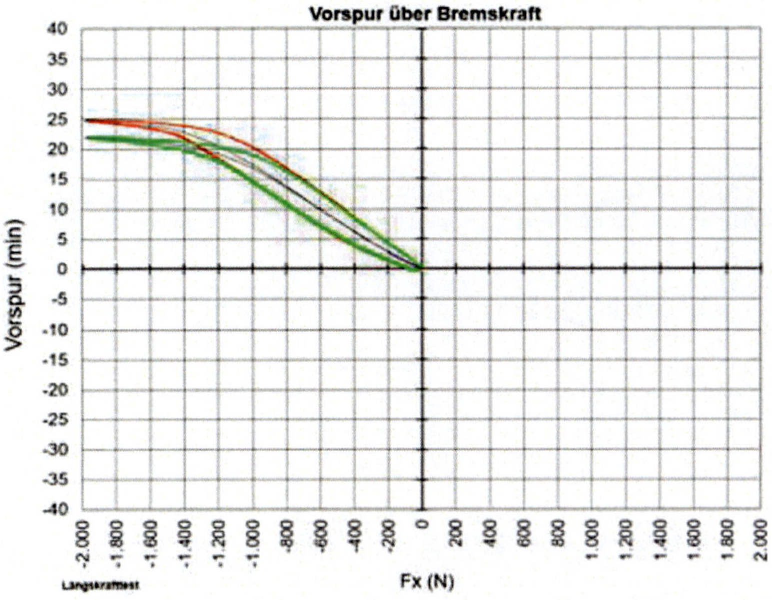

Fig. 9. Toe change over braking force

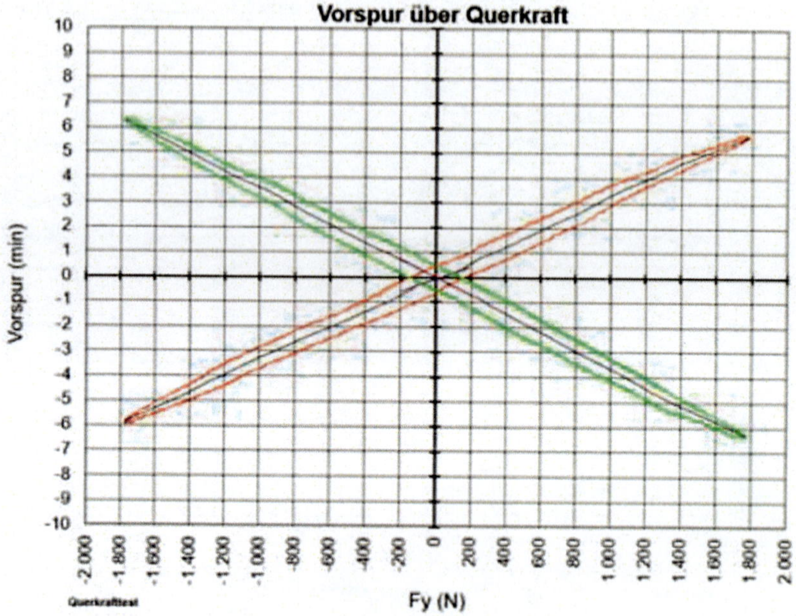

Fig. 10. Toe change over lateral force

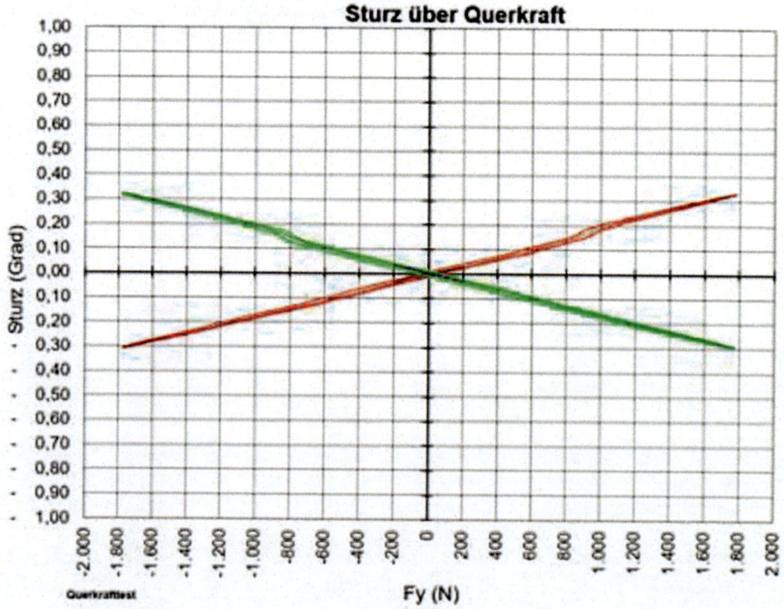

Fig. 11. Camber change over lateral force

Compared with the standard rear axle the spring/control arm rear axle shows similar characteristics, especially in bump and rebound. The understeering toe changes in bump and under longitudinal forces are more pronounced whereas under lateral load there is slightly less increment in toe in.

3.3 Road test

The road performance of the prototype vehicle with spring/control arm rear axle was evaluated on a restricted proving ground with high speed track, skidpad, handling track and several uneven surfaces with different qualities.

The findings of an experienced test driver with ample background in chassis and suspension series development were as follows:

- Stable vehicle behavior up to lateral acceleration limits, no reaction on power off/power on or braking during cornering, slight two-phase reaction on high-speed lane changes due to more pronounced understeer
- Significant improvement in ride comfort driving over small positive obstacles like edges of concrete slabs, lower driving noise on rough surfaces
- No sensitivity on lane grooves, free-rolling and with brakes applied

The road test results show the potential of improved NVH characteristics due to the insulating properties of the fiber-reinforced plastic spring/control arm. Furthermore, it is obviously possible to achieve a level of driving dynamics suitable for series release with this suspension concept after appropriate fine tuning. The evaluated prototype vehicle received no tuning at all and was tested as it came out of the work shop.

4 Summary and future prospects

A new suspension concept with a minimal number of parts saving required space especially for battery electric vehicles was introduced. Key component of the presented rear and front axles is a fiber-reinforced plastic spring/control arm which integrates the two functions wheel load support and wheel guidance into a single component. The only other wheel guiding component is the shock absorber, no other links are involved. After test rig and misuse testing a basic structural stability could be demonstrated as a starting point for the series release process. A road test with a prototype vehicle principally showed functional equivalence to the standard rear axle even without any fine tuning. A potential of improved ride comfort and NVH could be identified.

Cost and weight of the new suspension concept depend significantly on the design of the interfaces between spring/control arm and vehicle body. By optimization of these interface parts a cost and weight reduction compared with the standard axle seems to be possible. The introduction of through-holes in the FRP part might be an enabler for the replacement of the prototype milled aluminum parts by steel sheet metal during the series development process. These steps could be a viable way to realize the desired potentials.

References

1. Hacker, C., Hofmann, P., Krahn, S. et al.: Composite Leaf Spring Axle für Pkw. Lightweight Design 8, pp.14–19 (2015)
2. Matschinsky, W.: Radführungen der Straßenfahrzeuge, 3. Aufl. Springer, Heidelberg (2007)
3. Simon, M., Gerhards, T., Frantzen, M., David, W.: Das Fahrwerk des Ford Focus RS 500. ATZ 10, 764–768 (2010)

Utilizing artificial neural networks and design solution spaces to cope with the complexity in subframe design

Martin Kundla[1], Dieter Schramm[2], Markus Zimmermann[3], Nils Justen[1], Florian Scheirl[1], Tim Lindner[1], Florian Till[1] and Jaffrey Hudson Immanuel Jeyakumar[1]

[1] BMW Group, Germany
[2] University Duisburg-Essen, Germany
[3] Technical University of Munich, Germany

Abstract. Vehicle subframes are architecturally significant and highly integrated structural components of the chassis. They have a major influence on functional chains at the overall vehicle level and are characterized by a large number of interfaces. The conflicting objectives of increased component complexity versus cross-platform design modularization require new methods and concepts aimed at more efficient development and evaluation of subframe designs across multiple stages of product maturity. This paper presents a method to cope with the complexity and uncertainties in the early development phase of structural components and to carry out a multidisciplinary as well as functional and geometric design at the detail level. First, the complexity is reduced by using automatically generated knowledge-based CAD models and corresponding FEM models, which can cover different degrees of maturity in the design process within the early development phase. Secondly, automated processes are used for all relevant requirement disciplines in order to generate data in a fast and efficient manner to train neural networks, which can map the complex and often non-linear relationships between the design parameters and the characteristic values. Thirdly, epistemic uncertainties are considered with the help of the solution space engineering framework, in which maximized solution intervals for the individual design parameters are determined, to increase robustness and flexibility for the following design process. The effectiveness of the approach is demonstrated using a rear subframe as an example problem.

Keywords: Subframe Design, Solution Space Engineering, Robust Design, Early Phase Development, Knowledge-based Engineering, CAD/CAE, High-dimensional System, Nonlinear System, Multidisciplinary Optimization.

1 Introduction and state of the art

New challenges, like the growing number of competitors as a result of increasing globalization as well as the development of new technologies, e.g. in the field of electromobility and autonomous driving, force automotive companies to constantly optimize their processes and products. To maintain or improve the market position, they are forced to a more time and cost efficient development in accordance with increasing requirements

P. E. Pfeffer (ed.), *11th International Munich Chassis Symposium 2020*, Proceedings, https://doi.org/10.1007/978-3-662-63193-5_21

of customers, legislators and due to new innovations. The resulting increase in the number of vehicle models, drive systems and equipment options, coupled with shorter development times and the consideration of cross-product modular strategies utilized by most automotive companies, require further development towards more efficient and robust design processes. This is particularly necessary in the early phase of development in order to avoid expensive and time-consuming design iteration loops later on. However, numerous requirements and boundary conditions that have to be considered during component development are subject to great uncertainties due to a lack of knowledge about the development status of the entire vehicle. These uncertainties lead to increased complexity, especially in systems with many interacting variables, which may have non-linear influences on many different design goals.

In [1], Zimmermann et al. presented a design framework, called solution space engineering, which enables a systematic approach to the design of these high-dimensional, multidisciplinary and uncertain systems as represented by the vehicle. The proposed framework is based on the V-model theory (e.g. [2]), in which a complex overall system can be broken down in several steps into subsystems and finally into their individual components. The objective requirements of the overall system are then broken down to the lowest level along the resulting V (see Fig. 1). The design process does not prescribe a specific solution, but the solution must fulfill the given requirements.

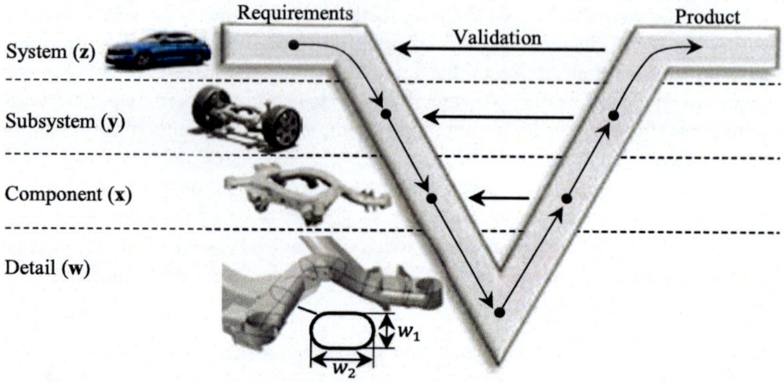

Fig. 1. System design according to the V-Model theory (see e.g. [2]).

Solution space engineering combines the V-model theory with clear design methods and comprises three main steps. First, the system structure is modeled in the form of a dependency graph, which represents the relationships between the respective design variables and the objective target values, thus creating a concrete structure for the design process. Secondly, physical or mathematical models are generated to describe these dependencies in order to obtain quantitative bottom-up representations. Third, quantitative top-down mappings are used to determine areas of acceptable designs, so-called design solution spaces. This approach differs from the classic point-based design process, in which a single design is optimized to meet the requirements of different departments or disciplines step by step. Design solution spaces, on the other hand, are selected as large as possible while taking all requirements into account in order to allow maximum flexibility and robustness in the later design process.

A single design that meets all requirements is called a good design and is part of the complete solution space. The solution space can take arbitrary non-linear, non-convex and non-connected forms, which makes the treatment and calculation of the mostly high-dimensional solution spaces very difficult. To address this problem, in [3] a stochastic algorithm was presented, which instead determines d-dimensional box-shaped solution spaces using Monte Carlo sampling and Bayesian statistics. The goal of this algorithm is to find the solution box with the largest volume containing a specified fraction of good designs that meet all requirements. These solution boxes can be calculated more easily and allow the design parameters to be decoupled, which means that they have to fulfill independent target intervals and are therefore robust to changes and uncertainties (see Fig. 2). As long as the design parameters remain within their intervals, the requirements are fulfilled one level higher in the V-model despite changes.

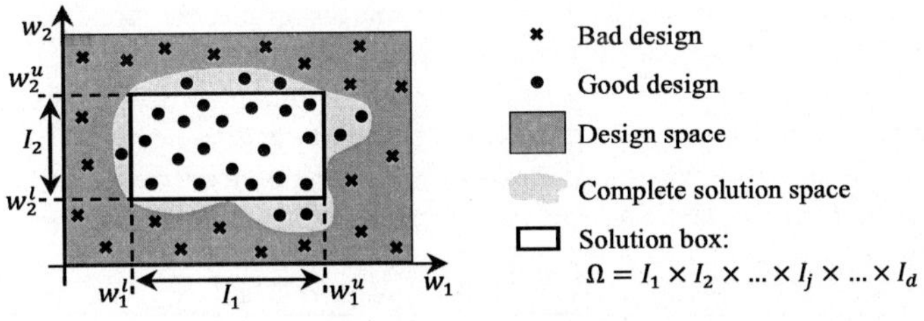

Fig. 2. Design and solution space projected on a 2D-map (according to [1] and [3]).

A large number of publications in recent years show the application of the solution space engineering method with box-shaped solution spaces to industrial problems. The main fields of interest are the crashworthiness of the vehicle (e.g. [3], [4] and [5]) and the chassis design (e.g. [6], [7], [8], [9] and [10]). Further developments on engine mounts (e.g. [11]) and control systems (e.g. [12] and [13]) show the wide range of applications of this method. It is significant however, that almost all publications deal with only one discipline and thus the multidisciplinary application is left out. Only in [11], which is also used in [1] as an example, is the consideration of several disciplines such as operational stability, comfort, acoustics and geometry carried out. In the example the geometry is described with few parameters due to the simple outer form of an engine mount. It is also noticeable that most publications remain at the functional design level and do not advance further to the geometric level, which is necessary to consider requirements relating to packaging, manufacturability, assembly, mass and costs.

In the field of chassis design, Wimmler et al. were concerned at least with the feasibility of designs determined from the solution space method. In [8], the solution intervals determined for functional tire characteristics are afterwards compared with real tire measurement data and thus checked whether these tires are feasible and which parameter intervals can be expected. In [10] an extended method is presented, where an intermediate layer in the V-model is introduced with so-called functional tire characteristics (FTCs). With the help of the solution space method and a support vector machine only feasible solution intervals for the FTCs are determined, which can be passed on to the tire manufacturer. The fulfillment of these requirements is then the task of the tire manufacturer and is not discussed further in the thesis, which again results in a strictly functional rather than a geometric consideration and design. In [9] and [10], a point-based feasibility optimization is carried out following the solution box optimization, in which intervals for kinematic properties such as camber and toe characteristics are determined under consideration of overall vehicle requirements. This checks whether certain axle systems (e.g. McPherson or double wishbone axle) with a combination of their varied hardpoints and bearing stiffnesses can fulfill the specified kinematic properties. However, whether the determined kinematic points are feasible in terms of package space or assembly is not taken into account, since a deeper geometric analysis is not done.

Only in [5] is an investigation of geometric parameters of a laminated car body rail carried out with the help of the solution space method. Here, an FEM model is divided into different segments and cross-sections by means of parameters, which are varied in a design of experiment (DoE) and the corresponding solution intervals are determined. However, as in most other publications, only one discipline, the crashworthiness, is considered. Topics such as package, manufacturability, assembly, mass and costs are not considered. But especially for complex structural components in the chassis, these disciplines play an important role in the early phase of development and must be considered in addition to the functional requirements during concept development.

In [14], Steber investigated and analyzed various processes and development models for structural components in the early phase and developed a gapless maturity-based design process, divided into five degrees of maturity (DoM), shown in Fig. 3. At the end of each phase, a maturity model with a certain level of detail is available, which must meet the respective requirements and on which the subsequent model is based.

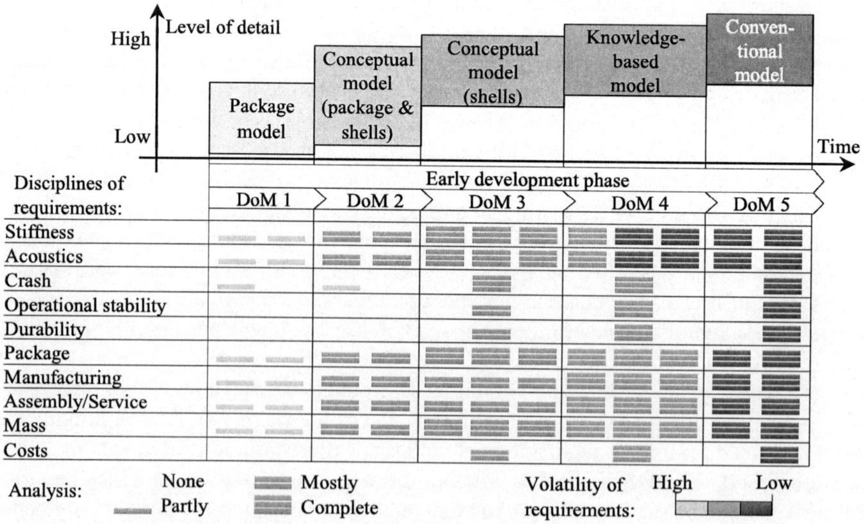

Fig. 3. Overview of the different degrees of maturity and requirements (according to [14]).

It becomes obvious that at maturity level 3 at the latest, all functional and geometric requirements must be considered mostly or completely, although they still have a medium volatility.

This paper presents a method to cope with the complexity and uncertainties in the early development phase of structural components, here using the example of the rear subframe, and to carry out a multidisciplinary as well as functional and geometric design at the detail level. On the one hand, the complexity is reduced by the fact that the knowledge-based CAD models used in this work can cover both the DoM 3 and 4 presented in [14]. On the other hand, uncertainties are considered with the help of the solution space engineering framework presented in [1] and solution intervals for the individual design parameters are determined using artificial neural networks to increase robustness and flexibility for the following design process.

2 New solution space subframe design

The high number of design variables, complex integration and numerous volatile requirements from a wide range of disciplines lead to major challenges in subframe development, concept decision and solution finding. Despite the already broken down requirements to the component level by the V-model, today's subframe development is usually still carried out with the help of classical design approaches, such as incremental or iterative development. In these approaches a single design is optimized step by step to meet all currently existing requirements that are specified in the respective maturity phase. This often leads to repeated concept and design changes in the later development phases and thus to expensive iteration loops (see Fig. 4). The subframe itself can thus be considered as a high-dimensional multidisciplinary system with uncertainties. The framework for the treatment of such systems presented in [1] is based on three enablers, which have already been briefly described in chapter 1 and which are applied to a rear subframe (RSF) connected to the body via rubber mounts in the following chapters.

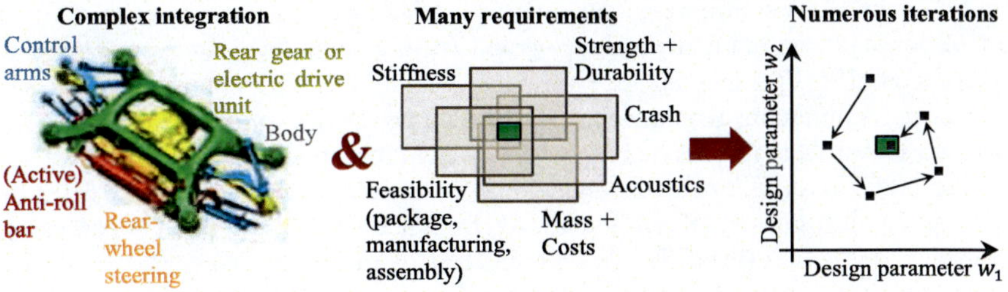

Fig. 4. Overview of the difficulties in subframe development resulting in iteration loops.

2.1 Dependency graph for the rear subframe

Due to the large number of objective characteristic values (CVs) at the component level and the large number of design parameters at the detail level, the dependency graph in Fig. 5 is simplified by overarching themes for better clarity. The d design parameters that are varied in this context can be divided into five sections, which in turn have varying degrees of influence on the individual m CVs. The CAD parameters, such as wall thicknesses, cross-sectional values or radii, and the amount, type and position of kinematic points or interfaces determine the geometric shape of the subframe and thus influence all disciplines considered. The material and associated weld properties also have a wide area of influence, whereas the influence of the bearing stiffness is less in the component calculation models used in the early phase. Calculation values, such as the number of units over the production period or the material price only influence the cost valuation. If, however, the variable assembly location is added, the tools derived from it have an effect on the assembly analysis. The CVs of each requirement discipline and their calculation are described in more detail in respective sections in chapter 2.2.

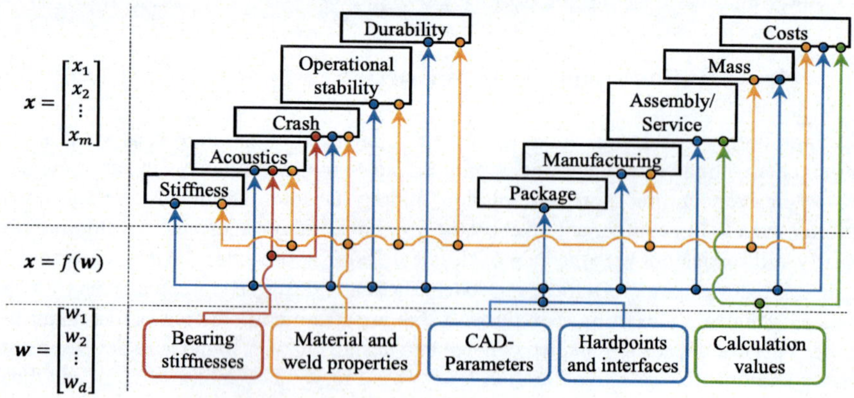

Fig. 5. Simplified dependency graph for the rear subframe.

2.2 Automated processes for the bottom-up mapping

Both CAD and FEM subframe models are required to determine all objective CVs of the disciplines shown in the dependency graph at component level. In the publication of Kundla et al. [15] a method was presented which allows an automated creation, meshing and simulation of knowledge-based parametric rear subframe models. The procedure shown there is extended in this paper, therefore it will be described briefly.

Based on an existing kinematics model and a component table previously filled in by the user, a subframe model is automatically built in the CAD program CATIA using so-called power copies (PCs). PCs are pre-designed components that can be fully parametrized and automatically assigned to new reference elements under defined boundary conditions when a CAD model is created. They can thus be used repeatedly for the creation of numerous different concepts, making complex design work worthwhile in order to increase their robustness and flexibility. In addition, they can be supplemented with information and knowledge to enable the automated execution of subsequent processes, such as the pre- and post-processing of simulations. The number and type of parameters used to vary the CAD model depends on the selection of components, which are differentiated into interfaces and design zones, type of design zone (cross zone, longitudinal zone, etc.) and type of manufacturing process (sheet metal forming, internal high pressure forming, etc.). After model creation, the automated FEM meshing takes place in two steps. In the first step, the information integrated in the knowledge-based PCs, such as middle surfaces, weld seams, bearings, connections and mesh alignment lines, is determined and used for an initial mesh creation in CATIA. After exporting the FEM model and other required information, the mesh quality is optimized in the FE-specialized pre-processor ANSA and individual details such as the assignment of node IDs and the creation of coordinate systems are carried out in a second step.

Due to the possibility of automatically meshing more detailed CAD models with this method, the utilized knowledge-based model and the corresponding FEM model can cover the DoM 3 and 4 in a single step. This reduces the complexity of the development process and the time required. The initial selection of the concept, the components and their parameter sets can be based, for example, on the previously determined information from the DoM 2 or on a possibly existing model of a predecessor vehicle. The method for automation presented in [15] is used and extended here. In addition to varying the CAD parameters and material, the hardpoints and their associated interfaces contained in the kinematics file can be varied, resulting in a more robust design with respect to subsequent kinematics and interface changes. A variation of the rubber mounts can be realized by a geometrical adjustment in the CAD model as well as a functional adjustment of the bearing stiffnesses in the input decks for the respective simulations. Furthermore, the calculation values which are used for the later cost calculation can be varied. All these parameter variations are implemented in the automated process and each of them has an influence on certain disciplines or CVs at the component level, as shown in the dependency graph in chapter 2.1. They serve as input and the respective CVs of the disciplines as output for the training of the neural networks (see chapter 2.3). The corresponding automated bottom-up mapping processes for the generation of output data are briefly described in the following sections.

Stiffness. The static stiffness of a component is relevant for the driving dynamics and performance of the entire vehicle. In the V-model, the vehicle dynamics requirements are broken down in such a way that a certain minimum component stiffness target must be achieved at the respective interface connections. Thus, objective stiffness CVs are obtained for each interface under defined constraints, rigidly modelled rubber mounts and applied static forces.

Acoustics. In order to achieve the high comfort goals of the overall vehicle and to isolate the high-frequency drive noise from the body, the rear axle transmission or electric drive unit is attached to the rear subframe, which in turn is attached to the body, both mounted using elastic bearings. The rear subframe itself receives requirements at the component level which can be determined with a frequency response analysis. First, sufficient local dynamic stiffnesses must be achieved at the drivetrain interfaces. To identify this, a harmonic dynamic force is applied in the X-, Y- and Z-directions at each connection point of the drive unit in order to determine the associated stiffnesses at different frequencies. This results in three local dynamic connection stiffness (LDCS) diagrams per interface of the drivetrain. Secondly, the transfer function (TF) between the drivetrain and body connection points is targeted. The acceleration (vector magnitude) caused by the applied dynamic force must remain below specified limit values at the body points. The number of diagrams to be considered results from the number of drive connections multiplied by three (X-, Y- and Z-direction) and multiplied by the number of body connections. To reduce the amount of CVs to be learned by the neural network, the respective diagrams are divided into frequency intervals and their corresponding worst case (WoCa) value is determined, as shown in Fig. 6 (see also [15]). The number of objective acoustic CVs per diagram results from the total frequency range to be considered and the number of chosen frequency intervals.

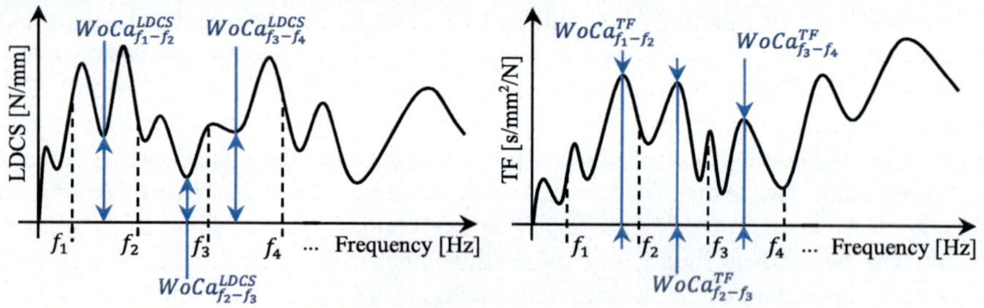

Fig. 6. Examples to illustrate the CVs for the LDCS and TF per frequency interval.

Crash. To generate a sufficient amount of data points for the training of the neural networks, a crash simulation with the entire vehicle would be too time consuming. Therefore, simplified static component substitute load cases are often used in crash simulations. However, this simplified analysis cannot represent certain properties accurately enough. Within the scope of this thesis a simplified substitute crash model for the rear subframe was developed, which considers not only the rear subframe but also other critical components in the load path (see Fig. 7).

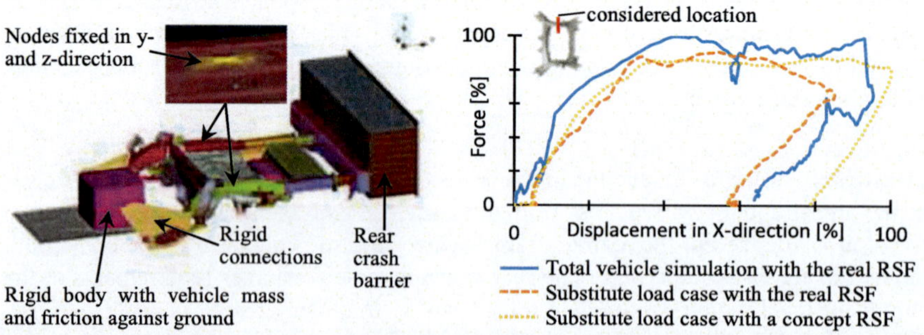

Fig. 7. Substitute crash model for the rear subframe (left) and comparison of the force-displacement between the total vehicle simulation and the surrogate load case (right).

The starting point is a complete FEM vehicle model, from which the substitute model with the relevant components is derived semi-automatically and needs to be completed once. This model achieves sufficient predictive quality for the early phase and reduces the simulation time by more than 90% compared to the overall vehicle simulation. During the DoE, the rear subframes are continuously and fully automatically replaced and the objective CVs are calculated after each crash. The derived crashworthiness requirements for the subframe are specified in the form of corridors for the force-displacement curves at defined points (see [1] and [3]). Additional CVs, such as the absorbed energy or the maximum plastic deformation, can also be determined and taken into account.

Operational stability. The stability analysis serves as a first estimation of the component capability. Particularly in the early phase, there is no comprehensive load vs. time data available for the vehicle to calculate the more complex and accurate durability. For this reason, the first stress evaluation is based on the simulation of various static driving conditions using a full vehicle multi-body system. The resulting forces at the respective interfaces from the load cases are then applied linear-statically to the subframe and the resulting stresses are calculated. However, this results in unrealistically increased stress peaks, showed in Fig. 8, at restraints with rigid body elements of the form 2 (RBE-2) and at the welds, which are modeled in simplified form as shell elements.

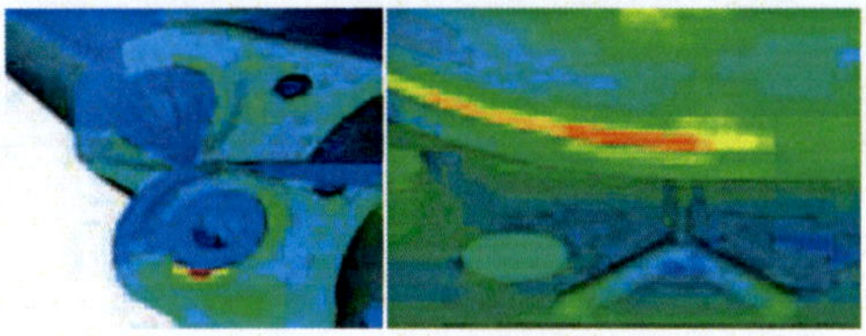

Fig. 8. Unrealistic stress peaks (red) at a RBE-2 restraint (left) and a weld seam (right).

These peaks are usually assessed by experienced engineers as non-critical and are therefore partly ignored or considered in more detail in the subsequent durability simulation. For this reason, an automated process for filtering these stress peaks has been developed within the scope of this work. For this purpose, the information extracted during the FEM meshing in CATIA is used. The knowledge-based CAD model provides information regarding which PID was assigned to welds or RBE-2 restraints during meshing. This information is used to determine the elements associated with these PIDs in post-processing and to exclude them from the maximum stress analysis. In addition, their associated nodes and the adjacent elements of the remaining PIDs are determined and also included in the filtering process. Thus, the automatically filtered stress maxima per load case and interface or design zone are used as CVs to train the neural networks.

Durability. In order to carry out the durability oriented design of vibration-stressed components, the assumed vehicle load is first compiled into a load collective with the help of classification methods (e.g. Rainflow method [16]) and the load vs. time data for the respective components are determined with detailed multi-body simulations.

Afterwards these are combined with previously performed unit load simulations and the total damage of the respective component is calculated taking into account the Wöhler curves determined from extensive material and weld seam experiments. In order to avoid a complex and exact modelling of the welds, programs for the calculation of the durability have predefined weld seam types and models which can be assigned to the individual simplified welds in the FEM model. However, this assignment is usually done manually and is therefore very time-consuming and not suitable for an automated process (see Fig. 9, top left). An alternative method for defining the welds can be implemented directly in the FEM model by assigning certain material properties to the elements of the weld and the adjacent elements and by clustering the corresponding nodes in defined sets. In the context of this work a process was developed for this purpose. With the help of the information derived from the CATIA mesh, the process can carry out this assignment automatically for all welds of the subframe (see Fig. 9, bottom left and right). This enables the software FEMFAT, which is used for the durability simulation, to read in all necessary data automatically and to perform the calculation. The automatically evaluated maximum damages per interface and design zone are used as objective CVs for the subsequent training of the neural networks.

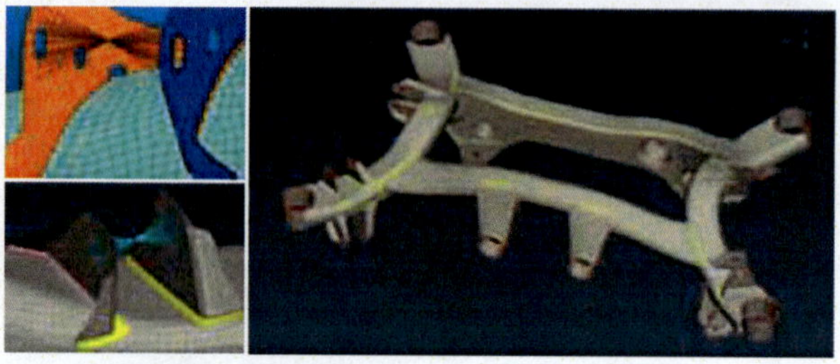

Fig. 9. Manual assignment of the weld seams in FEMFAT (top left) and automatic assignment of material cards to the weld seams and adjacent elements in ANSA (bottom left and right).

Package. Since a major part of the entire vehicle development is done in CAD, a model of the package space belonging to each component can be quickly generated at any time during the development phase. This model contains all surrounding components and possible envelope geometries. By means of a subsequent automated clash analysis, every parameter variation in the rear subframe CAD models which are created and varied in this context, can be checked for interferences or overlaps. As objective CVs, the values of the distances or overlaps to each individual component of the package model are determined for each interface and each design zone. The advantage of this method is that no restrictions to the CAD parameters regarding the package space have to be defined first. Limits for the CVs can be defined later for the solution space optimization, such that the desired packaging is taken into account. Initially, the limits can be ignored or extended in order to determine the space into which the optimized CAD model would like to enter. Thereby critical packaging restrictions can be identified for which it may be worth discussing an alternative solution in order to fulfill the requirements of the other disciplines in an optimal and mass-optimized manner.

Manufacturing. The type of feasibility analysis depends on the selected components and their manufacturing processes. As the current automated CAD generation and FEM meshing is still limited to thin-walled components, two manufacturing processes are mainly used: deep drawing of sheet metal parts and internal high pressure forming (IHF) of tubes, both classified under DIN 8580 as tension forming. There are two ways of taking component manufacturability into account. On the one hand, rules and formulas for individual parameters can already be stored in the CAD design, thereby preventing certain parameter combinations that cannot be manufactured. However, this method can very quickly lead to complex formulas and thus to undesired restrictions. In this paper, this method is therefore only used to a limited extent, such as for minimum bending radii, which are only dependent on the material and wall thickness. The second type of analysis is to subsequently examine the CAD components already created. Therefore, two automation processes were developed within the scope of this work, each of which analyzes the components and derives objective CVs for one of the two manufacturing processes used in the CAD models.

For the analysis of more complex sheet metal parts, the middle surfaces contained in the PCs are first extracted and a curvature dependent mesh is automatically generated in ANSA. Since the subsequent multi-step forming analysis used in the series process requires many component-specific user inputs, a one-step forming analysis is performed for the automation here. In this method, the sheet metal is flattened in one step and the resulting strains and stresses are calculated. This allows at least a rough estimation of whether cracks or folds are to be expected during the manufacturing process. To determine the respective critical areas, so-called forming limit curves (FLCs) are used, which describe the formability up to the failure limit for each material and sheet thickness and are determined by experiments. By adding an additional safety margin, the individual elements can be divided into three categories: cracks (red), danger of cracking (yellow) and safe (white/green), as shown in Fig. 10 at the top. Either the percentage or the absolute number of elements above the safety margin are used as objective CVs.

To determine the manufacturability of the IHF tubes, the circumferential expansion along the entire tube is considered for the first estimation.

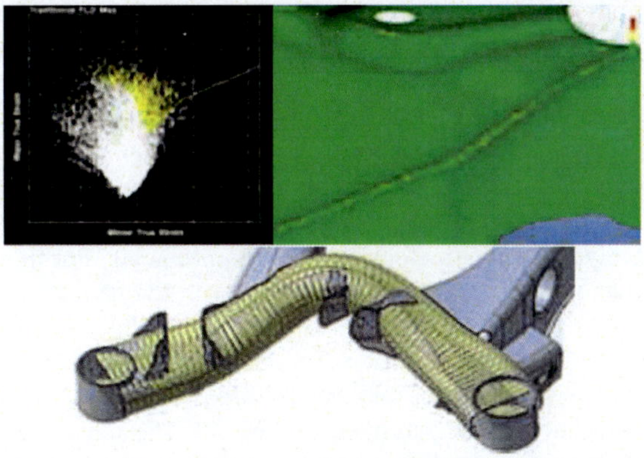

Fig. 10. Manufacturability analysis of sheet metal parts (top) and IHF tubes (bottom).

For this purpose, the circumferences of the respective component are determined automatically in CAD at small intervals (see Fig. 10, bottom). The tube is then divided into larger areas and the maximum expansion of every section is calculated in relation to the smallest circumference of the entire tube and used as a CV. For later optimization, the requirements for these CVs can be defined in such a way that the circumferential strains are lowest in the middle of the component and increase evenly towards the outside to a permitted maximum value.

Assembly/Service. The tools used for assembly are usually available as detailed CAD models. In the scope of this work, the tools used for the rear subframe and their envelopes were created as simplified PCs. These can thus be inserted automatically into the CAD model, depending on the assembly location and the type of interfaces selected. As objective CVs, the distances or overlaps of the individual subframe components with the assembly areas are then determined by means of a clash analysis, as already described in the subchapter on packaging constraints.

Mass. Using the selected material and the corresponding density, the mass per interface and design zone can be determined in CAD. The summed mass of the entire subframe is used as an objective CV.

Costs. In the scope of this work a simplified tool was developed for the cost calculation, which uses different information from the knowledge-based CAD model as input. This includes the number and type of different components, which are already divided into various categories during the PC construction, such as small, medium or large sheet metal parts and simple or complex tubes. On the other hand, the total length of existing welding seams and the mass of the subframe, which depends on the material, are determined. With additional information on the expected number of units over the production period and location, the manufacturing costs per unit are calculated with various factors approved by the department of cost calculation and used as an objective CV.

2.3 Using artificial neural networks for the top-down mapping

To determine the solution spaces and thus the regions of permissible designs, the existing requirements for the component CVs are mapped one level lower to parameters at the detail level. This step, however, requires a large number of function evaluations listed in chapter 2.2 and thus considerable optimization time. This time would have to be applied repeatedly, even for small changes in requirements. For this reason, a large amount of data is generated once at the beginning with the help of a DoE in order to create a substitute model for each objective CV, which can then be used repeatedly for a much faster solution space optimization. To generate the DoE input data, parameter intervals are defined and the Sobol sampling method is used, which generates samples uniformly in a multi-dimensional parameter space [17]. The substitute models are artificial neural networks with a feedforward structure, a single hidden layer, a sigmoid base function and 5 to 20 neurons (see e.g. [18]). Too complex network structures with many hidden layers or many neurons lead to overfitting, which means learning the data through memorization. For this reason, the generated DoE data is divided into training and test data, whereby the test data is used exclusively to check the quality of the substitute model. The corresponding coefficient of determination R^2 (see e.g. [19]) of the validation data should be as close as possible to the value 1.

3 Example problem for a rear subframe

The presented example deals with a rear subframe of a mid-size vehicle and can be placed in the development process at the beginning of the DoM 3. The goal of the new development is a significant improvement in acoustic properties at a minimal increase in mass without causing a deterioration of the other disciplines. The model of a predecessor vehicle is used as reference for the initial concept and the PC start parameters, allowing the DoM 1 and DoM 2 to be skipped. A total of 147 parameters are defined in the CAD model, which can be varied to various degrees from their respective starting values. The intervals are selected to cover as much design space as possible without causing many update errors during CAD design. The correspondence of the initial concept subframe with the original predecessor model is shown in Fig. 11.

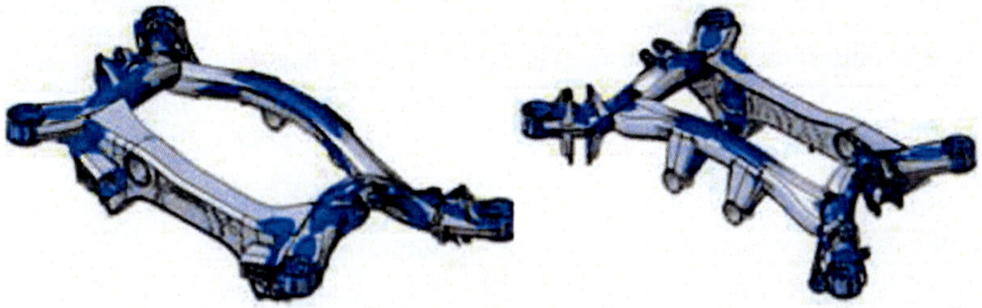

Fig. 11. Correspondence between the original RSF (blue) and the initial concept RSF (grey).

The parameters of the remaining sections at the detail level are not varied in this example, since they should remain unchanged from the predecessor model if possible. Only the number of units over the production period is varied in order to allow different cost scenarios later on, resulting in a total of 148 design inputs. This value also corresponds to the total dimensions of the design space. For requirements at the component level, 6 of the 10 disciplines are considered in this example. The crash and durability analysis are not performed for this large number of varying design parameters and associated intervals because of the increased simulation time compared to the other disciplines. These should only be included at a slightly later stage of development, when the design space has already been narrowed down by the preliminary investigations. The package and assembly criteria are also neglected at first. The subsequent optimization is intended to determine whether it would be advantageous to change the packaging of the new vehicle and the assembly tools used in favor of a new mass-optimized subframe design. With the other disciplines taken into account, the total number of CVs is 754.

Even for a very experienced engineer, this large number of parameters to be varied and CVs to be taken into account is hardly manageable with the methods used presently and would inevitably lead to many expensive and time-consuming iteration loops. For this reason, the new solution space subframe design method presented in this work will be applied to the described example problem. Table 1 gives an overview of the number of CVs, the amount of filtered data points generated from the DoE and the averaged quality of the trained neural networks per requirement discipline. The different number of data points results from various errors during the automated execution of the numerous programs and the subsequent filtering of outliers from the result data.

Table 1. Overview of the generated DoE data and quality of the artificial neural networks.

Discipline	Number of CVs	Amount of data points	Average R^2 (validation)
Stiffness	19	4485 - 4512	0,969
Acoustics	432	3911 - 4532	0,727
Operation. stability	289	3936 - 4111	0,835
Manufacturing	12	4523 - 5826	0,861
Mass	1	5802	0,998
Costs	1	5796	0,997

In the following section, two optimizations are carried out in which the acoustic properties will be improved, while setting as a requirement the fact that the component stiffness is at least equal to that of the original concept model and the filtered maximum stresses are below the yield strength of the material used. To determine a solution that is both mass-optimized and robust, a minimum mass is first determined with a particle swarm optimizer. Subsequently this mass is increased by 0.5 kg and used as an upper limit to the mass requirement for the following solution space optimization. For comparison purposes, a first optimization (Opt1) is used to find the largest possible solution box without considering manufacturability, a second optimization (Opt2) takes manufacturability into account. In both optimizations, the costs are set to a specified maximum limit. Displaying multi-dimensional solution spaces, 2D maps can be used to serve as a projection plane for the selected parameter intervals (see e.g. [1] and [10]). Fig. 12 shows an example of such a 2D map with two of the 148 possible parameters displayed on the axes for each of the two optimizations carried out.

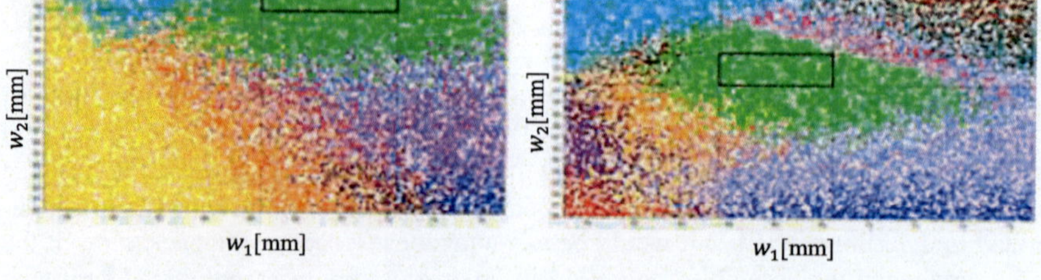

Fig. 12. Example of a 2D map after optimization 1 (left) and optimization 2 (right).

They represent the height and width of the cross-section between the middle and the end of the longitudinal IHF tube (shown in Fig. 1 as w_1 and w_2 in simplified form). Each point in the 2D map represents a subframe design with a certain combination of parameters. If the design meets all requirements, the point is green and belongs to the solution space. If it has a different color, at least one of the requirements is not met. It is evident that by considering the manufacturability in the second optimization the parameters are optimized to smaller values than in the first one. This results from the manufacturability requirement that the maximum circumferential expansion is limited and therefore the parameters cannot take on arbitrarily high values.

After the optimization is concluded, the determined solution intervals for the individual design parameters are derived. The parameter combination, in which all parameters lie in the respective center of their solution intervals, can be seen as a possible robust design for further development. Using the PCs and the automated concept creation, a design with the newly selected parameters can then quickly be created in CAD, as shown in Fig. 13. The two different sets of design requirements with regards to manufacturability thus result in two different models when compared to the initial design.

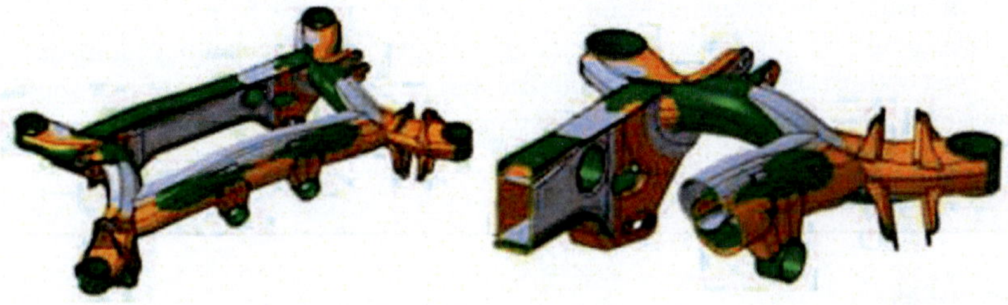

Fig. 13. CAD models of the initial design (grey), Opt1 (orange) and Opt2 (green).

Since the optimization is based on the trained neural networks, there may be deviations between the optimization results and the actual simulation data due to the imperfect substitute models. For this reason, the two identified designs from the respective optimizations are automatically simulated again and their results from the various disciplines are compared with the previously specified requirements. For the second optimization 718 of 754 CVs lie within the defined limits. The other CVs slightly miss the given requirements. Larger amounts of data are therefore necessary to train better neural networks. For the second optimized subframe Fig. 14 shows an example of the improved acoustic properties using two transfer functions.

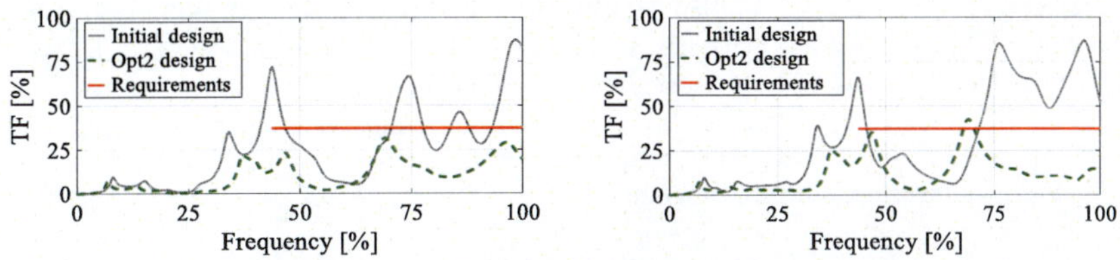

Fig. 14. Examples of the improved acoustic properties.

The optimization enabled the increased acoustic requirements to be achieved in a quick and efficient manner, taking the other disciplines and a mass-optimized structure into account. The design from the second optimization can be used for the next development phases, since the manufacturability has also been taken into account and therefore represents a realistic concept.

4 Conclusion

The solution space engineering design method supports the effective development of large and uncertain systems in the early phase of development. In this work, its application to the complex development of a rear subframe was demonstrated. With the help of the dependency graph the influences between the parameters at the detail level and the characteristic values at the component level can be clearly visualized. By using knowledge-based CAD models, powercopies and the developed automated processes, fast and consistent bottom-up mappings can be generated for all relevant requirement disciplines. The created data was used for the training of neural networks for a faster top-down mapping by means of solution space optimization. The presented new solution space subframe design demonstrates an effective way to carry out multi-disciplinary robust optimizations and a functional as well as geometric design of rear subframe concept models in the early phase of development. By adapting the bottom-up mappings accordingly, this method can be applied to other structural components in the future.

References

1. Zimmermann, M., Königs, S., Niemeyer, C., Fender, J., Zeherbauer, C., Vitale, R., Wahle, M.: On the design of large systems subject to uncertainty. Journal of Engineering Design, 233-254 (2017).
2. Haskins, C.: Systems Engineering Handbook. 3rd edn. INCOSE, San Diego, CA (2006).
3. Zimmermann, M., von Hoessle, J.E.: Computing solution spaces for robust design. International Journal for Numerical Methods in Engineering, 290-307 (2013).
4. Fender, J., Graff, L., Harbrecht, H., Zimmermann, M.: Identifying Key Parameters for Design Improvement in High-Dimensional Systems with Uncertainty. Journal of Mechanical Design 136(4), 041007 (2014).
5. Hesse, S.H., Lukaszewicz, D.H.-J.A., Duddeck, F.: A method to reduce design complexity of automotive composite structures with respect to crashworthiness. Composite Structures, 236-249 (2015).
6. Eichstetter, M., Redeker, C., Müller, S., Kvasnicka, P., Zimmermann, M.: Solution spaces for damper design in vehicle dynamics. In: Pfeffer, P. (eds.) 5th International Munich Chassis Symposium, pp. 107-132. Springer Vieweg, Wiesbaden (2014).
7. Münster, M., Lehner, M., Rixen, D., Zimmermann, M.: Vehicle steering design using solution spaces for decoupled dynamical subsystems. In: International Conference on Noise and Vibration Engineering, Leuven, Belgium (2014).

8. Wimmler, J., Schramm, D., Wahle, M., Zimmermann, M.: Concurrent design of vehicle tires and axles. In: Pfeffer, P. (eds.) 6th International Munich Chassis Symposium, pp. 839-851, Springer Vieweg, Wiesbaden (2015).
9. Wimmler, J., Wahle, M., Zimmermann, M., Schramm, D.: Optimizing Suspension Systems for Feasibility, Performance and Cost. Proceedings of the 13th International Symposium on Advanced Vehicle Control, CRC Press, Munich (2016).
10. Wimmler, J.: On the Design of Vehicle Architectures for Driving Dynamics subject to Uncertainty and Feasibility Restrictions. DuEPublico, Duisburg-Essen (2019).
11. Königs, S., Zimmermann, M.: Resolving conflicts of goals in complex design processes – application to the design of engine mount systems. In: Pfeffer, P. (eds.) 7th International Munich Chassis Symposium, pp. 125-141, Springer Vieweg, Wiesbaden (2017).
12. Zare, A., Michels, K., Rath-Maia, L., Zimmermann M.: On the design of actuators and control systems in early development stages. In: Pfeffer, P. (eds.) 8th International Munich Chassis Symposium, pp. 337-352, Springer Vieweg, Wiesbaden (2017).
13. Korus, J.-D., Ramos, P.G., Schütz, C., Zimmermann, M., Müller, S.: Top-down development of controllers for highly automated driving using solution spaces. In: Pfeffer, P. (eds.) 9th International Munich Chassis Symposium, pp. 325-342, Springer Vieweg, Wiesbaden (2018).
14. Steber G.: Reifegradbasierter Gestaltungsprozess für Strukturbauteile im Fahrzeugbau. TEWISS Verlag, Garbsen (2019).
15. Kundla, M., Heussner, T., Ye, X., Schramm, D.: Evaluation of competition and virtual rear subframes by means of the data envelopment analysis. In: Bargende, M., Reuss, H.C., Wagner, A., Wiedemann, J. (eds.) 19. Internationales Stuttgarter Symposium Automobil, pp. 1201-1217, Springer Vieweg, Wiesbaden (2019).
16. Radaj, D., Vormwald, M.: Ermüdungsfestigkeit. 3rd edn. Springer-Verlag, Berlin Heidelberg (2007).
17. Burhenne, S., Jacob, D., Henze, G.P.: Sampling based on Sobol' sequences for Monte Carlo techniques applied to building simulations. In: 12th Conference of International Building Performance Simulation Association, pp. 1816-1823 (2011).
18. Aggarwal, C.C.: Neural Networks and Deep Learning. Springer International Publishing, Cham, 2018.
19. Nagelkerke, N.J.: A note on a general definition of the coefficient of determination. Biometrika 78(3), pp. 691-692 (1991).

Virtual operational strength analysis for commercial vehicles

Robert Buchmann[1,2], Dr. Roland Krivachy[2], Prof. Dr. Alexander Lion[1]

[1] Bundeswehr University, Germany
[2] MAN Truck & Bus SE, Germany

Abstract. Customer demands and governmental emission regulations require new lightweight and electro mobility concepts in the commercial vehicle market. The necessary design changes have a significant impact on both, the structural strength and the dynamics of the chassis and the body. The evaluation of the chassis components and the body in terms of structural dynamics and operational strength at an early stage of the development process plays an important role for a time and cost effective validation process. Using simulation methods, the virtual vehicle can be analyzed long before a physical prototype is available which allows faster structural design optimizations. The virtual rough road testing combines multiple simulation disciplines and takes the full vehicle dynamics into account. Starting with the full finite element model for strength analysis, a superelement is generated using component mode synthesis. This superelement is incorporated in multibody rough road simulation to consider the elastic body deformations. The multiaxial loading history and the material data of all components are subsequently used in the operational strength analysis to determine the resulting hotspot regions and to derive structural optimizations. For a reliable operational strength forecast, an accurate prediction of the input load history is important. Therefore, the modelling of the front axle used in city busses is investigated. A flexible model of the axle is used for a stress comparison with real world test track data. In a further step, the necessity of the flexible body is evaluated with a comparison of the resulting bushing forces attached to the frame structure.

Keywords: Operational Strength, Flexible Bodies, Load Validation.

1 Introduction

One of the core keys of Europe´s economy is the efficient transport of goods and people. In the transportation sector, the customer demands in terms of cost effectiveness, comfort and driver-assistance systems as well as governmental emission regulations are rising. Together with a high number of variants along with comparable small production numbers the complexity of the development process for new vehicle generations increases. In order to fulfill the demands for future vehicles, new lightweight structures

P. E. Pfeffer (ed.), *11th International Munich Chassis Symposium 2020*,
Proceedings, https://doi.org/10.1007/978-3-662-63193-5_22

and powertrain concepts are developed. The design changes have a significant impact on both the structural strength and the dynamics of the chassis and the body. The evaluation of the chassis components and the body in terms of structural dynamics and operational strength at an early stage of the development process plays an important role for a time and cost effective validation process. Currently, rough road testing for reference vehicles and comparative simulations for variants are used for the final operational strength approval. Using simulation methods, the virtual vehicle can be analyzed long before a physical prototype is available which allows faster structural design optimizations. Virtual fatigue simulations hence offer the possibility of carrying out strength evaluations at an early stage of the development process, as physical rough road tests can only be carried out at a late stage when prototype vehicles are available.

2 Durability testing of vehicles

Repetitive loading of a metallic structure leads to fatigue of the material. The amount of cycles the material can withstand depends on the load characteristic, the geometry and the material properties. In order to prevent structural failures of vehicles in the field durability tests are performed that incorporate the customer service loads. In this chapter a brief overview over physical and virtual methods for determining the fatigue life of the entire vehicle structure is presented.

2.1 Physical durability testing

In order to accomplish the required number of loading events specified for a new vehicle in the mission profile, different approaches exist. The physical testing of structures is extremely time consuming because the accumulation of fatigue damage depends on the number of load cycles. Therefore, principles for accelerated testing are commonly used in order to avoid the accurate, but time consuming, long term testing under customer service conditions. Halfpenny [2] classified the methods in three categories.

- Compressed time
- Load amplification
- Combined

The rough road and test rig testing described in the following are a combined accelerated approach. They make use of amplified loads and disregard non-damaging load sequences.

Rough road testing. The rough road testing at MAN consists of multiple test tracks that are repeated several times in a strict order. The test tracks include various road conditions and obstacles, with exaggerated dimensions for a reduced number of cycles necessary. Non-damaging parts of real world customer service loads are omitted by the construction of the test tracks. The different test tracks at MAN Truck and Bus for durability testing are shown in figure 1.

Fig. 1. MAN test track seen from above [3]. Different road conditions and obstacles are used for the rough road testing.

Road simulation test rig. A further reduction in required time can be achieved with road simulation test rigs. The vehicle is therefore placed on a system of cylinders that reproduce the operational loads from the rough road test tracks. The operational loads are either measured signals from rough road testing or taken from simulation models. The test rig requires a minimal human involvement and makes a continuous testing possible. The advantages compared to rough road testing are reduced time and costs, with the challenging task of choosing the appropriate input signal. A full physical and virtual vehicle test rig used for busses and trucks is displayed in figure 2.

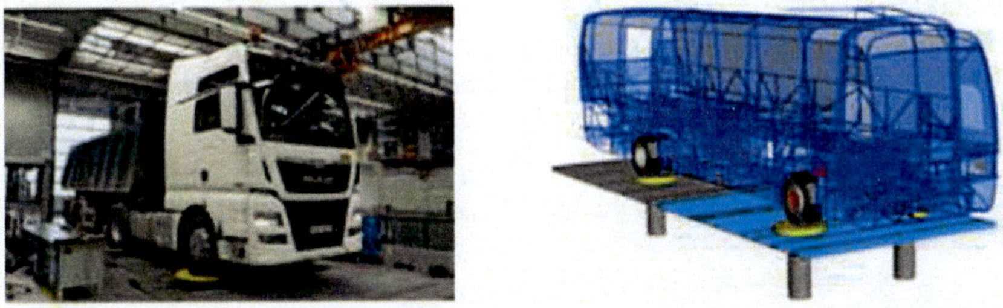

Fig. 2. Physical (left) and virtual (right) hydraulic test rig for trucks and busses. The virtual test rig uses a multibody simulation model with a flexible frame structure

2.2 Virtual durability testing

Virtual durability testing methods do not require physical prototypes and can thus be performed in earlier stages in development processes [1,4]. They rely on simulations models to analyze the lifetime of the structure and consist of three subsequent steps. In a first step, an appropriate load history needs to be generated. This can be done either from experience, experiments or simulation. In a second step, a stress analysis using a finite element model of the structure with the defined load history is performed. The evaluation of the fatigue strength can either directly rely on the stress distribution or on a subsequent fatigue life assessment taking both the load cycle information and the material fatigue properties into account.

Static and quasi-static load cases. Equivalent static load cases derived from real world measurements or dynamic simulations, combined with a vast experience, are a fast method to predict the fatigue of the vehicle structure. The load cases contain different scenarios during customer service such as braking, accelerating and cornering that are separately applied to the finite element model of the structure. Based on the calculated stresses, the durability is evaluated. The transformation of the load histories into equivalent static load cases is a challenging task, lacks accuracy and is not always possible as vibration phenomena and dynamic effects are not captured [4,5].

The static load cases can be extended to harmonic frequency excitations under consideration of the inertia relief method. This permits an insight into the basic dynamical behavior. The dynamic behavior under constant harmonic frequency excitation is studied to compute the stress distribution due to sinusoidal input loads. Both static and quasi-static load cases provide fast and, together with a vast experience, reliable results for classic vehicle concepts. New disruptive vehicle structures due to lightweight and electro mobility demands can only be assessed to a limited detail level since vibration phenomena are not captured and the load prediction is not accurate [4].

Virtual rough road simulation. For a detailed fatigue life evaluation considering dynamic effects and vibration phenomena of the vehicle structure, rough road testing is performed virtually. The virtual rough road testing is a multidisciplinary fatigue life assessment approach, incorporating finite element- and multibody-simulations, with a subsequent stress cycle-based fatigue life estimation. In this process, the digital twin of the test track of the real world rough road are used for an accurate load history prediction. The single steps of the process are illustrated in figure 3.

The finite element model used for the strength analysis of the vehicle structure is too large for a transient simulation in the time domain. Therefore, a superelement of it is derived using the technique of component mode synthesis. The reduced order model consists of Eigenmodes and additional static attachment modes. The use of additional attachment modes improves the static response of the structure to loads acting at the connection points of the structure to other components, such as the axles, the engine and the steering [6].

The superelement is then integrated in the multibody simulation model of the vehicle and connected to the drivetrain and axle models. In the multibody simulation domain, the superelement of the structure is called a flexible body. It permits elastic deformations of the structure and takes dynamic effects into account, such as vibration phenomena. This allows an accurate prediction of the time-dependent loads and stresses when driving over the digital test tracks. The stresses are calculated based on the modal participation factors from the multibody simulation and the unit load stresses obtained for each normal and attachment mode from the finite element analysis.

The calculated stress history is post processed with the Rainflow counting method to obtain a set of stress reversals. This set can be efficiently handled using the Miner´s rule to calculate the local fatigue life prediction. The local fatigue life values are then evaluated to optimize the structure. This can either be reinforcements to strengthen the structure in critical areas or lightweight modifications in uncritical areas.

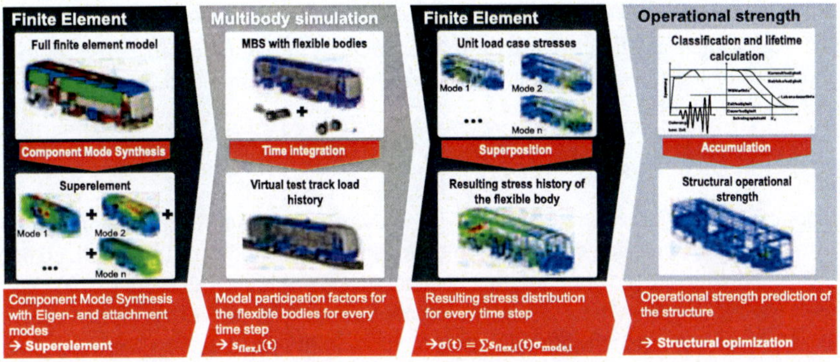

Fig. 3. Simulation process for virtual rough road testing

3 Flexible Bodies in multi body simulation

The classical multibody simulation approach assumes rigid bodies, meaning the deformability of the body is ignored. If the deformations of the body to be modelled are negligible or unimportant, this assumption provides a simple modelling and short computation times. If the deformations of the body become important, the elastic behavior has to be taken into account.

3.1 Flexible body representation

In this subsection, the general approach for a flexible representation of bodies in multibody simulation is shown. Further explanations can be found in [7,8,9]. The position of a volume element dV of a body can be described by a position vector for the undeformed state $\mathbf{u_r}$ and the elastic displacement $\mathbf{u_e}$. Both vectors depend on the location $\mathbf{c}$ of the volume element and time t.

$$\mathbf{u}(\mathbf{c},t) = \mathbf{u_r}(\mathbf{c},t) + \mathbf{u_e}(\mathbf{c},t) \tag{1}$$

Assuming small deformations, the elastic deformation can be approximated by a modal representation following the Ritz approach, where the displacements are expressed by a linear combination of mode shapes $\mathbf{u_j}(\mathbf{c})$ and time dependent scalar weighting functions $q_j(t)$.

$$\mathbf{u_e}(\mathbf{c},t) \approx \sum_{j=1}^{m} \mathbf{u_j}(\mathbf{c}) q_j(t) \tag{2}$$

The position of a volume element dV consequently depends on the rigid bod coordinates $\mathbf{u_r}$ and a displacement $\mathbf{u_e}$ that is approximated by m Eigenmodes and particular modes. The quality of the approximation depends on the number of mode shapes and their appropriateness to capture the dynamic boundary value problem. The vehicle structure in the multibody simulation model is subjected to inertia forces and reaction forces acting in the constraints and force elements attached to it. It is therefore necessary for the modal stress calculation to consider, in addition to Eigenmodes, particular mode shapes such as static or inertia relief modes. The approach to derive both the Eigenmodes of the free structure and the inertia relief attachment modes is outlined in the following section.

3.2 Component Mode Synthesis

Component mode synthesis reduces the system matrices of the full finite element model to a smaller set of interface degrees of freedom and a truncated set of generalized coordinates. A detailed derivation of the presented approach is found in [6,10]. For a damped system, the governing equation of the full model can be written in the following form with mass matrix **M**, stiffness matrix **K** and damping matrix **C**. The vector **f** contains the external forces acting on the structure.

$$\mathbf{M}\ddot{\mathbf{u}} + \mathbf{C}\dot{\mathbf{u}} + \mathbf{K}\mathbf{u} = \mathbf{f} \tag{3}$$

The physical displacement coordinates **u** are approximated in terms of generalized coordinates **q** using the Ritz approach in component mode synthesis.

$$\mathbf{u} = \mathbf{T}\mathbf{q} \tag{4}$$

Equation (3) can then be written as

$$\mathbf{M}\mathbf{T}\ddot{\mathbf{q}} + \mathbf{C}\mathbf{T}\dot{\mathbf{q}} + \mathbf{K}\mathbf{T}\mathbf{q} = \mathbf{f} + \mathbf{r}. \tag{5}$$

The residual force **r** results from the approximation made with the transformation if only a truncated set of generalized coordinates is used and can be eliminated with the Galerkin projection.

$$\mathbf{T}^{\mathbf{T}}\mathbf{r} = \mathbf{0} \tag{6}$$

Equation (3) then becomes

$$\widehat{\mathbf{M}}\ddot{\mathbf{q}} + \widehat{\mathbf{C}}\dot{\mathbf{q}} + \widehat{\mathbf{K}}\mathbf{q} = \widehat{\mathbf{f}} \tag{7}$$

with the reduced mass matrix $\widehat{\mathbf{M}} = \mathbf{T}^{\mathbf{T}}\mathbf{M}\mathbf{T}$, the reduced damping matrix $\widehat{\mathbf{C}} = \mathbf{T}^{\mathbf{T}}\mathbf{C}\mathbf{T}$, the reduced stiffness matrix $\widehat{\mathbf{K}} = \mathbf{T}^{\mathbf{T}}\mathbf{K}\mathbf{T}$ and the transformed load vector $\widehat{\mathbf{f}} = \mathbf{T}^{\mathbf{T}}\mathbf{f}$.

Component modes. The component mode synthesis consists of different sets of modes: normal modes, attachment or constraint modes and rigid body modes. A constraint mode is the static deformation of the structure due to a unit displacement of one interface node while restraining the other interface nodes and keeping all other nodes force-free. Attachment modes are defined as the displacement vector due to a unit force applied to a single interface node. A rigid body mode describes the undeformed configuration of the structure due to a unit displacement with all forces on the component set to zero and can be seen as a special case of a constraint mode. The normal modes are the Eigenmodes of the structure and are classified according to the interface boundary condition. For the fixed method these are the fixed-interface normal modes and for the free method these are the free-interface normal modes. The free-interface normal mode $\boldsymbol{\Phi}_j$ is calculated from

$$\left(\mathbf{K} - \omega_j^2 \mathbf{M}\right)\boldsymbol{\Phi}_j = \mathbf{0} \; j = 1,2, \ldots \mathrm{N_f} \tag{8}$$

with the corresponding Eigenfrequency ω_j up to a specified number $\mathrm{N_f}$ of modes or a frequency limit. Only a small subset of normal modes is used to approximate the dynamic behavior in model order reduction schemes. This approach is called mode truncation, assuming that the description of the dynamic properties of the component is still sufficient with the small subset.

Free-interface method. In this work, the free-interface method is used with inertia relief attachment modes. The derivation of the transformation matrix is presented in the following.

The nodal displacement vector u can be partitioned in a set of master degrees of freedom $\mathbf{u_m}$ and internal slave degrees of freedom $\mathbf{u_s}$. External forces are only applied to the master degrees of freedom.

$$\mathbf{u} = \begin{bmatrix} \mathbf{u_m} \\ \mathbf{u_s} \end{bmatrix} \tag{9}$$

Neglecting the damping forces equation (3) can be written as

$$\begin{bmatrix} \mathbf{M_{mm}} & \mathbf{M_{ms}} \\ \mathbf{M_{sm}} & \mathbf{M_{ss}} \end{bmatrix} \begin{bmatrix} \ddot{\mathbf{u}}_\mathbf{m} \\ \ddot{\mathbf{u}}_\mathbf{s} \end{bmatrix} (\mathrm{t}) + \begin{bmatrix} \mathbf{K_{mm}} & \mathbf{K_{ms}} \\ \mathbf{K_{sm}} & \mathbf{K_{ss}} \end{bmatrix} \begin{bmatrix} \mathbf{u_m} \\ \mathbf{u_s} \end{bmatrix} (\mathrm{t}) = \begin{bmatrix} \mathbf{f_m} \\ \mathbf{f_s} \end{bmatrix} (\mathrm{t}). \tag{10}$$

In the free-interface method in ANSYS [11], the generalized coordinates contains three sets of modes: constraint modes $\mathbf{p_r}$, inertia relief attachment modes $\mathbf{p_d}$ and free-interface normal modes $\mathbf{p_k}$. Thus, the transformation matrix for the free-interface method takes the form

$$\mathbf{u} = \begin{bmatrix} \mathbf{u_m} \\ \mathbf{u_s} \end{bmatrix} = \begin{bmatrix} \mathbf{I} & \mathbf{0} & \mathbf{0} \\ \mathbf{G_{sm}} & \boldsymbol{\Phi}_\mathbf{sr} & \hat{\boldsymbol{\Phi}}_\mathbf{s} \end{bmatrix} \begin{bmatrix} \mathbf{p_r} \\ \mathbf{p_d} \\ \mathbf{p_k} \end{bmatrix} \tag{11}$$

As pointed out in [6], rigid body modes are a special case of constraint modes. So, the first column of the transformation matrix **T** contains the constraint mode data. In the second column, the inertia relief mode data is stored and the third column contains the normal mode data. The components of the transformation matrix are briefly described, a more detailed explanation can be found in [6] and [11]. The matrix $\mathbf{\Phi}_{sr}$ contains the inertia relief attachment modes. The matrix $\widehat{\mathbf{\Phi}}_s$ is calculated from the slave partition, meaning the interior nodes of the free-interface normal modes $\mathbf{\Phi}_s$, the master partition (interface nodes) of the free-interface normal modes $\mathbf{\Phi}_m$ and the constraint mode matrix $\mathbf{G}_{sm}$.

$$\widehat{\mathbf{\Phi}}_s = \mathbf{\Phi}_s - \mathbf{G}_{sm}\mathbf{\Phi}_m \qquad (12)$$

The inclusion of the inertia relief attachment modes is important for the accuracy of the reduced order model if the free method is applied and only a truncated set of free-interface modes is used [6,10].

4 Flexible model of the front axle

One important aspect for a reliable simulation of the fatigue life is a good estimate of the input load history. While for the multibody simulation model of the rear axle of the bus a validation of the input loads was already conducted, the input loads from the front axle were so far not yet validated in detail. The measurement of the bushing forces during the rough road sections is a challenging task, as their mechanical behavior is highly nonlinear. For the assessment of the front axle components measurements with strain gauges were conducted to approve the axle components fatigue strength. These measurements are used in the following for a validation of the front axle input loads. A flexible model of the front axle allows a direct evaluation of the component stress history on a test track simulation. Assuming that the stress distribution in the hardware and modelled component matches, the resulting forces acting on the structure are comparable.

4.1 Front axle modelling

For an improved ride comfort and good maneuverability, a double wishbone axle is used for the new city bus generation Lion´s City at MAN. The lower arm is a single wishbone whereas the upper arm consists of two single arms connected with a bolt. The lower and upper wishbones are connected with bearings to a knuckle. The wheel carrier is connected to the knuckle to allow the steering. The front axle is displayed in figure 4 and was so far modelled with rigid bodies in the MBS connected with bushings to the frame structure. For the validation of the stresses in the axle components, a flexible modelling of the wishbones, the connecting bolts and the knuckle is introduced. In figure 4, the lower wishbone is shown in blue, the upper wishbone in yellow and the knuckle in red. The axle components were modelled with second order tetra elements and an average mesh size of 8 mm.

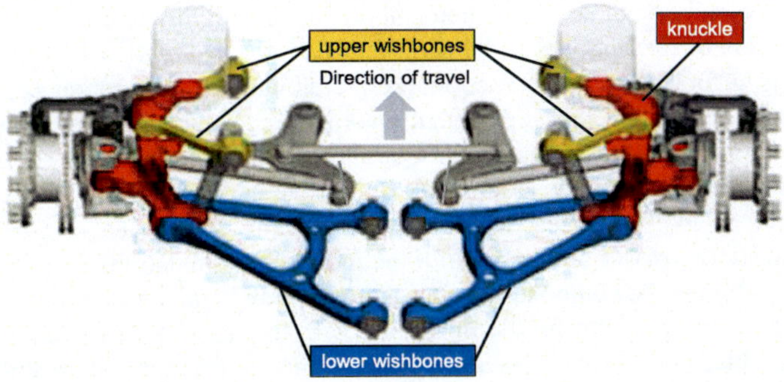

Fig. 4. Multibody Simulation Model of the front axle with wishbones and knuckles

4.2 Model variants

Besides the flexible modelling of the frame, the flexible modelling of the front axle leads to longer simulation times and increases the complexity of the model. For a full validation of the axle, the wishbones and knuckles are modelled as flexible bodies to compare the stress results with the measurements. The effect of the flexibility of the components on the resulting input loads is investigated with different variants shown in table 1. Starting with the rigid variant the lower arm is modelled as a flexible body in variant 1. Variant 2 also includes deformable upper wishbones and variant 4 is extended to a flexible knuckle.

Table 1. Investigated variants of the front axle

Variant	lower wishbone	upper wishbone	knuckle
1	rigid	rigid	rigid
2	flexible	rigid	rigid
3	flexible	flexible	rigid
4	flexible	flexible	flexible

4.3 Application of virtual stain gauges

For a validation with measured strains via strain gauges during the rough road testing, a similar approach is needed in the virtual world. Therefore, the flexible bodies were equipped with virtual strain gauges at the measurement locations in the MBS software SIMPACK. In a first step, the location of the strain gauges on the physical components were taken from a measurement map and pictures. Based on that, the location an orientation of the strain gauges is determined and the node closest to the corresponding position is selected to a set in the finite element model. The nodes contained in the set can later be used as masternodes during the model order reduction process. The selection as masternodes during the reduction process allows the creation of a deformed marker in the MBS at this position. A deformed marker moves with the respective body such that its position and orientation are dependent on the body deformation. For small elastic deformations of the body, also the use of an undeformed marker with the correct orientation is sufficient as a reference marker. This undeformed marker captures the rigid body motion, but the elastic deformation is ignored. The general process is illustrated in figure 5. In the physical testing, the strain gauges are applied to the components and then the whole front axle is assembled. Before the measurement is started, the vehicle is placed on an even surface and all mounted strain gauge signals are set to zero, such that the static stresses are ignored. In the multibody simulation, the virtual strain gauges are applied to the components to measure the static and dynamic parts. In order to remove the static part and for a comparison with the test data, an additional static maneuver is added. The static stresses are then subtracted from the test track maneuver simulations.

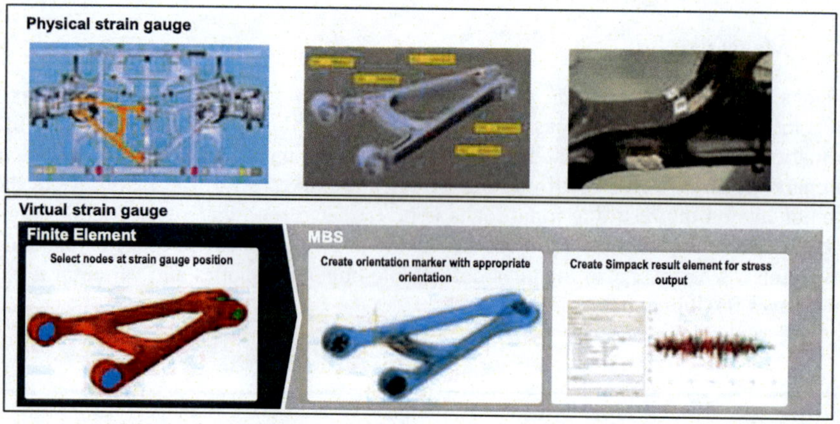

Fig. 5. Process for the application of virtual strain gauges in the multibody simulation for a comparison with real world strain gauge results

5 Results

In this section, a comparison of the simulated stresses with measured ones is presented. For a better interpretation of the measurement results, the measured strains are converted to stresses using the Young´s moduli of the components. Therefore, the comparison will be made using stress values. Furthermore, the necessity to model the axle with flexible bodies is investigated, comparing the bushing forces at the frame.

5.1 Comparison of simulated stress with measurements

In this subsection, the stress results for different test track sections and strain gauges are presented. Analyzing the location and measured strain histories, not all of the available strain gauges from the measurement can be used for a validation. If the strain gauge is close to a singularity in the finite element model, the stress values are very sensitive to a small positioning error and also to geometric deviations between the casted hardware component and the virtual model derived from CAD. In this subsection, the stresses from the physical and virtual strain gauges are compared for different levels of modelling according to table 1. The evaluation of the stresses in the simulation requires a flexible modelling of the component in the MBS environment. Consequently, not all variants can be compared for all components (e.g. stresses from the knuckle can only be evaluated for a flexible model of the knuckle).

In the physical testing, a human driver rides over the tracks multiple times in a given order. This leads to slightly different excitations in every run. For the validation of the front axle model, test tracks are chosen that show only a small sensitivity with regards to velocity and lateral guidance of the bus. In figure 6, the comparison between simulated stress and measured stress from multiple runs is shown for the lower wishbone.

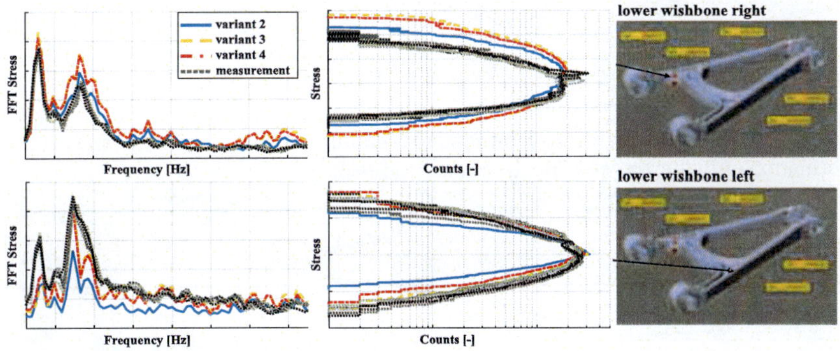

Fig. 6. Comparison of the FFT and level crossing for strain gauge 26 (top) of the right lower wishbone and strain gauge 25 (bottom) of the left lower wishbone for variant 2, 3 and 4 with measurements from various cycles.

A general improvement of both the frequency response and the load collective can be observed with increasing flexibility of the model. Especially for the strain gauge 25, it can be seen in the frequency analysis and level crossing, that the amplitude matches the measurements well if the deformation of the upper arms is taken into account.

In figure 7, a comparison of the stresses from simulation and measurement is shown for the upper arm. The frequency analysis of the stress history shows a good correlation of the peak frequencies. The peak amplitudes are underestimated in the simulation. The difference coming from a flexible modelling of the knuckle is negligible.

The comparison of the measured and simulated stress for the knuckle is depicted in figure 8. A good correlation for the peak frequencies can be observed. Looking at the stress level crossing, an offset between measurement and simulation can be seen. It results from a static pre-stress in the physical test. While the static part in the simulation can be removed with a zero load maneuver, in the real world test achieving a steady state condition without any pretension is a difficult task. Taking this into account also

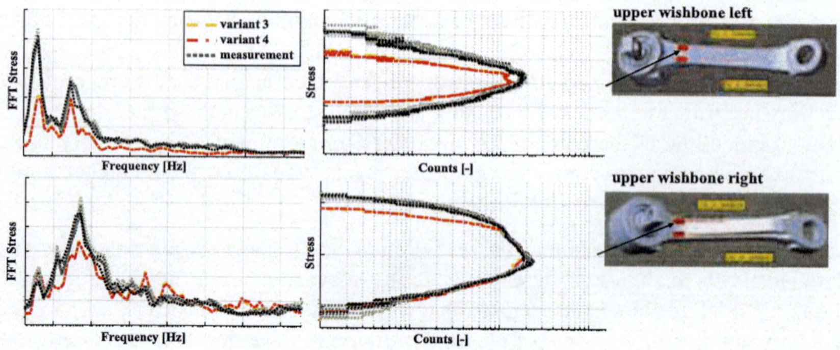

Fig. 7. Comparison of the FFT and level crossing for strain gauge 6 (top) of the left upper wishbone and strain gauge 16 (bottom) of the right upper wishbone for variant 3 and 4 with measurements from various cycles.

a good correlation for the amplitudes is given. The knuckle is a relative stiff part of the axle, which results in high Eigenfrequencies of the structure. The lowest Eigenfrequency is by far higher than the major input excitation Hence, for the description of the deformation only inertia relief modes are considered.

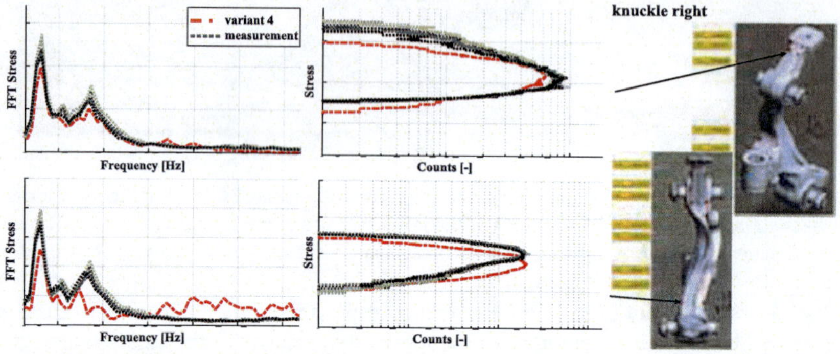

Fig. 8. Comparison of the FFT and level crossing for strain gauge 57 (top) and strain gauge 72 (bottom) of the right knuckle for variant 4 with measurements from various cycles.

5.2 Position sensitivity

The exact position of the strain gauges on the complex geometries is sometimes difficult to measure, especially when there is no landmark close to it. Furthermore, there might not exactly be a node in the finite element model at the measurement point. In this section, the sensitivity of the virtual strain gauges with respect to the position is investigated. In the multibody simulation, the virtual strain gauge elements were therefore used for the estimated position of the measurement position and the surrounding nodes. The orientation marker is kept the same. The sensitivity to the position of the strain gauge strongly depends on the stress gradient at this point. A high stress gradient, occurring for example close to sharp edges, leads to a high sensitivity with respect to the position. The used strain gauges for the measurement are uniaxial strain gauges that only capture the stress in one direction. Consequently, it is also important to match the direction of the strain gauge. In the physical testing, the orientation of the strain gauges is aligned with the geometry, which helps to find the appropriate direction for the virtual strain gauges. In figure 9, a strain gauge with a high stress gradient in direction north-south is shown. One can observe a high dependency of the resulting stress from the evaluated node position. Other strain gauges show only a minor influence to a small shift in position.

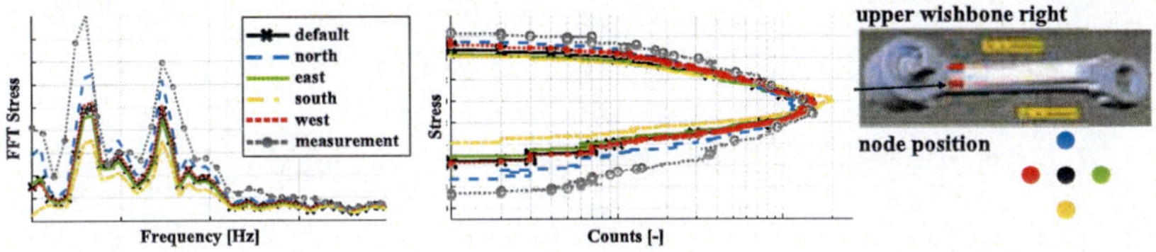

Fig. 9. Comparison of the stress FFT and level crossing for strain gauge 3 of the right upper wishbone for variant 4. The stress of the virtual strain gauge was evaluated at the nodes next to the default position.

5.3 Frame structure input loads

The comparison of the measured stress responses of the front axle parts showed a good correlation with the simulated stress responses. Only the version with the flexible bodies of the lower and upper wishbones with deformable knuckles allowed a full comparison with test data. This version also leads to the highest modelling and computational burden. Such a modelling depth is only required if the resulting forces in the bushings that connect the components to the frame show a change in comparison with the simpler front axle models.

In figure 10, the resulting bushing forces of the lower and upper wishbone for the investigated variants can be seen. For the lower arm, the force in x-direction depends mostly on the flexibility of the lower wishbone. A deformable model leads to a decrease in the force amplitude. The forces in y-direction at lower frequencies increase with added flexibility. In z-direction, the force amplitude at higher frequencies is lower if the lower arm is deformable. The straight black line in the plots shows the current cutoff frequency for the modal reduction of the frame structure. As mentioned in [12], a cutoff frequency of at least 1.5 times the excitation frequency is suggested. For the forces in z-direction this suggestion is satisfied, the forces in x- and y-direction would require a

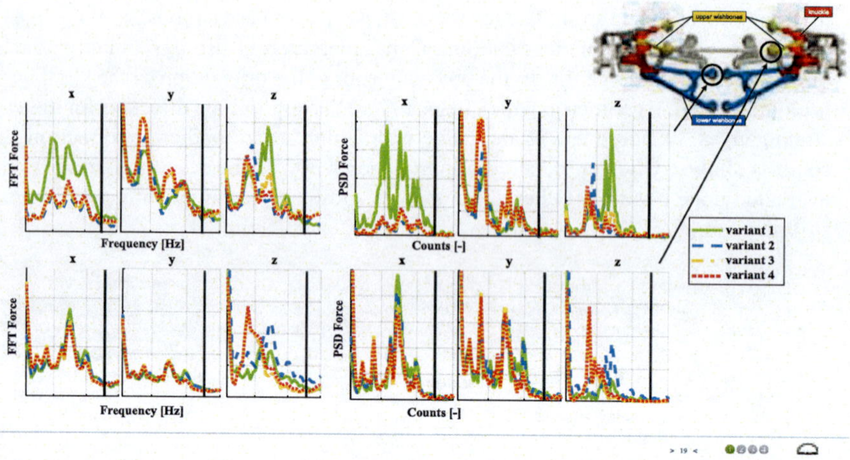

Fig. 10. Bushing forces of the left lower wishbone (top) and right upper wishbone (bottom).

slightly higher cutoff frequency, as can be seen from the fast Fourier transform and power spectral density plots. But it also has to be considered that the stress amplitude in x-direction is approximately only one tenth of the stress amplitude in z-direction.

For the upper arm, the major influence of the flexible modelling can be observed in z- direction, the influence with regard to the x- and y-directions is negligible. Looking at the cutoff frequency shown with the straight black line, one can observe that the suggestion of 1.5 times the excitation frequency holds for y- and z-direction. In x-direction, the forces are also present at slightly higher frequencies. As for the lower arm, the force amplitude in x-direction is about one dimension smaller than the amplitudes in y- and z-direction.

An accurate prediction of the transferred forces from the front axle to the frame structure requires the flexible modelling of the lower and upper wishbone. The influence of the flexibility of the knuckle on the resulting forces is negligible and consequently a rigid knuckle model can be used for an improved computation time.

6 Conclusion

The reliable prediction of the operational strength using virtual methods relies on the estimation of the input loads. While static load cases lack of accuracy for the fatigue strength determination of a vehicle structure subjected to multiaxial loading, a virtual proving ground simulation provides a more realistic prediction of the lifetime. The virtual proving ground incorporates a simulation of the vehicle response when driving over the digital twins of the real world test tracks. For an accurate lifetime simulation of the frame structure, a good prediction of the transferred loads from the road to the structure is necessary. In this work, the front axle model and loads are validated with test data. A direct measurement of the bushing forces was not available, so the validation was performed using the strain gauge signals from axle components. Therefore, a flexible modelling of those components was required with an application of virtual strain gauges. The components used are the lower and upper wishbones and the knuckles. A good correlation between the simulated stresses and measurement values could be shown. The flexible modelling of the front axle means an increase of complexity and computation time. The necessity of the flexible modelling was investigated with a comparison of the bushing for connected to the frame structure. For an accurate prediction of the bushing forces, the deformation of the lower and upper wishbones has to be taken into account. The flexibility of the knuckle has a negligible influence on the simulated forces.

References

1. Ferry, W. B., Frise, P. R., Andrews, G. T. & Malik, M. A.: Combining virtual simulation and physical vehicle test data to optimize durability testing. In: Fatigue & Fracture of Engineering Materials & Structures, Volume 13, Issue 12, pp. 1172-1134 (2002)
2. Halfpenny, Andrew.: Methods for accelerating dynamic durability tests. In: 9th International Conference on Recent Advances in Structural Dynamics. Southampton (2006)
3. Geoportal Bayern, https://geoportal.bayern.de/bayernatlas/?lang=de&topic=ba &bgLayer=atkis&catalogNodes=11,122&E=683535.27&N=5342708.49&zoom=13&layers=luftbild, last accessed 2016/11/21.
4. Johannesson, P. & Speckert, M.: Guide to load analysis for durability in vehicle engineering, John Wiley & Sons Ltd., Chichester (2014)
5. M. Haiba, D.C. Barton, P.C. Brooks, M.C. Levesley, Review of life assessment techniques applied to dynamically loaded automotive components. Computers & Structures, Volume 80, Issues 5–6, 481-494 (2002)
6. Craig RR, Kurdila A: Fundamentals of structural dynamics, 2nd edn. John Wiley, Hoboken N.J. (2006)
7. Shabana A.A.: Modeling of Flexible Bodies. In: Pfeiffer F., Bremer H. (eds) The Art of Modeling Mechanical Systems. CISM International Centre for Mechanical Sciences (Courses and Lectures), vol 570. Springer, Cham (2017)
8. Lion A.: Application of Elastic Bodies in the Durability Simulation of Vehicles: Methods, Examples and Open Questions. In: NAFEMS Seminar: Analyse von Mehrkörpersystemen mit FEM und MKS (2004)
9. Dietz S Vibration and fatigue analysis of vehicle systems using component modes. Zugl.: Berlin, Techn. Univ., Diss. Fortschritt-Berichte VDI Reihe 12, Verkehrstechnik/Fahrzeugtechnik, vol 401. VDI-Verl., Düsseldorf (1999)
10. Herting, D.N.: A general purpose, multi-stage, component modal synthesis method, Finite Elements in Analysis and Design, Volume 1, Issue 2, pp. 153-164 (1985)
11. ANSYS Help, https://ansyshelp.ansys.com/account/secured?returnurl=/Views/Secured/corp/v191/ans_thry/thy_anproc6.html, last accessed 2020/05/04.
12. Géradin M., Cardona A.: Flexible Multibody Dynamics: A Finite Element Approach, John Wiley, Chichester (2001)

Independent Design of Yaw Rate Gain and Phase Characteristics in SBW Vehicle

Kei Kitahara[1] and Hiroshi Mouri[2]

[1] JTEKT CORPORATION, Japan
[2] Tokyo University of Agriculture and Technology, Japan

Abstract. This article proposes a method to independently control the gain and the phase characteristics of the vehicle frequency response. In steer-by-wire vehicle, it is possible to design the characteristics of vehicle dynamics like the yaw response. However, the gain and phase of the frequency response of linear systems are coupled which limits the tunability. Therefore, when one characteristic is changed, the other is changed too. To deal with this issue, we propose a control method for enabling independent gain and phase characteristics of the yaw rate dynamics on a steer-by-wire system. The construction of a NARX model with a neural network enables generating the nonlinear characteristics. Furthermore, a LQR controller calculates the front wheel angle to track the nonlinear target yaw rate. The proposed control is validated on a steer-by-wire equipped vehicle. The control strategy is described and experimental data demonstrate the improvement on maneuverability and stability.

Keywords: Steer-by-wire, Vehicle dynamics, Motion control.

1 Introduction

The gear ratio of steering system (named steering ratio), defined as the ratio of the steering wheel angle over the wheel angle, can be modified mechanically with variable gear ratio steering system or artificially with steer-by-wire (SBW) system. For example, low steering ratio can be set at low speed for reduction steering amount when parking [1]. However, vehicles with low steering ratio are characterized by low damping of their yaw response to a steering input at low driving speed. A method to tune the frequency response of the vehicle dynamics has been attempted to solve the above issue.

It is known that the gain and phase characteristics of the frequency response of linear systems are coupled [2]. Decoupling them would enable extended tuning capability of vehicle yaw rate dynamics. The degree-of-freedom gained from steer-by-wire system permits to modify in real time the relation between the steering wheel angle and a wheel angle. This article proposes to decouple the gain and phase characteristics of the yaw rate dynamics by controlling the wheel angle independently of the steering wheel angle on a steer-by-wire system.

P. E. Pfeffer (ed.), *11th International Munich Chassis Symposium 2020*, Proceedings, https://doi.org/10.1007/978-3-662-63193-5_23

This paper is structured as follows. Section 2 recalls some basics of vehicle dynamics at low speed, describes the problem when using a low steering ratio and sets development goals. The control method proposed using a non-linear filter is explained in Chapter 3. Chapter 4 presents the experiment results used for validation. The developed control is applied to the steer-by-wire vehicle with low steering ratio and the experiment at low driving speed condition carried out. Further experiments were carried out at middle to high speed driving condition with higher steering ratio to demonstrate the wide scope of applications of the proposed control. At last, conclusions are given in Chapter 5.

2 Vehicle dynamics at low speed range

In this section, the performance reduction of the vehicle yaw rate characteristics at low speed when a low steering ratio is applied and the proposed solution are described.

2.1 Relaxation length

There is a dynamic relation between the slip angle and the lateral force of tires. This dynamics is often ignored but it dominates the tire lateral behavior, particularly at low speeds. It is described using the relaxation length as a speed-dependent first order lag [3]. Simulation of the dynamics of a two-wheeled vehicle model at low speed is made to confirm the effect of this behavior. (1), (2) represents the equation of motion of that model, respectively the lateral and the yaw movements. The symbols used are defined on Table 1.

Table 1. Definition of parameters

name	Symbol	Unit
Vehicle mass	m	kg
Moment of inertia of vehicle body in yaw axis	I_z	$\mathrm{kg \cdot m^2}$
Wheel base	l	m
Distance from the front axis to vehicle center of gravity	l_f	m
Distance from the rear axis to vehicle center of gravity	l_r	m
Cornering stiffness of front tires	C_f	N/rad
Cornering stiffness of rear tires	C_r	N/rad
Vehicle velocity	V	m/s
Front wheel angle	δ	rad
Slip angle of vehicle body	β	rad
Yaw rate	γ	rad/s

$$mV(\gamma + \dot{\beta}) = C_f\left(\delta - \beta - \frac{l_f\gamma}{V}\right) + C_r\left(-\beta + \frac{l_r\gamma}{V}\right) \quad (1)$$

$$I_z\dot{\gamma} = l_f C_f\left(\delta - \beta - \frac{l_f\gamma}{V}\right) - l_r C_r\left(-\beta + \frac{l_r\gamma}{V}\right) \quad (2)$$

C_f, C_r are represented as speed-dependent first-order lag.

$$C_f = \frac{c_f}{\tau s+1} \quad (3)$$

$$C_r = \frac{c_r}{\tau s+1} \quad (4)$$

The time constant t is defined as the ratio between the relaxation length and the vehicle speed. Assuming a relaxation length of 0.6 m, a low and a high speed time constant are defined for vehicle speed of respectively 20 and 100 km/h.

$$\tau_L = \sigma/V_L = 0.6/5.6 \cong 0.1 \quad (5)$$

$$\tau_H = \sigma/V_H = 0.6/27.8 \cong 0.02 \quad (6)$$

Figure 1 shows the frequency responses from steering wheel angle to yaw rate of the simulated model at low and high speeds. At 100 km/h, the peak gain increases of about 17% from the steady state gain. By contrast, this goes over 90% in the case of a vehicle speed of 20 km/h. The steering feel is not impaired by the low damping observed at the low driving speed. This is because a steady state gain is low.

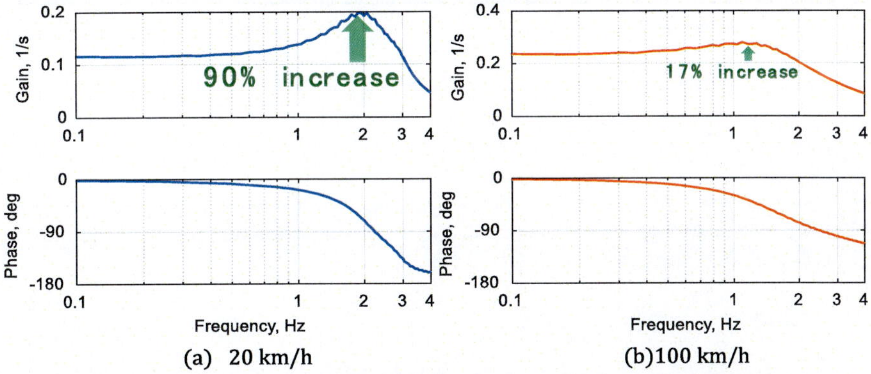

Fig. 1. Bode diagram from steering wheel angle to yaw rate at low and high speeds for a conventional vehicle with a steering ratio of 16.5.

2.2 Problem and solution

Figure 2 shows the frequency response from the steering wheel angle to the yaw rate at 20 km/h. The blue line represents the response of a conventional vehicle whose steering ratio is of 16.5 (equivalent to that shown in Fig.1 (a)). The orange line represents the response of a test vehicle which has a low steering ratio of 5. The damping is significantly lower for low steering ratio. This results in a less stable vehicle behavior.

To overcome this problem, increasing the damping would results in better vehicle performance especially at low ratio. Typically, a linear filter, such as a low pass filter, is applied but causes a phase lag as shown by the red curve in Fig. 3. Because of this coupled effect of linear filter on the gain and phase, the response of Fig. 3 is more stable but slow. If the damping could be increased while maintaining the phase characteristics, the vehicle stability would be guaranteed without degrading the response time. From this circumstance, the objective of this research is to develop the vehicle motion control strategy which can independently tune the gain and phase characteristics of the yaw rate dynamics.

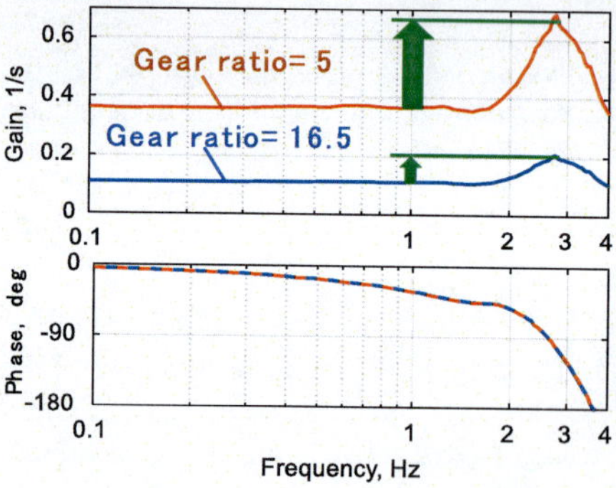

Fig. 2. Bode diagram from steering wheel angle to yaw rate for two different gear ratios

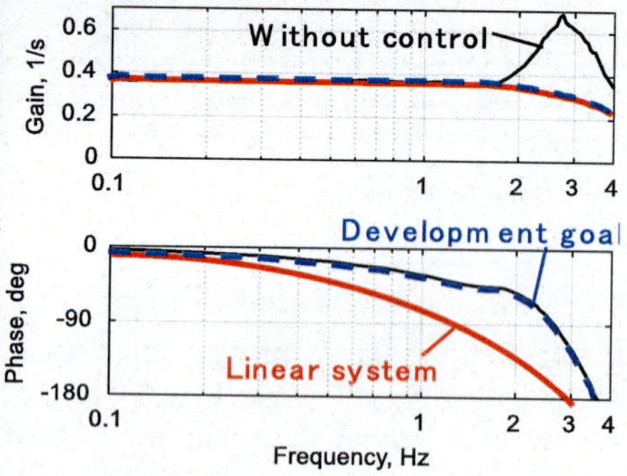

Fig. 3. Bode diagram from steering wheel angle to yaw rate

3 Method

In this Section, a vehicle control method for independent design of the yaw rate gain and the phase characteristics is described along with experimental result of the implementation of this tuned yaw rate dynamics.

3.1 Control structure

The control structure proposed for SBW which enables to independently design the gain and phase characteristics of the yaw rate dynamics is shown in Fig. 4. First, the target yaw rate γ^* is calculated from the steering angle input θ. In this step, a non-linear filter is used to compute a yaw rate which have independent gain and phase characteristics. The target front wheel angle δ^* is then generated based on the previously computed target yaw rate. Finally, the front wheel angle δ_m is controlled by the feedback controller. In the followings, the target yaw rate generator and the front wheel angle calculator are presented.

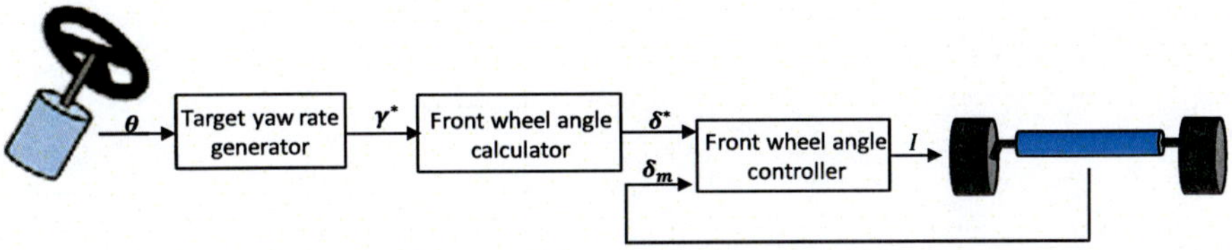

Fig. 4. Structure of the nonlinear yaw rate control system

3.2 Target yaw rate generator

This target generator computes a yaw rate which have independent gain and phase characteristics by using a non-linear filter based on a NARX (Non-linear Auto Regressive eXogenous) model [4]. The NARX model is defined as equation (7). The current output signal $y(t)$ is predicted by the input $u(t)$, the past inputs $u(t-1),\cdots,u(t-k_u)$ and the past outputs $y(t-1),\cdots,y(t-k_y)$.

$$y(t) = f\{u(t-1),\cdots,u(t-k_u),y(t-1),\cdots,y(t-k_y)\} \tag{7}$$

In this research, the steering wheel angle is used as the control input and the yaw rate as the output.

$$\gamma^*(t) = f\{\theta(t-1),\cdots,\theta(t-k_\theta),\gamma^*(t-1),\cdots\gamma^*(t-k_\gamma)\} \tag{8}$$

The function f is chosen as a neural network. NARX model build a mathematical model based on training data. This training data is computed from a data set of input steering angle filtered by a second order transfer function to set the phase characteristics and by a zero phase filter to adjust the gain characteristics. The zero phase filter is a non-causal (offline) filter that enables the modification of the gain characteristics without affecting the lag. The resulting signal γ_T represents the training data that is used to train the NARX model. The pink part of the block diagram shown in Fig.5 corresponds to offline training.

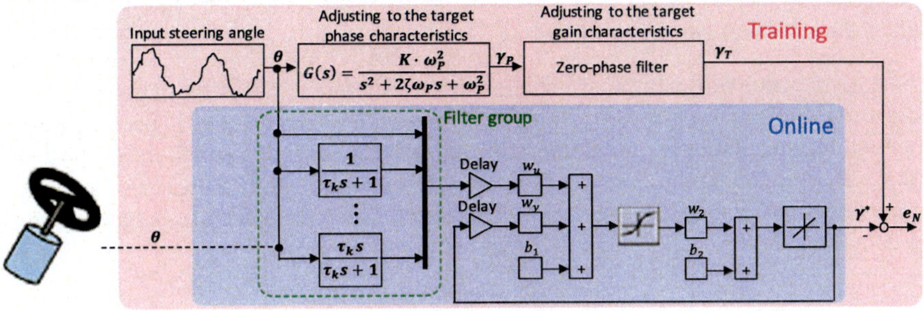

Fig. 5. Training structure of the NARX model

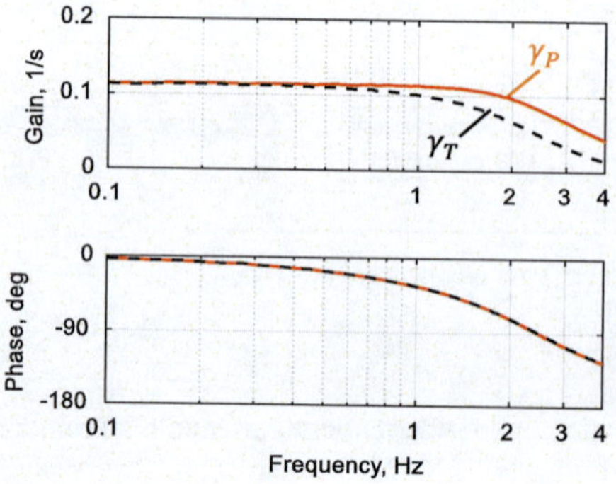

Fig. 6. Bode diagram from steering wheel angle to yaw rate from the training signal

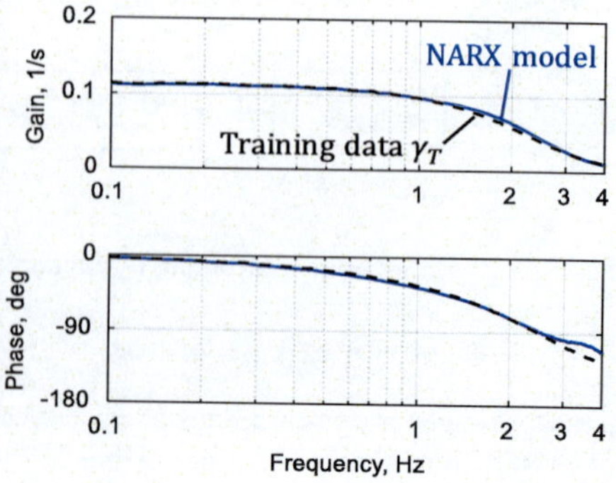

Fig. 7. Bode diagram from steering wheel angle to yaw rate of the NARX model output

Next, the NARX model is trained using the input steering angle and the training data computed as described above. Furthermore, to improve the accuracy not only the steering wheel angle is used as input but also its n-order derivatives (Filter group in Fig.5). The neural network of NARX model has parameters, weights w and the biases b, which are updated according to Backpropagation of the error e_N. This training process finishes, which means that the parameters of the NARX model are fixed, when the preset number of epoch is reached or when the error e_N becomes null.

After this offline training, the NARX model (blue part in Fig.5) can be implemented into the steering ECU to generate the desired nonlinear yaw response online.

The frequency response from the steering wheel angle to the yaw rate, as the output of the NARX model, is shown in Fig. 7. According to this figure, the output of the NARX model precisely follows the training signal. Therefore it is confirmed that the gain and phase characteristics can be independently designed online.

3.3 Front wheel angle calculator

The conversion from yaw rate to the front wheel angle is required in order to realize the characteristics of the target yaw rate on the actual vehicle. Figure 8 shows the feed forward control model used to calculate the target front wheel angle. The vehicle model should reproduce the dynamics of the actual vehicle for accurate tracking of the target characteristics. The front wheel angle δ^* is calculated using states of the model. The gain K_1, K_2, K_e is optimally set by a LQR controller.

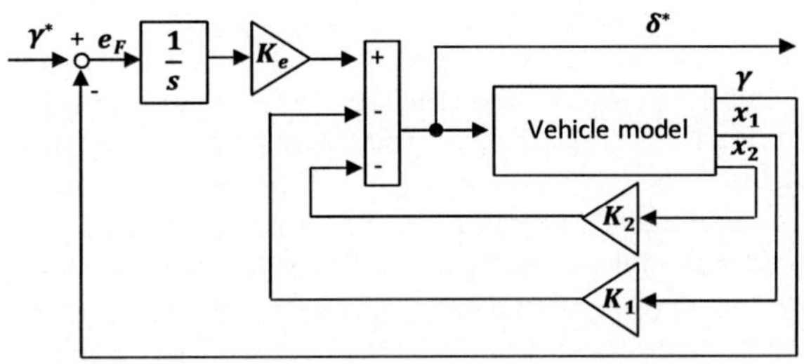

Fig. 8. Structure of the front wheel angle calculator

A second order transfer function is used to represent the yaw rate dynamics of the vehicle, which parameters are identified on the actual test vehicle. Equation (9) shows the transfer function, where K is a steady yaw rate gain.

$$G(s) = \frac{K \cdot \omega_n^2}{s^2 + 2\zeta\omega_n s + \omega_n^2} \tag{9}$$

Equation (10) and (11) are the state space formulation of (9).

$$\begin{bmatrix} \dot{x}_1 \\ \dot{x}_2 \end{bmatrix} = \begin{bmatrix} 0 & 1 \\ -\omega_n^2 & -2\zeta\omega_n \end{bmatrix} \begin{bmatrix} x_1 \\ x_2 \end{bmatrix} + \begin{bmatrix} 0 \\ K \end{bmatrix} \delta^* \tag{10}$$

$$\gamma = [K \cdot \omega_n^2 \quad 0] \begin{bmatrix} x_1 \\ x_2 \end{bmatrix} \tag{11}$$

The error e_F between the target yaw rate and the yaw rate of model output is (12).

$$e_F = \gamma^* - \gamma \tag{12}$$

Equation (13) is given by adding (12) to (10).

$$\begin{bmatrix} \dot{x}_1 \\ \dot{x}_2 \\ e_F \end{bmatrix} = \begin{bmatrix} 0 & 1 & 0 \\ -\omega_n^2 & -2\zeta\omega_n & 0 \\ -K \cdot \omega_n^2 & 0 & 0 \end{bmatrix} \begin{bmatrix} x_1 \\ x_2 \\ \int e_F \, dt \end{bmatrix} + \begin{bmatrix} 0 \\ K \\ 0 \end{bmatrix} \delta^* + \begin{bmatrix} 0 \\ 0 \\ 1 \end{bmatrix} \gamma^* \tag{13}$$

The front wheel angle δ^* is express using the gain K_1, K_2, K_e and the error e_F in (14).

$$\delta^* = K_1 x_1 + K_2 x_2 + K_e \int e_F \, dt \tag{14}$$

Equation (15) shows the cost function J when the weight for x_1, x_2, e_F, δ^* are respectively q_1, q_2, q_e, R. The cost function evaluates that the states x_1, x_2, e_F stabilize faster while the input δ^* remains bounded. The gain K_1, K_2, K_e are calculated so as to minimize the cost function J.

$$J = \int_0^\infty \left\{ [x_1 \quad x_2 \quad \int e_F \, dt] \begin{bmatrix} q_1 & 0 & 0 \\ 0 & q_2 & 0 \\ 0 & 0 & q_e \end{bmatrix} \begin{bmatrix} x_1 \\ x_2 \\ \int e_F \, dt \end{bmatrix} + R\delta^{*2} \right\} dt \tag{15}$$

3.4 Experimental validation of the yaw rate dynamics

The developed control was validated on the test vehicle equipped with a SBW system. Figure 9 shows a frequency response from the steering wheel angle to the yaw rate at a vehicle speed of 20 km/h. The measured gain characteristics (blue line) follows the target characteristics of the NARX model output (dashed black line) up to 2 Hz. However, an error occurs at higher frequency. In the phase characteristics, the measured value follows the target value with a small phase lag. Better accuracy is likely to be obtained with a more representative model of the vehicle used in the front wheel angle calculator.

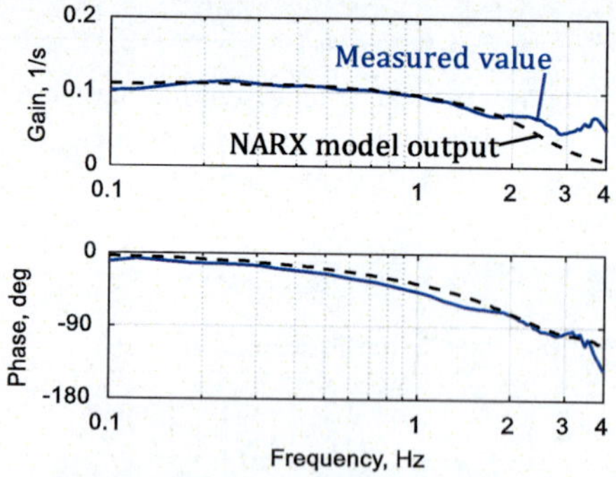

Fig. 9. Bode diagram from steering wheel angle to yaw rate

4 Validation of the control effects

The experimental validation is presented in this Section. At first, the confirmation of the effect on vehicle stability at low steering ratio and at low driving speed is carried out. Then, the maneuverability is validated on a left turn driving condition. Finally, lane change maneuver data are presented to confirm the contribution of the proposed control strategy at middle and high speed.

4.1 Vehicle stability at low driving speed

At first, the stability of the vehicle behavior at low speed is confirmed. An experiment was conducted in order to compare the three steering specifications that are shown in Table 2 and Fig. 10. Specification ① is the nominal specification of the test vehicle, with a steering wheel angle range of ±480 deg. Specification ② uses a lower steering ratio than Specification ①, which gives a range of steering wheel angle of ±150 deg. Specification ③ uses the same steering ratio than in ② and the developed control strategy which increases damping. The sine chirp shown on the upper diagram of Fig.11

was input as the steering wheel angle and the yaw rate at a vehicle speed of 20 km/h were monitored.

Table 2. Specification of steering wheel angle range and control type

	Range of steering wheel angle	Control
①	±480 deg	Without control
②	±150 deg	Without control
③	±150 deg	NARX

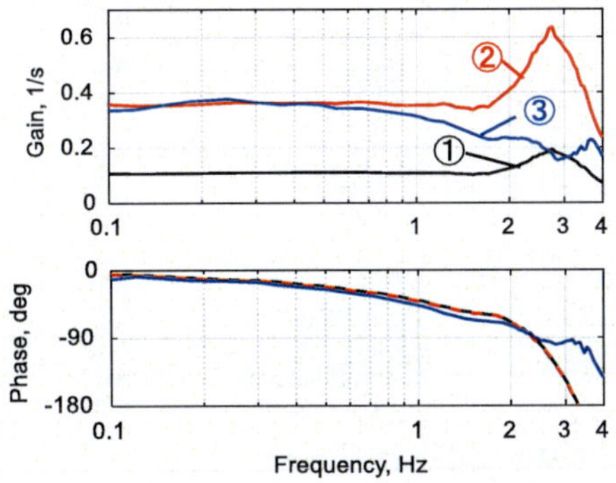

Fig. 10. Specification of experimental vehicle

The lower plot of Fig.11 presents the experimental results. The amplitude of Specification ② and ③ with low steering ratio is large at low frequency. It indicates that the vehicle can be operated with minimum effort at low speed maneuver. The amplitude of Specification ② increases at high frequency. This means that the vehicle is moved largely for high frequency input or the vehicle behavior is less stable for steering correction of high frequency. On the other hand, the amplitude of Specification ③ remains similar to that of Specification ① even at a high frequency. The improvement from Specification ② to ③ confirms the benefits obtained from the proposed control.

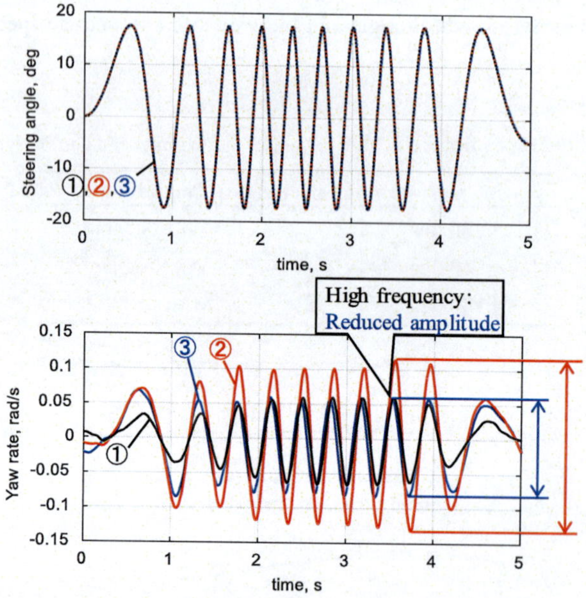

Fig. 11. The time series of input steering wheel angle and yaw rate

4.2 Maneuverability at low driving speed

Left turn experiments have been carried out to confirm the enhanced maneuverability. By maneuverability, it is mean small steering wheel amplitude, ideally enabling constant grip position during maneuver, while keeping stable vehicle performance. The layout of the path is illustrated in Fig. 12. The vehicle velocity is fixed at a constant speed of 20 km/h and subjects were required to operate the steering wheel angle only. The two specifications ① and ③ described in Table 2 and Fig.10 were used for the control of the steering system.

The time series data of the steering wheel angle and yaw rate are shown in Fig. 13. These plots represent averages trace over 5 runs. The plots show that the maximum steering angle could be reduced from about 240 deg to about 100 deg, while keeping similar yaw rate response. Hence, the maneuver is improved as the driver does not have to change the handle grip position without reduction of vehicle lateral performance.

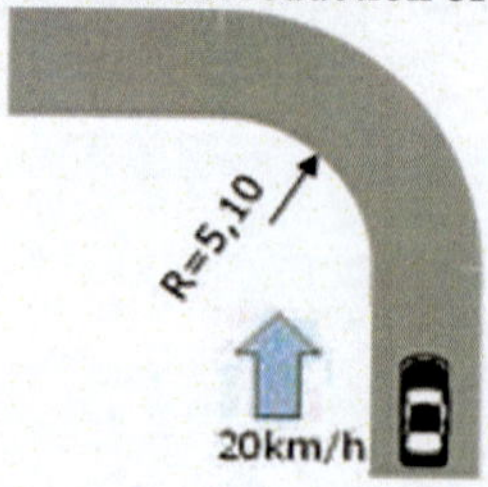

Fig. 12. Test course

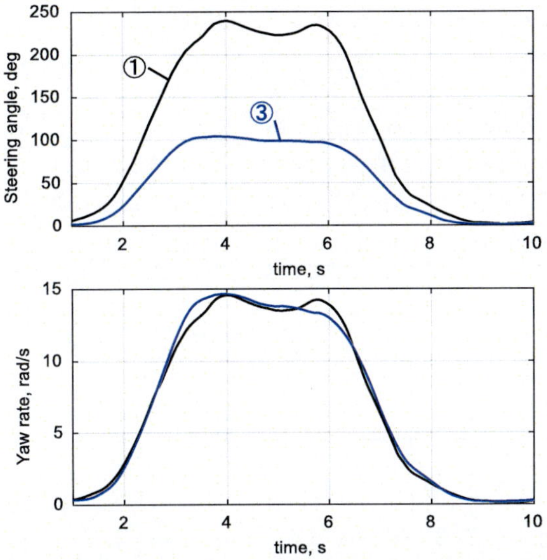

Fig. 13. The time series of steering wheel angle and yaw rate

4.3 Middle and high driving speed

The developed control strategy, which enables to independently change the yaw rate gain and the phase characteristics, is not only effective for low steering ratio at low driving speed condition but also for conventional steering ratio at higher driving speed condition. Generally, the smaller the phase lag of the vehicle frequency response is, the better is the vehicle performance [5]. However, the gain and phase characteristics are not able to be designed independently as explained previously. The damping reduces when the phase lag decreases. Therefore, the vehicle becomes less stable and steering overshoot is likely to occur. If the phase lag could be reduced while maintaining the gain characteristics, the vehicle maneuverability would be improved. Therefore, an experiment with the proposed strategy decreasing the phase lag only was implemented on the test vehicle and tested at 80 km/h as an example of middle and high speed range.

An experiment was conducted to compare the three new steering specifications given in Table 3 and Fig.14. These specifications considers the same steering ratio, given a steering wheel angle range of ±480 deg. Specification ④ is the nominal specification of the test vehicle without control. Specification ⑤ uses a linear-control resulting in a slight phase lag reduction. Here, the phase lag is reduced over the entire frequency bandwidth, however, the damping is reduced compared to ④. Specification ⑥ uses the proposed control resulting in further reduction of the lag while maintaining the peak gain as small as that of Specification ④. The test pattern is a lane change as illustrated in Fig.15. Vehicle speed was fixed at the constant value of 80 km/h, and the drivers were required to operate the steering wheel angle only.

Table 3. Specification of experimental vehicle

	Peak gain	Phase (at the peak gain)	Control
④	0.27	-75 deg	Without control
⑤	0.34	-45 deg	Linear control
⑥	0.27	-30 deg	NARX

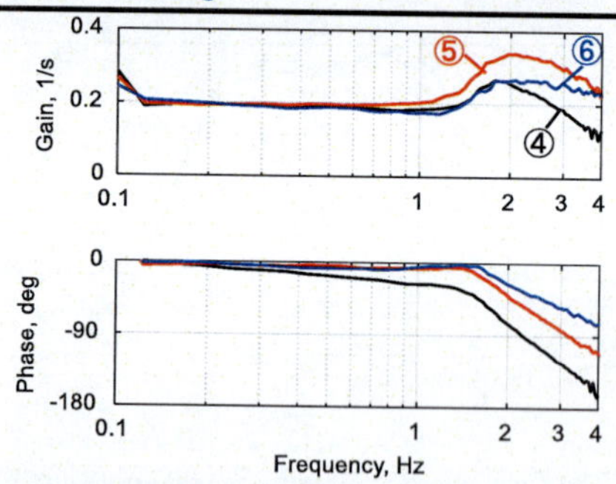

Fig. 14. Specification of experimental vehicle

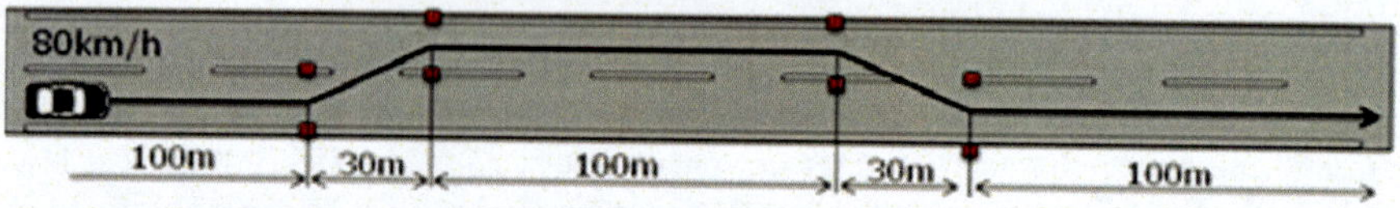

Fig. 15. Lane change pattern

The three time series data of the steering wheel angle are shown in Fig.16, which represent average plots over 5 runs. Compared to Specification ④, Specification ⑤ has smaller amplitude and more stable settling to 0 degree. However, some of the drivers commented out to have to pay attention to operate because the reaction of the vehicle for high frequency input is large. Specification ⑥ features the lowest amplitude and highest damping. Concerning Specification ⑥, almost all drivers commented that they could achieve the maneuver with less attention than with Specification ⑤. Therefore, it can be concluded that the proposed control improves the vehicle maneuverability at middle and high driving speed.

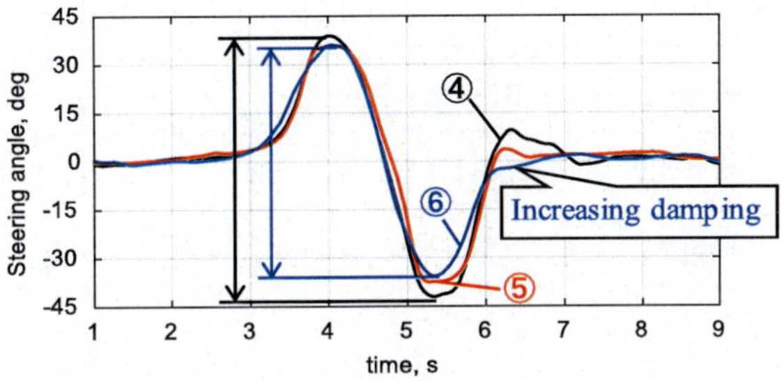

Fig. 16. Steering wheel angle during a lane change maneuver.

5 Conclusion

This research was carried out with the aim of improving stability of vehicle with low steering ratio. The cause of the degraded performance of vehicle with low steering ratio was explained with the relaxation length of the tire and illustrated with simulation examples. A method enabling to independently design the yaw rate gain and phase characteristics were presented as a solution to improve this behavior. This control strategy was applied to a vehicle with low steering ratio at low driving speed and experiments were carried out to confirm its effects. Furthermore, effectiveness of this control strategy was also presented at higher driving speed. The obtained knowledge is summarized below:

1. At low vehicle speed range, the damping of the yaw response reduces because of the effect of the relaxation length of the tire.
2. The vehicle behavior becomes underdamped on a vehicle with a low steering ratio.
3. A control method based on the combination of NARX model and steer-by-wire system enabling to design independently the gain and phase characteristics of the yaw rate dynamics was developed.
4. The yaw response of a vehicle with low steering ratio and at low driving speed could be improved by increasing the damping without impairing phase characteristics.
5. Decreasing the phase lag without increasing the gain at high frequency contributes to better maneuverability at middle and high driving speed.

References

1. Shimizu, Y. Kawai, T. Yuzuriha, J. and Takimoto, S.: A study of New Steering System with Variable Gear-ratio According to Vehicle Speed and Steering Wheel Angle. HONDA R&D Technical Review, pp.107-118 (1999).
2. Kazama, K. Haramiishi, Y. Mouri, H.: A Method to Design a Target Vehicle Dynamics by Using Variable Feedback Gain. Transaction of the Japan Society of Mechanical Engineers. Series C, Vol.79, No.806, pp320-329 (2013).

3. Shiang-Lung, K. Fanping, B. Han-Shue, T. and Tomizuka, M.: Vehicle Steering Control under the Impact of Low-Speed Tire Characteristics. Proceeding of the 2006 American Control Conference, pp.4676-4681 (2006).
4. Tsumuraya, Y. Mouri, H.: An investigation on the relationship between vehicle dynamics and drivability using NARX model. Transaction of the JSME. Vol.83, No.854, pp1-12 (2017).
5. Bergman, W., Relationships of Certain Vehicle Handling Parameters to Subjective Ratings of Ease of Vehicle Control, Proceeding of 16th FISITA Congress, Vol.16 (1976), pp.6.81-6.86

Ford steer-by-wire technology development using the example of vehicle ingress and egress

Florian Rath[1], Alexander Ein Waldt[1] and Oliver Nehls[1]

[1] Ford-Werke GmbH, Germany

Abstract. Steer-by-Wire (SbW) is a newly emerging technology entering the market in the next decade. The additional degree of freedom in the steering system leads to not only new and enhanced features, but also opens up numerous questions how the system needs to operate in specific situations to satisfy the driver. These situations encompass events where the separation of steering wheel and road wheels may lead to unique behaviours when compared to traditional steering systems. To better understand these use-cases, Ford Motor Company conducted multiple customer clinics as well as expert studies to define the requirements of the SbW system. One key question to be answered was, if a SbW system requires an electro-mechanic locking device to prevent misalignments between steering wheel and road wheel position. Other items of interest are feedback actuator sizing and strategies to mitigate hand and road wheel misalignment. This paper describes how a preliminary set of requirements were derived for Steer-by-Wire system to function without a feedback actuator locking device. Special attention was given to minimize system sizing and power consumption while achieving customer satisfaction.

Keywords: Steering, Steer-by-Wire, X-by-Wire.

P. E. Pfeffer (ed.), *11th International Munich Chassis Symposium 2020*, Proceedings, https://doi.org/10.1007/978-3-662-63193-5_24

1 Introduction

Building on experience with Electric Power Assisted Steering (EPAS) and Adaptive Front Steering (AFS) development, Ford is investigating Steer-by-Wire (SbW) as the next evolutionary step in steering technology.
Several pull factors are leading the industry in this direction, including, new opportunities for interior design, package, and manufacturing, as well as advantages for Advanced Driver Assistance Features (ADAS) and autonomous driving. While AFS added the ability to change steering ratios and superimposes feature angle requests, Steer-by-Wire mechanically decouples steering wheel and road wheels allowing the completely independent design of the torque feedback and vehicle response [1][2].

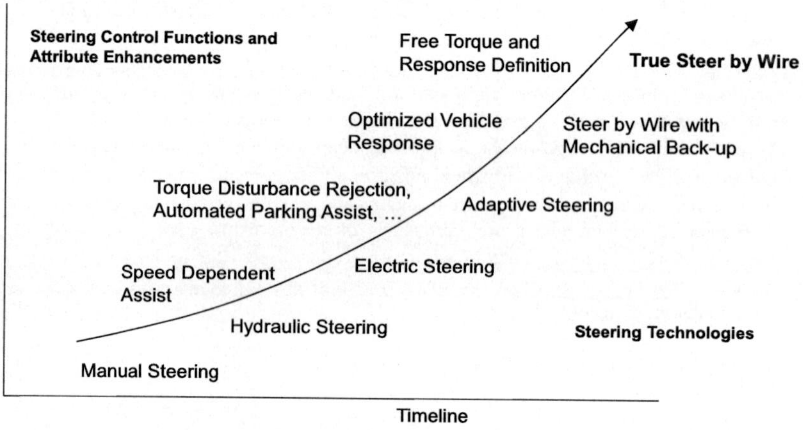

Fig. 1. Evolution of steering systems

A SbW System replaces a traditional steering column, intermediate shaft and rack with a Feedback Actuator (FBA) connected to the Steering Wheel and a Road Wheel Actuator (RWA) controlling the road wheel angles, the FBA and RWA are connected only by electrical wiring. The main tasks of the FBA are the measurement of the driver angle request and generation of a feedback torque, while the RWA estimates the actual rack force and steers the wheels of the vehicle into the requested position.

Ford Research and Chassis Engineering has a long history of developing steering systems and their features, including ADAS and comfort functions [3]. The experience and techniques that evolved in the development of these other technologies are now being used for the development of requirements and specific in-house software functions for Steer-by-Wire technology.

2 Problem Statement

In a traditional steering system, differences in the steering angle commanded by the steering wheel and the road wheels may deviate in one of two manners, a fixed offset due to mechanical misalignment, or a transient misalignment due to compliance of the mechanical elements of the steering system. As these error states are consistent, the driver typically adapts to them either consciously as with a clear-vision error, or unconsciously as when adjusting steering inputs to account for steering compliance.

Lacking this physical connection however, Steer-by-Wire systems are subject to induced offsets, and a variety of use-cases where such deviations may be introduced into the system, from minor deviations similar to a mechanical misalignment, or to gross errors up to the full range of motion of the steering wheel.

An offset is defined here as the difference between the actual roadwheel position and the driver desired position as calculated based on the steering wheel angle.

This paper is only treating those offsets that result from the additional degree of freedom between FBA and RWA in a Steer-by-Wire system.

An offset is any deviation between handwheel angle and road wheel angle, this may or may not be explicitly requested by the driver. For example, hand wheel is offset upon ingress or egress by the driver manipulating the handwheel and inadvertently inducing an offset. Large offsets can lead to the driver being uncertain where the road wheels are and how the vehicle reacts. Small offsets are acceptable to the driver and potential quality error states only.

The possibility to generate offsets is dependent whether the SbW system includes an electro-mechanical locking mechanism for the FBA. Such a locking mechanism locks the FBA in place, typically during engine off / vehicle asleep situations, disabling any rotational movement of the FBA and therefore of the Steering Wheel.

The functionality of a locking device is similar to a column lock with significantly increased performance and safety requirements. The system must be capable of locking and unlocking at any angle and have sufficient holding torque to prevent the driver from steering.

While there are advantages of such a locking device, due to the added complexity and weight, this paper examines how select use cases can be accommodated by a SbW system without a locking mechanism.

2.1 Considered Use-Cases

The following paragraphs will describe some example use-cases during startup and shutdown of the vehicle in which angular offsets between the driver requested and actual road wheel angles might occur. These situations will be referred to as offset events. The vehicle ingress/egress use-cases take place when the driver enters/exists the car and starts the engine whereas end-stop and curb interaction are related to situations where the car is running, and the steering system is operational.

Vehicle Ingress/Egress

The first and last interaction of the driver with a steering system occurs during vehicle ingress and egress. Drivers often use the steering wheel as a support handle to get in and out of the vehicle, leading to a torque input on the steering wheel. In today's vehicles with the engine off and therefore no steering assist, the steering wheel input of the driver results in a windup of the steering system, suspension and tire leading to a ramp-up of the torque until friction is overcome and the road wheels begin to turn. In a SbW system, the FBA must actively support the driver to emulate a similar behavior, providing sufficient counter-torque, as well as managing any existing or emerging offsets.

Figure 2 depicts one possible variation of the interaction of the driver and the vehicle during ingress. After a wake-up phase, the system checks for an offset and compensates for it by rotating the steering wheel, provided there is no intervention by the driver. Next, depending on the ergonomics, driver and vehicle type, the driver might use the steering wheel as a handle to support himself while getting into the seat. At this point, the FBA needs to ensure sufficient torsional stiffness to support the driver entering the vehicle. Once the driver is seated and starts the engine, the system will try to rotate the steering wheel to compensate any residual offset. A similar interaction with the steering wheel occurs during egress from the vehicle.

Multiple alternative variants of the ingress scenario have been analyzed, such as the car window being open while the system is switched off, or drivers who remain in the car after key-off.

Several nuances to the sequence of events bear consideration when developing functional strategies and performance specifications. For example, to minimize potential injury risks, the steering wheel should be rotated rather than the road wheels for system alignment. Should the system attempt to align the road wheels with the vehicle on a hoist or a roller dyno, the risk to those nearby are appreciable.

Similarly, to minimize risk inside the vehicle, hand wheel rotation needs to be limited in angular speed and maximum torque level to avoid any injuries to the driver's hands. Alternatively, during ingress/egress, a freely rotating steering wheel with no resisting torque could not only lead to large offsets but physical injury as the driver slips and falls when anticipated support from the wheel is not present.

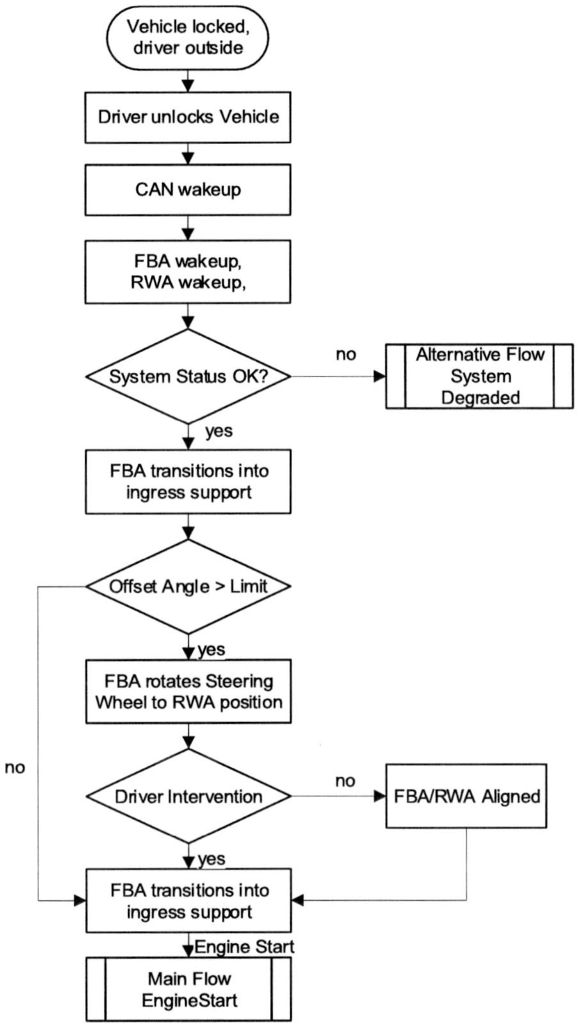

Fig. 2. Simplified main flow for the use-case v*ehicle ingress*

End-Stop and Curb Interaction

During normal driving the FBA has the task to provide situation dependent torque feedback to the driver. In conventional cars, the steering wheel movement is stopped by the rack travel end-stop. In a SbW vehicle this end-stop needs to be simulated to prevent the driver from over-traveling the hand wheel and inducing an offset. Such induced offsets may occur when the vehicle is stopped, or during maneuvers. Moreover, as in AFS, this end-stop position will vary with vehicle speed as the steering ratio varies.

This end-stop simulation needs to consist of the following items:

- Ramp-up of the FBA torque to inhibit over travel
- Controlled ramp-out of that additional torque during return to center
- Mechanism which consistently removes offsets due to overtravel

Typical torque levels during normal driving are 2-4 Nm, therefore the simulated end-stop torque level needs to be significantly higher. The optimal level needed is subject of the analysis described below.

A similar functionality is needed for curb situations; a curb situation is given when at least one of the road wheels is limited in its motion e.g. due to a high curb or similar obstacle. Complicating this situation compared to the above over-travel case; the curb contact can occur at any steering wheel angle in arbitrary road wheel positions and must to be detected by the roadwheel actuator at first.

2.2 Research Goal

In a system without a locking device, only the torque capability of the feedback motor is counteracting the driver. Sufficiently high levels of feedback torque would obviously lead to very low numbers and magnitude of offsets, but for a feasible system design, a compromise between package space, system sizing and performance must be found. Therefore, the authors sought to understand if a feedback actuator can be designed and operated without an electro-mechanical locking device with minimized system sizing. To achieve this goal, it was broken down into the following tasks:

- Ensure recognition of end-stops by the driver and hinder the driver movement
- Minimize the number of overtravel events
- Minimize the magnitude in case of overtravel (offsets generated)
- Achieve natural and consistent end-stop feel

3 Methodology

Steering feel attribute ratings are highly subjective, as such CAE methods are of limited utility. Therefore, Ford uses two methods in these kinds of investigations, expert studies and customer clinics.

The investigation requires a high-fidelity prototype system that ideally can exceed the performance requirements that are to be established. The test vehicle's feedback actuator must therefore have sufficient torque capabilities to exceed the anticipated upper bound of possible customer/vehicle needs to accurately define the true required performance limits.

3.1 Expert Study

Specific performance requirements may be determined by utilizing the knowledge of subject matter experts and conducting drive evaluations. Such subjective evaluations are typically conducted by a limited number of skilled drivers. The drivers are highly trained to control and evaluate the vehicle and its investigated attributes, using standardized maneuvers and a standardized rating scheme [4]. Most of the time, the experts are more sensitive in their rating compared to normal drivers. Typical examples for evaluation by specialists are controllability and fault insertion studies. Expert drives are effective for very specific driving scenarios but have limitations when evaluating new features or non-driving related attributes.

As an example, the required sensor accuracy and maximum allowable actuator delays to achieve acceptable steering performance with a SbW system has been established by expert drives.

For scenarios where a subject matter expert may be overly critical, their expertise may be alternatively leveraged to define the use-cases and for a customer clinic evaluation; defining test sequences, narrowing down parameters and boundary conditions for the requirement being evaluated.

The following investigations have been performed:

- Steering wheel torque level and shape for ingress/egress
- Potential user behavior at ingress/egress for multiple vehicle ride heights
- Maximum alignment speed of the steering wheel and maximum torque level during alignment

3.2 Customer Clinic

Whenever the interaction between driver and vehicle or a system within the vehicle is unknown, a customer clinic can be utilized. Goals can be to validate use-cases or finding performance requirements as well as to test customer acceptance or subjective preferences towards variants. A statistically relevant group of representative participants is exposed to different aspects of a new system in specially selected scenarios or given a specific task to execute. An interview during the study, directly asking questions to determine acceptance and preferences can be conducted.
To avoid customer bias and to allow a natural reaction of the customer, often cover stories are invented to distract from the real intent of the study.
The number of items to be investigated must be small enough, not to exceed the cognitive capacity of the test subject. Upon selection of the participants it must be decided whether demographic aspects or experience are relevant for the result. If not, an evenly distributed representative group of the average customer is favorable.

A customer clinic can help resolving diverging expert ratings. That is typically a sign that the subject under investigation is not yet completely understood and therefore needs more input from the customer base.

3.3 Test Vehicle

A demonstrator vehicle was built where the steering system was replaced by a fully fail operational prototype Steer-by-Wire system **without** mechanical fallback. Prototype safety was ensured by using redundancy throughout all subcomponents of the system.

The Feedback Actuator utilizes a brushless DC motor and a belt drive to reach sufficient torque levels while maintaining low friction. In addition, a locking device can be connected. The component is complemented by three angle sensors of different type. An absolute angle sensor is used to initialize the position of the steering wheel during startup. High resolution incremental sensors are then used to measure the driver angle request.

As Road Wheel Actuator, a belt drive rack EPAS with two power packs and redundant sensor set was used.
A redundant dSpace Autobox setup ensures fail operational capability. Two independent power supplies are used to be completely independent from the vehicle power supply. Communication lines between components are at least dual redundant.

Table 1. Steer-by-Wire Prototype System Performance

Max. Static FBA Torque	40Nm
Max. Dynamic FBA Torque	20Nm
Steering Wheel Angle Sensor Accuracy	<0,01°
Road Wheel Position Sensor Accuracy	<0,02°

3.4 Feedback Actuator Test Rig

To accelerate software development, a test rig was built utilizing the same FBA as in the test vehicle, connected to an RCP system. Scope of the test rig was software development and initial calibration. In early stages of development, a simple setup like this has been proven to be advantageous before validation in the vehicle.

4 Customer Clinic Setup

In total, four customer clinics were performed to answer the open questions in the investigated use-cases. They were staged to build on top of the results of the previous ones to go from general to more detailed questions and attributes.
The sample size ranges between 35 and 40 participants to achieve at least 30 acceptable data points. The population was almost evenly distributed between male and female as well as a wide spread over age.

Generally, all were conducted in a similar manner:

- Introduce the proband to the task to be conducted while giving a cover story about the background of the testing to ensure that there is no bias. Also, a test drive to familiarize the proband with the parkour was done initially.
- Three runs through the same test parkour with different settings (Scenario A-C) deemed to be an acceptable number for the participants to not overload their receptiveness.
- The order of the runs is randomized within the participant group to avoid bias due to the specific order.
- Finally, a questionnaire is filled-out to rate the different aspects tested

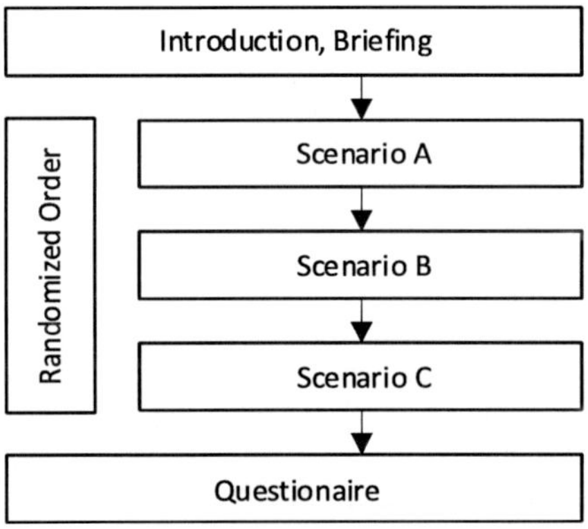

Fig. 3. Customer Clinic Execution

To ensure that the driver uses the car as naturally as possible and to distract him from the steering system back to the task to drive the vehicle, in each Scenario a subset of the tasks was repeated under stressful situation.

The analysis of the study was using the questionnaire (subjective ratings) as well as recorded objective measurement data (e.g. steering wheel angle, speed and torque). The following sections describe the investigated attributes.

4.1 End-Stop Shape

The test subjects were asked to drive a handling course including tight corners and parking maneuvers while steering wheel angle and torque were measured. The end-stop was simulated using a maximum torque of 20Nm. Three different end-stop shapes were tested: A replication of a conventional end-stop, a variant including additional damping torque and a variant where a high frequency vibration is added whenever the end-stop is exceeded.

4.2 End-Stop Torque Levels & Curb Interaction

To validate the previously investigated torque level against lower values, the second clinic featured two end-stop and curb simulations with a maximum torque of 12Nm against one with 20Nm. In addition, the test vehicle was parked close to a curb, that blocked the road wheels from moving when steering into it. The blockage was simulated with an end-stop like torque behavior.

4.3 Vehicle Ingress/Egress

The first clinic was conducted to identify different behaviors between medium and high ride height vehicles. In addition, preferences towards different levels of torque support compared to a freely rotating steering wheel where investigated.

The second clinic focused on the characteristics of the algorithm for torque support and the acceptance of different alignment strategies. Details are described in the following subsections.

Steering Wheel Utilization

Goal of the study was to understand if and how customers are using the steering wheel to enter and exit medium and high ride height vehicles. The preliminary assumption was, that these vehicles would provoke the most usage of the steering wheel by the driver pulling himself into the car.

A group of 49 test subjects were asked to enter/exit a mid-size SUV and a commercial vehicle. The steering wheel provided three different levels of maximum support torque:

- Freely Rotating
- Held in position by various levels of torque

A bystander observed the behavior during ingress and egress.

Alignment Strategies

In a second clinic focusing on ingress/egress, the steering wheel alignment type and timing acceptance was tested. Initially, the steering system was prepared with a significant offset before the subject arrived, while the roadwheel were in straight-ahead position. The subject then had to ingress in a narrow environment (parked close to a wall) to provoke usage of the steering wheel while entering. One lap on a small course was driven to distract the driver, then he had to exit the car again.

Three different triggers were used to initiate the alignment:

- Opening the door
- Grab and release of the steering wheel by the driver
- Engine start

Rotational speed for steering wheel alignment was determined in a prior expert clinic, considering maximum alignment duration and safety aspects. Maximum torque level as well as steering wheel speed were limited considering potential injury. Upon interaction with the steering wheel or blockage due to a hand or objects obstructing free rotation, the alignment would be stopped.

In addition to the alignment timing, two different ingress/egress torque behavior variants were mixed into the study. When the driver supports himself while entering the car by pulling at the steering wheel, based on his torque input, the steering wheel will be moved to a different position. Once he releases the steering wheel, two different deflection algorithms were used:

- the steering wheel stays at the deflected position. ("Damping")
- the steering wheel slowly rotates back to the original position. ("Spring")

In all scenarios, in case an offset is still present when the vehicle is launched, a quick alignment of the road wheels to the desired roadwheel position is triggered.

5 Results

5.1 End-Stop Shape

All three investigated torque shapes for the end-stop resulted in the same level of recognition, acceptance and amount of over travel. The torque shape of the ramp-in into the end-stop had no significant impact on customer perception as well the amount of overtravel. Solely the magnitude of maximum torque seemed to be the key factor to limit overtravel.

5.2 End-Stop Torque Levels & Curb Interaction

Fig. 4 summarizes the occurrence of overtravel events relative to all end-stop or curb interactions. For the 12Nm variants around half of all interactions result into an overtravel event.

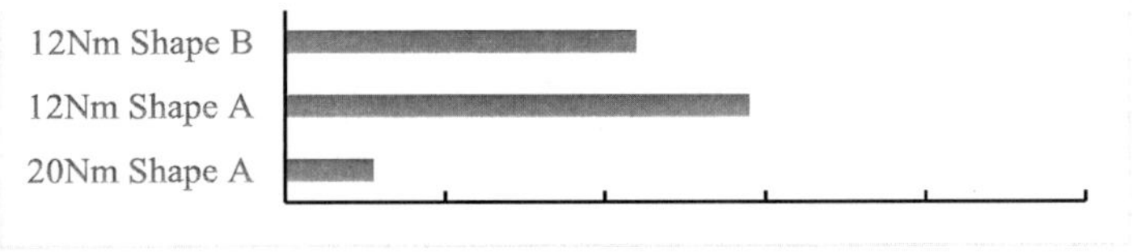

Fig. 4. Percentage of overtravel events during end-stop / curb interaction

In contrast to end-stop simulation, where 20 Nm are adequate to achieve customer satisfaction, for curb situations an additional torque is required. Most likely this is due to the different steering wheel position as well as the unpredictable occurrence which leads to higher steering speeds encountering the curb.

5.3 Vehicle Ingress/Egress

Steering Wheel Utilization

The results from the observed behavior of the test subjects are summarized in Fig 5. Observations from the second clinic are included for comparison.

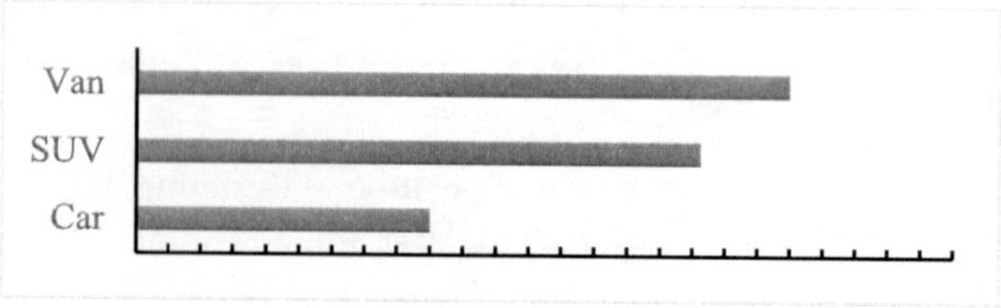

Fig. 5. Percentage of subjects who utilized the steering wheel on ingress/egress

Significantly more people used the steering wheel for entry support in the commercial car compared to the SUV and passenger car. But even in in a passenger car, more than a third of the test subjects used the steering wheel as a handle.

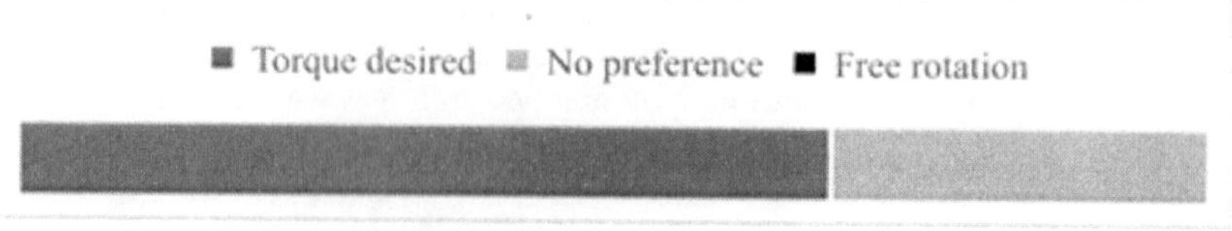

Fig. 6. Preference towards torque supported steering wheel on ingress/egress

Fig. 6 summarizes the customer preference for torque supported steering wheel for ingress and egress. Most customers expressed a preference for torque support, while some had no preference. However, there was no preference for a freely rotating steering wheel.

Alignment Strategies

During the realignment process, while the steering wheel was rotating to the desired position equivalent to the actual roadwheel position, most drivers did not disrupt the steering wheel movement. This was true for both realignment triggers, while opening the door as well as when starting the vehicle. Triggering the alignment upon grab & release was not successful. Partially, drivers did not grab the steering wheel at all or did release after ingress, so an alignment was not possible. In these cases the alignment was triggered upon starting the engine. Generally, all alignment strategies are acceptable for the driver. The strategy to rotate the steering wheel while opening the door was preferred by most customers as the alignment was finished when the driver was seated, and the vehicle is ready to drive away without any further delay.

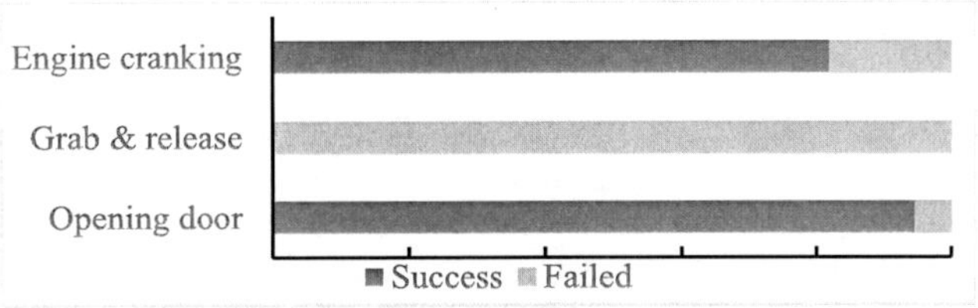

Fig. 7. Success rates of different of alignment triggers

Of the two deflection algorithms, both spring as well as damping were accepted by the customers. Most customers preferred the spring behavior as this prevent any misalignment of the steering system.

6 Conclusion

The expert studies proved beneficial to identify some core parameters of the feedback actuator and led to an expedient design of the customer clinics. Nevertheless, customer clinics indicated that some of the initial expert driver ratings were not necessarily aligned with the actual customer usage and acceptance, amplifying that for new features and uncommon behaviors expert drives have to be carefully planned, conducted and analyzed.

The number of drivers using the steering wheel to support entry of the vehicle is strongly dependent on vehicle type. Moreover, significantly more people use the steering wheel to support their exit of the vehicle compared to the entry.

A feedback torque of 20 Nm was found to be the minimum requirement to appropriately simulate end-stops. It prevents excessive over-travel in most cases. The investigated haptic warning in form of high frequency vibrations are more of a driver annoyance and can lead to misinterpretation.

Driver support for ingress and egress can be successfully demonstrated without a locking device, here the minimum assist torque needs to be at least 6-8Nm to support driver getting in and out of the vehicle. Generally, the higher the assist torque levels are, the higher the driver satisfaction. Here mainly the limited current draw is the restricting factor.

Overall, in all investigated use-cases, a SbW system can be designed to meet customer acceptance without the need of a physical locking device in the feedback actuator.

References

1. Höhfeld, M., Lodewijk, W., Nehls, O.: Enhancing the Driving Experience – Ford Adaptive Steering, Tag des Fahrwerks (2016)
2. Pfeffer, P., Harrer, M.: Lenkungshandbuch: Lenksysteme, Lenkgefühl, Fahrdynamik von Kraftfahrzeugen
3. Hestermeyer, Th., Bongarth, W., Dornhege, J.: Steering features of the new Ford Focus – 9th International Munich Chassis Symposium (2018)
4. Neukum, A. & Krüger, H.-P. (2003). Fahrerreaktionen bei Lenksystemstörungen – Untersuchungsmethoden und Bewertungskriterien. VDI-Berichte, 1791, pp. 297-318.

Steering Feel of Lane Keeping Assist Systems

Matthias Becker[1[0000-0002-2921-1667]], Manuel Höfer[2] and Peter E. Pfeffer[3[0000-0002-7026-4170]]

[1] MdynamiX AG, Germany
[2] Dr. Ing. h.c. F. Porsche AG, Germany
[3] Munich University of Applied Sciences, Germany

Abstract. Contemporary partially automated passenger cars assist the driver with the aid of longitudinal and lateral control systems without claiming the entire driving authority. The driver has the possibility to regain full control over the vehicle and relies on consistent feedback of the vehicle's driving state through haptic, optic, acoustic and kinesthetic sensing channels.

The steering system is the key element for shared lateral motion control with its steering wheel as the driver interface and its steering assist unit as the interface for lateral control systems such as lane keeping assistance. Depending on driving scenario as well as configuration and parametrization of the lane keeping assistance system, the driver can feel the system's presence and activity to a greater or lesser extent compared to steering without lane keeping assistance system.

Even if the actual lane keeping performance is improved by the assistance system, the driver might react skeptical or reject the technology due to an unpleasant steering feel. In order to solve this dilemma, an optimal cooperative assisted steering feel must be identified specifically for partial automation.

This research introduces an objective characterization method based on key performance indicators for assisted steering feel in the context of lane keeping assisted steering. Subjective assessment criteria and driving scenarios for individual evaluation are defined. Vehicle measurement data from driving maneuvers served as a basis for the extraction of characteristic key performance indicators. This method can support the development of a pleasant assisted steering feel and therefore improve vehicle safety.

Keywords: Assisted Steering Feel, Lane Keeping Assist, Objective Metrics.

1 Introduction

The great trend towards automated driving has facilitated a spread of advanced driver assistance systems on roads all over the world. The impact reaches from commercial logistics over public transport up to personal individual mobility. However further expansion especially in the latter area might be depending less on the availability or maturity of automated driving features and much more on users' acceptance. Even though the promising gains in road safety have already led to promotive steps in the international legislation and public image, the success of automated personal mobility lies mainly in the hands of the users. If the users do not approve these advanced driving features, they will not use or ultimately buy them at all, which in turn will not increase

P. E. Pfeffer (ed.), *11th International Munich Chassis Symposium 2020*, Proceedings, https://doi.org/10.1007/978-3-662-63193-5_25

road safety the way expected. To face this challenge, the development must concentrate on the driving experience from the users' point of view.

Lateral guidance as one of the core features of automated driving will be literally taken out of the driver's hands, thus it is essential to establish trust between the driver and the systems during the steps before. Today's prevalent lateral control systems basically share common operation principles, but their individual characteristics regarding availability, performance and driver interaction differ greatly. Reasons are varying interpretations or philosophies of the respective vehicle manufacturers. Some systems are calibrated to make decent interventions only when necessary while others are intended to intervene permanently and well noticeably by the driver. This is of course also wielded by manufacturers to define brand-specific behavior. Human perception on the other side is also individual as well as expectations on automated systems.

These circumstances altogether kept leading to low acceptance of lateral assistance features like Lane Keeping Assist Systems (LKAS) so far [1]. Although there are many influences on the drivers' perception when driving under assistance, the steering feel plays an important role in interactions between driver, LKAS, vehicle and environment. Untransparent or unnatural steering feel is considered a main contributor to drivers' skeptical attitudes towards LKAS.

Remarkable improvements are unlikely to be achieved if there is no method to identify the quality of steering feel systematically. A systematic approach will be addressed in this paper including detailed comprehension of the users' subjective preferences, a translation into corresponding verifiable objective key performance indicators (KPIs) and therefore a method to quantify the current state of development. In other words, a characterization of "assisted steering feel" in an objective manner.

2 Background and Literature

When addressing an objective characterization method for assisted steering feel several fields of research are involved. Driver and vehicle interaction used to be a subject typically investigated in vehicle dynamics with focus on the vehicle as well as in ergonomics and human factors research with focus on the driver. In the present context all the aforesaid disciplines contribute to the chosen approach. The following sections outline the most important knowledge and methods from the respective fields.

2.1 Partially Automated Driving

Vehicles with partial automation represent level 2 of the SAE J3016 classification scheme [2]. That means the human driver is responsible for monitoring the driving environment while being assisted in longitudinal and lateral vehicle guidance by one or more control systems. To ensure the awareness of the current driving situation the driver must keep his hands on the steering wheel and feet close to the pedals. In this way the driver can always overrule the control systems and acts as the fallback level in case of fault.

Level 2 lateral guidance systems can be divided into two variants. One variant is to refrain from intervening until the vehicle is close to the edge of the lane while the other variant is to guide the vehicle along the lane's center line, which is also called lane centering [3]. The latter system is of greater interest for this research due to its more frequent activity during operation.

Both variants share the same operating conditions as follows. The driver's hands remain on the steering wheel while a front camera module captures and processes the lane geometry and structure of the road ahead. When the vehicle leaves the desired path, a corrective control demand is transmitted to the steering control unit, where the steering system's assistance motor implements the action. A more detailed LKAS topology is shown in **Fig. 1**.

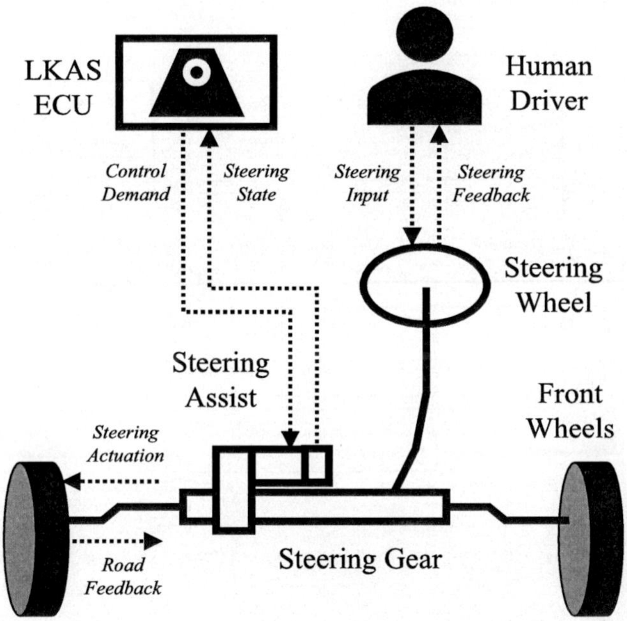

Fig. 1. Steering interaction during shared lateral control

2.2 Steering Feel

Since the driver is still constantly involved in the vehicle control loop during LKAS operation, he judges the whole driving situation by the feedback he receives through his senses. In this constellation he experiences steering feel in a conventional way. Various definitions of steering feel from the literature point out a common understanding. Steering feel describes an interaction between the driver and the vehicle. Hereby the driver causes a steering input according to his driving intention, for example following a road or taking a turn, which results in a vehicle reaction. This dynamic vehicle response is perceived and processed by the driver through haptic, visual, kinesthetic and vestibular sensory impressions. As an effect the driver draws conclusions and adapts his further driving intention accordingly [4].

The definition of Wolf [5] distinguishes steering feel in a wider sense involving the entire driver-vehicle interaction loop and in a narrower sense concentrating on the direct steering interface between the driver's hand-arm-system and the steering wheel. Primarily the latter scope is chosen in the present research because it allows a specific observation of the driver's haptic steering interaction. Assisted steering feel is therefore the steering feel in the narrower sense with LKAS switched on. Additionally, the steering wheel is assumed to reveal the biggest influence of LKAS on the driver, even though LKAS is performing the driving task. **Fig. 2** contains the detailed interaction scheme of assisted steering feel.

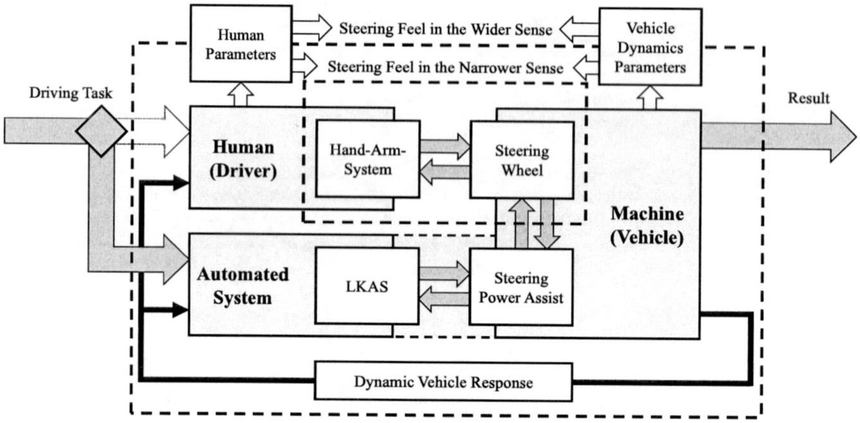

Fig. 2. Assisted steering feel in the narrower and wider sense based on [5]

2.3 Subjective and Objective Steering Evaluation

Systematic evaluation of vehicle behavior as a whole and of steering behavior in particular can be achieved by two types of established approaches. On the one hand, there are subjective assessment methods which utilize qualitative ratings of vehicle or steering attributes provided by test drivers. Common means are questionnaires and numerical rating systems that allow a systematic post-processing of a test collective. The advantages of this method are the low requirements for resources and the high coherence between the subjective human process of perception and the rating scheme. The method is however affected by the inter- and intraindividual variance of human judgement [4].

On the other hand, there are objective characterization procedures where the vehicle is set up with measurement equipment to capture relevant vehicle states during specific driving maneuvers. The recorded test data is processed and summarized into descriptive parameters or diagrams. Although more resources are needed for this method, it produces highly reproducible and comparable results, but the significance and coherence to subjective impressions from a driver's point of view is normally not given and the validity is restricted to the performed maneuvers [4].

To establish a translation between results from both methods several approaches have been published for vehicle and steering evaluation based on e.g. correlation analysis [6]. Most recent research on assessment of advanced driver assistance systems like LKAS and adaptive cruise control has introduced a correlation model based on a matrix, which assigns subjective criteria in different levels of detail to corresponding measurement channels and KPIs. The degree of correlation was hereby determined by empirical expert knowledge [7].

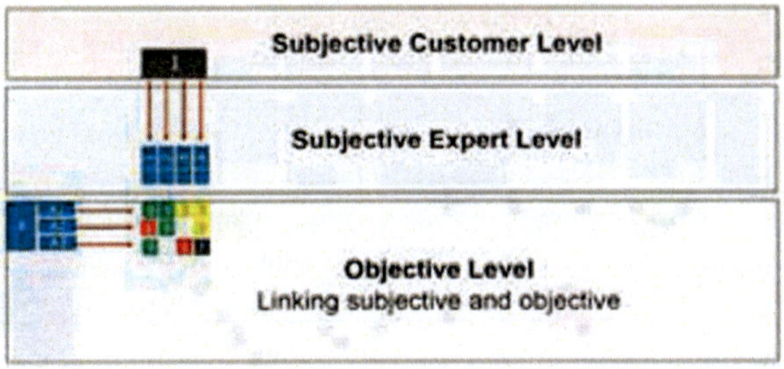

Fig. 3. Multi-level assignment model [7]

An analogical approach was also chosen in this research to transfer subjective criteria to objective attributes, which is described in the following chapter.

3 Characterization of Assisted Steering Feel

The main question to be addressed in this paper is how to quantify assisted steering feel. In this regard, quantification means characterization by objective metrics. As mentioned in the previous chapter, the approach originated from a collection of subjective assessment categories and criteria which reflect the assisted steering feel in a qualitative manner.

On this basis, driving situations that allow an isolated examination of the different subjective assessment criteria needed to be found and derived into repeatable driving maneuvers. These maneuvers were then performed while recording relevant vehicle states and steering related driver interaction quantities with the aim to extract objective KPIs correlating with the original subjective assessment criteria ratings.

3.1 Subjective Assisted Steering Feel Criteria

Previous research on systematic LKAS evaluation has already identified steering interaction as one of the most influential elements on overall driving experience from a customer's point of view [7]. The subjective steering evaluation was firstly limited to on-center steering feel and lane change behavior. Lately the assessment catalogue has been expanded by additional categories and criteria to reflect continuous and dynamic LKAS response, as well as hands-off behavior.

Table 1. Subjective assisted steering feel assessment categories and criteria

Categories	Criteria	Description
On-Center Feel	On Center Line Straight	Driving along the middle of a straight lane
	On Center Line Curve	Driving along the middle of a curved lane
	Off Center Line Straight	Driving along the edge of a straight lane
	Friction & Damping	Static and dynamic steering friction
Cooperative Steer	Steering Gap	Torque drop at return to center
	Returnability	Steering self-centering
Function Feedback	Maximum SWT	Peak torque of steering intervention
	SWT Increase	Torque build-up towards peak
	SWT Decrease	Torque relief beyond peak
Correcting Steer	Steadiness	Harmony of steering interventions
	Effort	Amplitudes of steering interventions
	Transition On-Off	Handover behavior upon deactivation
Perceived Security	Intervention Timing	Early or late intervention
	Intervention Distance	Intervention close or far from center line
	Transparency	Does the LKAS intervene reproducibly?
	Faulty Interventions	Number and severity of faulty interventions
	Road Response	Is the steering frequency response altered?
	Accuracy	Matches the LKAS path the driver's desire?

3.2 Driving Maneuver Design

When there are limited resources in terms of time or test capacity, it is practically impossible to characterize assisted steering feel of a vehicle completely for all possible situations through road testing. For that reason, the driving situation coverage was narrowed down to one significant driving scenario for each assisted steering feel criterium in favor of testing effort. Eight relevant open- and closed-loop driving maneuvers are introduced and explained in the following passage.

Weave Test. A well-known test in classical vehicle dynamics is a sinusoidal steering input for the analysis of on-center steering behavior [8]. This procedure was transferred to LKAS application with adaptions regarding test conditions and execution. Since the vehicle's lateral distance to the lane boundaries is relevant for LKAS it must be considered in the test design.

Driving in the center of a straight road with constant longitudinal vehicle velocity is the initial condition of the maneuver sequence. The weave motion is achieved by driving a sinusoidal path on the road around the center line (see **Fig. 4**). This is a notable difference to the classic vehicle dynamics weave tests where the steering wheel angle excitation has a sinusoidal shape which is not necessarily the case in this variant.

Replacing the straight road segment with a curve of constant radius and driving the sinusoidal path along the curve's center line makes it possible to examine the center feel under lateral preload. A third variant is created if the weave test along a straight road is laterally shifted towards the left or right boundary line, but with smaller amplitudes in order to stay in the corridor between center line and boundary line.

Lane Change. When driving over the lane boundary without turning on the indicator the LKAS will counteract in a certain characteristic way. Here it is crucial to suppress dynamic steering movements from the driver because the steering response would distort the characteristic counteraction force profile from the LKAS. Therefore, once the vehicle heading is brought out of center and towards the neighbor lane, the steering wheel angle should be kept constant to ensure harmonic lateral vehicle movement while driving over the line. This maneuver can be executed in both directions (right to left lane and left to right lane) and is finished as soon as the LKAS stops active operation and passes full driving authority back to the driver.

Stationary Curve. Driving with constant speeds on the road center along a curve with constant radius offers stationary cornering conditions similar to steady state circular driving [9]. This enables clear differentiations between assisted and manual driving in terms of e.g. steering effort. The stationary driving condition also allows the observation of deactivation of LKAS and thus hand-over to the driver when disabling the LKAS intentionally at any moment.

Steering Pulse. Dynamic steering responses can be provoked by quick manual steering pulses where the steering wheel is released after reaching the maximum pulse amplitude. The subsequent transient response reflects the effects of e.g. unintended steering wheel disturbances similar to ISO 17288-2 [10]. A straight road and stationary initial conditions are preliminary.

Curve Entry. Frequently encountered sections on public roads are straight segments followed by curves or vice versa. This maneuver and the following one represent scenarios on such combinations. In a curve entry, where a straight segment leads to a curve, the line crossing behavior can be examined in a different approach.

The vehicle's heading and velocity is fixed facing in the straight direction and held constant after entering the curve for as long as the LKAS is still operating. While the lane change maneuver is conducted on a straight road, the curve entry incorporates an increasing road curvature on which some LKAS may act sensitively.

Curve Exit. Leaving a curve towards a straight road segment makes it possible to observe the steering returnability. Referring to [11] this can be achieved by releasing the steering wheel in the moment the curve ends. This maneuver and the returnability in general are, however, relevant at speeds lower than ca. 60 km/h, which is below the minimum operating velocity of most LKAS available today.

Free Run. While all previous maneuvers had their focus on particular driving scenarios, the Free Run maneuver is intended to examine the steering feel in a statistical manner. Random or predefined road section samples are driven twice on the same lane with the same speed and by the same driver, but once with LKAS activated and once without. All constrains are chosen to allow a differential examination of both runs.

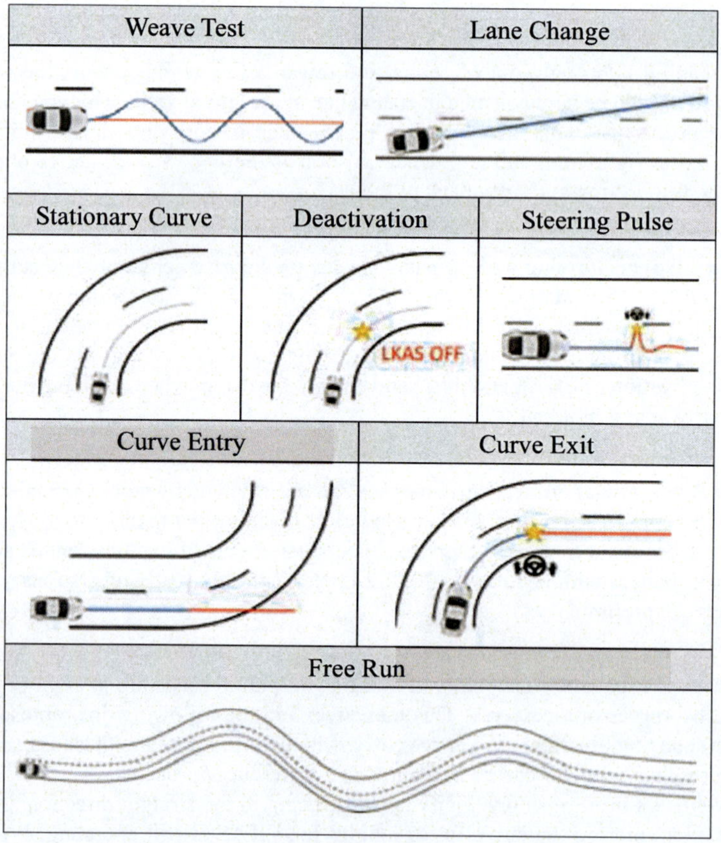

Fig. 4. Driving maneuver schematic overview

Maneuver Assignment. The purpose of all driving maneuvers is to capture situations that allow an assessment of corresponding subjective criteria and finally enable the extraction of characteristic objective data. The following illustration shows which criteria correspond to which maneuver.

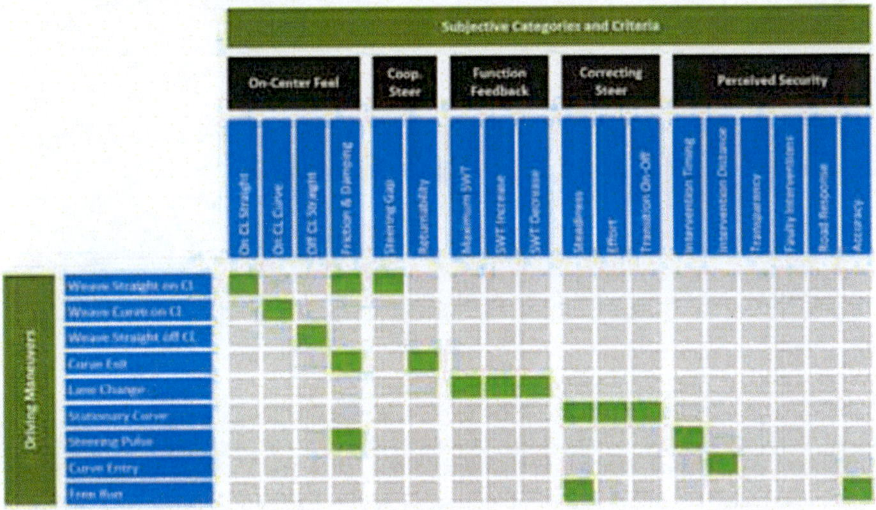

Fig. 5. Assignment maneuvers to subjective criteria (CL acronym for center line)

Some subjective criteria remain unassigned because suitable assessment procedures do not exist yet or require high expenses to develop corresponding quantification methods.

3.3 Vehicle Measurement

The introduced approach requires measurement data from the driven maneuvers for the calculation of reliable objective metrics or KPIs. A special set of measuring equipment is necessary for steering feel testing of standard LKAS equipped passenger cars. Even though many relevant signals can be taken from the vehicle's built in sensors via on-board communication buses, some channels are not sufficiently accurate for the upcoming evaluations and need additional external measuring devices. Furthermore, comparability of LKAS assisted steering feel between different vehicles is only ensured with identical measuring setups.

Great attention was drawn on the acquisition of the steering wheel torque (SWT), steering wheel angle (SWA) and its derivative steering wheel angular velocity (SWV) as they represent the feedback from the assisted steering procedure to the driver and the driver's steering intention, respectively. For this purpose, a universal measurement steering wheel hub was inserted between the original steering wheel and the upper steering column end as displayed in **Fig. 6**. This setup makes the steering wheel torque and angle measurement independent of the steering system configuration where internal torque and angle sensor readings might not contain friction and compliance from the mechanical parts between the driver and the sensor.

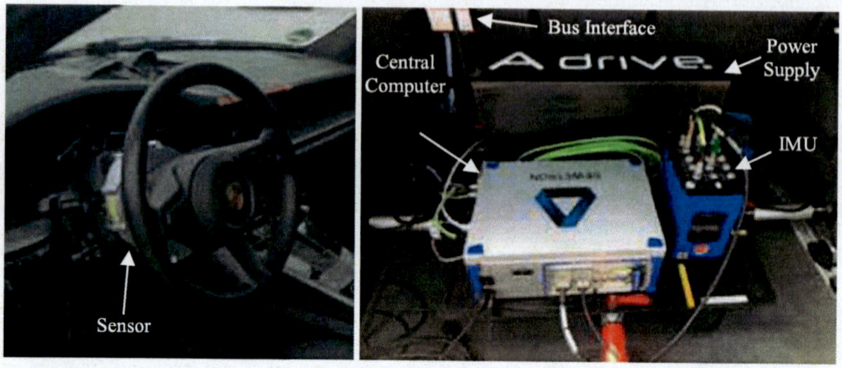

Fig. 6. Measurement steering wheel (left) and portable setup for ground truth positioning (right)

For most of the introduced driving maneuvers it is essential to know the exact position of the vehicle on the road with respect to lane properties such as center line, e.g. the distance to center line (D2CL), or boundary lines. Even though the on-board LKAS camera normally calculates and communicates relevant lateral offsets by itself in order to serve its function, validity of vehicle positioning was assured by a ground-truth approach as described in [12].

Emphasis was put on simplicity during design of the characterization maneuvers, so that all steering procedures can be conducted by hand. Yet it yields increases in measurement quality and repeatability when maneuvers like Lane Change or Weave Test are performed with the aid of a steering robot. The model in use offered a path-following feature via vehicle ground position control, where the initial vehicle heading can be modulated by sinusoidal patterns or yaw steps.

3.4 Objective Evaluation

The final step towards a quantitative evaluation method was the extraction of objective parameters from the measurements in the form of KPIs. For the Weave Test and other maneuvers related to international standards, the therein proposed evaluations have been taken into consideration as well as evaluations from the literature [6]. The necessary KPI calculations were performed in the MXeval software framework [13].

While most KPIs are self-descriptive due to being based on simple calculations such as maxima, minima, average or effective values of measurement signals or their time derivatives, certain indicators require a more detailed description as follows.

Steering Stiffness. Several definitions for steering stiffness describe a steering wheel torque gradient with respect to steering wheel angle. An example definition is the average torque gradient in Nm/deg within ±10% of the maximum steering wheel angle around the steering center (0 deg).

The herein used stiffness KPIs such as steering or lateral stiffness refer to the respective gradients from maximum to minimum angle or D2CL value. This definition was chosen considering all previously introduced variants of weave test where measurement precision was not given for finer steering stiffness intervals.

Push and Pull. Dynamics of a manually guided steering wheel can be abstracted into two non-static conditions. When the steering wheel torque acts in the same direction as the current direction of motion in terms of angular velocity it is physically accelerated and thus the driver senses it as a pulling force regardless of the current angular position. The opposite state, namely push force, is a decelerating torque which makes the driver push against it in order to maintain the current steering wheel motion. The proportions and alternation between both states have proven to act as important indicators for assisted on-center steering feel.

Table 2. Assisted steering feel KPIs

Maneuver	Domain	KPIs
Weave Test	SWT vs. SWA	Steering stiffness SWA and SWT dead bands SWA and SWT mean value (center point) Push and pull sections Hysteresis area
	SWT vs. D2CL	Lateral stiffness D2CL and SWT dead bands D2CL and SWT mean value (center point) Hysteresis area
Lane Change	SWT vs. D2CL	D2CL dead band width SWT build-up gradient SWT plateau location SWT drop-off gradient
Stationary Curve	SWA vs. Time	SWA mean value and standard deviation
	SWT vs. Time	SWT mean value and standard deviation SWT drop-off height and rate during deactivation
Curve Entry	SWT vs. Travel	Distance from curve start to SWT threshold
Curve Exit	SWA vs. Time	Remaining SWA
	SWV vs. Time	Max. SWV during returning
Steering Pulse	SWA vs. Time	Ratio SWA peak to peak
	SWV vs. Time	Max. SWV during returning Max. steering wheel acceleration
Free Run	SWA vs. Time	SWA standard deviation
	SWV vs. Time	SWV standard deviation
	SWT vs. Time	SWT standard deviation

4 Results from Vehicle Test

The previously introduced objective evaluation method for assisted steering feel was developed for any kind of test environment, either virtual vehicle dynamics simulation, real world road test or intermediate stages such as software- or hardware-in-the-loop (SiL/HiL). The following result visualizations however were taken from different road test sessions with the aim to outline the meanings of different chosen assisted steering feel KPIs.

4.1 Weave Test

Analysis of exemplary straight road weave tests has shown significant differences in steering hysteresis shapes shown in **Fig. 7** between LKAS off and on. While lying in the acceptable ranges for conventional steering feel KPIs [4] without LKAS, the intervention torque of an active system causes pull phases to last far beyond the steering center and leading to a shorter but steeper push phase shortly before the turning point. This observation corresponds well to the subjective impression of a steering gap.

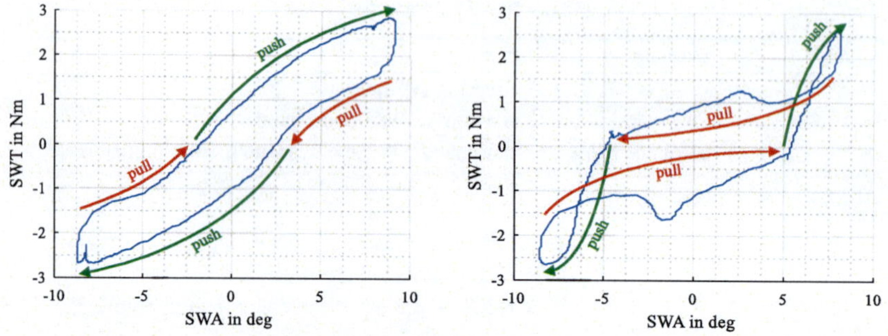

Fig. 7. Weave test SWT vs. SWA diagrams with LKAS off (left) and on (right)

4.2 Lane Change

A lane change procedure as introduced in the previous chapter creates a characteristic torque profile containing three characteristic phases: build-up, plateau and drop-off shown in **Fig. 8**. The shape of the build-up curve is not necessarily linear due to the underlying lane keeping algorithm and its calibration. During the plateau phase the maximum torque level is especially relevant for overruling the LKAS. Typical plateau heights range from ±1 Nm to ±3 Nm where the upper limit has compulsory requirements by law [14]. Relevant for the subsequent drop-off phase is the location with respect to the boundary line and the torque gradient before the LKAS stops operation or locks into the neighbor lane if there is any.

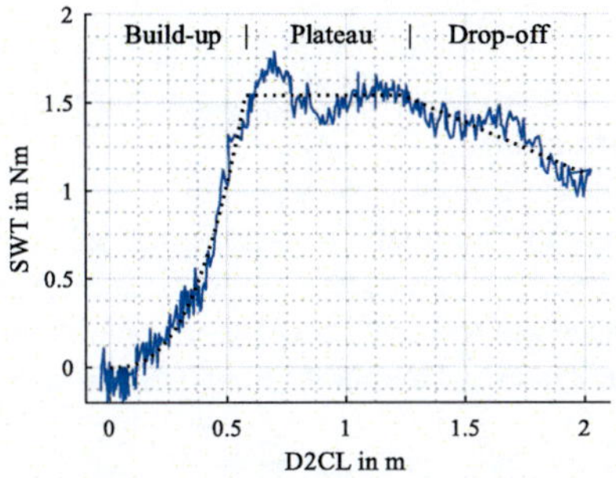

Fig. 8. Lane Change SWT vs. D2CL diagram

4.3 Stationary Curve

Average torque and angle effort of manual and assisted steering are the main KPIs evaluated by the stationary curve maneuver. The fluctuations of both quantities bring additional information about steadiness and harmony of steering interventions at static external conditions. The steering wheel torque diagrams in **Fig. 9** originate from a bend to the right with a radius of approximately 1000 m. The left diagram shows the steering wheel torque signal and mean values for both manual steering and with active LKAS. The right diagram contains the transient torque progression in the event of a driver induced LKAS deactivation out of stationary conditions, where the difference in torque levels and the gradient in between are characteristics for every system.

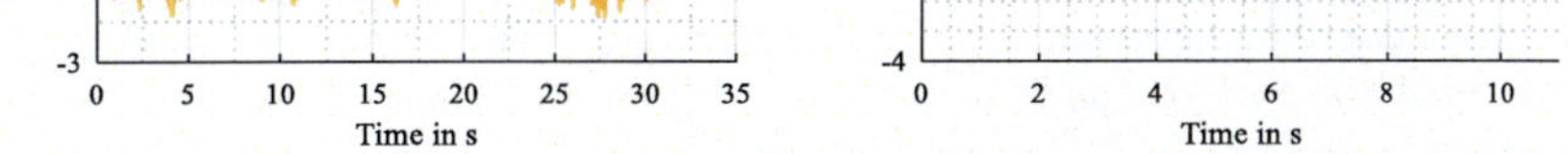

Fig. 9. Stationary Curve SWT diagrams continuous (left) and with LKAS deactivation (right)

4.4 Free Run

From all evaluation maneuvers the free run is the most subjectively affected one, because it requires a human driver to follow the same route twice, once with enabled LKAS and once disabled. It is intended to quantify the subjective steering accuracy which usually depends heavily on the individual driver's route following and corner cutting preferences. A driver who prefers to drive in the lane center will endorse an

LKAS which is designed in this manner, whereas a curve cutting or aggressive driver will dislike it and vice versa. Nevertheless, the connection to the subjective steering accuracy rating will not be impaired by this approach, but the validity will always be limited to certain groups of drivers.

Evaluated measurement data for a pair of free runs contains the characteristic information less in the absolute amplitudes of e.g. steering wheel angle or torque but much more in the respective deviations. Given the angle sensor recordings in **Fig. 10**, the fluctuations due to corrections are lower with LKAS on, which means the driver agrees with the path driven by the system. This result matched the perceived positively rated steering accuracy.

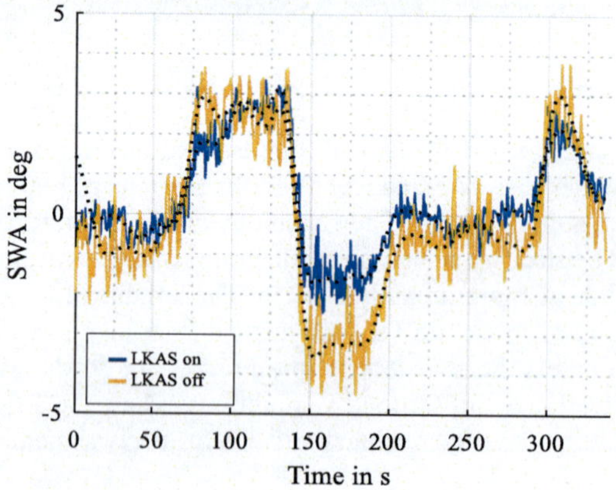

Fig. 10. Free Run SWA diagram with active and disabled LKAS

5 Conclusion and Outlook

In this paper a KPI based characterization method for assisted steering feel was introduced and exemplified by experimental vehicle measurement data. The research firstly defined subjective assessment criteria and the derivation of driving scenarios for individual evaluation. Measurement data from the driving maneuvers was then used to extract characteristic objective attributes. After multiple applications of the characterization method on different vehicles, most of the introduced KPIs have been proven to reflect the subjective assessment result accurately.

Further enhancements in the method are e.g. the consideration of adaptive systems and even the application on SAE level 3+ systems when assessing steering handover or takeover scenarios. A fundamental correlation analysis should be conducted with sufficient subjective and objective assessment data to further validate the selected KPIs.

Having validated objective target values of KPIs in general will lead to an effective employment of virtual methods during the otherwise extraordinarily time consuming and expensive development process of automated driving features. In terms of LKAS this research should contribute to this goal.

6 References

1. Aydogdu, S., Schick, B., Wolf, M.: Claim and Reality? Lane Keeping Assistant – the Conflict between Expectation and Customer Experience. In: 27th Aachen Colloquium Automobile and Engine Technology 2018, Aachen (2018).
2. SAE J3016™ JUN2018 – Taxonomy and Definitions for Terms Related to Driving Automation Systems for On-Road Motor Vehicles. SAE International (2018).
3. Winner, H., Hakuli, S., Lotz, F., Singer, Ch.: Handbuch Fahrerassistenzsysteme. 3rd edn. Springer Vieweg, Wiesbaden (2015).
4. Pfeffer, P., Harrer, M.: Lenkungshandbuch. 2nd edn. Springer Vieweg, Wiesbaden (2013).
5. Wolf, H.: Ergonomische Untersuchung des Lenkgefühls an Personenkraftwagen: Dissertationsschrift. Südwestdeutscher Verlag für Hochschulschriften (2010).
6. Harrer, M.: Characterisation of Steering Feel. Thesis, University of Bath, Department of Mechanical Engineering, Bath (2007).
7. Schick, B., Fuhr, F., Höfer, M., Pfeffer, P.: Eigenschaftsbasierte Entwicklung von Fahrerassistenzsystemen. Automobiltechnische Zeitschrift 121(4), 70–75 (2019)
8. International Standard ISO 13674-1:2010 – Road vehicles – Test method for the quantification of on-centre handling – Part 1: Weave test. International Organization for Standardization, Geneva (2010).
9. International Standard ISO 4138:2012 – Passenger cars – Steady-state circular driving behaviour – Open-loop test methods. International Organization for Standardization, Geneva (2012).
10. International Standard ISO 17288-2:2011 – Passenger cars – Free-steer behaviour – Part 2: Steering-pulse open-loop test method. International Organization for Standardization, Geneva (2011).
11. International Standard ISO 17288-1:2011 – Passenger cars – Free-steer behaviour – Part 1: Steering-release open-loop test method. International Organization for Standardization, Geneva (2011).
12. Schneider, D., Huber, B., Lategahn, H., Schick, B.: Measuring method for function and quality of automated lateral control based on high-precision digital Ground Truth maps. 34. VDI/VW Gemeinschaftstagung: Fahrerassistenzsysteme und automatisiertes Fahren 2018, Wolfsburg (2018)
13. MXeval Product Sheet, https://mdynamix.de/en/mx-products/mxeval/, last accessed 2020/05/11
14. UN Regulation No. 79 – Uniform provisions concerning the approval of vehicles with regard to steering equipment. Revision 3, United Nations (2017)

Enabling flexible Interiors for existing vehicle platforms with a Steering Wheel integrated Steer-by-Wire Force Feedback Actuator

Dr.-Ing. Christian Strümpler[1], Roland Grimm1 and Daniel Alt[1]

[1] Joyson Safety Systems, Germany

Abstract. Current solutions for steer by wire force feedback actuators (FFA) based on steering column-integrated actuators require a huge effort for integration as design spaces used by head up display, climate systems or knee airbags are infringed. Additionally the crash element in the steering column limits the available retraction length of the steering wheel. A solution highly integrated in the steering wheel based on Joyson Safety Systems (JSS) active front steering concept enables the introduction of Steer-by-Wire (SbW) for any vehicle platform and automation level along with a maximum interior flexibility: Ranging from the pure integration of SbW-Systems in existing vehicle platforms to retractable or foldable and stowable steering wheels.

Keywords: Steer-by-Wire, force feedback, steering wheel, integration, stowable.

1 Introduction

As for safety reasons SAE Level 3-5 autonomous vehicles and future advanced driver assistant systems (ADAS) require more and more redundant electric drives, communication interfaces and power supply for brake or steering actuators and control units, one of the major cost drivers of Steer-by-Wire (SbW) systems will be introduced in future vehicles. Thus, the introduction of SbW Systems is also becoming more likely. Furthermore, SbW enables flexible interiors featuring retractable, foldable or stowable steering wheels, as these systems require a decoupling of the steering wheel rotation while driving autonomous. These are major end-customer benefits of automated driving as space for the driver is generated enabling him to pursue side tasks like eating, reading, working or relaxing.

Current solutions for force feedback actuators (FFA) based on steering-column-integrated actuators require a huge effort for integration as design spaces used by head up display, climate systems or knee airbags are infringed. Additionally the crash element in the steering column limits the available retraction length of the steering wheel. A FFA highly integrated in the steering wheel based on JSS's active steering wheel (see Figure 1) enables the introduction of SbW for any vehicle platform and automation level along with a maximum interior flexibility.

P. E. Pfeffer (ed.), *11th International Munich Chassis Symposium 2020*, Proceedings, https://doi.org/10.1007/978-3-662-63193-5_26

2 Steering Wheel Integrated Force Feedback Actuator

With the Active Steering Wheel (ASTW) [1] Joyson Safety System introduced, an ASIL D compliant fail-safe steering actuator integrated into the steering wheel (see Figure 1). This active steering unit (ASTU) consists of a brushless DC (BLDC) motor with a worm shaft, fail-safe locking unit, electronic control unit (ECU) and worm wheel. The bearing support surfaces on a skeleton inlay are casted in the steering wheel skeleton, which is then part of the gear housing. The motor is rotating with the steering wheel enabling the basic function of a steering angle overlay. This means the steering column angle is the sum of the steering wheel angle set by the driver and the actuator angle set by the ASTU. The Active Steering wheel supports variable gear ratio and further ADAS features like, vehicle stabilization, side wind compensation or lane keeping in vehicles with hydraulic and electric power steering systems. Furthermore, an ASIL D certified fail-safe absolute steering wheel angle sensor (SAS) is part of the active steering wheel. It is located in the housing of the Clockspring Module.

By a fixation of the worm wheel hub the Active Steering Unit can be used as an force feedback actuator for SbW Systems. It is now possible to apply a feedback torque to the steering wheel with the BLDC motor of the active steering unit. As the Active Steering Wheel is a fail-safe system, which is locking the gear when a failure is detected, several requirements change when turning the system to a fail-operational FFA for SbW Systems:

- Maximum actuator torque requirement decreases,
- Gear set misuse torque requirements decrease,
- Maximum actuator speed requirement increases,
- The locking unit needs to be replaced by a mechanical multi-turn end stop to limit steering wheel travel,
- The steering wheel angle sensor needs redundancy,
- A fallback for torque generation is required,
- Redundancy in power supply and communication with the road wheel actuator (RWA) are required.

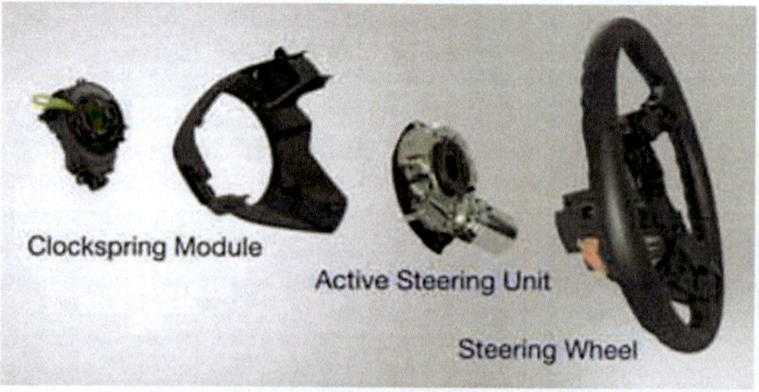

Fig. 1. Steering Wheel integrated Steering Actuator

These significant changes in the requirements lead to a complete redesign of the Active Steering Unit. The new gearbox design based on the decreased maximum torque requirements results in a significantly reduced actuator package (see Figure 2). Hereby the integration of the actuator in current contemporary steering wheel designs is facilitated.

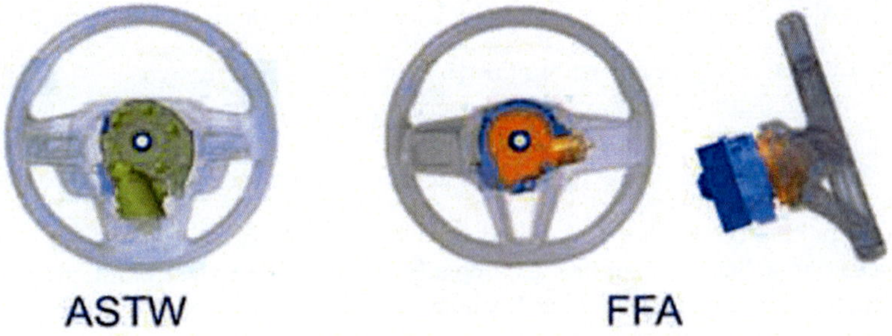

Fig. 2. SbW FFA Package

A multi turn end stop is part of the gear set and the spatially separated actuator ECU inside the steering wheel delivers a second ASIL D certified steering wheel angle signal. Consequently, a fail-safe steering wheel angle signal is two times redundant available (1 out of 2 fail-safe concept). Within the Clockspring Module an optional steering torque sensor can be implemented for safety or control purpose.

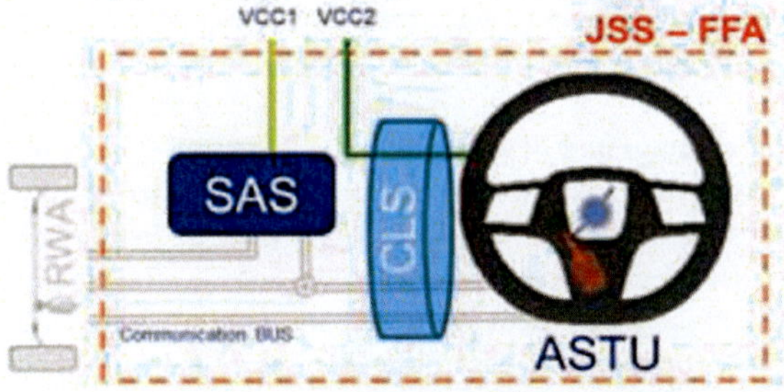

Fig. 3. SbW FFA System Layout

To meet fail-operational requirements each ECU has a separate redundant power supply. The high current power supply of the ASTU is transferred via a high power clockspring (CLS) as well as the signals of the torque sensor and two redundant communication channels (see Figure 3). In case of a failure in torque generation, the system can rely on either a passive feedback torque or a redundant motor design.

Thus, the system is fail-operational relying on

- Redundant power supply
- Spatially separated independent ECUs,

- Redundant communication channels for each ECU,
- Redundant fail-safe multi turn steering wheel angle signals,
- Either a passive feedback torque or a redundant motor design.

3 Integration in existing Vehicle Platforms

Based on the complete integration of the FFA into the steering wheel the effort to implement a SbW System in an existing vehicle platform is significantly lower, as

- No additional design space within the dashboard is necessary,
- Existing mechanical interfaces of the steering column to the dashboard can be used,
- A rotating steering column hub is no longer needed,
- Any kind of fail-operational road wheel actuator can be used.

Besides any kind of steering rack based road wheel actuator also frequently in small cars used column based actuators are compatible with a steering wheel integrated feedback actuator (see Figure 4) as the FFA does not use the bespoke design space within the dashboard.

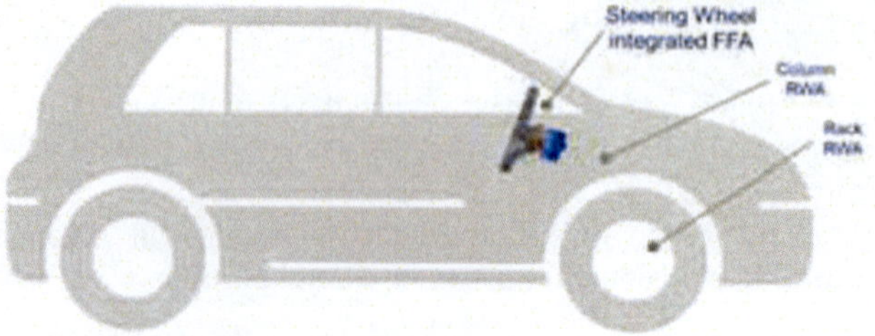

Fig. 4. SbW Integration options with Steering Wheel integrated FFA

The steering wheel integrated FFA requires for integration some smaller changes on the vehicle side: A reinforcement of the instrument panel carrier may be necessary to control the natural frequency behavior of the steering wheel. While the missing steering linkage reduces structure borne oscillation transmission to the steering wheel, the increased weight of the steering wheel decreases the natural frequency behavior of the steering wheel and steering column assembly. The increased weight can partly be compensated by a decrease in weight of the modified steering column. Nevertheless, natural frequency behavior and necessary measures must be reviewed for the concrete vehicle environment. Furthermore adaptations of the existing steering column control module including the stalk switches are required to package the non-rotating Clockspring Module of the FFA.

As these smaller modifications do not require any changes to major structural parts or major surrounding components a SbW system with steering wheel integrated FFA can be optionally assembled on the same production line to any kind of vehicle. This enables an assembly of SbW for new vehicles based on existing platforms if required by optionally available ADAS, automated driving systems, or selected by the customer.

4 Flexible Interiors

Having in mind the above-mentioned integration advantages of a steering wheel integrated FFA for existing platforms the design space occupied by todays steering columns can be reused for a steering wheel retraction or stowing mechanism. Based on the SAE definition of autonomous vehicles [2] two different applications and functions are in focus:

1. Highly automated driving (SAE Level 3): The driver is still responsible to drive after a system request. The Steering Wheel as one of the major control elements for manual driving [3] must still be reachable. Thus, a retraction of the steering wheel to free space for driver's side tasks and comfort is possible but limited.
2. Fully automated driving (SAE Level 4): The driver is not responsible to take over manual control, as the system can operate safely in a certain domain. In this case a complete stowing of the steering wheel is possible to free a maximum of space for the driver.

4.1 Retractable Steering Wheel

In case of a retractable steering wheel, the retraction length is limited to the reachability of the steering wheel in case of a necessary takeover of the driver. This means the adjustment speed of the steering wheel is relevant for the maximum travel in the retracted position. Assuming a take over time of 10 seconds and doubled adjustment speed of today's steering column adjustments (30 mm/s) a maximum retraction length of 300 mm can be realized. To use this theoretical retraction length in most of the vehicles changes in dashboard design are necessary to avoid a collision between dashboard surfaces and steering wheel or stalk switches. However, in order to be compatible in a highly dense package environment of the instrument panel, a complete redesign of the steering column is necessary. The new design must not only fulfill the requirements of the increased longitudinal adjustment and adjustment speed but also current requirements on energy absorption in case of a vehicle crash. The steering wheel integrated FFA allows a maximum usable design space for the new retractable steering column as the major part of the actuator is integrated in the steering wheel and additional design space used for hub joints are not necessary. Thus, in comparison to other FFA concepts the complete design space of today's adjustable steering column is available for the redesign.

4.2 Stowable Steering Wheel

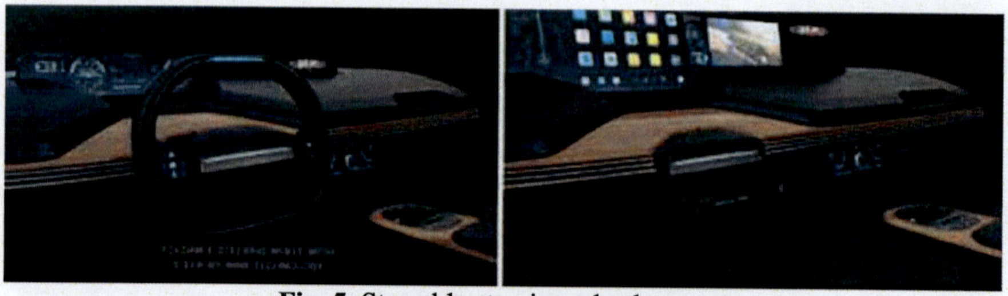

Fig. 5. Stowable steering wheel concept.

For fully automated vehicles, the steering wheel can be completely stowed in or underneath the dashboard as the unfolding is not limited to take over times. To minimize the necessary design space in the stowed position a rotation of the steering wheel rim or the complete steering wheel including the FFA by 90° is preferred (see Figure 5). In order to minimize the necessary impact on the dashboard package it is recommended to stow the steering wheel underneath the dashboard in the driver's foot well. This requires an additional increase of steering wheel retraction by half of the steering wheel height. Assuming the maximum retraction length of 300 mm a complete maximum longitudinal travel of approximately 480 mm is necessary.

Again, the design space of a steering column can be reused with a new design to enable the stowing of the steering wheel without major infringements of design spaces of head-up displays, climate control systems or other components located in the instrument panels.

While the rotation of the steering wheel rim by 90° is very steering column and dashboard package efficient it requires additional gear sets and joints inside the steering wheel. The rotation of the complete FFA to stow the FFA and steering wheel base inside the steering column is less complex but requires more effort with regards to steering column and dashboard package. Both solutions require a significant effort to realize the stiffness of the steering wheel and steering wheel column in the driving position.

5 Summary

With a steering wheel integrated force feedback actuator for steer by wire systems Joyson Safety Systems offers a highly flexible solution for the integration of SbW Systems and flexible interior concepts for automated vehicles. A consequent redesign of the Active Steering Unit along SbW requirements reduces system size and weight facilitating the integration in current contemporary steering wheel designs. As the feedback actuator gear is completely integrated into the steering wheel no additional design space within the dashboard is required. Thus, an integration in existing vehicle platforms with rack or column based road wheel actuators requires only smaller changes for the instrument panel support and steering column. Furthermore retractable or completely stowable steering wheels are enabled. Based on the steering wheel integrated FFA possibilities to integrate the retraction or stowing mechanism exist by reusing the given design space of the steering column in current vehicle platforms.

References

1. Grimm R. (2014) Active Steering Wheel, a new solution to implement active steering functionality into a vehicle. In: Pfeffer P. (eds) 5th International Munich Chassis Symposium 2014. Proceedings. Springer Vieweg, Wiesbaden
2. SAE International: J3016 - Taxonomy and Definitions for Terms Related to Driving Automation Systems for On-Road Motor Vehicles, June 2018
3. Strümpler C., Alt D., Ruck H. (2020) Future steering wheels for automated vehicles. In: Pfeffer P. (eds) 10th International Munich Chassis Symposium 2019. Proceedings. Springer Vieweg, Wiesbaden

Objective and Subjective Steering Feel HiL Testing on a New High Performance Steering HiL with Virtual Vehicle Prototypes

-

Objektive und subjektive Lenkgefühluntersuchungen auf einem neuartigen Hardware-in-the-Loop Hochleistungs-Lenksystemprüfstand mit virtuellen Fahrzeugprototypen

Markus Lemmen [1], Michael Peperhowe [2], Olav Lange [3], Michael Moczala [4]

[1] Bochum University of Applied Sciences Campus Velbert/Heiligenhaus, Germany
[2] dSPACE GmbH, Germany
[3] Ford Werke GmbH, Germany
[4] ZF Automotive Germany GmbH, Germany

Abstract. HiL testing as a standardized methodology takes real components into consideration, such as entire steering systems. The capabilities and possible use cases of these systems depend on the performance of the test bench and the maturity of the underlying real-time model. This paper describes a new level of performance steering test benches including the possibility of subjective steering feel tests.

Keywords: Steering Test Bench, HiL, Objective Test, Subjective Test, Real-Time Simulation, Steer-Feel.

1 Introduction

Nowadays, HiL testing is a state-of-the-art technology for ensuring a safe and robust development of software functions, mechanical components, and their integration into mechatronic steering systems of current vehicles. Particularly, steering system test benches are used for release tests on complete steering systems. In the V development process for mechatronic systems, the tests performed on a steering system HiL generally serve as a last assessment before the system is released and signed off for in-vehicle testing and finally for series production. Fig. 1 depicts this type of situation: Automotive system suppliers first try to validate subsystems as complex as possible before performing final validation tests on a full system-level HiL simulators, e.g. like steering system test benches as discussed in this paper. These HiL simulators that test steering subsystems (from top to down) prior to the steering system HiL simulator are called

P. E. Pfeffer (ed.), *11th International Munich Chassis Symposium 2020*, Proceedings, https://doi.org/10.1007/978-3-662-63193-5_27

EPPiL (Electric-Power-Pack-in the-Loop) and ECUiL (Electronic-Control-Unit-in-the-Loop). Even before the HiL simulation, the software is tested in an earlier validation phase, e.g., via SiL (Software-in-the-Loop) simulation.

In general, the following rule applies: The earlier in the validation or development phase, the more tests will be performed up to final sign-off though less hardware is available. This implies that it is necessary to model more of the hardware components. As soon as more system or sub-system hardware is available, HiL simulators testing these increased complex hardware assemblies can be used to validate and optimize the models, which are later required in the validation and development phase. Therefore, a HiL simulator, such as a steering system HiL simulator, is not only used as the last/final validation testing unit but also to obtain and validate models used for other HiL or SiL simulations.

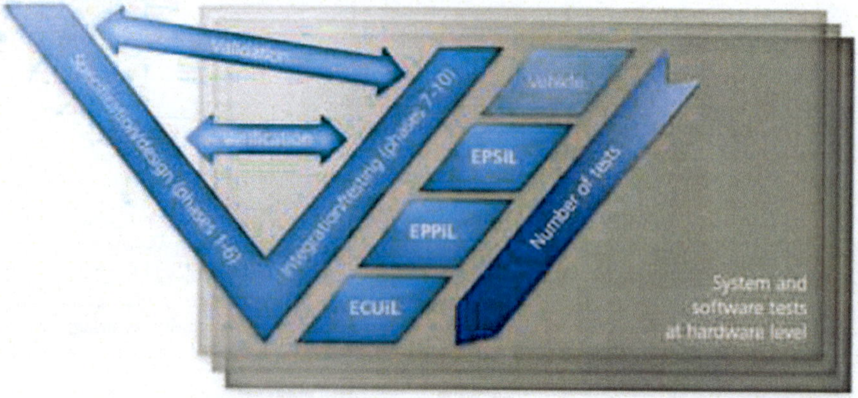

Fig. 1. Different HiL test benches used for testing various hardware complexity levels of components [1].

This paper presents a new generation of high-performance steering system test benches for closed-loop HiL operations (

Fig. 2) located at the Campus Velbert/Heiligenhaus (CVH) of the University of Applied Sciences, Bochum (Germany), referred to as CVH steering system HiL from now on in this paper. In particular, the system allows for subjective assessments prior to vehicle testing. While the general setup of these kinds of steering system HiL simulators currently is such that the basic layout consists of a wheel actuator (electric motor) and one or two high-load linear actuators (linear magnetic drives), this new HiL system also consists of an additional linear actuator of significantly lower mass/inertia. Furthermore, this paper also provides information on the first experiences gained with significantly improved friction compensation control for the linear actuators of the HiL and its impact with respect to steering system testing and steering system model parametrization.

The above-mentioned additional linear drive enables investigating devices under test with different actuators, which can significantly improve the knowledge gained during test bench measurements thanks to a better distinction between the dynamics of the device under test and the test bench (actuator) dynamics. This can help improve the modeling capabilities for future developments.

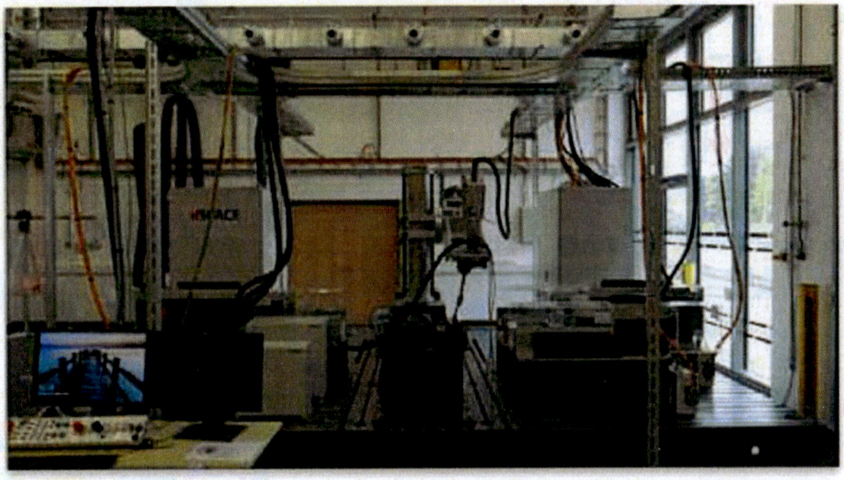

Fig. 2. The new CVH steering system HiL from dSPACE.

The additional smaller linear actuator also allows for performing subjective assessments of the device under test by a human 'test driver' on the HiL simulator. For this 'HiL test drive' use case, the device under test (DuT) can also be integrated into a full vehicle simulation.

One of the main goals of this paper is to present the newly achieved performances, especially in comparison to state-of-the-art test benches and to give an outlook on additional potential fields of operation that are new to the market. For example, changes to prototypes of the steering system can be assessed not only by means of measured data but also subjectively. Even different vehicles or vehicle prototype models can be assessed with respect to the subjective steering performance.

2 CVH Steering System Test Bench

Steering system HiL simulators are intended for testing complete steering systems and, if possible, subassemblies of an entire steering system, such as steering columns or steering gears. Consequently, steering system HiL simulators ideally consist of actuators generating the driver input or driver load and one or more actuators generating the rack loads that represent the forces coming from the road-tire contact through the suspension into the steering system.

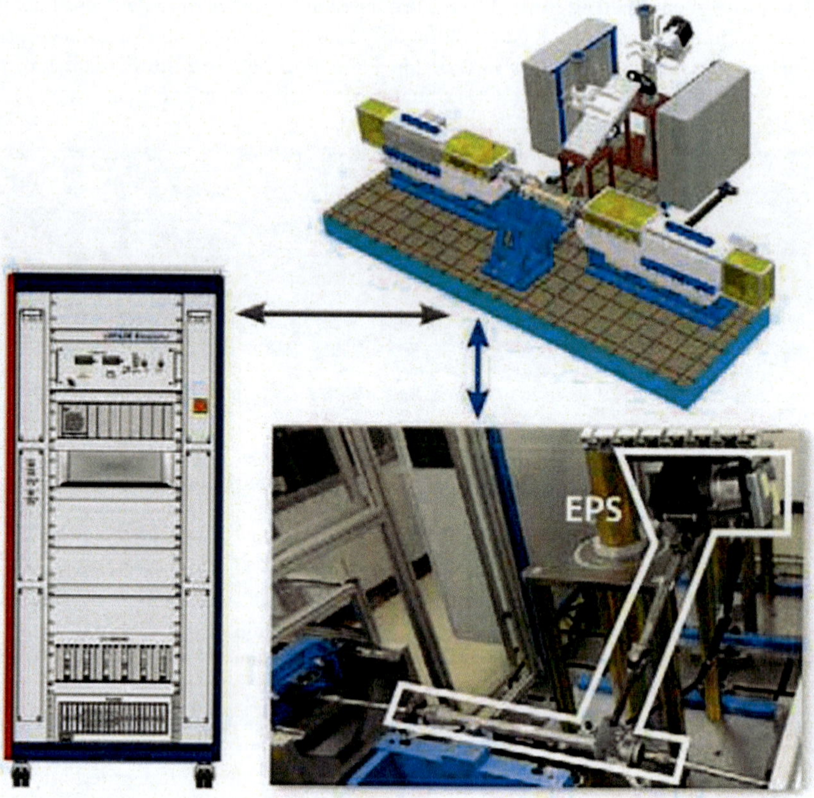

Fig. 3. State-of-the-art steering system HiL setup (e.g., [1],[6]).

Therefore, many steering system HiL simulators are set up in a structure that is identical or similar to the setup shown in Fig. 3: One or two steering gear rack load actuators provide the road loads, and a rotary drive provides the steering wheel loads. The setup is illustrated in Fig. 4 and can be found in different steering system HiL simulators at different OEMs and suppliers as well at research centers and universities. For convenience, the left (tie) rod actuator will be referred to as RLA, the right one as RRA, and the steering wheel actuator as WA. Usually, all these actuators can be controlled in different operation modes to offer force or torque control and position/speed or angle/steer speed control. Sometimes, even more sophisticated and more complex operation modes can be implemented and automated. The actual forces are often measured with force cells and the steering torque is usually also tracked. Additionally, actuator positions are documented, and the related speed is derived from these measurements.

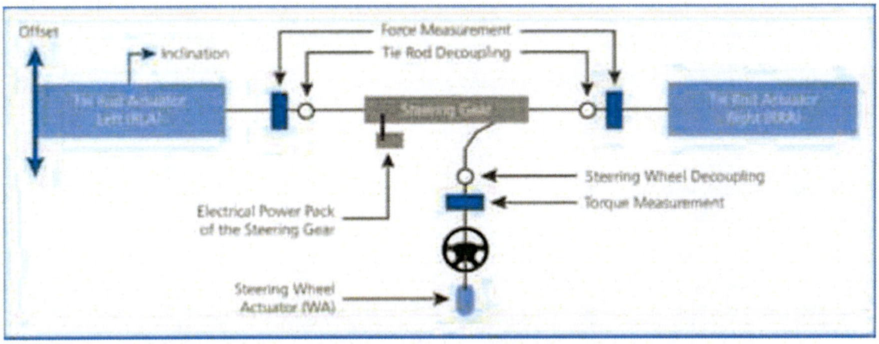

Fig. 4. Sketch of state-of-the-art steering system HiL simulators.

Basically, in this state-of-the-art setup the device under test (DuT) can be altered allowing for individual adjustments of the relative position and inclination between RLA and RRA with respect to the steering system. If the focus is on the reproducibility of measurements (measurement/test data quality) and reduction of machine setup time, the preferable test bench setup will be as shown in Fig. 4: The RLA, RRA, and steering gear are assembled in the same effective force axis. If the focus is on a more realistic scenario as seen in a real car, the assembly will have an assembly offset in the vehicle's direction of travel and height, and an inclination angle in orientation might also be considered. These assembly options can already be implemented in some of the state-of-the-art steering system HiL simulators. However, usually the focus is on the reproducibility and reduction of machine setup time and the setup shown in Fig. 4 is the standard setup found in almost every steering system HiL simulator. In current state-of-the-art steering system HiL simulators, the steering wheel depicted in Fig. 4 is exchanged by the steering wheel actuator (WA) to generate the driver's steering wheel input.

The new CVH steering system HiL simulator offers a number of features and details that exceed the state-of-the-art possibilities that are used as a standard: First, the new test bench offers three instead of two linear rack/road load actuators. The additional actuator is a smaller electric linear drive that supplies a max. force of 3.5 kN with significantly decreased inertia compared to the larger linear drives, which deliver max. loads of 20 kN. The load cells for force measurement are adopted to enable higher measurement accuracy for the smaller load case. In this context, moving the force measurement closer to the DuT, i.e., between the decoupling of the actuator and the DuT, also allows for a more accurately controlled force load for the intended tests. Identical to the state-of-the-art steering system HiL simulators, the steering wheel is exchanged with the steering wheel actuator (WA), which provides a max. 90 Nm if the larger force actuators RLA and/or RRA are used. However, when the smaller linear drive RLA_S is used, it is possible to keep the steering wheel and a real driver is allowed to manually generate the driver's steering wheel input up to max. steering wheel torque of 5 Nm for safety reasons. Therefore, Driver-in-the-Loop tests can also be performed on the same test bench. The new CVH steering system HiL simulator makes subjective assessments on the same bench possible for the very first time and provides very early insight into the subjective perception of the steering system and/or calibration. It also provides information on driver acceptance of the steering system and/or new functions, e.g., driver assist functions or investigations regarding the hand-over acceptance of automated driving vs. manual driving, etc.

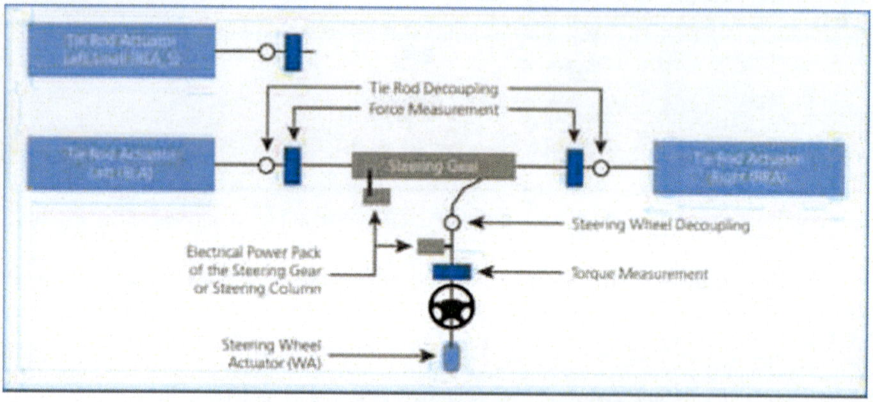

Fig. 5. Sketch of the new CVH steering system HiLs

The real-time computer runs all control tasks with 1 kHz (i.e., at a 1 ms sample time) for force, speed, position, torque, steer speed or steering angle control, and full vehicle model as well a vehicle rest bus simulation. However, the frequency converters are run in real time with 8 kHz and 16 kHz, all using synchronized signals. These measures result in the overall situation that, e.g., the large 20-kN electric linear force drives can be run in a closed-loop force control at high loads of more than 35 Hz with an error margin less than 3 dB. This can be achieved up to almost full load. Therefore, the steering system HiL simulator is considered as being at the high-performance end of the state-of-the-art steering system HiL simulators in terms of functionality and performance. In particular, the test bench capabilities for generating higher frequency loads are expected to become increasingly important, since, e.g., an insufficient (closed-loop) stability margin due to inappropriate calibrations very often result in steering torque oscillations in the region between 20 Hz and 35 Hz. As many more functions will be developed and calibrated by different distributed teams (and sometimes even developed by different companies) with regard to driver assist functionalities and automated driving/steering, the integration task and validation of suitable software releases and appropriate models that are valid for these frequencies might become crucial in the next years.

For reasons of completeness, we would like to mention that not only Electrically Powered Steering (EPS) systems can be tested on the same test bench but also Hydraulically Powered Steering (HPS) and Electrically Powered Hydraulic Steering (EPHS) systems. As a result, it is possible to obtain very reliable back-to-back measurements and comparisons.

Furthermore, the test bench supplier dSPACE developed a new friction compensation technique that is implemented in the electric frequency converters running the linear electric drives RRA, RLA and RLAS. This paper also investigates this particular technique's impact on the test performance and the obtained results for the DuT.

3 Component and Steering System Testing and Modeling

As indicated in the previous sections, aside from release and validation tasks, the steering system HiL simulator is often used to obtain models and model parameters of steering systems with different model depths that depend on the intended use of the models. Therefore, this section describes several state-of-the-art approaches to obtain certain important model parameters of steering systems, for example, stiffness or friction. Additionally, this paper also discusses the effect of using different control settings for the steering system HiL simulator, e.g., friction compensation or different actuator sizes with respect to the obtained measurements and the resulting model parameters.

3.1 Basic Idea

Steering Gear Model and Its Parameters

The hardware-in-the-loop technology is used by ZF to tackle challenges related to functional performance and safety-related release processes for today's steering systems. Based on the system integration strategy, the HiL test instances contain different fractions of the investigated hardware [1]. The EPSiL (Electrically-Powered-Steering-in-the-Loop) architecture corresponds to the high-performance steering HiL simulator presented in this publication and includes the complete steering gear hardware as the device under test (see Fig. 6). If only the ECU functionality is the focus, a complete simulation-based ECUiL (Electronic-Control-Unit-in-the-Loop) environment can be used (see Fig. 7). All other mechatronic components of the steering system, such as the mechanical power transmission elements or the steering sensor are virtual. The functionality of the steering servo motor is mapped by emulation. The emulation combines a model of the electric motor behavior with electronic load and power elements that are connected to the motor controller, which is an integrated part of the ECU. In both environments, EPSiL and ECUiL, a full vehicle dynamics model is integrated.

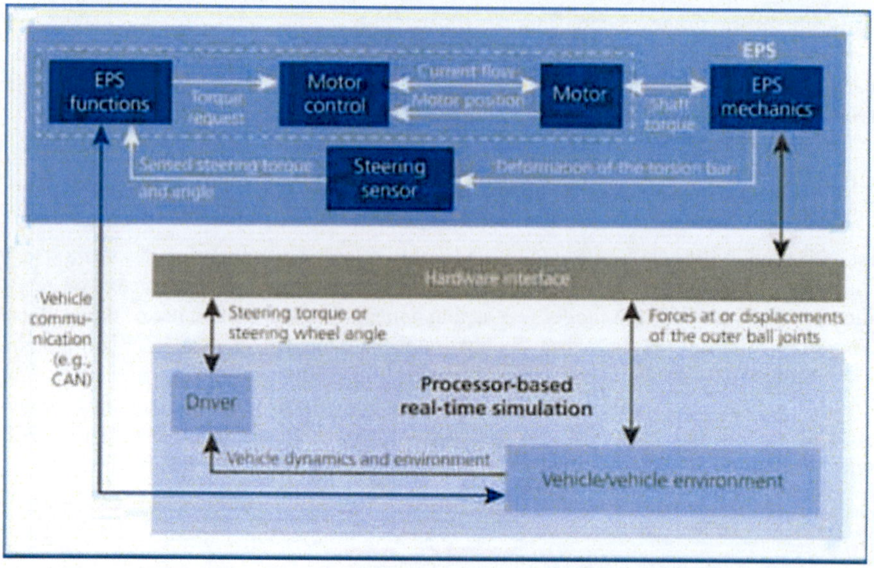

Fig. 6. HiL architecture with steering gear as hardware (EPSiL).

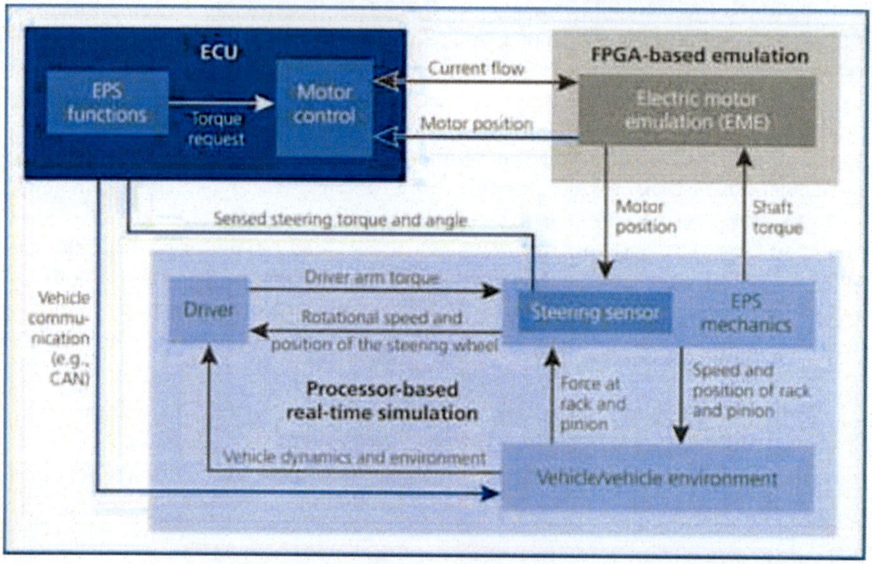

Fig. 7. HiL architecture with steering gear as a model (ECUiL).

Aside from the capability to map the realistic dynamic and functional system behavior, the most important requirement for a plant model in a HiL simulation is a low computational effort for underlying equations. This property is essential for the operability of the entire numerical environment on a real-time computer.

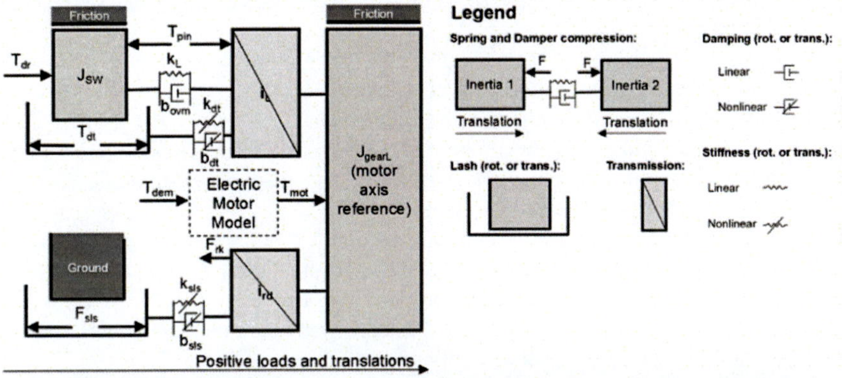

Short form	Meaning	Short form	Meaning
b	Damping	M	Weight
dem	Demand	mot	Electric Motor
dr	Driver	ovm	Overmould
dt	Dog-Tang	pin	Pinion
F	Force	rd	Rack Drive (Belt Drive or Dual Pinion)
i	Transmission Ratio	rk	Rack
J	Rotational Inertia	sls	Soft Lock Stop
k	Stiffness	sw	Steering Wheel
L	Lumped (Combined Properties)	T	Torque

Fig. 8. Rack-drive steering gear real-time model.

For the steering part, a customized two-degrees-of-freedom model has been designed (see Fig. 8), which is integrated in the full vehicle dynamics model later on (see section 4.1). In this schematic representation the steering wheel is connected to the input shaft of the steering sensor via the column shaft. Inside the steering sensor, a tensional bar links the input shaft to the steering pinion. The product of relative rotational displacement between the input shaft and the steering pinion and the torsion bar stiffness is a measure for the transmitted steering torque. In the rack and pinion gear, the rotational pinion movement is translated to the axial rack movement. Two tie rods are connected to the rack ends by ball joints. To ensure that the steering system is not damaged by running into the maximal stroke of the rack, two soft locks are implemented for the contact between the ball joints and the rack and pinion housing. To avoid high distortions of the torsion bar and subsequent damage, a mechanical rotational end stop system (Dog-Tang) reduces the possible rotational displacement between the input shaft and the steering pinion shaft. When the relative rotation reaches a certain limit value, contact between the two parts is made and a continued distortion of the torsion bar is restricted.

Even a simple model as this requires many parameters to represent the realistic behavior of the system. Depending on the mathematical description of the modeled physical property, the parameter identification can be broken down into three cases:

1. The physical property is defined by a single parameter, which can be determined directly by a simple measurement or analysis of CAD data of the single component:
   - Stiffness of basic elements and interaction contacts
   - Mass or rotational inertia of components
   - Transmission ratios of a simple gear
2. The interaction between components of the real system is described by parameters, which can be obtained analytically from a simple system measurement:
   - Nonlinear system stiffness
3. The system quantity is described by a complex mathematical formulation that specifies which parameters cannot be determined directly from the system test or for which analytical determination is very costly:
   - System transmission ratio considering cardan shaft effects and load dependent gear separation
   - System damping
   - Friction model

For the latter, the parameters must be identified in an alternative way. For this numerical optimization is very useful [3].

Parameter Refinement by Means of Numerical Optimization

The process of numerical optimization essentially consists of the following steps:

1. Identifying the influencing factors that determine the parameter value. For example, in the case of friction model parameters, this would be, e.g., the rack velocity or the transmitted torque if the model to be parameterized takes the effect of load-dependent friction into account.
2. Determining the relevant value range of the selected factors. To use our example of friction, this would be the range of rack speeds for which the steering gear is specified and the model is supposed to represent. In this case, of course, the capabilities of the test bench used must be considered.
3. Performing tests on the real component that cover the range of the values of the factors and, as a result, provide the value that is determined by the required parameter. For load-dependent friction, the factors are velocity and load, and the value measured in the test is friction force or torque.
4. Defining a target function that represents the deviation between the measurement result and the simulation result.
5. Carrying out the actual numerical optimization to find the values of the model parameters that guarantee the smallest magnitude of the objective function. Therefore, for this procedure the model parameters are varied until the results have an adequate quality and the deviation between the measurement and the simulation is reduced to a minimum.

The groundwork of the procedure used at ZF [4] is the Particle Swarm Optimization (PSO). PSO is a population-based stochastic method [5] and is grounded on observations of the behavior of flocks of birds that are looking for (optimal) feeding spots.

The selected objective function F_{LS} is defined by the sum of the squared errors between the measured results (y_{test}) and the simulation results (y_{sim}). In our example: A difference between the measured friction force and its value obtained from the simulation. The index i refers to the existing test data sets, i.e., to the variation of the influencing factors, e.g., rack velocity values. However, the test can be set up in such a way that i represents the time progression of the measurement:

$$F_{LS} = \sum_{i=1}^{n} \left(y_{test,i} - y_{sim,i}\right)^2 \quad (1)$$

The advantage of squaring the difference is that larger deviations are weighted more than smaller ones.

Test Execution – Measurement

The test procedure described in this document is based on the approach defined by Lin et al. [6] for identifying steering parameters.

It is very important to have sufficient technical knowledge about the model structure and the steering system to be characterized. The influence of the model parameters on system behavior must be sufficiently known. A sensitivity analysis can be carried out before examining the parameter influence. In addition, a suitable test bench must be available to carry out the tests. The tests must be planned in accordance with the capabilities of the plant model and of the used test bench.

As a reference to the high-performance HiL test bench presented in this document, the steering parameter identifications were investigated using an older generation of the EPSiL simulator available at ZF's Active Safety Technology division. The test bench had to be prepared accordingly for the parameter identification task. For example, the control logic of the simulator was changed so that individual actuators can be decoupled from the device under test for certain measurements. Special measuring devices were installed to enable a more precise recording of the friction forces. The high-performance CVH HiL simulator provides all these required features, already.

The schematic representation of the test bench configuration required for the implementation of the parameter identification procedure is the state-of-the-art schematic in Fig. 4. Depending on the test setup, the actuators are connected or decoupled from the device under test and working in the required control mode – position, torque, or force control.

3.2 Experimental Procedures

Two examples of automated parameter identification procedures are presented below. When determining the system rigidity, several measurements are statistically evaluated. The model parameter for the stiffness is then calculated analytically. When determining the values of the friction model parameters, the previously described optimization procedure was used.

Table 1. System stiffness test scheme

	Actuator		
	WA	RLA	RRA
Connection to DuT	coupled	coupled	decoupled
Control type	torque	position	none

System Stiffness

For the system rigidity identification tests, one of the linear actuators is decoupled from the tested steering gear. The reason for this is that the actuator transmits slight disturbances to the rack in force control mode, even if the setpoint for the force is 0 N. The new friction compensation for RLA and RRA does improve but not fully resolve the issue. The second linear actuator is controlled to a constant position. The steering wheel actuator is in torque control mode. The test scheme is shown as an example for a configuration with a decoupled right linear actuator in Table 1.

The steering torque setpoint is sinusoidal and contains different amplitudes. The amplitude depends, among other things, on the torsion bar stiffness. The aim is to twist the torsion bar until the mechanical rotational end stop in the steering sensor is closed. After the mechanical contact between the sensor input shaft and the output shaft is established, the total system stiffness increases.

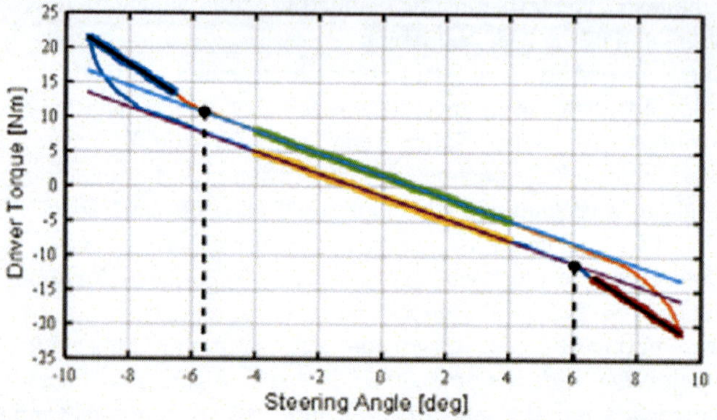

Fig. 9. System stiffness test: Steering wheel torque as a function of steering wheel angle.

Fig. 9 shows an example measurement of a stiffness test. There is a clear increase in the gradient at the marked points. At this angle, the maximum torsion bar twist angle is reached, and the Dog-Tang contact between the input shaft and the output shaft closes. This end stop angle must also be parameterized in the system model. However, it contains both, the torsion bar contortion and the deformation of the entire steering column.

Therefore, the maximum torsion bar contortion φ_{sen} is determined by the driver torque T_{dr} and the stiffness of the torsion bar k_{sen} :

$$\varphi_{sen} = \frac{T_{dr}}{k_{sen}} \qquad (2)$$

In this approach, the torsion bar stiffness is assumed to be known – see point 1 in "Steering Gear Model and Its Parameters". The torque measurement in the steering gear and the EPS assist torque is based on this parameter.

System Friction.

The friction test is used to identify the system friction model parameters. For this test, the steering wheel actuator is decoupled so that it is possible to move the rack freely. In order to rule out any interactions between the two linear actuators, one of them is decoupled in the same way as in the previously described rigidity / stiffness test. The test scheme is shown as an example for a configuration with decoupled right linear actuator in Table 2.

Table 2. Scheme of the friction test

	Actuator		
	WA	RLA	RRA
Connection to DuT	decoupled	coupled	decoupled
Control Type	none	position	none

As the underlying system model is defined in the EPP motor axis coordinate system, the friction forces measured in the rack axis must be converted to the rotational coordinate system of the motor using the related transmission ratios.

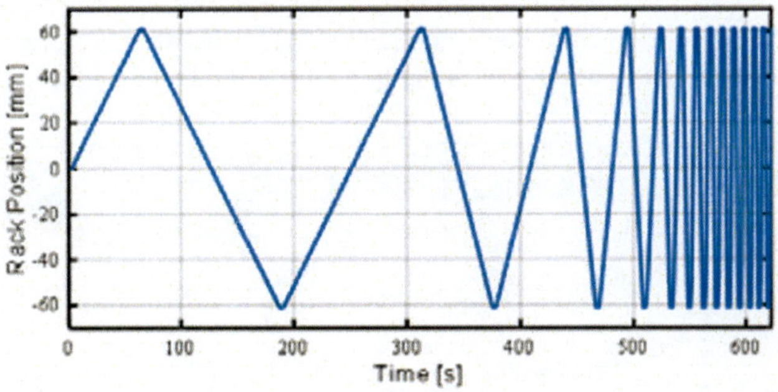

Fig. 10. Test vector: friction test.

The friction test is performed at a rack speed of up to 100 mm/s. This is a limitation of the test bench. The friction forces are gathered over almost the entire rack stroke. The measurements are carried out for different speed levels. The rack is moved at a constant speed during one test run. The individual test runs are performed successively. Fig. 10 shows the rack position over time. This vector serves as the setpoint value for the linear actuator.

Since the friction model parameters are determined by an optimization, the rack speed is the optimization factor and the friction force is the measured value, which is used to calculate the objective function in the optimization process. A swarm size of 50 iterations and a maximum of 500 iterations were selected for the optimization. The optimization error history is shown in Fig. 11.

The analysis of the measured data has shown that due to the mechanical structure of the steering gear, the levels of the measured frictional forces are dependent on the direction of rack travel. This behavior is independent of whether the rack is pushed or pulled. As the underlying steering model contains a friction formulation that is independent of the direction of movement, the measured forces were mapped to a symmetrical course.

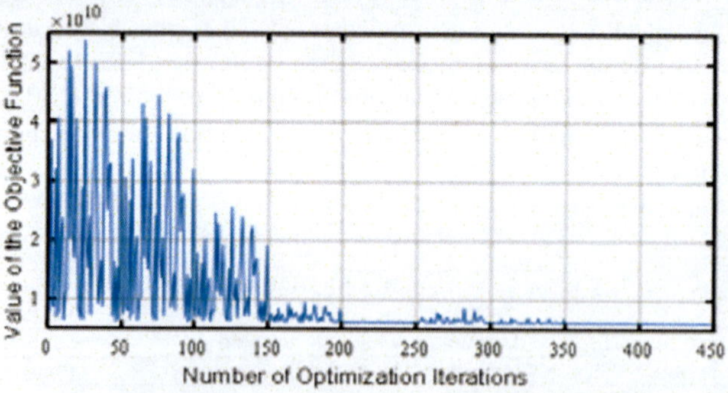

Fig. 11. Error history of the model parameter optimization

A comparison of the measured friction torque with results generated by the investigated friction model is shown in Fig. 12. The diagram contains the model results for both the initial parameters and their optimized values. This clearly shows that the real friction behavior of the steering gear is rather poorly reproduced by the starting parameter values.

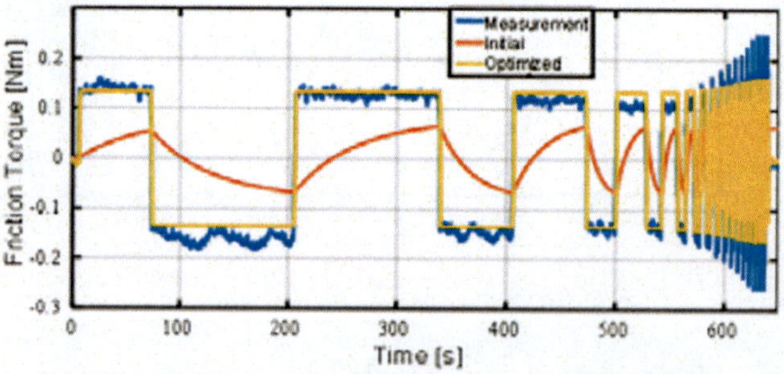

Fig. 12. Friction model – comparison of parameterization.

3.3 HiL Setup Variation and Its Impact on Measurement Results Used for Steering System Modeling

Since the low-level friction compensation control for the actuators RRA, RLA and RLAS is very new and has been developed and tested by dSPACE using the new CVH steering system HiL simulator, only limited experience is currently available. However, in this paper we would like to present first observations using this new HiL setting. Basically, it is not expected to find any changes in case (quasi-) static measurements are performed using closed loop force control for the HiL actuators RRA, RLA and RLAS, due the force measurements being performed near to the DuT (compare Fig. 5). Fig. 13 illustrates the first findings by comparing the measurement results with friction compensation activated (measurement shown in green) versus the case friction compensation being deactivated (measurement shown in red). Both tests were performed on the HiL simulator for generating the stiffness measurements, each by altering RLA and therefore altering the induced forces (plot on the left) and by altering the steering wheel actuator and therefore altering the steering wheel torques (plot on the right), respectively. The actuator that is not used to alter the loads is then switched to position control in order to 'stiffen' the steering system and to maintain a situation of 'driving straight'.

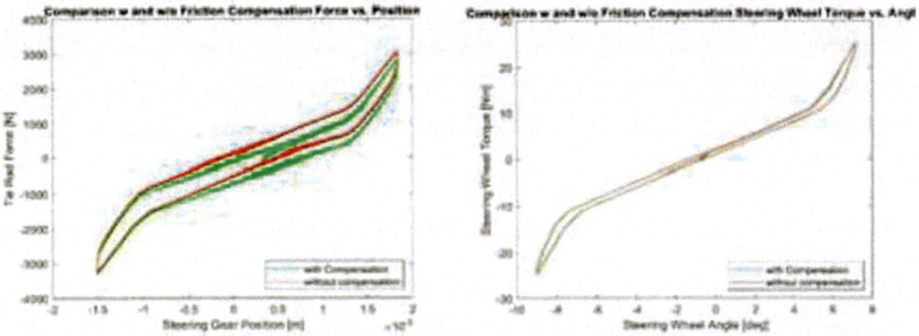

Fig. 13. Steering system stiffness test without HiL actuator friction compensation vs. friction compensation for actuators WA and RLA: Stiffness force vs. rack displacement [N/m] (left) and stiffness steering wheel torque vs. steering wheel angle [Nm/°] (right)

As can be seen from Fig. 13, the stiffness measurement during which the steering wheel loads are altered (plot right) shows no difference independent of whether the friction compensation is switched on or off. However, for the test setup in which the rack forces and position that controls the steering angle position are altered (plot left), the measurements show a difference. They even show on offset value and a difference in the gradients of the measurements which results in changes to the obtained/identified stiffness parameters representing the steering system. Since this result is in a certain way unexpected, the same test has been performed with the actuator RLAS instead of RLA. The result is depicted in Fig. 14. This second stiffness test shows results as expected: The (quasi-) static measurement of the stiffness presented in that type of cross-plot is the same independently if the friction compensation of the RLAS has been switched on or off; the controlled force (over time) might be more smooth for the case with friction compensation, however, the static cross-plot does not show (almost) any difference at all. The stiffness calculated from the cross-plot remains the same, independent of the choice of the low level controls of the actuator RLAS. Thus, the additional small actuator could be used in order to falsify an error-prone measurement shown in Fig. 13 (left).

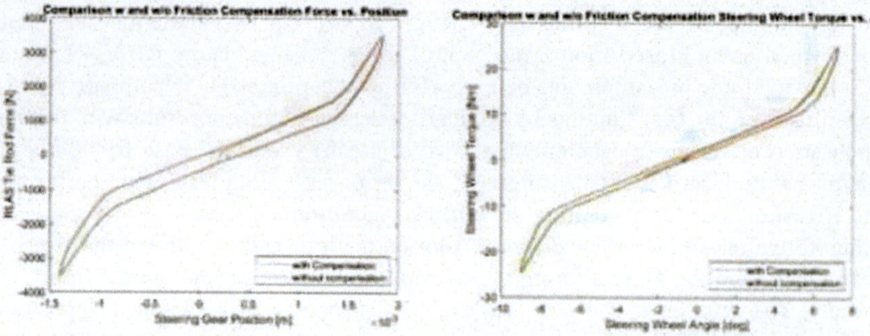

Fig. 14. Steering system stiffness test without HiL actuator friction compensation vs. friction compensation for actuators WA and RLAS: Stiffness force vs. rack displacement [N/m] (left) and stiffness steering wheel torque vs. steering wheel angle [Nm/°] (right)

4 Virtual Prototypes and Subjective Steering Feel Validation

4.1 Simulation Model

We chose the Automotive Simulation Models (ASM) from dSPACE as the underlying real-time vehicle dynamics model. The open and modular model structure allows for interfacing at any hierarchy level. Therefore, customized components can easily be integrated, and a complete closed-loop operation including the entire steering test bench is possible.

A virtual driver controls the vehicle and provides a reference steering wheel angle or torque, depending on the interfaces that are required at the steering wheel of the test bench (see Fig. 15). In the lower part of the test bench, reference signals for the counter motors (the linear drives RRA, RLA and RLAS) at the steering rods are generated by the model as a result from the torques and forces acting between the tires and road. These signals are translated to the degree of freedom of the test bench and controlled by the test bench motors. To close the loop, the measured displacement of the steering rod at the test bench is returned to the model to be considered in the suspension system, such as kinematics and compliances, and thus has an impact on the overall vehicle dynamics.

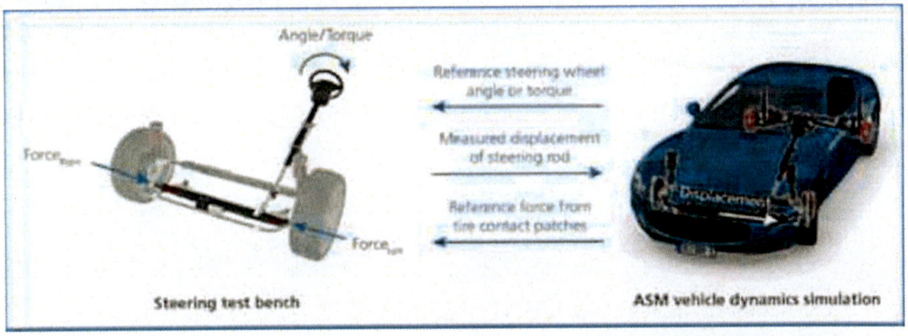

Fig. 15. Closed loop operation of the steering test bench with the ASM simulation

The previously mentioned friction and stiffness tests only take into account the steering system in an open-loop setup, whereas a sophisticated vehicle dynamics model helps drive realistic scenarios in an entirely closed-loop manner.

A set of multiple scenarios is available for putting the system into general operation, such as driving along a handling track with various curvature and height profiles, braking maneuvers in a curve, or different sinusoidal characteristics for the steering. For the later validation process and objective steering feel tests, additional profiles based on real maneuvers can also be imported.

The newly achieved performance level of the steering test bench allows for special stability investigations. Stimulations in a frequency range from 10 Hz up to 30 Hz are possible. This means that the interaction of the EPS ECU with other ECUs, e.g., for active safety or ADAS, can be analyzed.

What makes this steering test bench unique is the option to replace the virtual driver with a real driver in the loop as well as the ability to also execute subjective steering feel tests for haptic analysis, etc.

Furthermore, the interaction of the EPS ECU with other ECUs in the loop, such as ESC or ADAS controllers, can be investigated

4.2 Virtual Vehicle Prototypes

A major task for proper steering feel validations is the correct parameterization of the virtual vehicle, i.e., obtaining matching results from the simulated vehicle dynamics compared to the characteristics of the real target vehicle.

The parameterization process starts with the selection of a pre-parameterized vehicle class, such as an SUV or VAN. Based on this basic vehicle type, the available parameters from the target vehicle are set. There can be different sources where the data is coming from. For example, KnC data from hardware- or simulated test rigs. In most cases for HiL application the inputs are obtained from very detailed Multi-Body-Dynamic models, like ADAMS/Chassis or ADAMS/Car. Tire parameters can be used from real tire measurements in the standardized TIR file format. For this project, Ford provided a validated model from a production vehicle.

During the parameter validation process, virtual test runs are driven according to steering and speed profiles of real test runs (see Fig. 16). The generated simulation results are compared with the on road test results for the main vehicle characteristics, such as lateral acceleration, yaw rate, side slip angle, etc. (see Fig. 17).

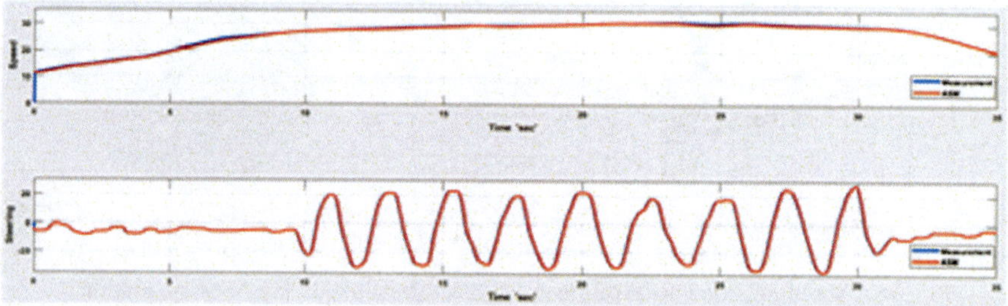

Fig. 16. Example stimulus signals (speed, steering) for the validation process.

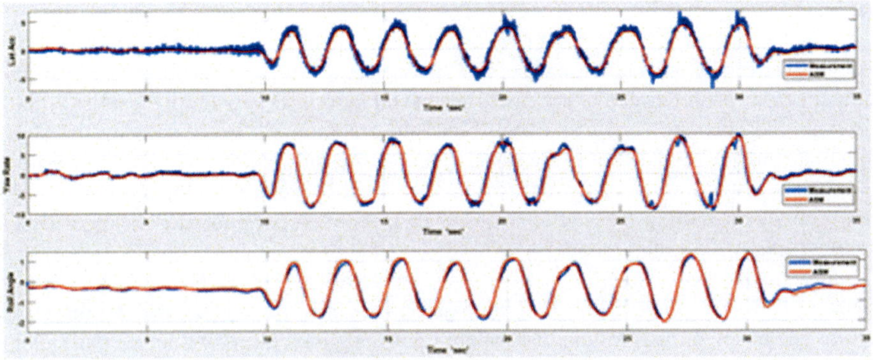

Fig. 17. Example validation results for lateral acceleration, yaw rate, and roll angle.

As the real test car differs from the series car in some aspects, parameters with tolerances can be varied for fine-tuning afterwards [7]. Based on this digital clone, new target vehicle characteristics can be defined to serve as a basis for an ongoing parameter variation. As a result, a digital prototype can be derived. This virtual prototype will now be used in combination with the steering test bench to evaluate both the objective and in particular the subjective steering feel behavior. To put it simply, single parameters from the digital clone, such as the mass of the vehicle or the center of gravity, can successively be varied to directly see the impact of a single vehicle parameter variation on the steering feel [8].

4.3 Subjective Steering Feel Tests

The new CVH steering system HiL simulator has been used to perform some first subjective assessments of a DuT. For this purpose, the ASM full vehicle model as described in section 4.1 and parametrized according to section 4.2, is simulated in (hard) real time, calculating the road loads/rack force for the RLAS actuator depending on the actual steering wheel input from the (real) test driver in the HiL simulator. Thus, this section documents this subjective assessment of the steering system on the HiL being part of the virtual ASM full vehicle model. For this purpose, the measured steering wheel angle input data is propagated in real time by the ASM simulation model for calculating the required desired forces for the RLAS, which will then control the forces accordingly. Since the closed loop force control is improved with the friction compensation for the actuator RLAS of the HiL, it is expected to perceive some difference when the friction compensation of RLAS is switched on or off.

The overall impression is that depending on whether the EPS is powered on or off, different vehicle speeds can be perceived. However, the subjective steering feel impression differs from the steering feel perceptions in a real vehicle. The reasons are manifold. Some of the reasons are not due to the HiL simulation itself, because the steering feel impression of a test driver inside a real vehicle is much more than 'only steering wheel torque and steering wheel angle'. The subjective impression of the test driver is significantly influenced by the subjective pressure felt in the inner ear, the perceived force with which the driver is pressed into the seat (and its support), the impression of the relative motion of the vehicle compared to the surrounding environment (particularly influenced by the inclination angle of the vehicle's A-pillar, since this significantly influences the possibility to either easily detect roll motion or yaw rate), the noise while driving, and other influencing factors. Thus, the steering feel impression on the HiL simulator cannot be identical to the steering feel impression inside a vehicle. Nevertheless, the investigation is useful and reasonable, because we are not looking for 'absolute' ratings of steering feel but for relative changes of steering feel for different DuT settings and/or the vehicle model.

Consequently, in this paper we focus only on certain aspects of how to improve the steering feel generation in the HiL simulator to reduce errors generated by the HiL setup. For this, the main residual issues are improving the vehicle model (in particular tire modeling and friction modeling of tires and suspension) and improving the HiL force actuation performance. In this paper we focus on the effect of improved friction compensation for the RLAS closed-loop force control situation, since this is purely HiL-related and improvements in this field will result in better overall performance and investigation acceptance for any type of investigation.

It is difficult to quantify the subjective impression. Therefore, we decided to perform the same sinusoidal on-center straight driving test for different configurations of the steering system and the HiL. This type of test is frequently performed in vehicle tests to gain what is known as steering hysteresis plots, for example. We performed these experiments in four different configurations: EPS on and off and also each time the new friction compensation of the RLAS HiL controller switched on and off. This adds up to four different scenarios tested during the same type of test drive. The results are as follows.

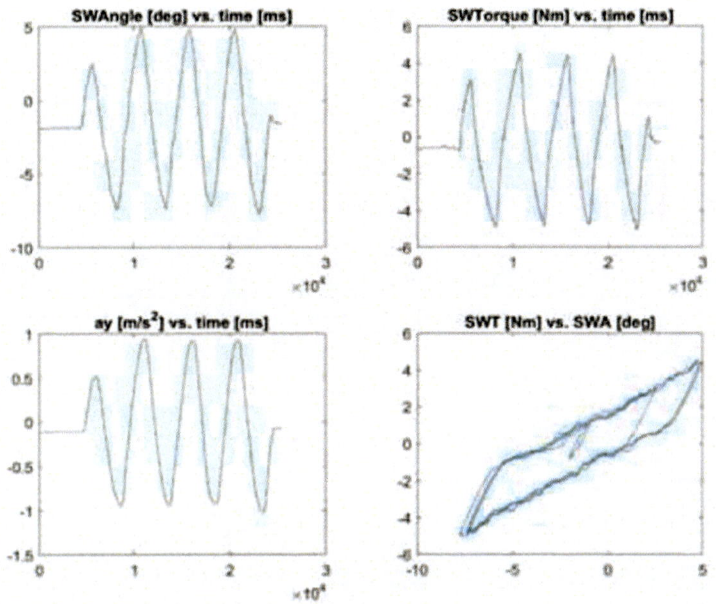

Fig. 18. On-center sine steering test with the configuration EPS off, friction compensation off.

Fig. 18 shows the test results of the on-center test performed in the EPS off and friction compensation off scenario. The sinusoidal manual input can easily be seen and has been adjusted such that the max. steering wheel angle inputs relate to approximately. 0.1 g (simulated) lateral vehicle acceleration. Since the steering system does not provide any type of assist (i.e., EPS off), the results clearly illustrate an issue, as this power-off-scenario is reaching the maximum allowed limit of steering wheel torque due to the safety concept implemented to obtain CE-conformity of the test bench. An increase of the allowed applicable manual steering wheel torques would be a welcome benefit aiming at increasing the amount of lateral acceleration of the simulated vehicle being tested on the HiL in the EPS power-off-scenarios. For this investigated vehicle model and simulated vehicle speed, already 0,1g is the upper limit of test to be performed on the HiL.

Fig. 19 illustrates the results obtained if the friction compensation is switched on for the low-level controls for the RLAS closed-loop force control mode. Only a small deviation from the results in Fig. 18 can be seen. This result indicates that the change might be helpful to improve the obtained test results. However more detailed investigations must be performed in future to get a more detailed answer to this question.

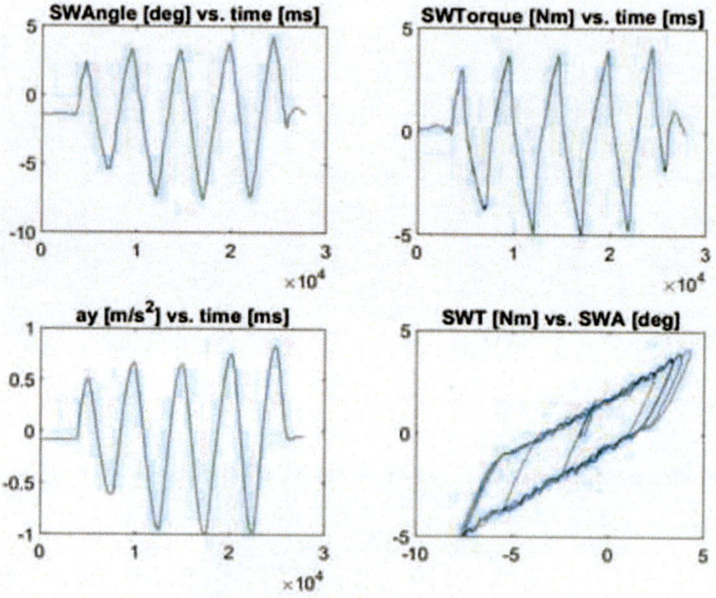

Fig. 19. On-center sine steering test with the configuration EPS off, friction compensation on.

Fig. 20 shows the result for the EPS on scenario with the friction compensation of the RLAS actuator switched off. First, it can be seen, that subjective assessments can very well be carried out on the CVH steering system HiL simulator, because the operating range of steering wheel torque remains significantly below the safety limit of approximately 5 Nm. The well-established hysteresis plots can be generated and used for development purposes. However, it must be kept in mind that the steering wheel torques obtained are a result of the underlying full vehicle model. We strongly recommend not to use these types of obtained results to replace vehicle tests but to enable, e.g., an early and 'easy' back-to-back comparison of different prototype or software levels of the DuT that was obtained in the same bench scenario instead.

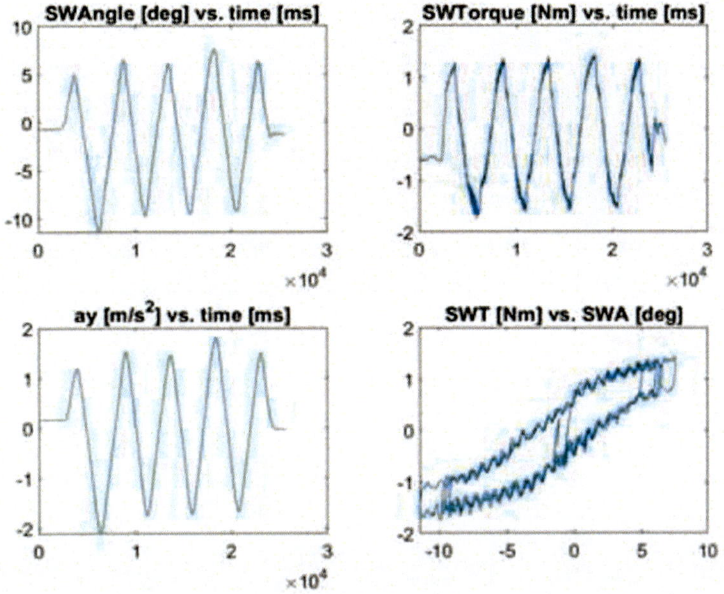

Fig. 20. On-center sine steering test with the configuration EPS ON Friction Compensation OFF

Fig. 21 depicts the results obtained for the EPS on scenario with friction compensation on. As can be seen, the safety limits for operation are still far away from the steering wheel torques required for the test (same result as obtained for the EPS on scenario with friction compensation off). However, it also can be seen, that the well-established hysteresis plots show slightly different values and shapes. Whether this effect is reflecting better or worse compared to the real in-vehicle-test-situation still has to be validated and the tests have not yet been performed. This part of the question must be answered in future work.

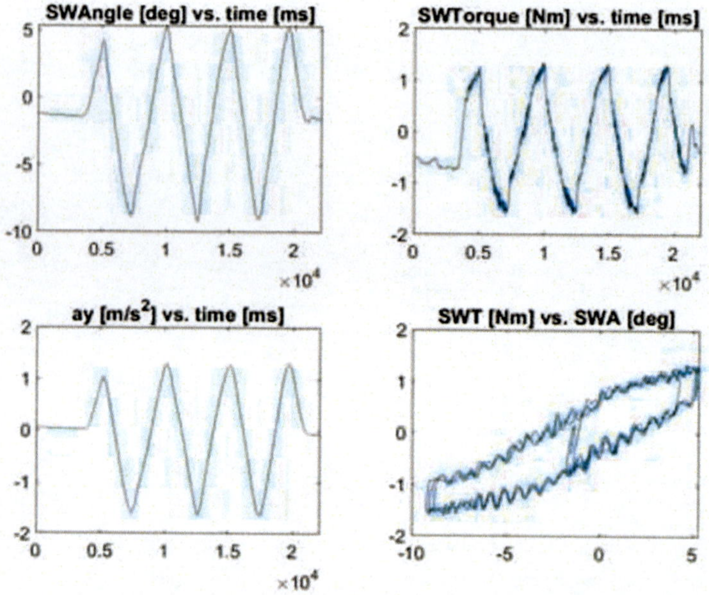

Fig. 21. On-center sine steering test with the configuration EPS on, friction compensation on.

5 Summary and Outlook

This paper gives an overview for which purposes different HiL types for steering systems are used during the development process according to the V-model for developing mechatronic systems and products in the automotive development context. In particular, steering system HiLs not only are being used for final release and validation tests but they are being used to obtain models and model parameters for other purposes, too. In this context, it has been shown, that the application of parameter identification procedure proposed by Kruse [3] and Lin et al. [6] implemented in ZF's HiL test environment is an efficient approach. Their functionality has been proven. The automated process clearly shows not only the room for improvement of the underlying plant model, but also the need to enhance the deployed test rig.

The new High-Performance CVH (Campus Velbert/Heiligenhaus) Steering HiL presented in this article addresses these shortcomings. The torque and force measurement accuracies as well as the position control of the load actuators have been enhanced with this new HiL. The overall inertia of the actuation systems has been significantly reduced, which further improves the general system dynamics. This HiL simulator is not only suitable for haptic investigation of the steer feel but it is also ideal for numerically based steering gear parameter identification.

Subjective assessments with the CVH steering system HiL are possible, however evasive steering maneuvers with high driver torques (e.g. loss-of-assist-tests) are hard to realize due to the limitation of a maximum of 5Nm driver torque.

The improved closed loop control force performance has been discussed in this paper. In particular, during the subjective assessment tests some changes in the results could be seen. Vehicle model simplifications e.g. with respect to suspension friction can distort the subjective impression during the HiL tests compared to the driving experience during on road tests. Thus, further investigations need to be carried out in future in order to perform the comparison of the tests documented in this paper with real on road assessments.

References

1. Moczala, M., Maur, T.: Lenkung auf dem Prüfstand, dSPACE-Magazin 03/2015, pp. 13-17
2. Moczala, M.: 2 DoF Rack Drive Steering Gear Model, ZF TRW Engineering Document, (2016)
3. Kruse, D.: Teilautomatisierte Parameteridentifikation für die Validierung von Dynamikmodellen im modellbasierten Entwurf mechatronischer Systeme, Dissertation, Universität Paderborn, Paderborn (2018)
4. Zelles, T. J.: Entwicklung einer prüfstandsbasierten Methode zur Lenkungsvermessung durch simulationsgestützte Parameteridentifikation am Beispiel eines ausgewählten PKW-Lenkungsmodells. Master Thesis, ZF Automotive Germany GmbH & Hochschule Niederrhein, Düsseldorf (2019)
5. Hu, X., Shi, Y., Eberhart, R.: Recent advances in particle swarm, Proceedings of the 2004 Congress on Evolutionary Computation, pp. 90-97, IEEE (2004)
6. Lin, J., Kloos, R., Pfeffer, P., E.: Steering Systems Models - An Efficient Approach for Parameter Identification and Steering System Optimization, 8th International Munich Chassis Symposium 2017, pp. 535-551, Springer Vieweg, Wiesbaden (2017)
7. Peperhowe, M., Schindler, W.: Efficient Vehicle Dynamics Development via Simulation-Based Prognosis Tools. chassis.tech (2013)
8. Heller, A.: In-the-Loop Methoden am Beispiel eines Lenkungsprüfstands, Entwicklungsprojekt im Master-Studiengang Mechatronik & Informationstechnologie (Masterproject Thesis), Campus Velbert/Heiligenhaus der Hochschule Bochum (2019)
9. Janorschke, M.: Bestimmung von Reibungskenndaten von Lenksystemen mithilfe von Prüfstandsmessungen, Laborprojekt im Master-Studiengang Mechatronik & Informationstechnologie (Laboratory Thesis), Campus Velbert/Heiligenhaus der Hochschule Bochum (2019)
10. Janorschke, M.: Experimentelle Modellbildung einer Lenkung mit Hilfe eines Lenkungssystemprüfstands, Entwicklungsprojekt im Master-Studiengang Mechatronik & Informationstechnologie (Masterproject Thesis), Campus Velbert/Heiligenhaus der Hochschule Bochum (2020)

Effects of Hub Bearing Integrated with Additional Wheel Steering System on Vehicle Agility and Stability

– Development of an independent front wheel hub-steering system –

Mitsunori Ishibashi[1], Makoto Yamakado[1], Yoshio Kano[1], Masato Abe[1], Atsushi Ito[2], Norio Ishihara[2] and Hirokazu Ohba[2]

[1] Kanagawa Institute of Technology, Japan
[2] NTN Corporation, Japan

Abstract. The hub-steer system by which the additional front wheel steer of the vehicle is avail-able with no modification of the external layout configurations of the existing steering and suspension systems has been developed. The control algorithms of the hub-steer for improving the vehicle handling characteristics during steady circular turning and also for improving the vehicle transient response characteristics to steer input are introduced. Following to the simulation studies, the experimental investigation into the effects of the hub-steer system with the control algorithms using a conventional passenger vehicle equipped with the hub-steer system are carried out on a proving ground.

Keywords: Hub-steer, Handling Characteristics, Ackermann Steer Angle.

1 Introduction

A wheel steering system installed inside the wheel hub is developed, with which an additional front wheel independent steer of the vehicle is available without any change of external configurations of conventional steering and suspension layout.

An additional front wheel active steering control law is proposed to improve the vehicle handling in wide range of the steady circular turning from low to high speed and the transient response of the vehicle to steering-input at high speed. The effect of the additional steer on the vehicle motion is theoretically investigated using a full vehicle model and the results are confirmed experimentally on a proving ground with the vehicle equipped with the hub-steering system developed.

In the control law, the left and right front steering angles are adjusted by the hub-steering system depending on the steering wheel angle input during circular turning. At low speed turning, an exact Ackermann steer angle is added to the left and right wheels respectively to keep the side slip angles of both wheels zero to secure an eminently smooth tight cornering with low speed. During turning at higher vehicle speed, the additional steer angle is added to the respective wheels in order to have the side slip angles of both wheels proportional to the estimated vertical loads of the wheels to make more

P. E. Pfeffer (ed.), *11th International Munich Chassis Symposium 2020*, Proceedings, https://doi.org/10.1007/978-3-662-63193-5_28

severe cornering at higher vehicle speed available. In order to improve the vehicle transient response to the steering wheel input by the driver especially at high vehicle speed, the steering wheel speed dependent additional steer is added to the front wheels by the hub steer system according to the feed-forward type of front active steering control algorithm.

The result obtained in the investigation shows us significant effects of the hub-steering system on improving the handling characteristics of the vehicle at the steady state turning and also the transient response of the vehicle motion to steer input at high speed [1], [2].

2 Hub-Steering system

The external view of the hub-steering system developed and its main components are visually shown in Fig.1 and the fundamental specifications of the system are as follows:

Total mass of the hub:	13.8kg (additional mass: 5.0kg)
Maximum steering torque:	350Nm
Power supply voltage:	24V
Maximum steering angle:	3.5deg
Maximum steering angle speed:	16.0deg/s

A human driver steers the knuckle connected to the rack/pinion through the steering wheel shaft. In accordance with the control algorithm described in the next chapter, the hub bearing with the tire wheel will be steered relatively to the knuckle around the additional steering axis perpendicular to the ground at the center of the tire contact patch by the trapezoidal screw thread attached to the motor on the knuckle.

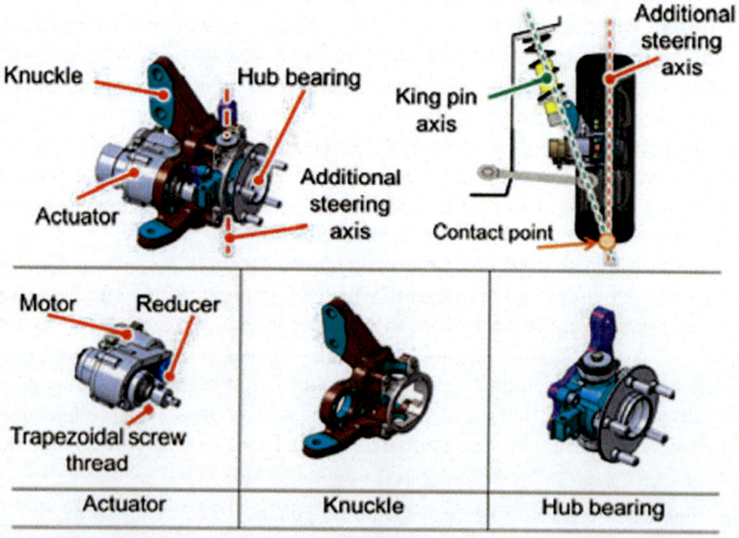

Fig. 1. External view and main components of the hub-steering system

3 Control algorithm of additional front wheel steer

3.1 Symbols

Following symbols are used in this study:

m : vehicle mass
l : wheel base
d : tread
l_f : longitudinal position of front wheel from vehicle center of gravity
l_r : longitudinal position of rear wheel from vehicle center of gravity
h : center of gravity height from the ground
K_f : front tire cornering stiffness
K_r : rear tire cornering stiffness
V : vehicle speed
δ : front wheel steer angle by steering wheel
δ_h : steering wheel angle
δ_{hb} : hub-steer angle (right +, left-)
a_y : lateral acceleration
r : yaw rate
β : side slip angle
n : steering gear ratio
W : vehicle weight
ΔW : transversal load transfer at front

3.2 Control algorithm for improving steady turning characteristics

If the steer angles of front right and left wheels given by the steering wheel are the same value, δ, the geometrical side slip angle of right and left wheels during steady turning are different caused by the difference of radius of curvature due to the vehicle tread. On the other hand, the vertical loads of the wheels differ between each which is caused by transversal load transfer due to lateral acceleration. Since the relation of tire lateral force to side slip angle is regarded to be similarly changed to the variation of vertical load, it is reasonable to keep the side slip angle of the wheel proportional to the vertical load of the wheel from the view point of effective use of the tire.

The figure 2 is a conceptual diagram to show that it is reasonable to use the tires under different vertical load with the side slip angles proportional to the respective vertical load rather than to use them with the same side slip angle.

The side slip angle of left and right wheels, β_1 and β_2, are described as follows, when both wheels have an additional hub-steer, δ_{hb}, in counter direction each other:

$$\beta_1 = \delta - \delta_{\mathrm{hb}} - \frac{\beta + l_f \frac{r}{V}}{1 - \frac{dr}{2V}} \tag{1}$$

$$\beta_2 = \delta + \delta_{\mathrm{hb}} - \frac{\beta + l_f \frac{r}{V}}{1 + \frac{dr}{2V}} \tag{2}$$

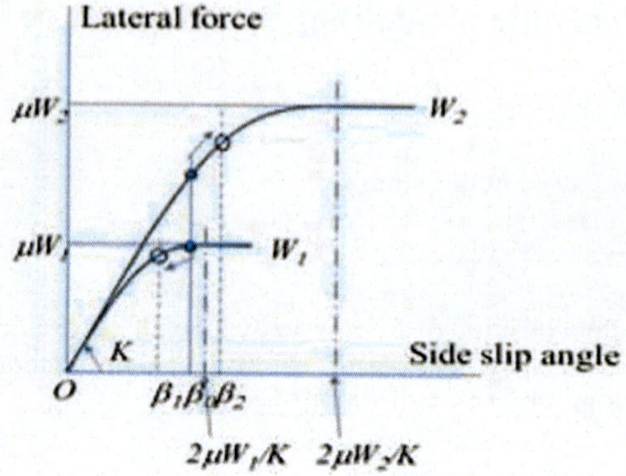

Fig. 2. Right and left tire characteristics with different vertical load due to load transfer

The side slip and the yaw rate, β and r, during steady turning are described as follows:

$$\beta = \frac{1-BV^2}{1+AV^2}\frac{l_r}{l}\delta,\ r = \frac{V}{1+AV^2}\frac{1}{l}\delta,\ \left(A = -\frac{m}{2l^2}\frac{l_fK_f - l_rK_r}{K_fK_r},\ B = \frac{Wl_f}{2ll_rK_r}\right) \quad (3)$$

Putting Eq.(3) into Eqs.(1) and (2) gives us:

$$\beta_1 = \left(1 - \frac{1-B\frac{l_r}{l}V^2}{1+AV^2-\frac{d}{2l}\delta}\right)\delta - \delta_{hb} = k_1\delta - \delta_{hb} \quad (4)$$

$$\beta_2 = \left(1 - \frac{1-B\frac{l_r}{l}V^2}{1+AV^2+\frac{d}{2l}\delta}\right)\delta + \delta_{hb} = k_2\delta + \delta_{hb} \quad (5)$$

Letting the side slip angle proportional to the vertical load:

$$\frac{\beta_2}{\beta_1} = \frac{k_1\delta+\delta_{hb}}{k_2\delta-\delta_{hb}} = \frac{W_2}{W_1} \rightarrow k_1W_2\delta - W_2\delta_{hb} = k_2W_1\delta + W_1\delta_{hb}$$

Then the followings are obtained as the additional steering control law during steady turning:

$$\therefore \delta_{hb} = \frac{k_1W_2 - k_2W_1}{W_1+W_2}\delta \quad (6)$$

where, W_1 and W_2 are the vertical load of the front left and right wheels at steady state cornering respectively and expressed by:

$$W_1 = \frac{l_r}{2l}W - \Delta W = \frac{l_r}{2l}W\left(1 - \alpha\frac{2lh}{l_rd}a_y\right), W_2 = \frac{l_r}{2l}W + \Delta W = \frac{l_r}{2l}W\left(1 + \alpha\frac{2lh}{l_rd}a_y\right)$$

The transversal load transfer at front axle, ΔW, is written as:

$$\Delta W = \alpha\frac{h}{d}Wa_y, a_y(s) = \frac{1}{gl}\frac{V^2}{1+AV^2}\frac{1}{1+T_Ds}\frac{\delta_h(s)}{n}\ (T_D = 0.1\sim0.2[s])$$

Here, α is a transversal load transfer rate at the front axle determined by roll stiffness distribution between front and rear and $a_y(s)$ is the estimated steady state lateral acceleration response to the steering input with the first order delay.

Putting the above W_1 and W_2 into Eq.(6) gives us the following description as a final form of the control law:

$$\delta_{\mathrm{hb}} = \left[\frac{1}{2}(k_1 - k_2) + \alpha(k_1 + k_2)\frac{lh}{l_r d}a_{\mathrm{y}}\right]\delta$$

$$= \underline{\frac{-\left(1-B\frac{l_r}{l}V^2\right)\frac{d}{2l}\delta}{(1+AV^2)^2-\left(\frac{d}{2l}\delta\right)^2}} + \underline{\alpha\frac{\left(A+B\frac{l_r}{l}\right)V^2(1+AV^2)-\left(\frac{d}{2l}\delta\right)^2}{(1+AV^2)^2-\left(\frac{d}{2l}\delta\right)^2}\frac{2lh}{l_r d}a_y\delta} \tag{7}$$

As the first underlined part and the second one become $-d\delta^2/2l$ and zero respectively when the vehicle speed becomes zero, the first underlined term corresponds to the Ackermann steering angle and the second one is for adjusting the side slip angle depending on α and a_y to be proportional to the vertical load. The figure 3 is a diagram to conceptually show how the right wheel hub, δ_{hb} , is additionally steered by Eq.(7) depending on the lateral acceleration, a_y , with increase of the vehicle speed at constant steering wheel angle of left turn.

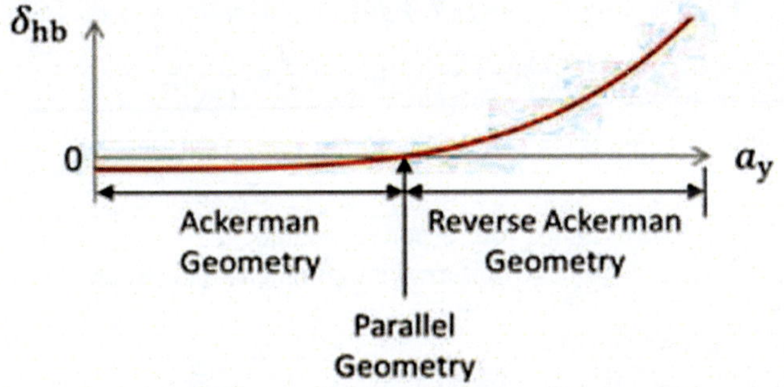

Fig. 3. Conceptual diagram of hub-steer with respect to lateral acceleration

3.3 Control algorithm for improving transient response to steer input

The side slip angle and yaw rate response functions to front wheel steer angle are described as follows based on the two degree of freedom planar bicycle model:

$$\beta(s) = G_\delta{}^\beta(0)\frac{1+T_\beta s}{1+\dfrac{2\zeta s}{\omega_n}+\dfrac{s^2}{{\omega_n}^2}}\delta(s),\ \ r(s) = G_\delta{}^r(0)\frac{1+T_r s}{1+\dfrac{2\zeta s}{\omega_n}+\dfrac{s^2}{{\omega_n}^2}}\delta(s)$$

Introducing the feed-forward type of active front wheel steer, $\delta/\delta_h(s)$, the above response functions are formally rewritten as:

$$\beta(s) = G_\delta{}^\beta(0)\frac{1 + T_\beta s}{1 + \frac{2\zeta s}{\omega_n} + \frac{s^2}{\omega_n{}^2}}\frac{\delta}{\delta_h}(s)\delta_h(s),$$

$$r(s) = G_\delta{}^r(0)\frac{1 + T_r s}{1 + \frac{2\zeta s}{\omega_n} + \frac{s^2}{\omega_n{}^2}}\frac{\delta}{\delta_h}(s)\delta_h(s)$$

As it is well known that the vehicle handling characteristics is predominantly influenced by the response parameters, natural frequency ω_n and damping ratio ζ, the following transfer function is introduced as a control law of the front active steering control:

$$\frac{\delta}{\delta_h}(s) = \frac{1}{n}\frac{1 + \frac{2\zeta}{\omega_n}s + \frac{1}{\omega_n{}^2}s^2}{1 + \frac{2\zeta(1+\alpha_1)}{\omega_n(1+\alpha_2)}s + \frac{1}{\omega_n{}^2(1+\alpha_2)}s^2}(|\alpha_1|, |\alpha_1| \ll 1.0) \quad (8)$$

Then, the side slip angle and yaw rate response with active front wheel steer becomes as follows:

$$\frac{\beta}{\delta_h}(s) = \frac{1}{n}G_\delta{}^\beta(0)\frac{1 + T_\beta s}{1 + \frac{2\zeta(1+\alpha_1)}{\omega_n(1+\alpha_2)}s + \frac{1}{\omega_n{}^2(1+\alpha_2)}s^2},$$

$$\frac{r}{\delta_h}(s) = \frac{1}{n}G_\delta{}^r(0)\frac{1 + T_r s}{1 + \frac{2\zeta(1+\alpha_1)}{\omega_n(1+\alpha_2)}s + \frac{1}{\omega_n{}^2(1+\alpha_2)}s^2}$$

Therefore, it is possible to adjust the response parameters, ω_n and ζ, by changing α_1 and α_2 appropriately so as to have preferred handling characteristics of the vehicle for a specific driver.

The equation (8) is rewritten as:

$$\frac{\delta}{\delta_h}(s) = \frac{1}{n}\left[1 + \frac{\frac{2\zeta}{\omega_n}\frac{\alpha_2 - \alpha_1}{1+\alpha_2} + \frac{1}{\omega_n{}^2}\frac{\alpha_2(2+\alpha_2)}{(1+\alpha_2)^2}s}{1 + \frac{2\zeta(1+\alpha_1)}{\omega_n(1+\alpha_2)}s + \frac{1}{\omega_n{}^2(1+\alpha_2)^2}s^2}\right] \quad (9)$$

Here, the first part of the right side of the equation presented by 1.0 is the steering angle directly steered by the driver through the steering wheel. The second part described by fractional expression is the additional control steer part to be steered by the hub-steering system depending on steering wheel angle speed.

In the meantime, the lateral acceleration response of the vehicle with an active front wheel steer is described as follows:

$$a_y(s) = G_\delta{}^{a_y}(0)\frac{1 + \frac{l_r}{V}s + \frac{I}{2lK_r}}{1 + \frac{2\zeta s}{\omega_n} + \frac{s^2}{\omega_n{}^2}}\frac{\delta}{\delta_h}(s)\delta_h(s)$$

A problem for conventional vehicles exists that there is a significantly larger delay in the phase lag of the lateral acceleration response compared with that of the yaw rate at high speed. This is due to the decrease of the first order lead time constant of the lateral acceleration, the coefficient of s in the numerator, l_r/V, with increase of vehicle speed, which is related to the side slip response acting in opposite direction to the steering angle at higher vehicle speed than $V_{\beta=0} = \sqrt{2l_r l K_r / m l_f}$.

In order to solve the above problem, the following front active steering control is introduced:

$$\frac{\delta}{\delta_h}(s) = \frac{1}{n}\frac{1 + \tau s + \frac{I}{2lK_r}s^2}{1 + \frac{l_r}{V}s + \frac{I}{2lK_r}s^2} \tag{10}$$

Then, the lateral acceleration response to steer input becomes as follows:

$$\frac{a_y}{\delta_h}(s) = \frac{1}{n}G_\delta{}^{a_y}(0)\frac{1 + \tau s + \frac{I}{2lK_r}s^2}{1 + \frac{2\zeta s}{\omega_n} + \frac{s^2}{\omega_n{}^2}}$$

Therefore the decrease of the first order lead time constant in the numerator is compensated by τ, which must be determined appropriately.

The figure 4 shows the relation of the first order lead time constant, τ, provided and the side slip gain constant, $G_\delta^\beta(0)$, and in order to avoid the excessive decrease of the lead time constant due to l_r/V, τ is kept constant at the vehicle speed higher than the speed at which the side slip response is zero to steering input.

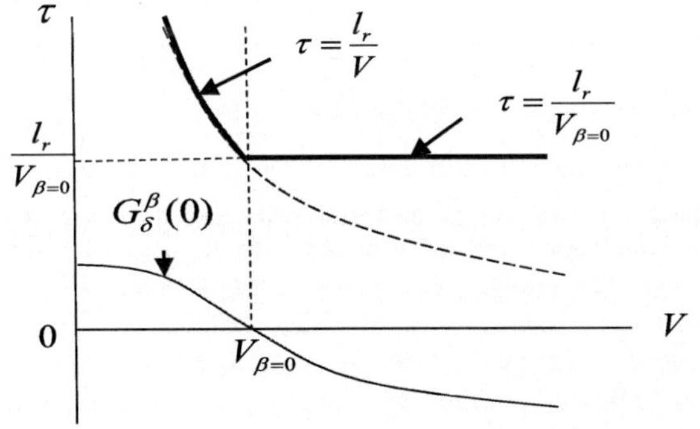

Fig. 4. The first order lead time constant in lateral acceleration

The equation (10) is rewritten as:

$$\frac{\delta}{\delta_h}(s) = \frac{1}{n}\left[1 + \frac{\left(\tau - \frac{l_r}{V}\right)s}{1 + \frac{l_r}{V}s + \frac{I}{2lK_r}s^2}\right] \tag{11}$$

The first term is corresponds with a steering angle directly steered by the drivers through the steering wheel and the remnant part described by the fractional expression is the additional control to be steered by the hub-steering system depending on the steering wheel speed.

When the adjustment control of the response parameters, ω_n and ζ, as well as the compensation control of the lead time constant in lateral acceleration are adopted at a time, the active front wheel steering control law is described as:

$$\frac{\delta}{\delta_h}(s) \approx = \frac{1}{n}\left[1 + \frac{\frac{2\zeta}{\omega_n}\frac{\alpha_2 - \alpha_1}{1+\alpha_2} + \frac{1}{\omega_n{}^2}\frac{\alpha_2(2+\alpha_2)}{(1+\alpha_2)^2}s}{1 + \frac{2\zeta(1+\alpha_1)}{\omega_n(1+\alpha_2)}s + \frac{1}{\omega_n{}^2(1+\alpha_2)^2}s^2}s + \frac{\left(\tau - \frac{l_r}{V}\right)}{1 + \frac{l_r}{V}s + \frac{I}{2lK_r}s^2}s\right] \tag{12}$$

The first term is a steering angle directly steered by the driver and the fractional expressions of the second and third term are the additional steering control by the hub-steering system.

4 Simulation

In order to confirm the effects of the hub-steer, δ_{hb}, of the left and right wheels in opposite direction each other on the steady circular turning, a simulation of the vehicle response to the large ramp-step steer input using full vehicle model is carried out before the experimental investigation. The results are shown in Fig.5. Though the inner wheel steer angle is larger than the outer wheel at lower vehicle speed due to Ackermann steer function, it is obvious as the effects of the hub-steer that the steer angle of the outer wheel becomes larger compared with that of the inner wheel with increase of the lateral acceleration and the radius of path curvature becomes smaller. This means that larger lateral force is provided by the hub-steer and tighter cornering is available at higher lateral acceleration for the vehicle with the hub-steer.

In order to confirm the effects of the active front wheel steer by the hub-steer on the transient response characteristics of the vehicle, the frequency response of the vehicle with the active control is analyzed by the simulation. The results are shown in Fig.6. It is easy to understand through this result that the vehicle response characteristics are available to be widely changed by changing the active steering control parameters. Therefore, it will be possible to tune the vehicle response characteristics preferred by the drivers by adjusting the control parameters of the active front wheel steer.

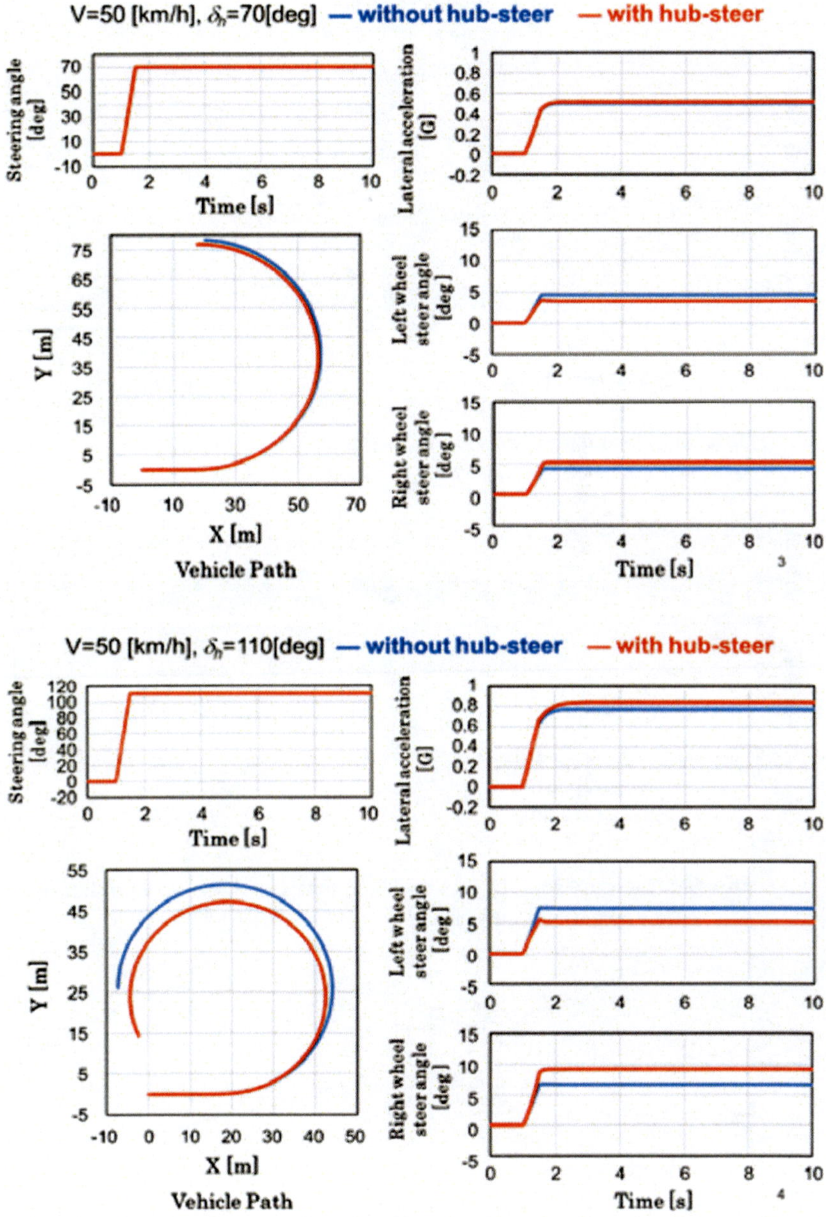

Fig. 5. Simulation results of the effects of hub-steer on vehicle response to ramp-step steer input

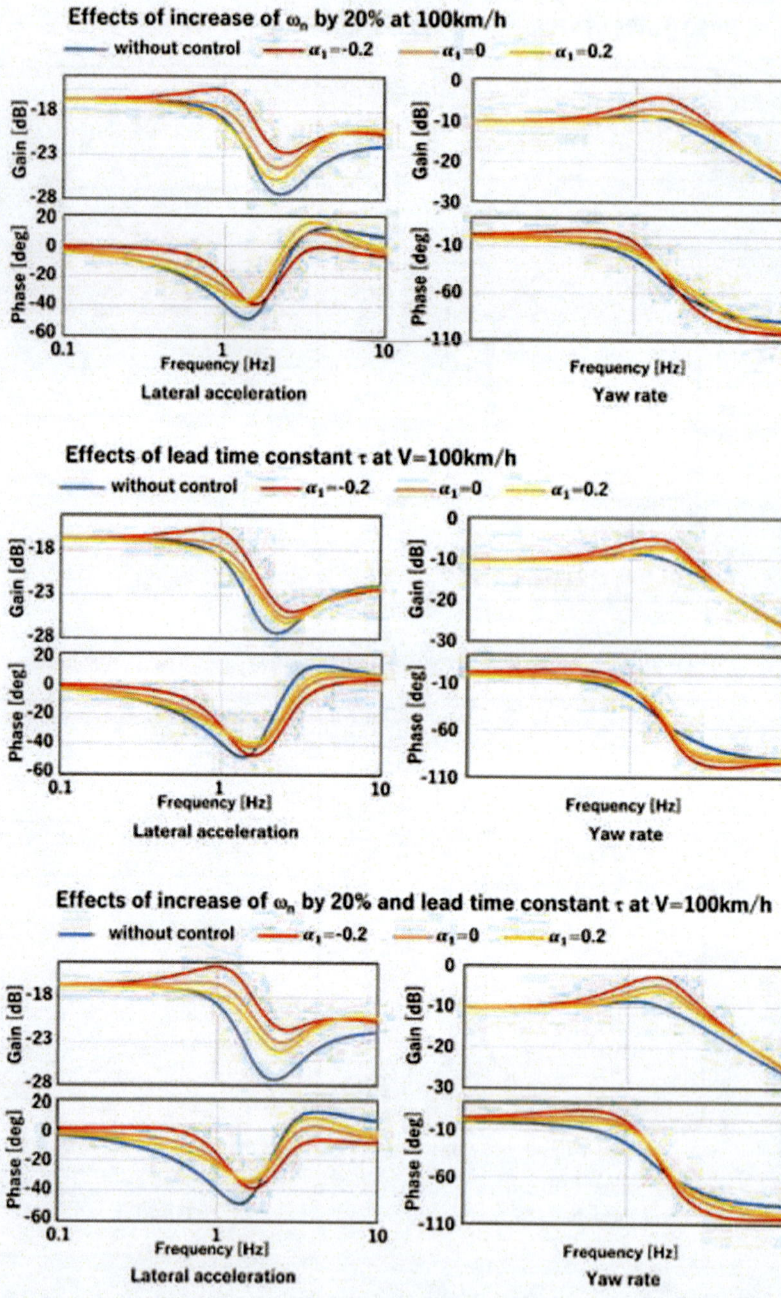

Fig. 6. Effects of active front wheel steer on vehicle frequency response

5 Experimental validation

An experimental validation of the effects of hub-steer control system on improving handling quality of the vehicle is carried out on a proving ground using a conventional passenger vehicle in which the developed hub-steer system is installed being substituted for the existing hub system.

5.1 Effects of the left and right wheel hub-steers in counter direction each other

The figure 7 (next page)shows an example of the effects of the hub-steers on the vehicle response to the ramp-step steer input same as shown in the simulation. It is shown that as the vehicle speed and steering angle increase, more severe cornering with smaller curve radius is available for the vehicle with the hub-steer compared with the vehicle without it.

An effect of the hub-steer on the vehicle motion during a slalom running is shown in Fig.8. It is obvious that the slalom running of smaller path amplitude is available with smaller amplitude of the steering wheel angle for the vehicle with the hub-steer and the amplitudes of both yaw rate and lateral acceleration also become smaller.

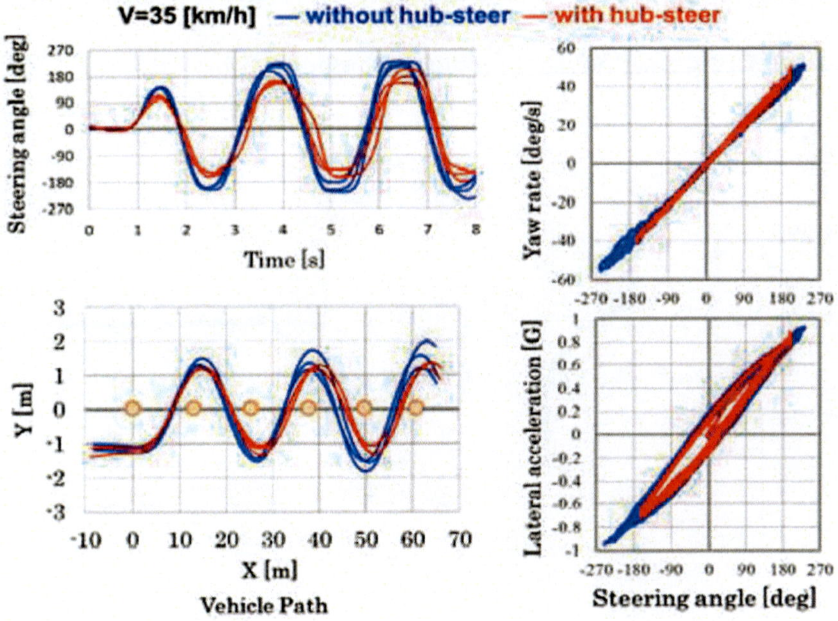

Fig. 7. Effects of the hub-steer on the vehicle slalom running

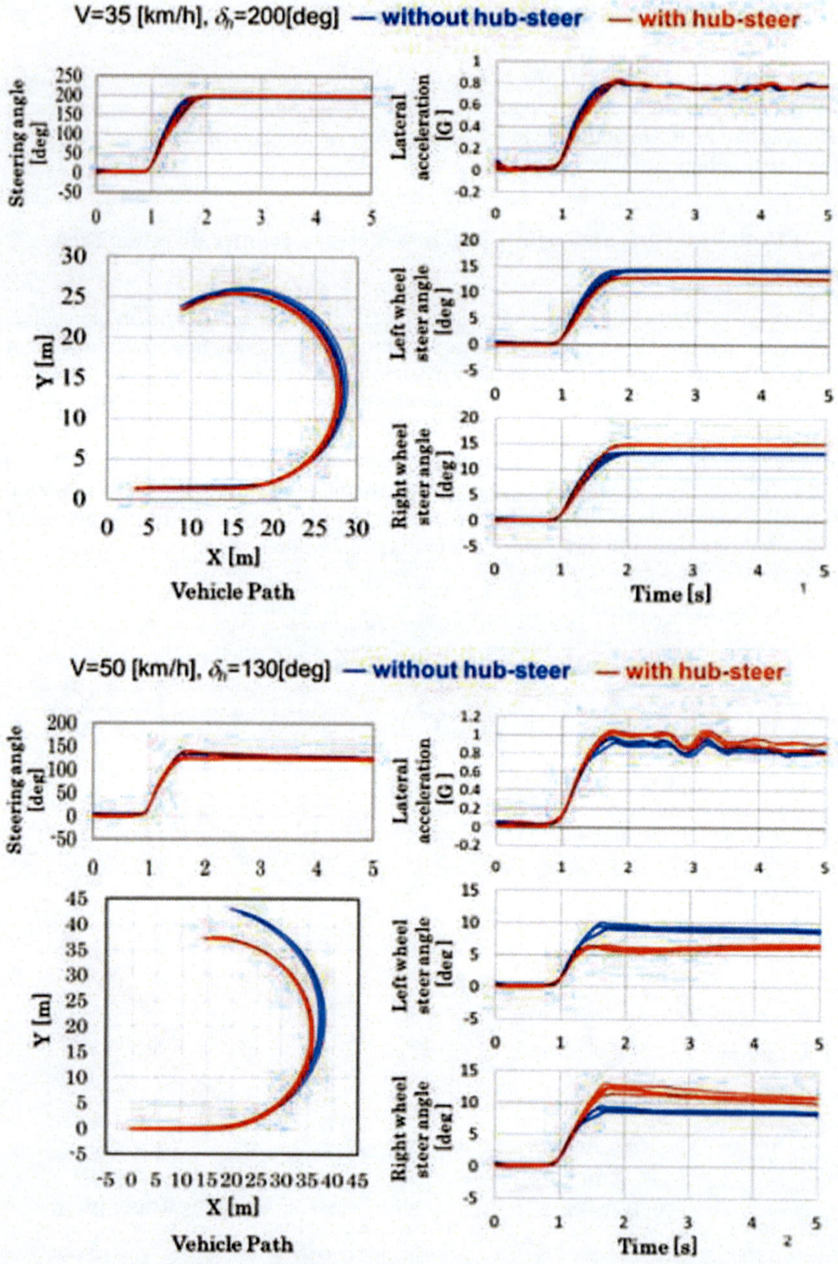

Fig. 8. Effects of the hub-steer on the vehicle response to ramp-step steer input

5.2 Effects of the active front wheel steer by the hub-steer

An effect of the active front wheel steer on the transient behavior of the driver-vehicle-system during a lane change is shown in Fig.9. Stabilizing effects of the active front wheel steer on the driver-vehicle behavior during the lane change is quite significant.

The figure 10 shows the effect of the control on the vehicle behavior during a slalom running. It is obvious that the width of lissajous curves of both steer angle - yaw rate and steer angle - lateral acceleration of the vehicle with the active steer control is narrower than that of the vehicle without control. This means that the front active steer control is contributive to reduce the response delay of the vehicle to steer input and to stabilize the driver-vehicle-system behavior, which corresponds to the effect of the active steer control on the phase advance in the vehicle frequency response to steer input shown in Fig.6 as a simulation result.

6 Conclusions

The hub-steer system by which the additional front wheel steer of the vehicle is available with no modification of the external layout configurations of the existing steering and suspension systems has been developed. The control algorithms of the hub-steer for improving the vehicle handling characteristics during steady circular turning and also for improving the vehicle transient response characteristics to steer input are introduced. Following to the simulation studies, the experimental investigation into the effects of the hub-steer system with the control algorithms using a conventional passenger vehicle equipped with the hub-steer system are carried out on a proving ground.

Followings are confirmed as the effect of the left and right wheel hub-steer in opposite direction each other to have each side slip angle proportional to the respective vertical load of the wheels. While at low speed turning the left and right wheels are provided with an exact Ackermann steer angle to have smooth tight corner turning with large steering wheel angle, during circular turning with higher vehicle speed, the steer angle of the outer wheel becomes larger than that of the inner one due to the hub-steer, which result in larger total lateral force at the front wheel of the vehicle and more tight and severe steady cornering with large lateral acceleration becomes available.

The feed-forward type of active front wheel steer control by the hub-steer is confirmed to bring us significant effects on improving the transient response characteristics of the vehicle to the steering input especially at relatively higher vehicle speed. Most preferable response characteristics is available by adjusting the natural frequency, the damping ratio and the first order lateral acceleration lead time constant of the vehicle response transfer function appropriately in the control algorithm.

Thus it is concluded that the hub-steer system developed is significantly effective in improving the vehicle agility and stability in wide range of vehicle maneuvering conditions such as from very low speed tight corner turning to high speed tight/severe circular turning with high lateral acceleration and during transient period of vehicle response to steer input especially at very high speed.

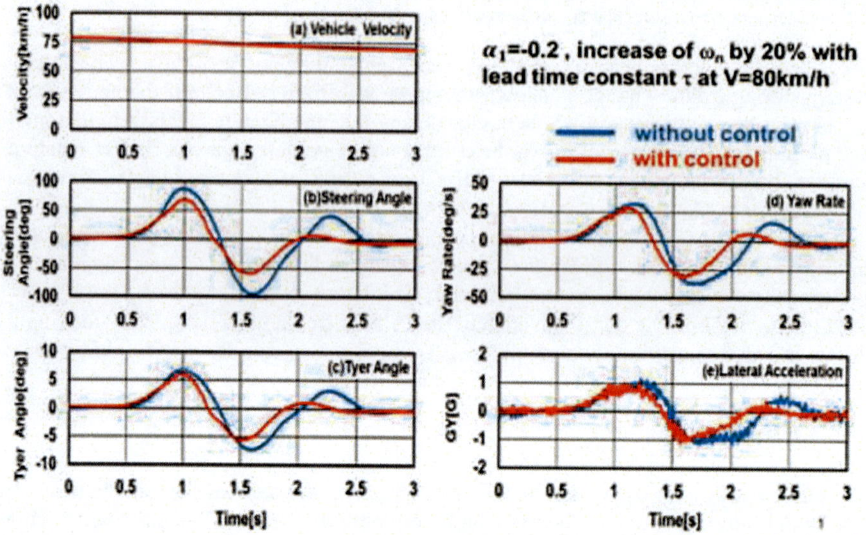

Fig. 9. Effect of the active front wheel steer on lane change behavior of the driver-vehicle system

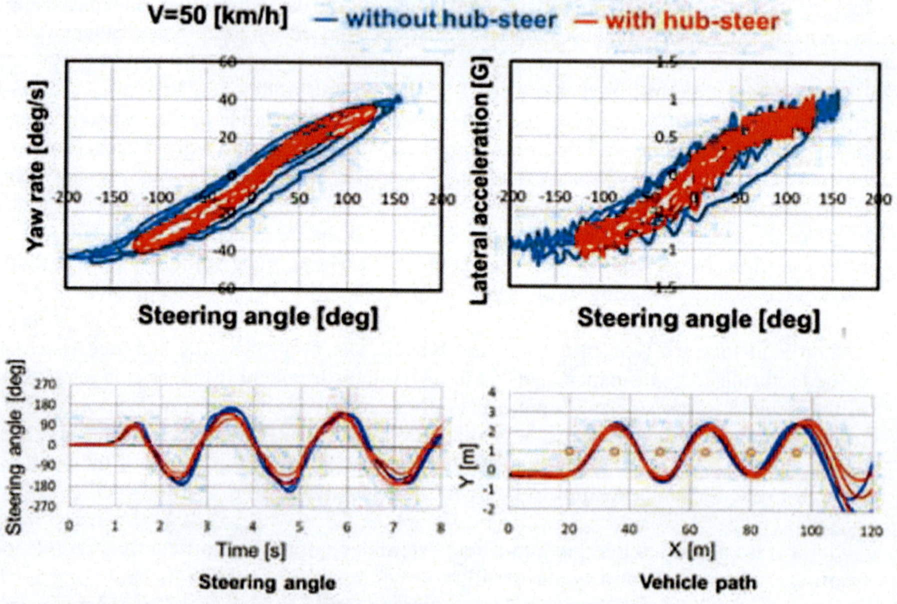

Fig. 10. Effects of the active front wheel steer on vehicle slalom running

References

1. N. Ishihara, M. Ishibashi et al. "Development of hub bearings integrated with active steering function for improvement of vehicle dynamics" Proceedings of 2018 JSAE annual congress (spring), May 2018, Yokohama, Japan (in Japanese)
2. A. Ito, M Ishibashi et al. "Hub bearing integrated with active steering function that improves vehicle dynamic performance" Proceedings of 2019 JSAE annual congress (spring), May 2019, Yokohama, Japan (in Japanese)

Steer-by-Wire in the Context of the Connected Vehicle: Opportunities for Future Steering Innovations

Joe Klesing

Nexteer Automotive Corp., USA

Abstract. Currently, the application cycle of a steering system goes together with the life cycle of a vehicle application. Significant changes in function and content are typically introduced through new vehicle applications and platform architectures. With the fast development of connected vehicle functions, new chassis applications and innovations can be introduced during the life cycle of a vehicle via over-the-air system updates.

What made the Smartphone so successful that we all have one? A powerful, flexible, software-upgradable platform architecture that comes to life with new downloadable applications is a major part of the answer.

Steer-by-Wire will have the same impact in a connected and software-defined vehicle: Elimination of the I-shaft eliminates constraints in packaging and functionality. It improves new functions like Automatic Emergency Steering and enables functions that we have not yet even thought of but that will develop over the coming years. The steering wheel is transforming into a Human Machine Interface by creating additional degrees of freedom for haptic interaction between driver and vehicle – like fly-by-wire in the aerospace industry.

The connected, software-defined vehicle will feature new software functions, updated over the air over the lifetime of the vehicle, that keep the product portfolio fresh. Therefore, the connected, steer-by-wire–equipped vehicle brings multiple trends together: connectivity and software.

Keywords: connected vehicle, by-wire, safety, multi-modal communication, platform architecture.

P. E. Pfeffer (ed.), *11th International Munich Chassis Symposium 2020*, Proceedings, https://doi.org/10.1007/978-3-662-63193-5_29

1 Trends and Lessons-Learned in Related Industries

1.1 By-wire

Fly-by-wire was introduced into mass-produced commercial and military planes in 1976. Originally, it was needed to enable novel breakthrough features that otherwise would not have been possible, like the supersonic flights of the Concorde or the stealth of the Lockheed F117. Other benefits followed, including fuel savings and new safety functions via automation. Lessons-learned: By-wire and automation go hand in hand, enabling brand new functionality.

Fig. 1. Early by-wire application examples in aerospace: Concorde, F117 [1]

An equivalent application in automotive steering is the retractable steering wheel in an autonomously driving vehicle. Without by-wire, the steering wheel cannot be prevented from rotating during the storing process (Fig. 2).

Fig. 2. Steer-by-Wire application in automotive

The introduction of automated driving in mass production on public roads will be determined by the pace of legislation, the efforts in validation to make this technology safe for public use, and the cost of the necessary sensor suite. The trend towards the connected vehicle provides a tailwind towards the introduction of Steer-by-Wire (SbW) and will be examined in the following sections.

1.2 Trends in life of vehicle vs. innovation cycles

According to the IHS Markit light vehicle sales forecast [2], the life of vehicle platforms is increasing from one generation to the next. This development is generating higher volumes per platform and therefore economies of scale (Fig. 3).

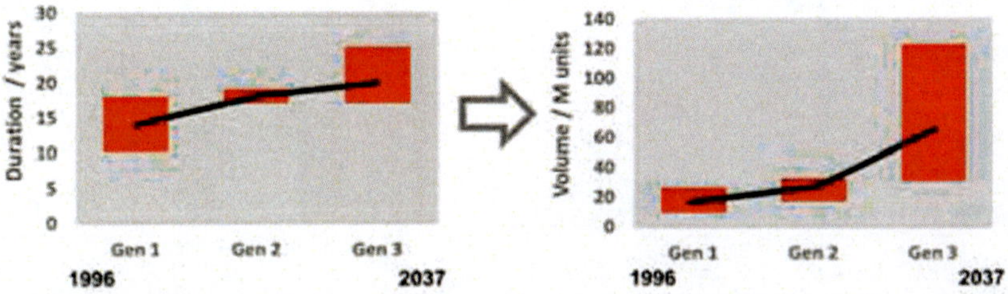

Fig. 3. Duration and production volume of vehicle platforms

At the same time, the vehicle cycles per platform and thus the introduction of new features for the consumer are becoming shorter (Fig. 4).

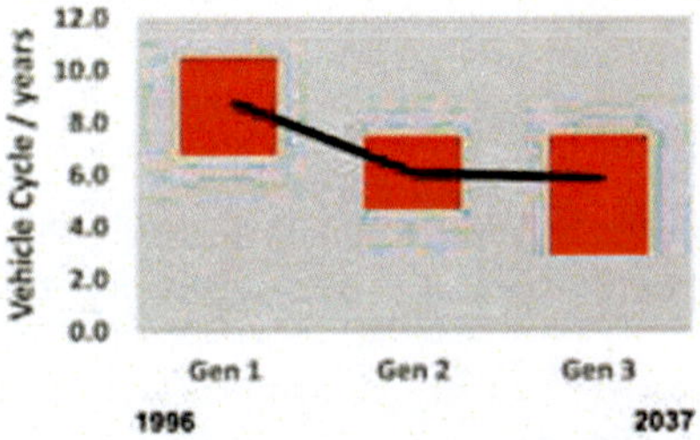

Fig. 4. Duration and production volume of vehicle platforms

This results in an increased "freshness" of the vehicle portfolio, enabled and driven by new software functions (Fig. 5). According to Accenture Research [3], in 2030 software will account for 40% of the total vehicle value.

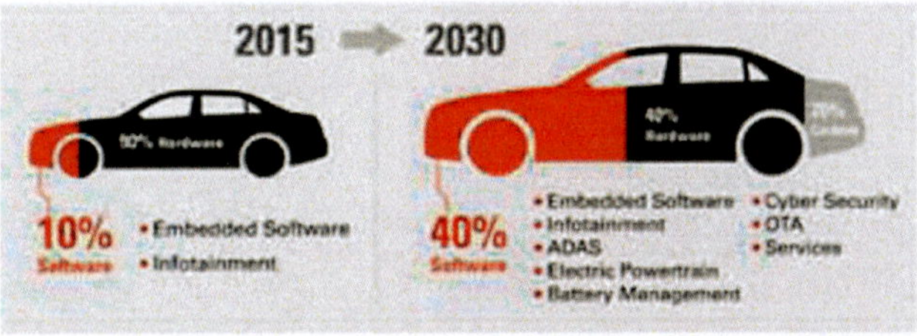

Fig. 5. The software-enabled vehicle

New functions will be launched not only at but also after start of production and can be sold separately during the life of a vehicle platform. Features, including trials, subscriptions, purchases, and on-demand features, can be either activated or downloaded from an App store in the cloud. Example features include drive modes, assist functions, location-based services, and range extension for electric vehicles and are accessible via a User Interface in the vehicle or on a mobile device (Fig. 6).

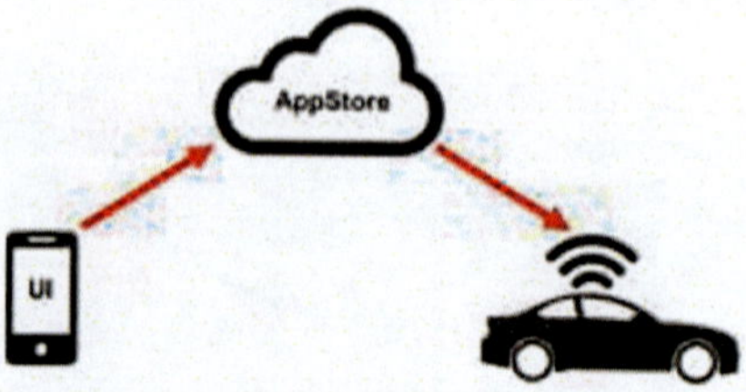

Fig. 6. Appstore enables feature upgrades

Tesla [4], for example, introduced a number of new and updated features, such as adaptive suspension, auto lane change, power increase, lane departure warning, and auto folding mirrors, in such a fashion. The software updates included beta versions and "sneak previews."

1.3 Lessons-learned From the Mobile App Market

What can we learn from connected, mobile applications in terms of monetization? Previous research [5] identified three phases in the development of the market over the period from 2008 to 2018:

- Experimentation: Mobile device owners discover and experiment with new Apps. Download numbers increase precipitously.
- Adoption: Habits start to form, and users identify their preferred Apps while engagement with the Apps climbs. This phase is characterized by a steep rise in usage of Apps.
- Ubiquity: This phase is characterized by increased consumer spending.

So, the lessons-learned from the mobile App market is that once the consumer became accustomed to the concept of an "App store," the monetization opportunities and thus the revenue for the App providers increased (Fig. 7).

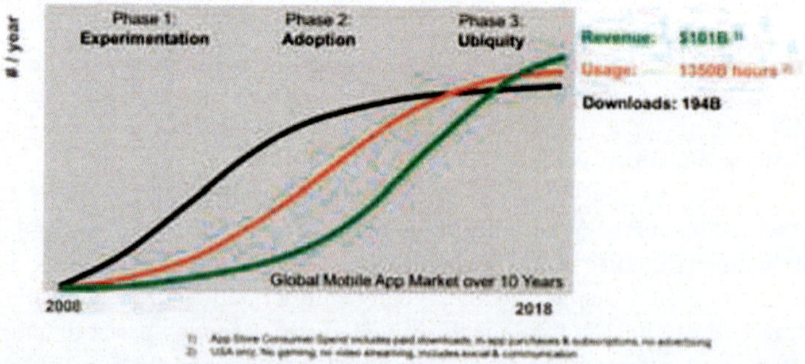

Fig. 7. Downloads, usage, and revenue in the global mobile app market

Consumers try out significantly more Apps than they actually keep on their mobile devices in the long term. Successful Apps are tailored to solving a specific problem for the user, have a lean feature set and great user experience [6] and are continuously replenished with fresh content [7].

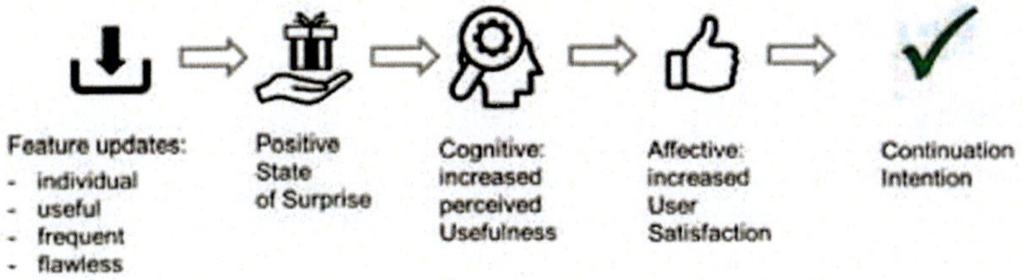

Fig. 8. Flawless updates with useful features increase customer loyalty

Another lesson from the mobile App market concerns the architecture of the mobile devices. A standardized, modular, and flexible platform architecture allows features to be more easily integrated into an existing system.

A platform architecture partitions a system into stable core components and variable components. It enables new features to be added or existing features to be removed in tailoring the products to market needs. By promoting the reuse of core components, such partitioning can reduce the cost of variety and innovation at the system level [8].

Upgradability with new software-enabled features drives new E/E architectures also in automotive applications. It consists of multiple tiers, including an actuator tier, Domain Controllers (DCU), and the cloud (Fig. 9).

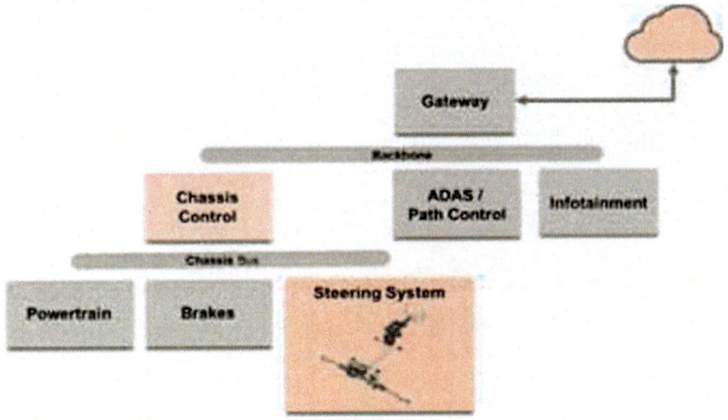

Fig. 9. Simplified view of the E/E architecture enabling new chassis functions

Functions such as time-critical motor controls are located in the Electronic Control Units (ECU) in the actuator tier. Customer-relevant features reside in the DCU and the cloud.

2 Chassis Apps

2.1 Safety

What kind of automotive Apps are consumers willing to spend money on? A 2019 Nexteer survey of 501 people confirmed an interest in safety-related Apps: 85% of respondents like to drive their car, and 77% either have been involved in an accident or are close to someone who has (Fig. 10). Thus, safety is on consumers' minds, but they want to continue to enjoy driving.

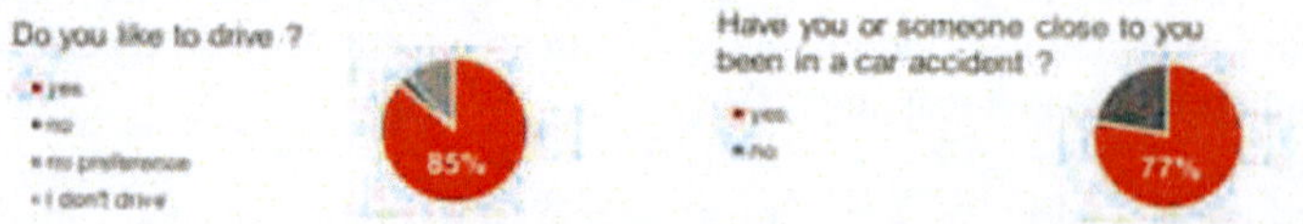

Fig. 10. Consumers like to drive but are safety conscious

A majority of the respondents (79%) regularly use Apps on mobile devices (Fig. 11). The familiarity with the concept of Apps will help with the introduction of Apps into cars and for chassis applications and might even be expected by the consumer.

Fig. 11. Consumers are familiar with the concept of Apps

Proposals to pay more for specific safety-related warnings or vehicle stability are met with more mixed results. Among the respondents to the Nexteer survey, 48% said they would be willing to pay for multimodal alerts of a hazardous situation and 52% responded that they would be willing to pay to eliminate the risk of their vehicle sliding or spinning (Fig. 12).

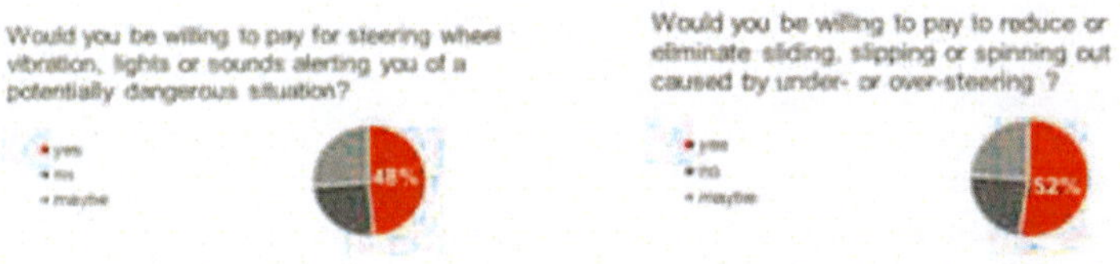

Fig. 12. Consumers' willingness to pay for warnings and stability features

2.2 A consumer input on safety applications

A Lead User analysis [9] provides further insights. Lead Users are consumers who face the needs that will become general needs in the marketplace and who would benefit from a solution. And, more importantly, they have ideas and volunteer them, e.g. by participating in a study and thus being part of the extended product design team [10].

As part of this approach, 1701 contributors interacted with the research team via a web portal. Over 3 stages of interaction, the team identified 30 Lead Users. The ideas generated from this process can be summarized as follows: Consumers are looking for practical helpers that they can experience every day and that give them peace of mind and help them avoid surprises. If a situation should arise that overwhelms consumers, they need the helper to be available at their fingertips (Fig. 13). The solution includes multimodal communication, including visual, acoustic, and haptic communication between the driver and the vehicle.

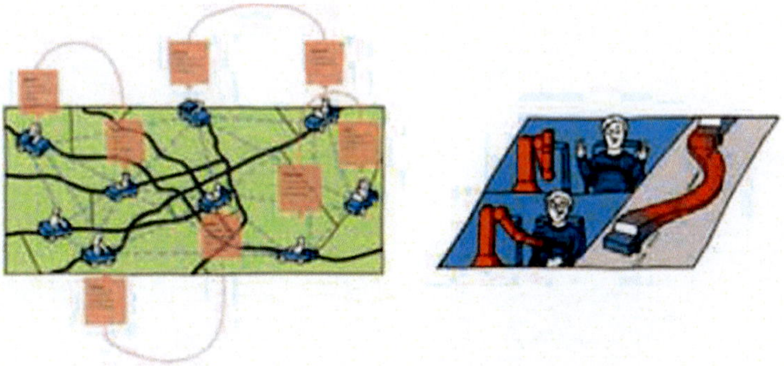

Fig. 13. Practical helpers for peace of mind and as a support net

3 Steer-by-Wire–enabled Safety Applications

3.1 Automatic Emergency Steering and Park Assist

Vehicle-initiated Automatic Emergency Steering (AES) and Park Assist are two functions that benefit from the introduction of SbW. In both applications, the driver can leave their hands on the steering wheel during the maneuver. At the same time, the vehicle communicates its intentions to the driver by giving directional cues via the reduced steering wheel angle (Fig. 14).

Fig. 14. SbW for vehicle-initiated AES and Park Assist

3.2 Shared Vehicle Control

But SbW can do even more to improve the safety of a vehicle. Its strength is the ability to communicate haptic cues to the driver that are independent from the actual steering angle. Thanks to this feature, shared haptic control concepts [11] become feasible and the steering wheel becomes a bi-directional Human Machine Interface (HMI) element.

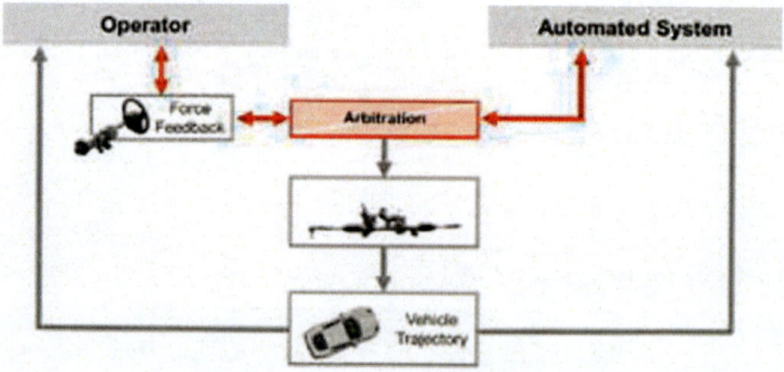

Fig. 15. Shared Vehicle Control enabled by SbW

A study by Nexteer [12] showed that such a control concept can help to manage issues of situational awareness, skills degradation, mode confusion, and workload fluctuation during the handover from automated to manual driving. These types of applications support keeping vehicle behavior predictable and safe for the average driver in all driving scenarios.

An example is a lateral acceleration limiter to prevent understeer on low friction surfaces (Fig. 16). A crowd-sourced preview of the friction coefficient enables the stability limit to be determined before the vehicle hits an icy patch of road and thus reduces the system's response time. SbW can then be applied to limit lateral acceleration and provide stability control. The driver gets torque feedback that keeps them safe, similar to the feedback with Lane Keep Assist, and prevents them from "fighting" the system.

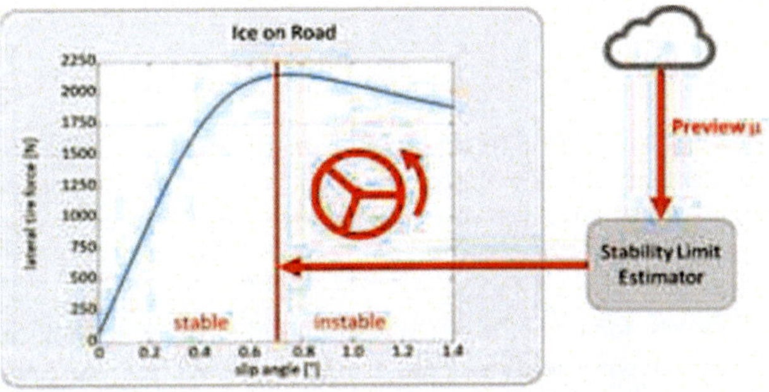

Fig. 16. SbW can keep vehicle behavior predictable and prevent understeer

As an outlook into the future, shared control algorithms will enable a guided automated driving mode (Fig. 17). Today, guided automated flight controls are applied in the aerospace industry to prevent the pilot from bringing the airplane into an unstable mode.

The character of the vehicle, its feel and behavior become entirely synthetic, i.e. customizable via software. The driver conveys their intent via the steering wheel and brakes, while the automation controls the vehicle in the safest possible way. The steering wheel offers additional degrees of freedom for the haptic communication between driver and vehicle.

This type of application is not yet ready for deployment. However, it was chosen to show that in a connected world, with the appropriate hardware architecture in place, including SbW, functions like this will be available for download and deployment without changing the hardware.

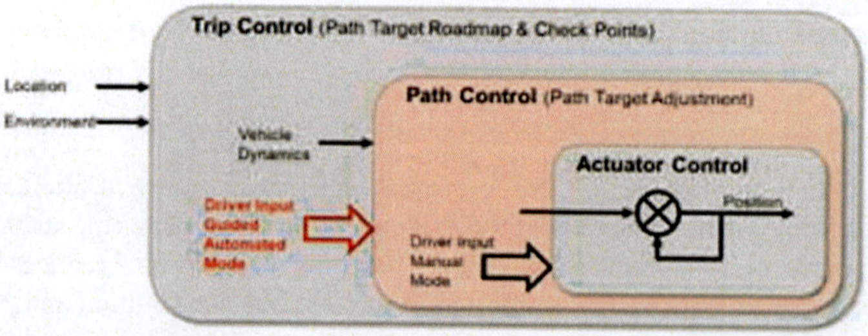

Fig. 17. SbW enables Guided Automated Driving mode

4 Summary

Connected vehicles will offer a new level of experience for the consumer. The vehicle will stay in permanent contact with its environment, the vehicle owner, and the manufacturer. The data and insights from those interactions will result in continuous innovation over the lifetime of a vehicle.

The ability to provide flawless updates of new features over the life of a connected vehicle builds consumer loyalty. Consumers value practical, safety-related features that result in peace-of-mind and real-time support if they should become overwhelmed in a driving situation, and thus these features offer monetization potential.

A standardized, flexible, modular platform architecture, which includes Steer-by-Wire, enables the upgrade and integration of new features that have not yet been developed, without changing the vehicle hardware.

Hence, in a connected, software-defined vehicle, the Steer-by-Wire platform enables unprecedented, future-proof customization and upgradability.

References

1. Images by Schoepal, Allmon from en.wikipedia.org
2. IHS Markit Automotive Industry, Light Vehicle Sales Forecast, December 2019
3. Accenture Research, 2019
4. https://teslascope.com/teslapedia/software
   https://teslafi.com/firmware/
   https://teslabout.com/firmware
5. https://www.appannie.com/en/insights/market-data, State of Mobile 2018 report
   https://www.businessofapps.com/data/app-statistics/#5
   https://www.statista.com/topics/983/mobile-app-monetization/
   https://www.statista.com/statistics/266210/number-of-available-applications-in-the-google-play-store

6. Clark, J.: Tap worthy: Designing Great iPhone Apps, 1st Edition, O'Reilly (2010)
7. Fleischman, M. et. al., Decision support systems: The role of software updates in information systems continuance — an experimental study from a user perspective, Elsevier (2016)
8. Baldwin, Carliss Y. and Woodard, Charles Jason, The Architecture of Platforms: A Unified View (September 8, 2008). Harvard Business School Finance Working Paper
9. Von Hippel, E.: Democratizing Innovation, MIT Press (2005)
10. Von Hippel, E.: https://evhippel.mit.edu/
11. David A. Abbink, Mark Mulder, and Erwin R. Boer. Haptic shared control: smoothly shifting control authority? Cognition, Technology & Work, 14(1):19–28, Mar 2012. ISSN 1435-5566.
12. Klesing, Zuraski, Rezaeian (2018), Steering on Demand for dual-mode vehicles, 9th International Munich Chassis Symposium 2018

Optimal design of a DC brushless motor for automotive brake actuation

Francesco CAMOZZI[1], Antonino DI GERLANDO[2],
Cristiano FISSORE[1], Massimiliano GOBBI[3], Giampiero MASTINU[3],
Paolo RIGHETTINI[4], Roberto STRADA[4], Damiano TODISCO[1],

[1] Brembo, Stezzano BG, Italy
[2] Politecnico di Milano, Dep. of Energy, Italy
[3] Politecnico di Milano, Dep. of Mechanical Engineering, Italy
[4] University of Bergamo, Italy

Abstract. A Consortium of Companies and Universities cooperated to design and manufacture electric motors for brake actuation. The project, in an Industry 4.0 framework, aimed to combine the design of both product and process.
The electric motors for brakes have been optimized to shrink as much as possible the geometric dimensions while keeping high output torque, with the constraints of reduced production cost and extremely high level of reliability and performance.
A smart production plant has been studied and optimized together with the electric motor design.
Multi-objective optimization has been used to design the brushless DC motor. Fifteen design variables were considered for the definition of stator and rotor geometry, pole pieces and permanent magnets. The performance indices are peak torque, efficiency, rotor mass and inertia. The design constraints refer to components stress levels and temperature thresholds.
The physical mathematical models used for optimal design refer to electromagnetic field and related currents computation, to thermo-fluid dynamics simulation, to local stress and vibration assessment. A surrogate model based on Artificial Intelligence has been used to speed-up the simulations. An Artificial Neural Network model, trained with an iterative procedure, was employed.
Pareto-optimal solutions resulting from the design process are presented. Significant improvements of the performance indices with respect to a reference solution have been found.

.**Keywords:** Electric motor, brake, optimization, Industry 4.0.

1 Introduction

The aim of the paper is to present how to construct effectively electric motors for vehicles. The normal automotive paradigm is to conceive both the product and the construction process at the same time. Since we have fully digitized such two activities, we also

P. E. Pfeffer (ed.), *11th International Munich Chassis Symposium 2020*,
Proceedings, https://doi.org/10.1007/978-3-662-63193-5_30

resort to Industry 4.0 paradigm. All of the aspects concerning virtual design, virtual production and virtual testing are addressed.
The market penetration of electric vehicles improves in an unpredictable way. KPMG states that "the rate of EV uptake is uncertain, it is unlikely to be slow and linear." [1]. The high purchase cost of EVs is one of the causes that prevents such vehicles to gain a sound penetration in the market.
Our commitment is to produce high quality electric motors at a relatively low cost. This requires mass production, which implies a proper design of the production process and a coincident proper design of the electric motor.
The paper deals with the construction of electric motors for actuating brakes. The original feature of such products is that they are exclusively conceived and manufactured for automotive applications, with safety, lightweight construction and performance as driving issues.
In the literature [2-5], a number of contributions have been given on the design and construction of electric motors for automotive applications. In this paper we will highlight the relevant features of both the design and manufacturing process of electric motors. This contribution seems original.
The paper is organized as follows. At first, the technical specifications and product impact analysis are briefly addressed. Then the design of both the product and the manufacturing process are described. The optimal design of electric motor is addressed. Then, the flexible production line is described.
At the very end, the prototypes are presented.

2 Product requirements and technical specifications analysis

Electrically actuated brakes allow brake-by-wire. The advantages of such technology have been described in [6]. Fig.1 shows the reference Brake-by-Wire system for the study, a distributed braking system concept where each actuator includes an electric motor controlled in order to press the brake pads on disks. Both front and rear brake actuators were considered, in order to define the respective sizes of electric motors.

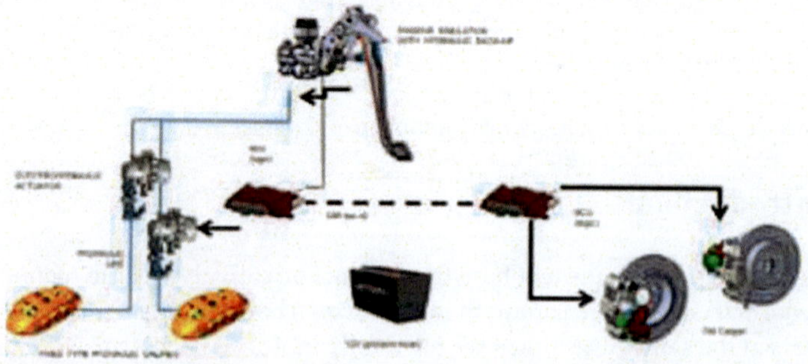

Fig.1. Layout of the Brake-By-Wire system with actuated calipers (Courtesy of Brembo S.p.A).

In order to address proper design of the electric motors, a system design approach was adopted, starting from the definition of vehicle functions and high-level requirements (including safety aspects). The electric motors requirements were used as input for the optimized motor design.

The front brake actuator motor is shown in Fig. 2. The rear brake actuator is shown in Fig. 3.

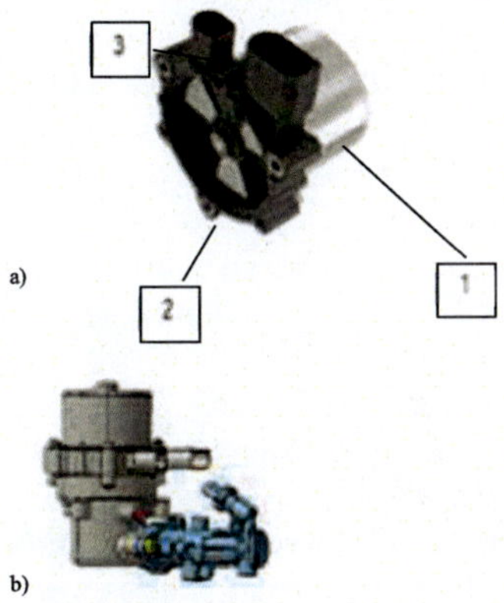

Fig. 2. Front brake hydro-electric motor. a) body of the front brake electric motor. 1: motor body, 2: mechanical interface, 3: electrical interface b) Electro-hydraulic actuator (courtesy of Brembo S.p.a.).

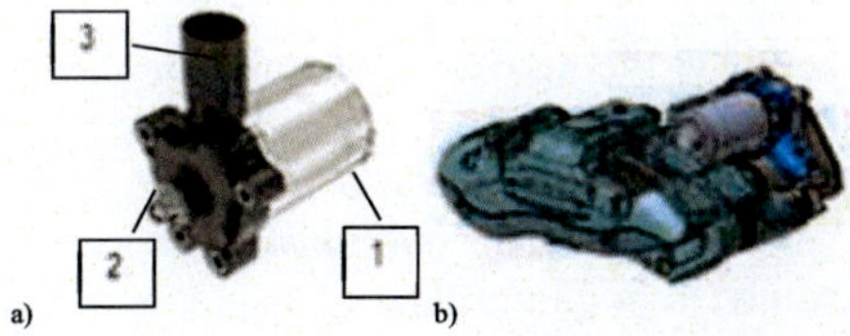

Fig. 3. a) rear brake electric motor. 1: motor body, 2: mechanical interface, 3: electrical interface. b) caliper and electric motor (courtesy of Brembo S.p.a.).

Main design parameters taken into account for the optimal design of the brake motor have been:

- Volume
- Mass
- Max Temp.
- Min Temp

- Max current
- Nominal Motor Voltage
- No load phase current
- No load speed
- Nominal power
- Max phase current
- Nominal Torque
- Battery Voltage range

3 Electric motor description (Brushless DC motor - BLDC)

The brushless DC motors in Figs. 2 and 3 have been optimized. Here the optimization of the motor in Fig.3 is dealt with.

Motor topology (stator and rotor) was selected in order to fit the available room. The rotor magnetic field is generated by surface-mounted permanent magnets (PM) and concentrated stator winding.

The permanent magnets are glued onto the rotor, but bandages are needed to counteract centrifugal forces. A metallic sleeve and a non-metallic bandage (carbon fiber, fiberglass, ….) have been studied.

The number of poles and slots have been set according known paradigms. The number of poles is an even number, the number of slots is a multiple of the number of phases. The winding factor is maximized. The number of poles is different from the number of slots, to avoid cogging torque. Cogging torque produces unacceptable vibrations.

The material used for the magnets is neodymium iron boron (NdFeB).

The motor voltage is 12V. The current input of the 3 phases is defined by the BCU (Brake Control Unit) model that considers the switch and diode characteristics. The actual simulation condition is with constant current value that allows shorter computational times.

The electric motor is tested at the maximum operative temperature required by the manufacturer, that is 120 °C.

The windings section geometry and placement is defined by using the copper slot fill factor. This factor is kept constant in the optimization process presented in the next sections. The windings pattern through the slots of the stator is fixed for all the motor configurations and equal to the reference design.

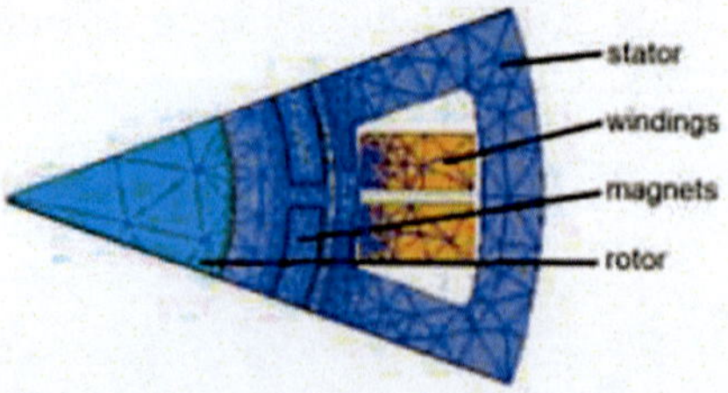

Fig.4. Preliminary design of the BLDC electric motor. Adapted from [12].

4 Physical models

The multi-physics mathematical models of the motor refer to electromagnetic field and related currents computation, to thermo-fluid dynamic simulation, to local stress and vibration assessment (see Fig. 5).

The electromagnetic model is a two-dimensional FE model developed in Motor-CAD®. The output torque, the cogging torque and the actual power consumption are computed. Operating temperatures are computed by a lumped parameter model. The dependence of magnetic and electrical characteristics are taken into consideration.

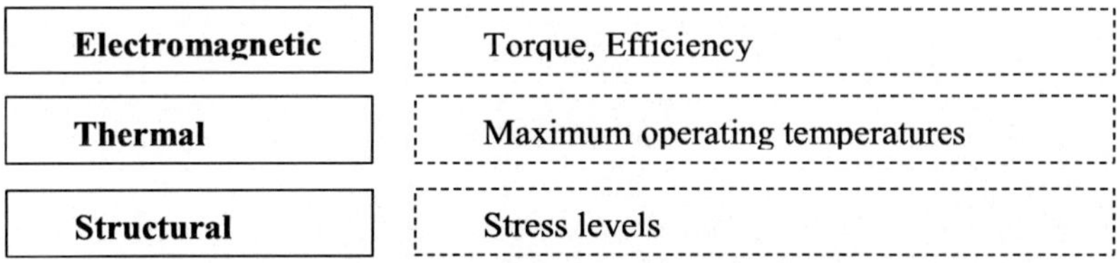

Fig.5. Physical models used in the optimization

A two-dimensional ABAQUS® FE model computes the stress levels. The rotor is the critical part. A Python script is used to update the geometry of the parts for a given set of parameters values and to take into account the mechanical and thermal loading of the components resulting from the Motor-CAD® simulations.

5 Optimization

The electric motor optimal design is formulated according to Multi-Objective Programming (MOP).

In a MOP problem, solutions are not unique. Solutions belong to the Pareto-optimal set. A preferred solution has to be chosen from this set. Pareto-optimal solutions are the best obtainable compromise among conflicting objectives [8].

6 Design variables

15 design variables pertaining to stator and rotor geometry, pole pieces and permanent magnets are considered. Fig. 6 shows the relevant geometrical design variables.

AutoCAD® is used to parametrize the geometry. An Excel® code is used to update the drawing parameters to feed analysis. A MATLAB® code was developed to import the geometry in Motor-CAD®. The electromagnetic and thermal analyses are performed as function of the many different geometries.

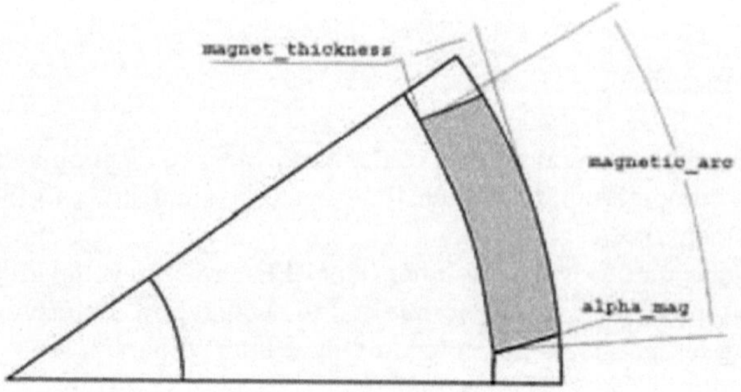

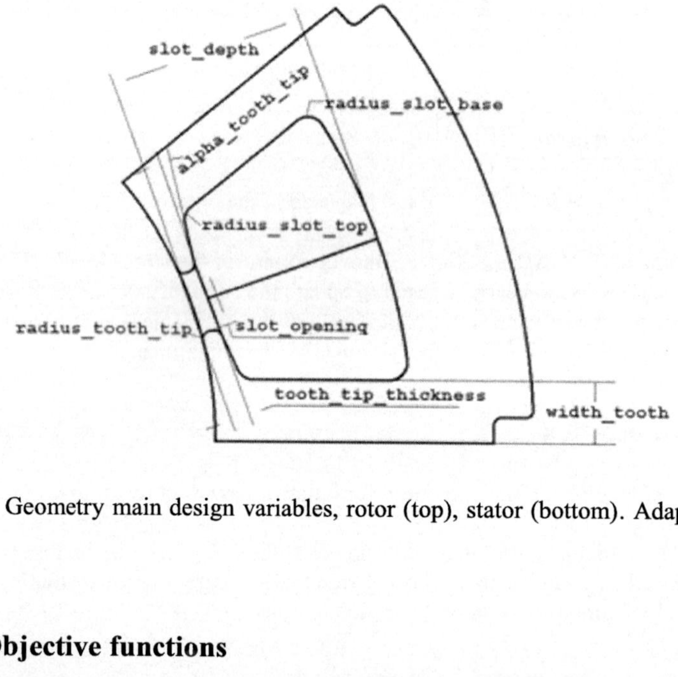

Fig.6. Geometry main design variables, rotor (top), stator (bottom). Adapted from [12].

7 Objective functions

The considered three objective functions are
T_M maximum holding torque (to be maximised),
I_{zz} rotor inertia (to be minimised)
P energy consumption while stalling at 1Nm
(efficiency index, to be maximised).

8 Constraints functions

5 geometry constraints have been introduced to avoid non feasible designs (rotor geometry, stator geometry, ..).

Additionally, design constraints that refer to components stress levels, temperature thresholds and maximum cogging torque are applied.

They read

$T_{windings} < T_{w\,max}$
$T_{magnets} < T_{mag\,max}$
$\sigma_{Von\,Mises,\,rotor} < \sigma_{yield}$
Cogging Torque < Admissible cogging torque

Artificial Intelligence is used reduce the optimization time. One thousand design variable sets are used to simulate the behavior of the electric motor. Physical models (Electromagnetic, Thermal and Structural) are used. The one thousand solutions are used as 'training set'. Supervised machine learning algorithms (SMLAs) are used and tested in terms of generalization capability, i.e. the errors of the SMLA have been computed on a 'validation set' constituted by 100 designs not belonging to the 'training set'. If the accuracy is not adequate, the process can be iterated by increasing the number of simulations used to train the SMLA and/or by varying the SMLA architecture.

The SMLA is thousands of times faster than the physical models used to train it (120s per design for the physical model, 0.01s for the SMLA).

The trained SMLA has been used to evaluate the performances of a large set of experiments (1 million different design solutions).

Further details on optimization are provided in [12]

9 Design of Experiments

When the dimensionality of the problem increases, the design variables domain volume increases so fast that the available data become sparse and the amount of data needed in the sampling phase grows exponentially with the dimensionality (curse of dimensionality). Several sampling techniques [12] can be employed in order to define "low discrepancy sequences" having almost the same discrepancy of a uniform grid but using a limited number of samples.

10 Supervised Machine Learning (Artificial Intelligence)

The supervised machine learning algorithm (SMLA) is based on Artificial Neural Networks (ANN). A multi-layer neural network was used [12]. A feedforward back-propagating neural network (ANN) with 15 inputs, 6 outputs and 2 hidden layers has been designed [12].

11 Search Method

The Parameter Space Investigation (PSI) method has been applied [12]. The feasible and Pareto optimal solutions are determined by applying the definition of Pareto-set. More details in [12].

12 Optimization results

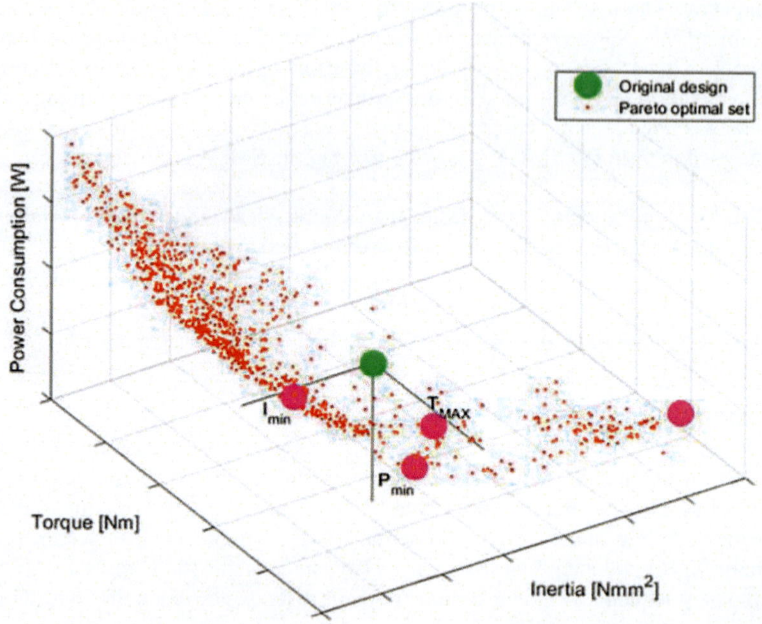

Fig. 7. Pareto Optimal solutions.

A preliminary design provided an electric motor with a very good performance, however the optimization allowed even better behavior. See Fig.7.

12.1 Max torque solution

The torque at low speeds is the most relevant index of performance.

All the design constraints (cogging torque, maximum stress levels, maximum operative temperature) are satisfied.

Referring to the design variables, the stator section has been widened especially at the tooth tip and stem region. This leads to a reduction in the magnetic saturation of the ferromagnetic material that constitutes the magnetic circuit, allowing for a larger magnetic flux value and thus more torque. The curvature radii at the tooth tip and slot corners are reduced, but are within the feasibility constraints. More details in [12].

12.2 Min Inertia

The rotor inertia is reduced by almost 35%. This is achieved by a 12% reduction of the stator bore radius. The torque characteristic is very close to the original one across all the speed ranges and the power consumption very close to the original one.

13 D-FMEA process

The brake electric motors that have been developed are safety components, so a comprehensive Design FMEA has been carried out from the initial stage of the development, in synergy with the buildup of the Process FMEA.

14 Flexible line design

One of the main innovative characteristics of the project is the integrated design of product and process, taking into account the effects that each product specification may have on the relevant phase of the manufacturing process. Since the flexibility of the production line is paramount, the opportunities that its complete digitalization offers have been used to achieve the following targets [11]:

1. The processing of a batch of products also with the possibility of a production schedule customizable according to the needs;

2. The processing of a single B-PMM or even only one of its subassemblies with the same smart and flexible approach of the previous mode.

What has been stated introduces an innovative work method that enables the line to be adequate with all the different phases of a product development, in particular with the phases of:

1. Prototyping;
2. Series production.

Focusing on the first one, the prototyping requires the management of a variety of parameters and, for each of them, of different values. For this reason, the flexibility has been also designed to satisfy the maximum range of values that may be used during the testing activities.

14.1 Process steps and layout design

The integration of design and manufacturing development, has allowed to analyze the production technology in order to scale the machine performance and optimize the B-PMM components design by applying "Design for manufacturing" methods.

According to what has been previously mentioned and since the line produces a variety of B-PMM, the process of manufacturing has been designed to work in parallel with the following main subassemblies

1. Stator;
2. Rotor;
3. Flange;
4. Motor.

While considering the requirements of production of each component, the layout design has been developed in order to achieve an accurate space-saving, reducing the surface needed and improving the process capability.

The flow chart is shown in the following Fig.8 in which the main subassemblies have been reported.

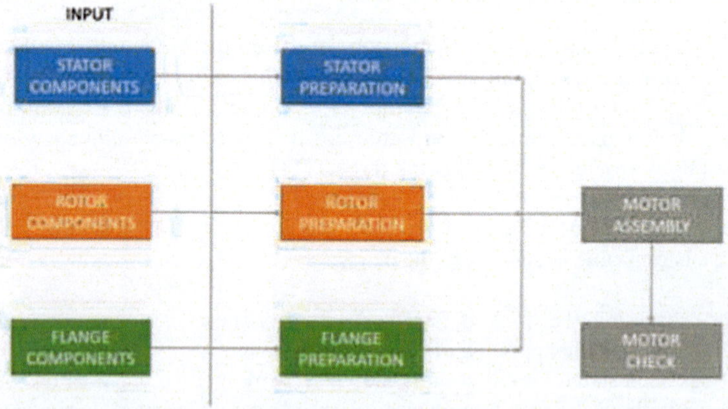

Fig. 8: Production flow chart

15 PFMEA of B-PMM products

Since the B-PMM requires a strict compliance with automotive safety requirements, a scientific and quantitative approach has been introduced and applied during all the steps of the project in order to guarantee each phase of manufacturing to work according to what has been declared [11]. For these reasons, the predictive approach known as FMEA, acronym of Failure Mode and Effects Analysis, has been selected to evaluate the effectiveness of the choices stated on the real case of manufacturing, and for this reason called FMEA of Process (PFMEA).

More specifically, the PFMEA tool has been largely applied to check and avoid the failures that may occur during the production [9][10].

An integration of FMEA's, Design and Process, allows to align the functional requirements of B-PMM from the design to the manufacturing, in order to get, through such synergic approach, the best performance of the product and the quality level required [9][10].

The output of this analysis is a unique parameter (Risk Priority Number RPN) useful to check the functions; if they have this value higher than a stated limit of acceptance, it means that they need an improvement effort.

An example of the relationships existing between causes and failures is provided in Fig.9. In the same figure, the different levels above mentioned are shown [11].

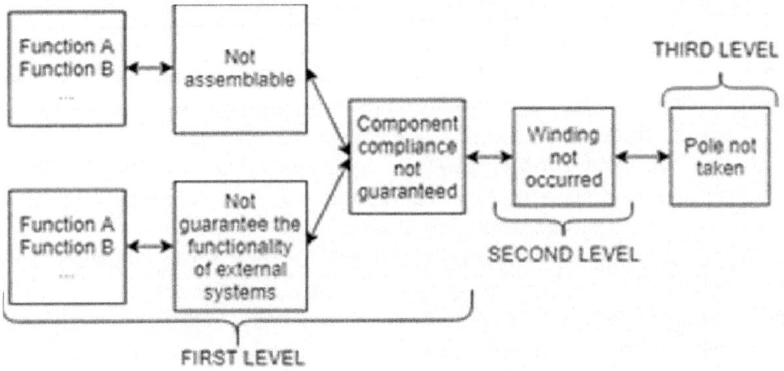

Fig. 9. Example of the connections between failures

15.1 Production process checks and quality design

As above mentioned, in order to comply with the requirements, the product must be checked during the process of manufacturing.

The design of each manufacturing step is made with the Total Quality Management approach.

Based on PFMEA, causes and effects failures analysis, the controls are tailored on the functional characteristics of process and design.

Related to the most critical phases of work, checks based on different technologies are performed, with the final scope to monitor the process and to store information.

All manufacturing steps are monitored by means of one or more parameters. The reason for each control is to evaluate the manufacturing process and to monitor the process steps in order to prevent scrap generation; moreover, such controls allow to guarantee the functional requirement stated during the design phase. A positive result of this control enables the component to proceed according to its flow chart, otherwise it may be considered to be reworked or, at most, a waste.

Furthermore, the phase of quality control of the B-PMM is the final step of the process of manufacturing and its target is to verify the compliance with its main specifications. According to this, the specifications must be pointed out as measurements needed to be performed by the workstation of motor check and guaranteed through an adequate phase of design for the integration of the different measurement systems.

As a result, the output of the "production report" is a detailed report that contains all the information picked up by sensors and the final statement about quality and history of each B-PMM produced on the line.

15.2 IT architecture and IoT integration

Since the principles of the 4.0 Industry have been introduced, the traceability of each single item in term of part number as well as of the working phases must be guaranteed. In order to allow this, a complete digitalization of the entire process of manufacturing through an adequate structure of data management has been designed.

In this instance, the production is monitored by a Supervisor System Unit (SUU), linked to the stations by a specific logic and architecture. The communication of all data is done by standard protocol applied for each machine.

The data acquired by sensors are managed by the SSU that can act on the parameters achieving the results while preventing risks and avoiding damages [11].

This implies that the SSU must be provided with an integrated knowledge about the variety of products and recipes available, that implies different data acquired that must be stored in different formats in order to enable the next phases of Data Analysis.

As a result, the SSU allows, on one hand, the control of the real time production pointing out the actual state of completion of the product and the relevant performance achieved by the line, and, on the other, provides the data for the detailed historical report of the B-PMM already made.

16 Conclusions

Two electric motors, fully engineered with their own production lines, have been presented. The two motors allow brake-by-wire and are devoted to the front and rear brakes of a reference car, respectively. All of the electric motors have been designed and manufactured after an in-depth optimization process. Such a process referred to both product design and production line engineering. The electric motors will contribute to the penetration of electric vehicles into the market.

References

1. KPMG, https://assets.kpmg/content/dam/kpmg/au/pdf/2018/electric-vehicles-is-the-energy-sector-ready.pdf
2. Momen, F., Rahman, K., Son, Y., and Savagian, P., "Electric Motor Design of General Motors' Chevrolet Bolt Electric Vehicle," SAE Int. J. Alt. Power. 5(2):286-293, 2016, https://doi.org/10.4271/2016-01-1228.
3. A Giorgetti, L Cavestro, M Rampazzo, Design of a Lightweight Braking System for Electric Cars, SAE Technical Paper 920649 doi: https://doi.org/10.4271/920649
4. Ramakrishnan, K., Romanazzi, P., Zarko, D., Mastinu, G. et al., "Improved Analytical Model of an Outer Rotor Surface Permanent Magnet Machine for Efficiency Calculation with Thermal Effect," SAE Int. J. Alt. Power. 6(1):34-46, 2017, https://doi.org/10.4271/2017-01-0185.
5. A Carriero, M Locatelli, K Ramakrishnan, G Mastinu, and M Gobbi, A Review of the State of the Art of Electric Traction Motors Cooling Techniques, SAE Technical Paper2018-01-0057 ISSN: 0148-7191, e-ISSN: 2688-3627, doi: https://doi.org /10.4271/2018-01-0057
6. Sundar, M. and Plunkett, D., "Brake-by-Wire, Motivation and Engineering - GM Sequel," SAE Technical Paper 2006-01-3194, 2006, https://doi.org/10.4271/2006-01-3194.
7. Mastinu, G. Ploechl, M, Road and off-road Vehicle System Dynamics Handbook, CRC Press, Boca Raton, New York, USA, 2014
8. Mastinu, G., Gobbi, M., Miano, C, Optimal Design of Complex Mechanical Systems, Springer, Berlin 2006
9. "Potential Failure Mode and Effects Analysis (FMEA) – Reference Manual Fourth Editon", AIAG (Automotive Industry Action Group), April 1993.
10. "SAE AS13004 - Process Failure Mode and Effects Analysis (PFMEA) and Control Plans", 2017.
11. Righettini, P., Strada, R., Gosatti, A., Togni, S., Camozzi, F., Fissore, C., "Smart Mobility: a modern approach to automotive product development for vehicle electrification", Proceedings of IEEE RTSI 2019 (5th International Forum on Research and Technologies for Society and Industry), Firenze, 9-12 september 2019.
12. Di Gerlando, A., Gobbi, M., Mastinu, G., and Miotto, A., "Electric Motor for Brakes - Optimal Design," SAE Technical Paper 2020-01-0919, 2020, doi:10.4271/2020-01-0919.

Acknowledgments

Lombardy Region Government has partially funded the research with POR/FESR INPROVES grant (ID 241363). The Italian Ministry of Education, University and Research is acknowledged for the support provided through the Project "Department of Excellence LIS4.0 - Lightweight and Smart Structures for Industry 4.0" of the Department of Mechanical Engineering of Politecnico di Milano.

Brake-by-Wire Actuator for Electromechanical Disc Brake

Dr. Carsten Kügeler[1], Dominic Scopacasa[2], Dr. Joachim Funke[1], Dr. Ingo Geue[1], Jens Hunecke[1]

[1] HELLA GmbH & Co. KGaA, Germany
[2] HELLA Corporate Center USA, Inc., USA

Abstract. Well-known for its brake system components, e.g., the vacuum pump for brake systems, HELLA develops a digitally controlled actuator for brake-by-wire applications with a small design environment, high dynamics, and extended diagnostic functionality. Besides the functionalities in the brake system like ABS or EPB, the actuator can also provide Park Brake Actuator functionalities in a vehicle system.
Keywords: Brake-by-wire, electromechanical braking, disc brake actuator.

1 Introduction

The transformation of individual mobility has already started. Public transport, car sharing, last-mile vehicles, etc. show that comfort and environmental concerns become more and more important for individuals. It is assumed that the future will see the continued electrification of passenger vehicles and the improvement of vehicle's ability to drive autonomously. The necessary technical modifications require a safe and redundant power supply as well as reliable real-time data processing. With these preconditions, the implementation of drive-by-wire functions such as steer- and brake-by-wire seem to be straight forward and will enable more sophisticated driving operations. Focusing on electromechanical brake (EMB) systems, several advantages are indicated. Since the EMB is driven by electric means, there is no need for the installation of hydraulic lines or to do service on a hydraulic system. Also, the space requirements in the engine compartment are reduced, and the control software might be integrated into any suitable vehicle control with an appropriate safety level and enough computational power. In the case of autonomous driving, EMB is considered to be faster than traditional hydraulic systems, and so more time is available for the computation of autonomous driving functions. Finally, since EMB allows for true zero drag, it will be beneficial for the driving range of electric vehicles and the reduction of CO_2. Please note, electrified brake calipers are already a reality today: The use of electric parking brakes is expected to reach more than 40% of the total market in 2021, and the introduction of electromechanical brakes is expected in 2027 [1].

P. E. Pfeffer (ed.), *11th International Munich Chassis Symposium 2020*, Proceedings, https://doi.org/10.1007/978-3-662-63193-5_31

2 Components for Brake Systems

2.1 System Overview

Traditional brake systems translate the driver's demand for braking from the brake pedal to the wheels via a hydraulic master cylinder, hydraulic pipes, and hydraulically actuated brake calipers. In order to enhance the vehicle's braking performance, an additional booster (either vacuum-based or electrohydraulically driven) may be provided to support the driver (Fig.1).

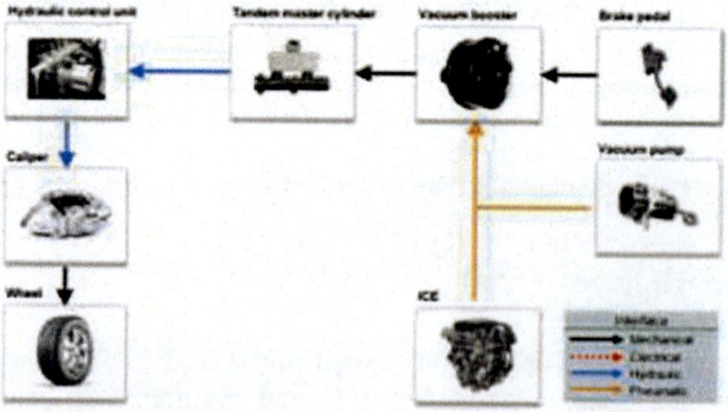

Fig. 1. Simplified overview of a vacuum-based brake system.

HELLA has a long history of experience with electric vacuum pumps associated with traditional brake circuits (Fig. 2). HELLA offers a variety of pumps for all engine types and performance levels (from compact to high-power pumps), ensuring compliance with the latest safety standard and reliable operation of the vehicle.

a) b)

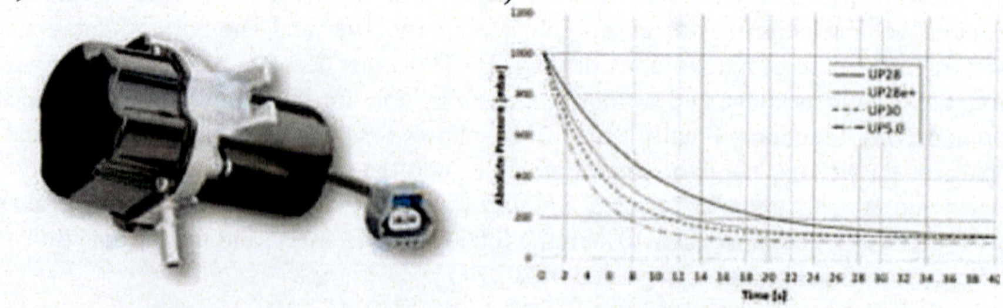

Fig. 2. a) HELLA vacuum pump UP5.0. b) Performance of the vacuum pump family (booster volume: 4 l, supply voltage: 13 V, room temperature).

Autonomous driving vehicles (L3-5) do not rely on the brake request from the driver. Further, recuperation braking alters the demand to the friction brakes and the subsequent charging state of the battery. Therefore, it seems like a logical evolution to introduce brake-by-wire (BbW) systems where a potential driver is mechanically decoupled from the brake calipers. The simplified overview of an electromechanical brake system (Fig. 3) shows a couple of changes.

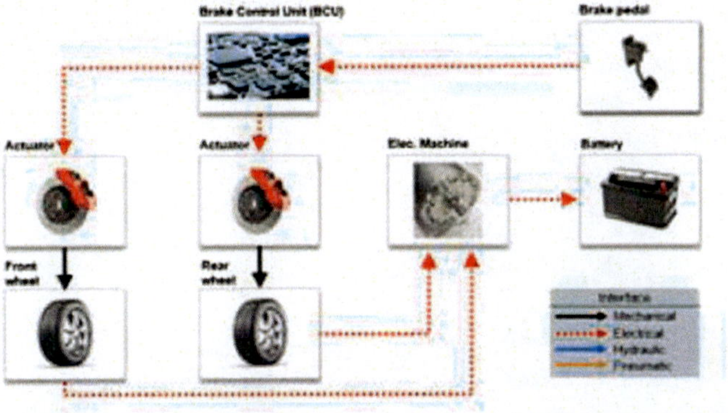

Fig. 3. Simplified overview of a brake-by-wire system for (P)HEV and BEV.

First, all hydraulic pipes and cylinders are omitted and replaced by electrical wires. If a driver is in the control loop (L3,4) he is mechanically decoupled, and a new kind of brake pedal is needed, which provides the brake demand by reliable position sensing and gives mechanical feedback according to the brake performance. Especially while braking with a hybrid friction braking and recuperation braking system, the BbW system does not need the adjustment of the brake feeling by using an additional pressure simulator. HELLA is developing an electric brake pedal.

Second, an electronic control unit with sufficient software algorithms is introduced. While the control algorithms can be implemented on any control unit, the ECU needs to provide a high level of functional safety features to ensure the appropriate ASIL level is achieved. The safety level, timing requirements, and power demands are very similar to those in electric steering systems. Because HELLA has been in this market with steering ECUs for some time, it can transfer this know-how to braking systems.

From a high-level view, the required modifications of the brake caliper are more an evolution than a revolution. The brake pad arrangement will remain unchanged, but the hydraulic cylinder will be exchanged with an electromechanical actuator, and exact and robust sensing needs to be added to gain exact control of the braking event. Inductive position sensors typically fulfill the sensing requirements.

2.2 Electric Brake Pedal

The electric brake pedal (Fig. 4a) simulates the pedal "feel" of a conventional brake system. It transfers the driver's intent electrically to the brake control unit (BCU). Two independent and fail redundant sensors provide an instantaneous signal which can be adjusted to platform-specific requirements. Also, the force-travel characteristic of the pedal can be adjusted individually (Fig. 4b).

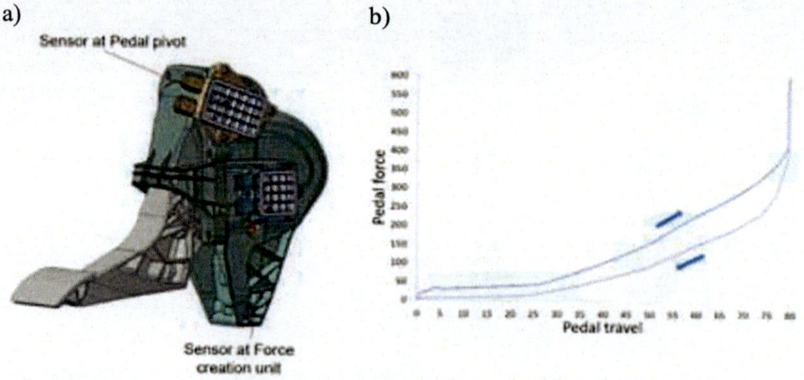

Fig. 4. a) Electric brake pedal. b) Pedal force characteristic at 9 mm/s.

Whereas the pedal "feel" is not a safety goal, but very important for driver acceptance, the avoidance of submitting a wrong brake request (too high, too low, unintended) is very critical. This issue is resolved by using a synchronized monitoring of three, up to six, independent output signals based on the system architecture of the brake system.

The use of the electric brake pedal without any mechanical link into the engine compartment also comes with advantages in size, weight (reduced pedal arm length leads to smaller dimensions and will allow the use of plastic materials) and crash behavior (no mechanical link between EHB unit and brake pedal required) In addition, the electric brake pedal can easily be combined with the accelerator pedal leading to a standardized cost-optimized module.

2.3 Steering / Braking ECU

Since 2010, HELLA has produced electronic control units (ECU) for electrical power steering (EPS) applications to support and promote fuel efficiency as well as autonomous driving. EPS devices allow needs-based steering-assistance using an electronically controlled electric motor: the steering control module (SCM) calculates the required steering support based on the incoming signals from the sensor system. The

SCM also controls the EPS system's electric motor power supply to deliver the appropriate level of assistance. HELLA continuously develops its ECUs and offers them in a fail-operational version (Fig. 5). The fail-operational version ensures that in the event of a fault in the electronics, the system will not be shut down and the EPS will continue to work. The SCM can drive either 3- or 6-phase motors, provide interfaces for redundant torque and angle sensors, and is ready for dual battery connection. With the fail-operational function, HELLA has laid a necessary foundation that allows the ECU to be suitable for SAE level 4 automated driving. From the point of the control modules, the requirements for future steering and electromechanical brake systems are similar.

Fig. 5: HELLA fail-operational steering control module.

2.4 Contact-less Inductive Position Sensor (CIPOS)

The main advantages of inductive sensors for automotive applications are their mechanical variability (linear, rotational, arc segment, through-hole and end of shaft installation etc.), high-temperature range, simplicity and robustness against electromagnetic interference. The inductive contact-less sensor consists of a stator and a rotor, or cursor. (Fig. 6) The stator contains an excitation coil (which is part of an LC resonance network), receiver coils and electronics for signal processing. The rotor or cursor is a passive element with one closed winding, which is designed with a particular geometry.

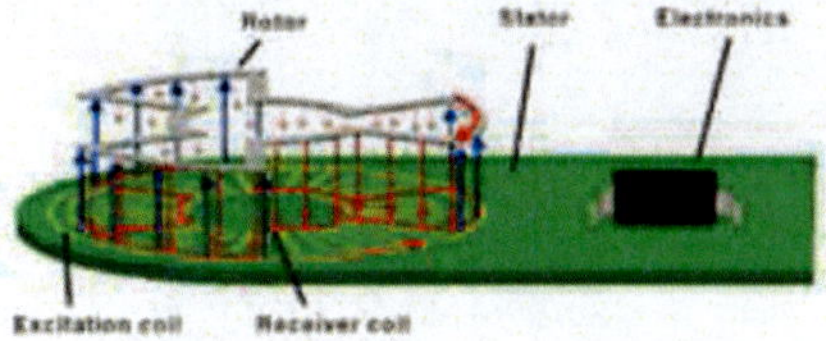

Fig. 6 Principle of rotational inductive position sensor [2].

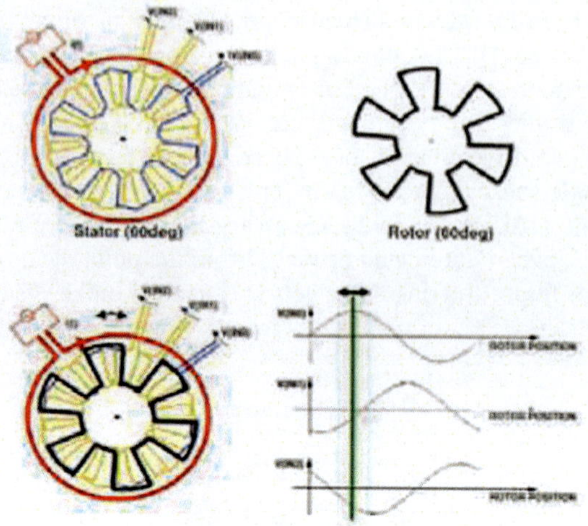

Fig. 7 Output signals of CIPOS technology [2].

The inductive coupling between the excitation coil, the rotor/cursor, and the three receiver coils lead to three ratio-metric signals that are related to the angular rotor. Through the use of analog and digital signal processing, a position for the rotor/cursor is extracted from this signal. Note that the signal is linearly dependent on the rotor/cursor position (Fig. 7).

Position sensors based on the CIPOS principle are widely used in the automotive industry e.g., acceleration pedals, engine compartment actuators, body actuators, or motor position sensors. Fig. 8 shows the rotor and stator implemented in a throttle flap actuator.

Fig. 8. CIPOS-Sensors in throttle-flap actuator [2].

3 Disc brake actuator

3.1 Requirements

Brake calipers are mounted in an extremely harsh environment with stringent requirements regarding vibration, thermal shock, and durability. The electromechanical actuator, which moves the brake pads, must fulfill the requirements as well. The most critical parameter is the dynamic speed of the brake. Achieving a clamping force of 16 kN in 200 ms is the baseline for the development of the motor-gear combination.

3.2 Gear design

The gear configuration is at the core of the HELLA EMB design. The gear train's relevance to the system is shown in Fig 9a. The actuator includes the gear, a brushless motor with a controller, a position sensor, and in some cases, a parking brake interlock mechanism; however, the integration with the caliper and brake ECU into the vehicle will be the responsibility of the brake system supplier. The gear design was derived from the technical requirements. After the evaluation of several options, a multi-stage gear with a combination of a spur gear and planetary gear, integrated with a permanent-magnet synchronous machine provided the best performance. The gear train has a ratio of 1:75 and is calculated to achieve ~ 12 Nm at an output speed of 48 rpm. Details of the calculation are given in Fig. 9b.

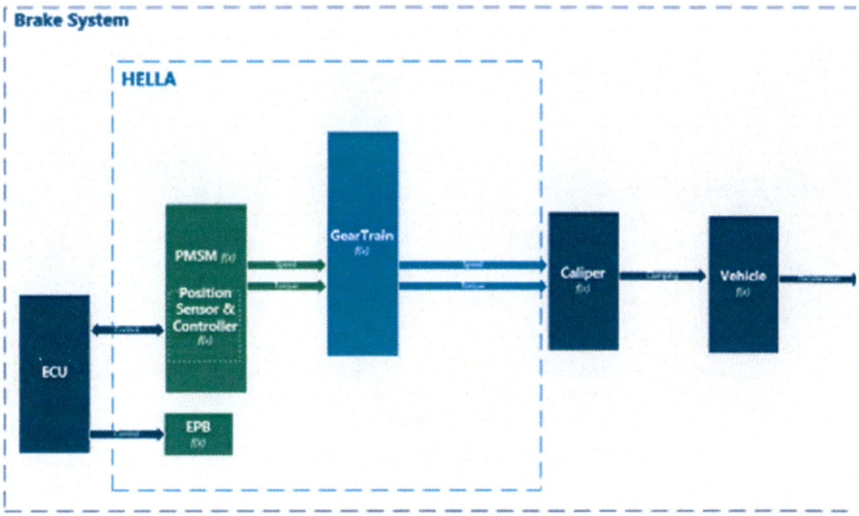

Fig. 9a: System overview.

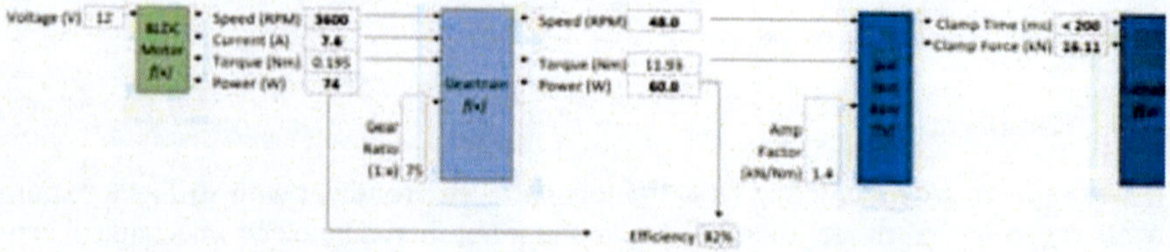

Fig. 9b: Gear Calculation with the use of an electronically commutated motor.

3.3 Motor design

A permanent magnet synchronous machine was designed based on the performance requirements and the gear train calculations. In order to reach the operating point of 0.195 Nm at 3600 rpm, it was determined that a motor with nine slots and six magnetic poles was required. For the given temperature profile, NdFeB magnets were selected as they offer an efficient and compact motor design. The dimensions of the motor are expected to be 40 mm in diameter and 32 mm in axial length. The calculated motor characteristics are presented in Fig. 10. The operating point at 3600 rpm comes with an input current of about 7.6 A and 73.6 W output power at 20°C, which results in a motor efficiency of 86.3%. The motor is designed so that operating points up to 254 mNm can be reached, and it can withstand even higher short-time overload conditions. A margin of safety is built into this design. Note that there is an efficiency drop down to 81.2% at 120°C, but the operating points can still be reached.

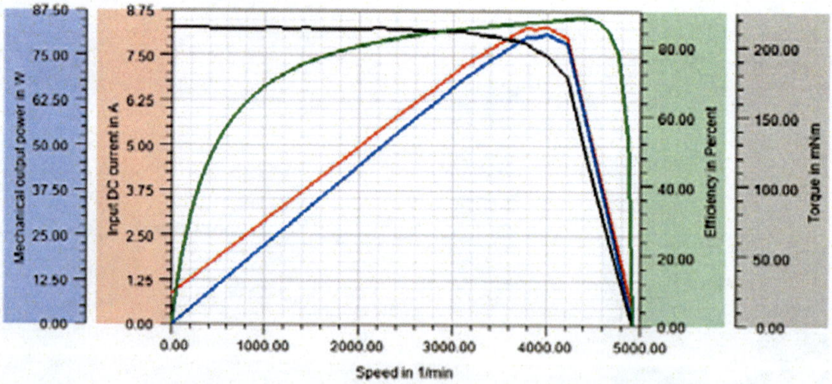

Fig. 10: Calculated motor characteristics at 20 °C.

Fig. 11 shows the finite-element based flux-density distribution in the stator and rotor structures. The distribution indicates that the dimensioning of the slots and stator yoke is suitable to prevent magnetic saturation and the subsequently reduced efficiency even under full load operation.

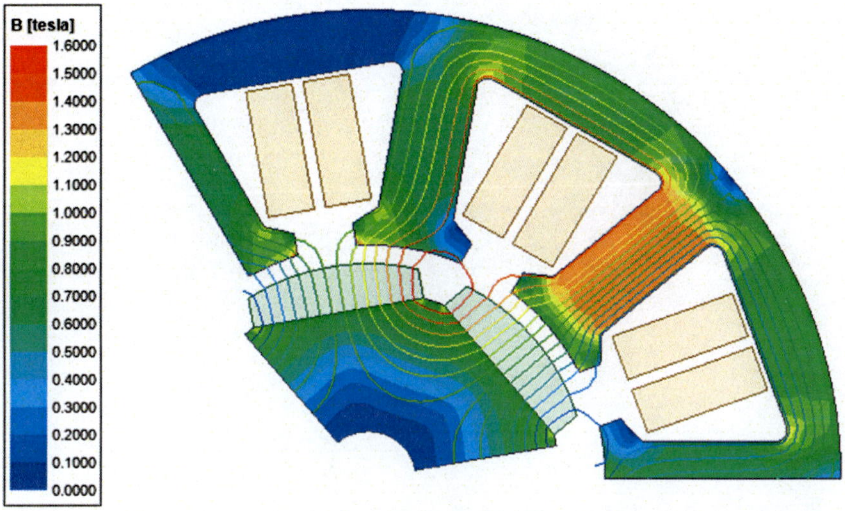

Fig. 11: Flux-density in the motor and flux distribution.

3.4 Actuator incl. position sensor and force measurement

Based on the packaging requirements, the motor definition and the gear calculation, the actuator layout shown in Fig. 12-14 was selected. Fig. 12 displays the CAD design, including the motor position sensor on top of the motor in an end-of-shaft arrangement. The first gear stage is designed as a spur gear and the next gear stage is in a planetary gear arrangement with a direct connection to the spindle in the brake caliper. The cross-section in Fig. 13 gives more details. In this image, the arrangement of the CIPOS-based motor position sensor is clearly displayed. The rotor is directly mounted at the end of the motor shaft. The PCB, with the sending and receiving coils, is integrated into the cover. Please note that some space is reserved for the parking brake interlock mechanism.

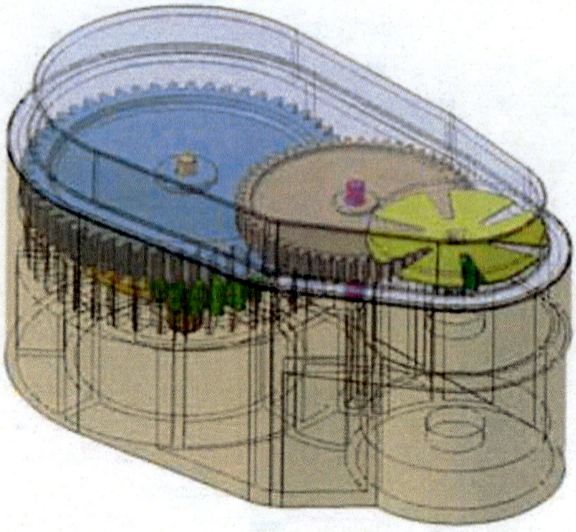

Fig. 12: CAD outline of the actuator prototype.

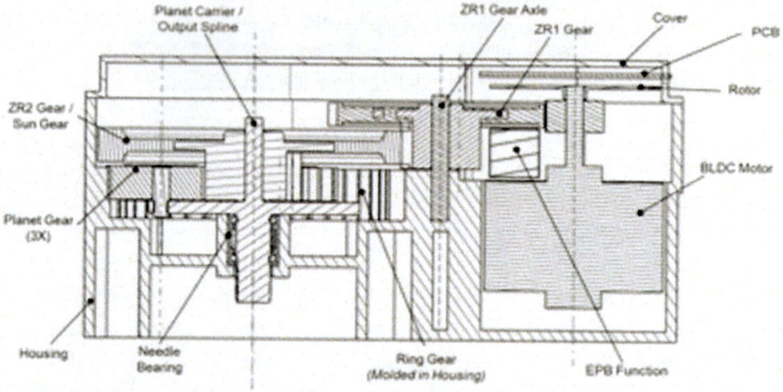

Fig. 13: Cross-section of the actuator prototype.

Fig. 14 shows two pictures of the actuator prototype. The housing is made of aluminum for prototype purposes only. Although the motor specifications have been precisely calculated for a specific custom motor design, this prototype makes use of an off-the-shelf motor together with an internal MPS and control ECU. In addition, to accelerate the development, the prototype was manufactured without a CIPOS-based motor position sensor. Fig. 15 presents a model of the actuator attached to the disc brake caliper.

a) b)

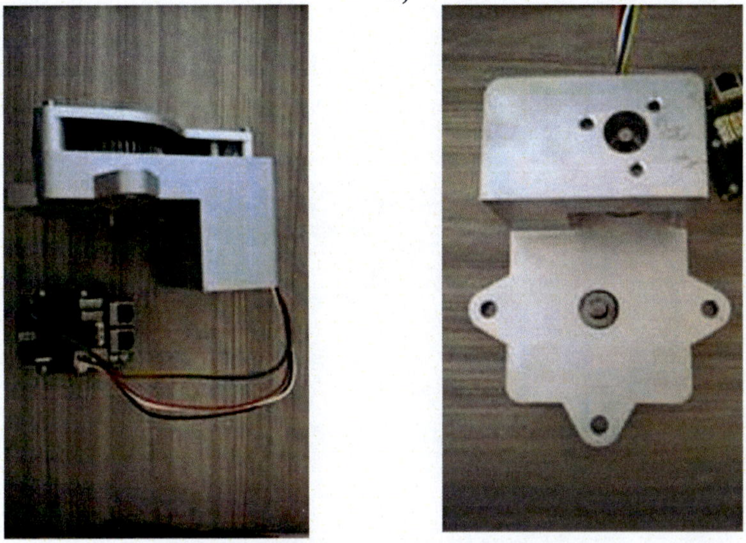

Fig. 14: Picture of actuator prototype a) side view b) top view.

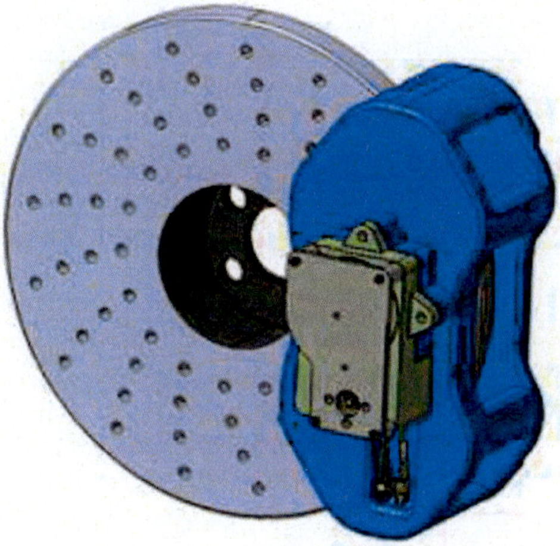

Fig. 15: Sketch of the actuator attached to the disc brake caliper.

4 Test results

4.1 Service brake

Fig. 15 shows the actuator mounted on a test bench and attached to a dynamometer with a maximum torque of 28 Nm.

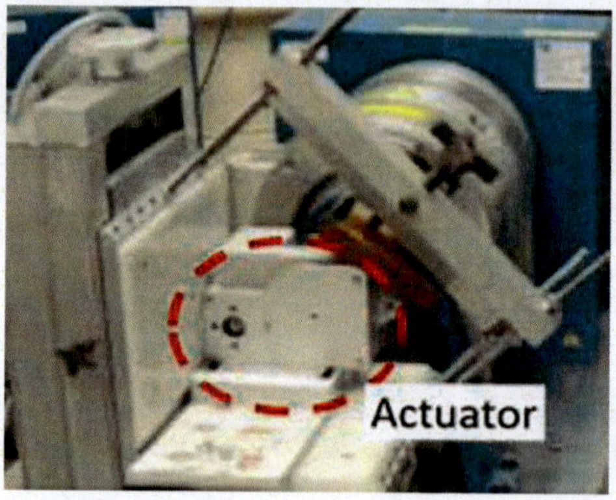

Fig. 15: Actuator mounted on the test bench.

The torque-speed curve test measures the torque versus the speed of the actuator throughout its entire operating range. The actuator is loaded from free run (zero load) to stall (full load). The test uses a method where open-loop current to the dynamometer brake is ramped from 0% to 50%. This range was determined from the pull-in torque test to be sufficient current to stall the actuator. As the torque versus current characteristic of the dynamometer's hysteresis brake is non-linear, the start of the curve was incremented quickly from 0% to 20% in one second, and 20% to 50% in ten seconds. This bypasses the lower portion of the hysteresis curve, where large changes in current produce small changes in torque. Fig. 16 displays the measurement result of the device under test and shows that the actuator reaches a torque of more than 20 Nm before it stalls.

The pull-in torque test is to determine the maximum torque (load) the actuator can start rotating against. The dynamometer was set to an open-loop load that was close to the desired pull-in torque of the actuator. The actuator was powered on and data was recorded. If the actuator started and continued to rotate, the open-loop value was increased slightly, and the process repeated. This was continued until the actuator failed to rotate against the applied load. Fig 17 shows that in less than 200 ms a torque of 16 Nm was surpassed, and therefore the actuator fulfills the initial requirement target for service brake function. The required torque output of 18 Nm to support the EPB function is also achieved in less than 250 ms.

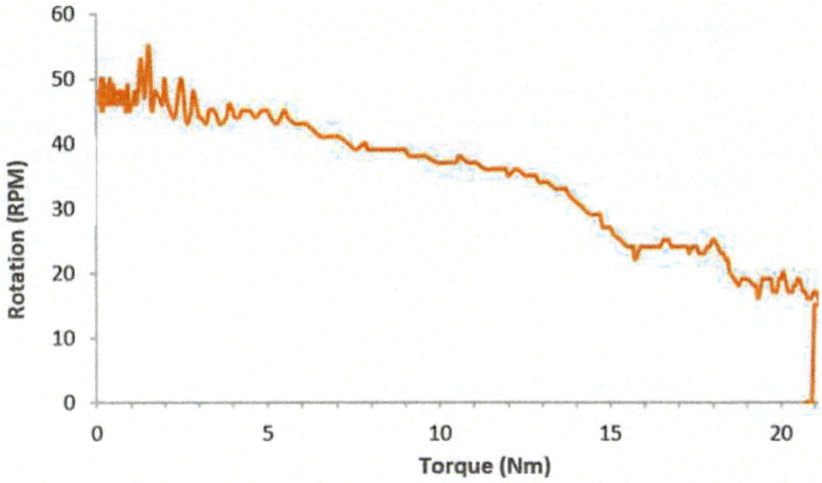

Fig. 16: Measurement of torque-speed cure test.

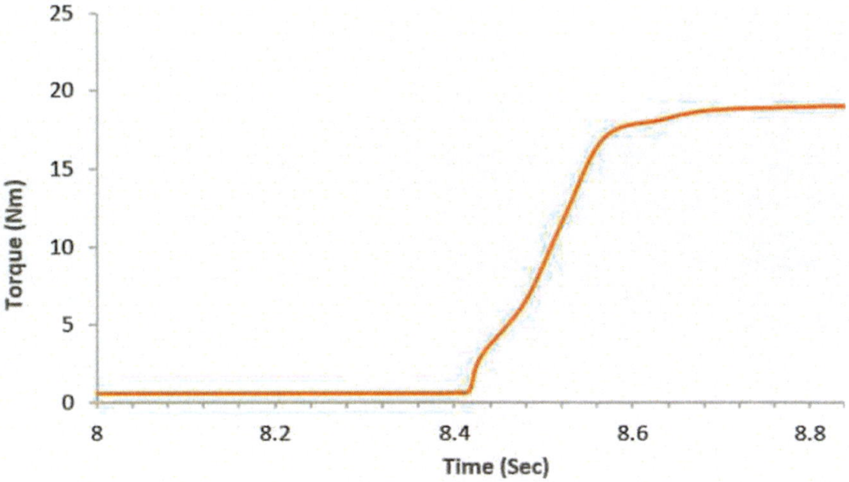

Fig. 17: Measurement of pull-in test.

5 Summary

The introduction of electromechanical braking is beneficial regarding electrified, autonomously driving vehicles. Therefore, a motor gear combination was developed, which can be attached to a disc brake caliper and fulfills the requirements regarding speed and torque of the service brake.

Acknowledgments

The authors would like to thank Dennis Burke, Christoph Schulte and Bodo Bücker for their support for this paper.

References

1. Barbale P.: The chassis market and technology trends – understanding the impact of automotive megatrends on the chassis sector. In: Pfeffer P. (ed.) 10th International Munich Chassis Symposium 2019, pp. 13-20, Springer Fachmedien Wiesbaden (2019).
2. Kamenicky P., Horsky P., H. Irle, An Inductive Position Sensor ASIC. In: Casier H., Steyaert M., Van Roermund A.H.M. (eds) Analog Circuit Design., pp. 33-53. Springer, Dordrecht (2008).

Wheel individual longitudinal control systems

Prof. Dr. Eberhard Drechsel[1], Maximilian Böhle, M. Sc.[2] and Miso Kvesic, M. Sc.[2]

[1] formerly Munich University of Applied Sciences, München
[2] formerly municHMotorsport e.V., München

Abstract. The paper investigates the importance of wheel-individual drive and brake systems for the development of vehicle dynamics control systems for automated road vehicles. Based on the findings from the Formula Student competition, system requirements for wheel-individual drives for automated driving are developed using the example of vehicles from the Munich University of Applied Sciences in the categories Electric (FSE) and Driverless (FSD). The conceptual system design as well as the technical implementation of the competition vehicles will be reported in detail. A further focus lies on software requirements and the structure of the control systems. Special aspects of safety and comfort in automated driving are discussed using the system description of the trajectory tracking control of the FSD vehicle. In addition, simulation results of driving operation are presented in order to show the safety gain due to the stability reserve set by means of continuous wheel-individual drive and brake control.

Keywords: Wheel individual drives, Automated driving, Electric wheel drives

P. E. Pfeffer (ed.), *11th International Munich Chassis Symposium 2020*, Proceedings, https://doi.org/10.1007/978-3-662-63193-5_32

1 Trend towards wheel individual systems for drives, brakes and steering

The motion behaviour of conventional road vehicles is determined by the wheel forces applied by the driver through the actuation of the drive, braking and steering systems. In the case of automated vehicles, on the other hand, a system of sensors, controllers and actuators must ensure that the direction of the speed vector at the vehicle's center of gravity is brought into agreement with the tangent to the planned motion trajectory. Besides additional requirements resulting from individual traffic situations and regulations as well as comfort demands, control over wheel individual tire potentials is necessary, especially for autonomous driving, in order to be able to maintain a safe distance from the physical limits. In the context of this safety reserve, driving comfort and the range of environmental perception are closely related.

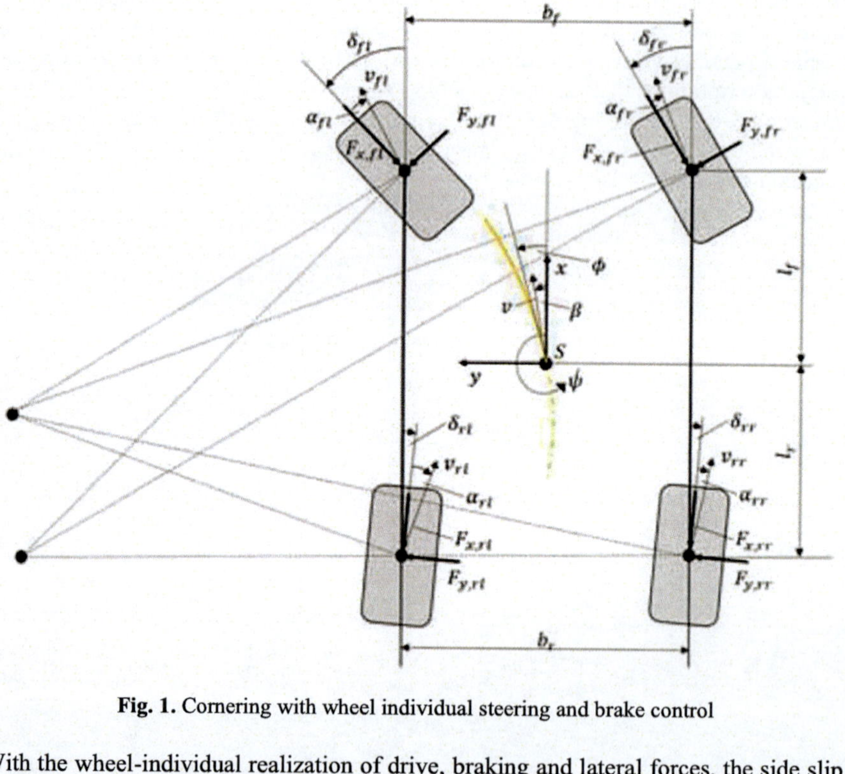

Fig. 1. Cornering with wheel individual steering and brake control

With the wheel-individual realization of drive, braking and lateral forces, the side slip angle required to follow a given trajectory can be achieved with low latency while avoiding overshoots. The research project UNICARagil [1], which is being carried out in cooperation with eight German universities, is a good example for a current use of this solution. This project for the development of four modular driverless vehicle concepts plans the integration of drives, brakes and steering systems that can be controlled individually for each wheel (see also [2]). In addition to the advantage in overall efficiency due to recuperative braking, a considerable improvement in efficiency can be achieved by adjusting the longitudinal and lateral slip during normal driving operation individually for each wheel to avoid unnecessary wheel slip differences [3]. These and similar effects will not be discussed further in this report.

Current systems only partially fulfil the demand for controlled use of individual wheel potentials. Drive systems are usually configured axle by axle, wheel individual torque control before reaching the stability limits is hardly common. Drive slip control only occurs at the frictional limit, brake systems are usually controlled axle by axle and likewise feature individual wheel control (ABS) only at the frictional limit. This is also used for driving stability control in critical situations of lateral dynamics (ESP). A rigid mechanical coupling of steering systems prevents the optimum wheel position for all steering angles and speeds.

2 System requirements for longitudinal dynamics control regarding autonomous driving

For these reasons, there is an increasing trend towards the development of wheel-individual systems. An effective system for wheel-individual four-wheel drive and braking is presented below. It was developed within the framework of the international design competition "Formula Student Electric" (FSE) in the form of wheel-integrated high-performance motors with integrated planetary gears. Since 2015, the competition vehicles of the FSE team of the Munich University of Applied Sciences have been equipped with such a system, which is subject to continuous development.

Fig. 2. FSE vehicle from municHMotorsport at the Formula Student Germany [12]

Since 2017, this system has also been used by the same team in the category for automated driving, the "Formula Student Driverless" (FSD). It enables the control of drive, braking and driving stability over the entire operating range in autonomous driving mode.

2.1 System design using the example of a 4WD FSE competition vehicle

The advantages of wheel-individual drives have also been of great importance for Formula Student competition vehicles for some years now, above all the possibility of making the greatest possible use of individual tire potentials and the ability to adjust the yaw behaviour according to subjective driver wishes or established stability limits. For their drive configuration, the electric vehicles from municHMotorsport rely on four electric motors in combination with 1.5-stage planetary gears integrated into the wheel carriers. A permanently excited synchronous machine developed in-house is used for each wheel, which delivers a maximum torque of 29 Nm up to a threshold speed of 10,600 rpm and an output of 35 kW at a maximum speed of 20,000 rpm.

Fig. 3. System design of the wheel drives of the FSE vehicle

On a conceptual level, this system design offers advantages that are of great relevance for vehicles with electric drive systems, even beyond their use in a competition vehicle. For example, a permanently excited synchronous machine does not deliver the theoretically highest possible efficiency in principle, but it is clearly superior to alternatives such as reluctance motors in this application in crucial points for road vehicles such as noise behaviour and ripple of the drive torque. As it has been proven that potential customers have higher comfort requirements for electric vehicles than for vehicles with conventional drives in these areas, synchronous machines are to be preferred here. In order to make use of the package related advantages of the motors being positioned close to the wheels, the integration of high transmission ratios in combination with liquid cooling is also of decisive importance.

In this way, significantly higher volumetric and mass-related power densities can be achieved compared to other motor concepts, which on the one hand enables the realization of a purely electric drive over a wider speed range, and on the other hand has a great influence on the known disadvantages due to the increase in unsprung masses as well as the package advantages of an electric drive close to the wheel. Thus, the defining characteristics of the present drive on design level are:

- high speed synchronous motors in combination with efficiency optimized planetary transmissions,
- liquid cooling,
- functional design integration of gearbox, wheel hubs, uprights and suspension mounting points

resulting in a power-to-mass ratio of more than 5 kW/kg at a total weight of 7 kg.

On a software level, torque vectoring and model-based traction control in municHMotorsport's electric vehicles allow for improved utilisation of the tire potential and increased system efficiency through continuous control of the power recuperated by the electric motors. The maximum transferable wheel forces are determined by an extended two-track model while driving. In combination with a scaling factor to take into account the current track conditions by the driver - the software does not estimate the friction coefficient - the engine torque requested by the torque vectoring can then be limited by the traction control system to the calculated maximum transferable tire torque. The yaw rate and longitudinal slips are used as the control variables. In the FSD vehicle, the yaw angle difference to the desired trajectory is also taken into account. A more detailed description of the underlying system can be found in [4].

With regard to automated driving, the influence of individual wheel drives gains additional importance. The autonomously driving vehicle from municHMotorsport, which competes in the FSD, uses the same vehicle dynamics software for the basic functions such as torque vectoring, traction control and recuperation. However, the implementation of a trajectory planning and a motion controller that ensures sufficient tracking of the target trajectory enables the automated vehicle to make better use of the system potential. The choice of optimization targets and boundary conditions for the selection of the trajectory can already influence the utilization of tire potentials and recuperation. In the case of a global trajectory planning, for example, a constant safety margin to the stability limits can be set. Considering the necessity of local motion planning for both comfort and safety-oriented driving on roads with traffic, which is further explained in the following, this factor is of utmost importance. In this way, the required prediction range for a driven speed can be reduced by wheel-individual drive and brake control.

2.2 System integrated braking

The high recuperative braking power available due to electric four-wheel drive covers the range of the brake system's normal operating conditions to a large extent. Instabilities, especially on roads with low friction coefficients, can be compensated for highly dynamically over the entire operating range. For emergency braking, especially from high speeds, and for braking to a standstill, blending with an electronically controlled conventional braking system is indispensable for road vehicles.

Essential boundary conditions here are perception range, driving speed, reaction and response times as well as comfort criteria. In conventional vehicles, the prediction horizon, reaction time and comfort requirements are specified by the driver. In automated driving, these variables must be realized situationally by evaluating sensor systems and predefined control strategies. The prediction horizon here depends primarily on a realistic recognition of the environment under all circumstances but can be extended by car-to-car communication compared to conventional vehicles [5] [6].

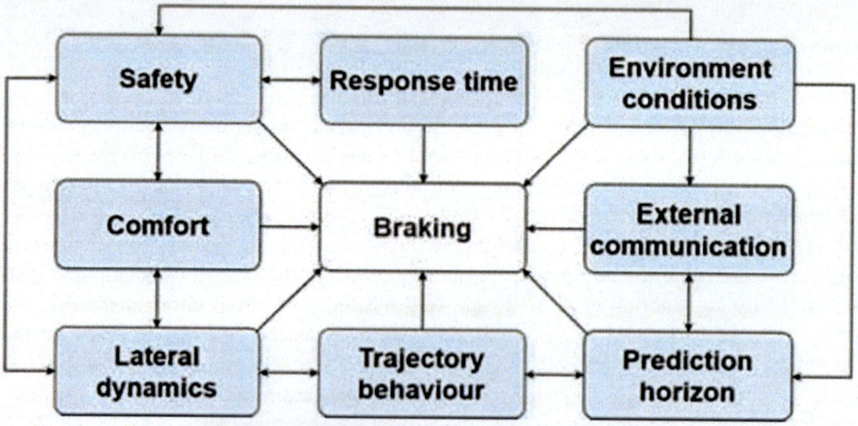

Fig. 4. Influencing factors on the brake strategy

The larger the reliably generated prediction horizon, the more comfort aspects can be taken into account during braking. A decisive advantage in automated driving is that the reaction time is close to zero. As can be seen in figure 5, the distance travelled during this time is thus available as braking distance and can be used for a higher safety reserve or improved braking comfort, depending on the situation and the selected driving profile.

Starting from a straight line travel at 100 km/h, the comparison is shown between emergency braking delayed by the human reaction time of one second with a constant maximum deceleration of -10 m/s² and the brake behaviour resulting from a constant jerk of -1, -2 and -3 m/s³. Additionally, a reference braking is displayed, such as is carried out by current commercially available driver assistance systems for automated emergency braking (AEB). The dotted lines mark constant decelerations of 2, 4, 6 and 8 m/s². It is clearly recognizable that the achievable driving comfort can be significantly increased by eliminating human delay in a fully automated motion planning system.

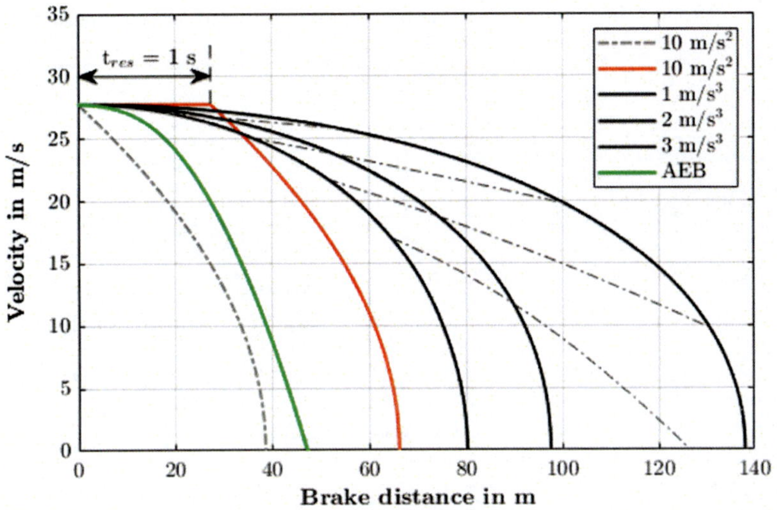

Fig. 5. Brake distance behaviour for constant deceleration, constant jerk and an AEB profile

At the same time, automated driving requires the minimization of delay between the controller's braking request and the resulting wheel deceleration through mechanical and hydraulic latencies. It is also in this context that the possibility of recuperative all-wheel braking offers a clear advantage through significantly improved performance in comparison with current hydraulic systems. If obstacles suddenly appear within the detection horizon during a normal braking maneuver, it is necessary to switch from comfort mode to safety mode as quickly as possible. In this situation once more, automated driving has the advantage of a very short reaction time. A specific implementation of this requirement is shown in the following chapters using a control concept based on the vehicle dynamics systems of the FSD vehicle and the simulation results of its driving operation.

2.3 System architecture of an autonomous vehicle

The system architecture of an autonomous vehicle can be subdivided in different ways depending on the task at hand and the algorithms specifically selected. In the case of the FSD vehicles from municHMotorsport, the task is to ensure both manual and autonomous driving operation. In addition, due to the resources available in the development team, it is desirable to adopt findings and specific software modules from the FSE vehicle. For these reasons, a modular structure is presented, which is divided into five subsystems. Thus, the boundary conditions of the system are similar to the requirements for motion planning in road vehicles. The subsystem *Environment Recognition* processes the sensor data from camera, lidar and radar, which are transferred to the subsystem *Localization and Mapping*. All vehicle-specific state variables are generated in the *Vehicle State* block using sensor data and vehicle dynamics models. The *Motion Planning* and *Vehicle Control* blocks represent the most relevant subsystems for the following explanations and are therefore explained in more detail.

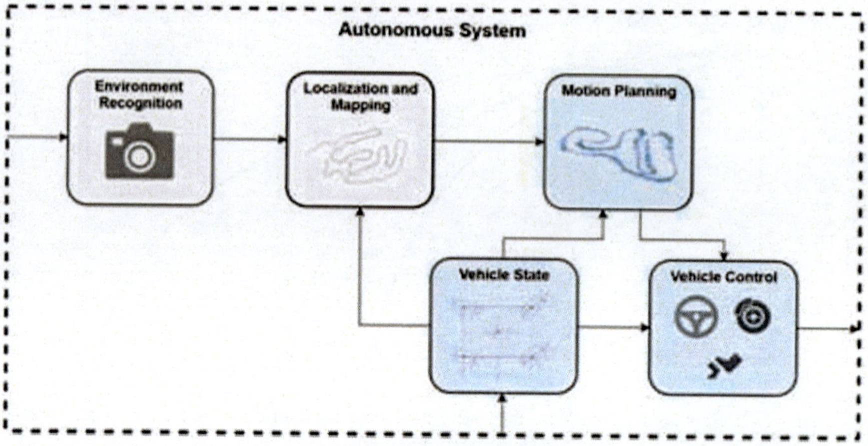

Fig. 6. Autonomous System architecture

The structure of the motion planning system of the autonomous vehicle PWd 2.18 is shown in figure 7. The actual motion planning requires a preceding tactical planning, which deals with the fulfillment of a basic goal, the mission. A mission represents, for example, following the instructions of a navigation system, selecting a parking space or simply keeping a speed limit. Such tactical planning significantly reduces the exploration space for actual motion planning and allows the use of computationally intensive planning algorithms.

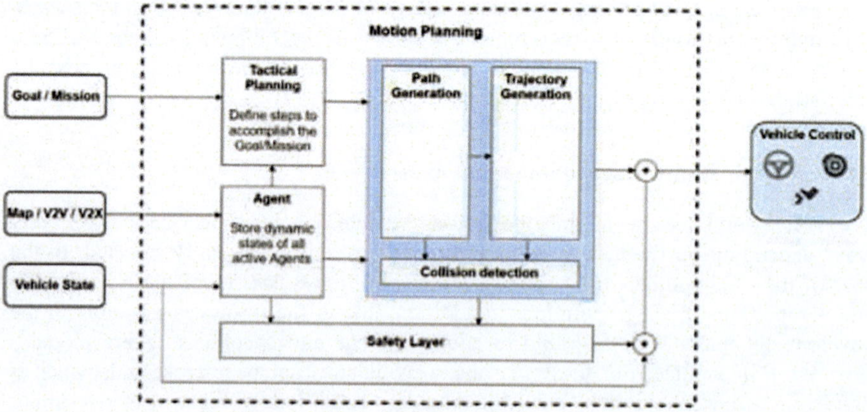

Fig. 7. Motion Planning system architecture

In this paper the mission represents the following of a given route. Based on this, a global, static path is calculated and transferred to the local planning as a reference line. The local planning block (blue block in figure 7) uses a sample-based procedure in which motion primitives are generated from quintic G2 splines [7] based on a substitute model. In addition, speed profiles are calculated using a forward-backward approach [8] in order to dynamically map the vehicle's dynamic properties. Within the solution space, a cost function determines the optimal trajectory, which is passed on to the *Vehicle Control* subsystem. The safety layer represents the monitoring element and checks whether logical solutions are generated. This information is passed on to the Vehicle Control Subsystem to bring the vehicle into a safe state if necessary. The *Vehicle Control* environment calculates the manipulated variables to control the vehicle movement in order to follow a given trajectory in a stable manner. For this purpose, the subsystem is divided into three further submodules:

High-Level-Controller - *Low-Level-Controller* - *Low-Level-DAS-Controller* (Driving Assistance System)

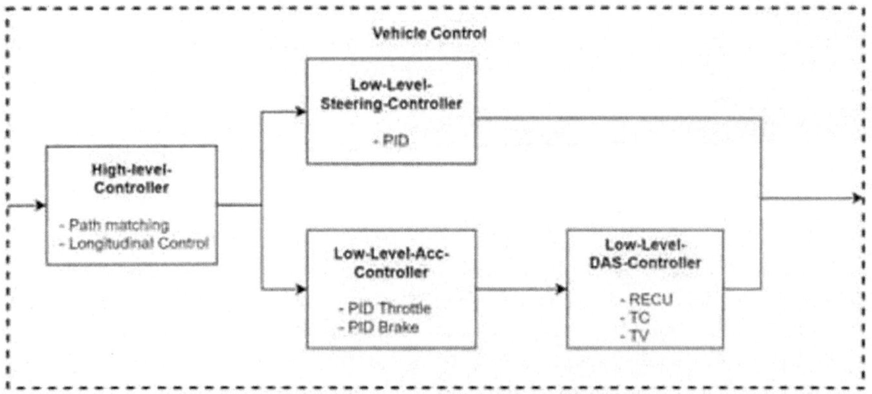

Fig. 8. Vehicle Control system architecture

The high-level controller follows a parallel approach, as it is divided into lateral and longitudinal control. A combinatorial approach is not necessary for this problem, since vehicle stability is guaranteed by the motion planning that takes place beforehand. The quintic splines guarantee control of lateral deviation with minimal jerk, while the speed profile takes the acceleration potential of the car into account. The vehicle's steering response and thus its tendency to under- and oversteer are taken into account in the lateral controller (see also [8]).

The low-level controllers generate the signals to control the steering and braking actuators as well as the motors of the vehicle. The DAS environment includes the control systems Torque Vectoring, Traction Control and Recuperation, which provide the immediate stability control of the vehicle.

Control interventions that are perceptible to the vehicle occupants are generally perceived as more unpleasant during automated driving than in conventional vehicles. For this reason, control systems should be capable of actively intervening over the vehicle's entire operating range, especially for automated driving functions. In contrast to conventional systems, the DAS environment described above does not only intervene at the stability limit, which means that the jerk in both acceleration and braking can be adjusted to fit the selected system strategy and individual driver requirements. This allows the implementation of comfort focused motion, distance keeping systems as well as safety braking.

As a matter of principle, motion planning for road vehicles should avoid reaching dynamic limits in automated driving operation. If the forces required for control (e.g. braking forces for driving stability control) have a negative impact on the passengers' perception of comfort and safety, this can be avoided by applying compensation forces to those wheels within the stability limits using wheel-specific drive control. Slip and driving stability control as well as torque vectoring thus achieve a new level of control quality necessary for automated driving by means of wheel-individual all-wheel drive and braking.

3 Simulation of driving operation

3.1 Simulation setup

In order to demonstrate the function of the presented architecture, the controller structure of the FSD vehicle of municHMotorsport are extended by the described software components for the realization of an emergency maneuver to avoid an obstacle unknown to global planning. On the basis of a validated vehicle model, a simulation is carried out which operates the models of the controllers used by the vehicle together with the additional components of the emergency layer as software-in-the-loop. Delays caused by the latency of the conventional brake system are mapped by a simple second order delay acting as a substitute model of the hydraulic system. Since neither the type of common blending approaches nor the detailed design of a corresponding brake deceleration controller is the subject of this paper, in this model the recuperative brake intervention by the longitudinal controller is also assigned an electrically induced latency and superimposed on the effect of the conventional brake without further signal processing.

3.2 Driving maneuvers

In order to illustrate the dynamic advantages of the described concepts, the vehicle behavior when initiating an evasive maneuver during cornering is monitored. After braking from 100 km/h according to a speed profile generated on the basis of global path planning, an online generated trajectory is selected when a significantly slower vehicle in front is detected in order to avoid the necessity of emergency braking for collision avoidance. This maneuver allows the evaluation of the utilization of the resulting tire potentials as a key factor for the possible safety distance from the friction limits as well as the dynamic response of the side slip angle in the transient steering phase.

3.3 Results

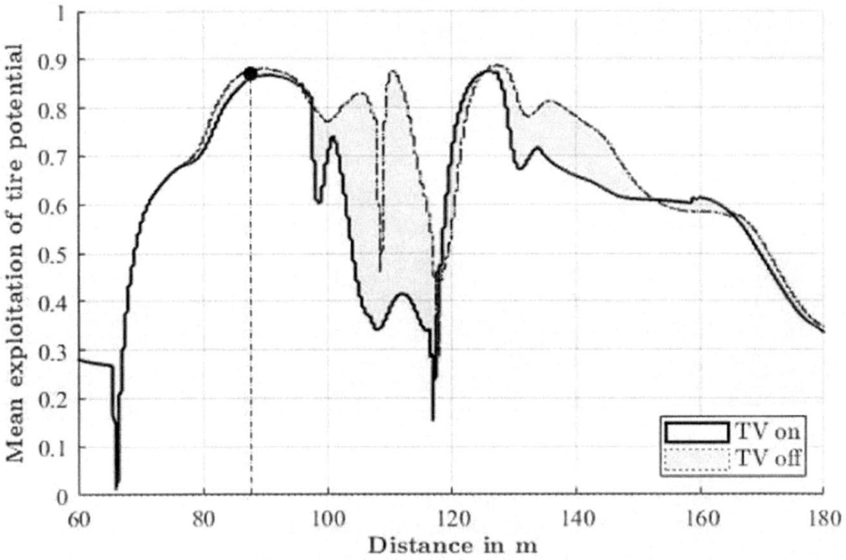

Fig. 9. Mean utilization of tire potential during evasive manoeuvre

The advantageous increase in the frictional safety buffer is shown in figure 9. The marking indicates the beginning of the evasive manoeuvre. Especially in the quasi-stationary phases, the influence of the wheel-individual torque distribution on the increase of the remaining lateral force potential can be seen. As described at the beginning, the coefficient of friction available due to environmental conditions is assumed to be known and constant, the calculation of the maximum tire potential is therefore based on the wheel forces calculated in the tire model. A further decisive advantage of wheel-selective control over the entire operating range is shown in figure 10. The continuous control of the yaw moment, taking into account the angular deviation from the target trajectory as well as individual tire potentials, considerably reduces the amplitude maximum of the side slip angle overshoot and shortens the transient phase. In the vehicle without wheel-individual torque control, the side-slip angle clearly exceeds the vehicle-specific limit value of 7 degrees, which would require stability control by brake intervention in order to be able to continue to follow the target trajectory with sufficient accuracy. In the vehicle with continuous support of the trajectory tracking control by means of torque vectoring and model-based limitation of the individual wheel torques, on the other hand, the reaching of the stability limits can be effectively prevented by maintaining a frictional buffer.

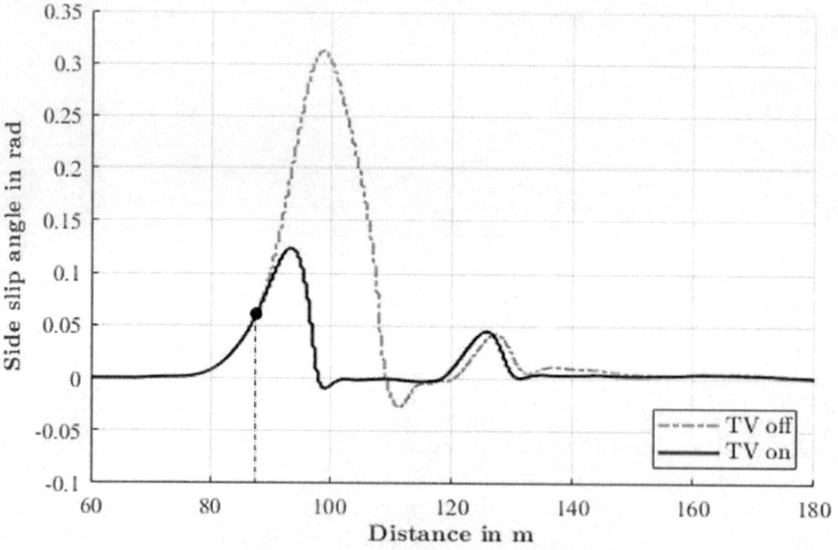

Fig. 10. Side slip angle response during evasive maneuver

If the driving profile on which the motion planning is based allows even higher utilization of the wheel potential, the vehicle with conventional drive configuration can no longer follow the trajectory required for collision avoidance due to the higher target speed, while the vehicle with individual wheel drive can still successfully perform the evasive maneuver. This case is less relevant for a normal ride in an automated road vehicle with a high coefficient of friction, since a basic safety design must not allow a planned use of more than 90% of tire potential. In the lateral direction this value is usually in the range of less than 25% due to common comfort requirements to limit maximum lateral accelerations to less than 2 m/s^2. However, since the motion planning in longitudinal and transverse direction is often carried out separately [9], common control approaches can result in significantly higher utilisation of friction potentials through combined slip. In the present case, this results in combined loads of up to 45% of absolute tire potential. Conversely, this also means that the lateral boundary conditions leave little room for variations regarding the planned use of dynamic wheel potential, for example to realise different driver characteristics. This is particularly true in the case of uncertainties in the friction coefficient information. Thus, even small friction coefficient differences, which under certain circumstances cannot be detected by car-to-x communication installed for this purpose, for example due to dirt on the road surface or local weather changes, result in a similar problem with the trajectory tracking. This is illustrated by the following simulation. Based on an external friction coefficient assumption, a driving profile with medium overall dynamics is created. For comfort reasons, a maximum of 25% of the wheel potential is used both longitudinally and laterally. With a local friction coefficient reduction of 20%, the swerve maneuver from the previous tests is now to be carried out.

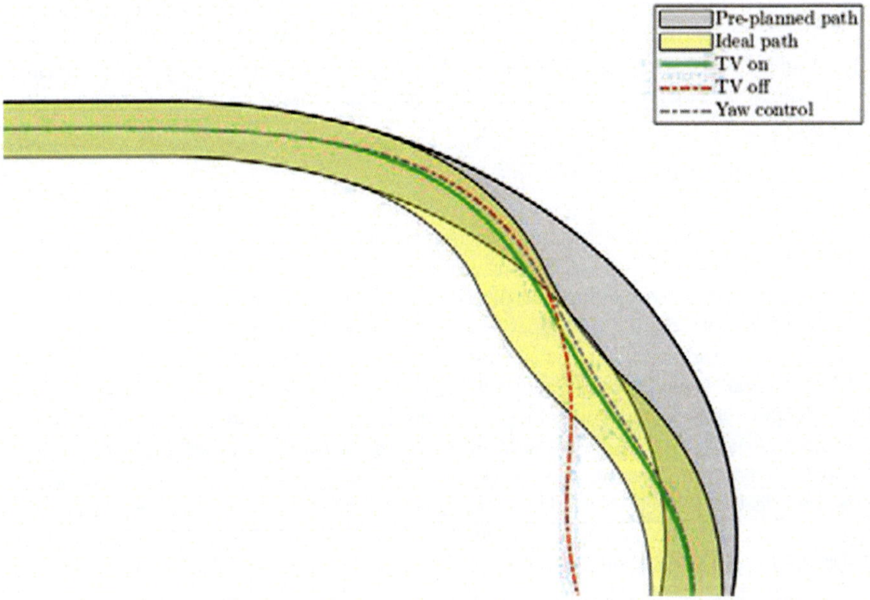

Fig. 11. Trajectory tracking with local reduction of friction coefficient

The ideal path describes an evasive maneuver as requested by the motion planner assuming the higher friction coefficient and a sufficiently early information on obstacle position. In this case, it should be noted that the prediction horizon also has a decisive influence on the success of such a maneuver, taking into account external environmental information provided by car-to-x systems. It can be seen that in the phase, which is covered by global motion planning, all drive controllers show a similar performance. However, at the beginning of the additionally required steering input due to the increase in curvature requirement, an increase in the control deviation can be immediately detected in the vehicle with conventional torque distribution. The significantly increased resulting side slip angles cause a reduction of yaw gain, making higher steering angles necessary. As a result, considerably higher frictional loads arise in comparison to a continuously controlled vehicle. In the present simulation, the vehicle is unable to follow the evasive maneuver without control intervention despite the low utilization of tire potential on the part of the driving profile and leaves the lane. For comparison, the trajectory for a vehicle with pure yaw rate control to improve driving stability is also shown. This behaviour is similar to the effects of a permanent ESP intervention with the addition of compensating torques on wheels with lower frictional load. Thus, the torque distribution is adjusted individually for each wheel in favour of driving stability to follow the yaw rate progression of a linear two-track model, but without taking into account the deviation of the trajectory tracking control. In contrast to an ESP intervention, the selected control structure has the consequence that the control intervention does not become noticeable for the occupants by an unintentional reduction of the longitudinal speed. However, the disadvantages with regard to the control deviation are also apparent with this approach in both longitudinal and lateral direction. By combining reduced response times with the concept-related advantages of the discussed type of motion control, the perception of driving safety, which is decisive for customer acceptance, can thus be significantly improved.

4 Conclusion

It could be shown that wheel-individual longitudinal control systems can improve the driving dynamics, safety and comfort of road vehicles. This is of particular importance for automated driving. As the simulation of different vehicle reactions to a traffic obstacle at country road speed demonstrated, the fundamental findings from the FSD competition are also relevant for the development of road vehicles.

The essential aspects for the successful integration of the presented drive concept are high speed motors with planetary gears and liquid cooling as well as strong function integration on component design level. In this way, the system can use the advantages of continuous control of drive and braking forces over the entire operating range, while the concept-related disadvantages of drives close to the wheel can be largely avoided. For example, the possibility of a wheel-individual control of frictional tire load means that a safety buffer to the stability limits can already be set during operation within stability limits, without retaining unused friction reserve. It is also possible to continuously adjust the desired steering response. Similarly, compensation for disturbing influences such as cross winds, road inclination or differences in driving resistance between lanes is also possible. The simulation of an evasive manoeuvre not covered by the global driving profile showed that this approach provides advantages in driving stability as well as in the quality of the trajectory tracking with a higher resulting force reserve by avoiding overshooting of the side slip angle. In contrast to conventional systems, the necessary intervention of the stability control in the area of physical limit states can be implemented with reduced jerk by compensating control forces on wheels with uncritical friction conditions. Since the driving comfort of automated driving functions is the decisive criterion for acceptance by the customer, the concept presented here offers an opportunity to significantly improve the performance of automated driving in road vehicles.

References

[1] T. E. L. Woopen, „Neue Architekturen für die Mobilität der Zukunft“. In: *ATZ/MTZ 24, 22–27,* 2019.

[2] M. Krug, „Level 3 ist (fast) erreicht“. In: *Technik in Bayern, Heft, S. 6,* 02 2020.

[3] S. D. S. B. O. e. a. Köhler, „Torque Vectoring Systems To Improve Energy-Efficiency and Vehicle Dynamics“. In: *ATZ Elektron Worldw 9, 46–53,* 2014.

[4] M. G. M. Böhle, „Milestones of the Drivetrain Development in the FSE“. In: *ATZextra Worldw 23, 50–53,* 2018.

[5] G. F. Reichart, C. Dorrer, H. Rieker, E. Drechsel und G. Wermuth, „Potentials of BMW Driver Assistance to Improve Fuel Economy“. In: *Paper FISITA Congress Paris,* 1998.

[6] K. Bengler, „Psychologie des Automatisierten Fahrens“ . In: *Technik in Bayern, Heft, S. 12,* 02 2020.

[7] A. &. G. L. B. Piazzi, „Quintic G2-splines for trajectory planning of autonomous vehicles“. In: *IEEE Intelligent Vehicles Symposium*, 2000.

[8] J. S. a. J. C. Gerdes, „Autonomous vehicle control for emergency maneuvers: The effect of topography“. In: *American Control Conference*, 2015.

[9] M. Werling, Ein neues Konzept für die Trajektoriengenerierung und -stabilisierung in zeitkritischen Verkehrsszenarien, KIT Scientific Publishing, Karlsruhe.

[10] T. Süd, *Firmenzeitschrift ABOUT TRUST,* 01 2019.

[11] M. Schindlmeier und M. Mainusch, „Selbst entwickeltes Planetengetriebe für ein Formula-Student-Rennfahrzeug“ . In: *ATZextra,* 11 2016.

[12] „Formula Student Germany Website“
Available: https://media.formulastudent.de/2018/Hockenheim/20180809-Thursday/i-MfSshpm/A, last accessed 2020/05/13.

Test Procedure Development and Validation for Brake Wet Fading in the FKFS Thermal Wind Tunnel

Domenik Schramm, M.Sc. [1]

Adrian Bernhard, M.Sc. [1]

Dr.-Ing. Timo Kuthada [1]

Prof. Dr.-Ing. Jochen Wiedemann [1]

Prof. Dr.-Ing. Andreas Wagner [1]

Lars Altseimer, B.Sc. [2]

[1] Research Institute of Automotive Engineering and Vehicle Engines Stuttgart (FKFS), Germany
[2] University of Stuttgart, Germany

Abstract. The main function of a vehicle braking system is to decelerate the vehicle predictable and reliable in any driving condition. Overheating of brake components can cause undesirable effects, such as brake fading or boiling of brake fluid. To ensure safe operation, aerodynamic concepts such as brake ducts, deflectors or brake disc ventilation have been introduced and steadily improved. These measures ensure sufficient heat-rejection but can reduce brake performance under rainy conditions, due to water ingress towards the brake. This phenomenon is called "brake wet fading" and results in a decrease of the friction coefficient between brake pad and disc.

During the vehicle development process, brake components (pad and disc materials) are tested on component test benches. However, realistic environmental and aerodynamic boundary conditions in the vehicle cannot be applied during these tests. In order to obtain more realistic results, measurements for brake wet fading are taken on the test track–in later development stages, although environmental conditions such as side wind and ambient temperature, rarely remain constant. This makes direct comparisons between different test configurations difficult.

For these reasons, FKFS developed a new test procedure for brake wet fading in the FKFS Thermal Wind Tunnel. It provides a robust test environment in which aerodynamic measures regarding brake wet fading can be evaluated. This paper presents the brake wet fading test procedure and correlation measurements with road tests. Results of the new brake test procedure show good agreement with those from on-road measurements, with a deviation of less than 3% in friction coefficient. With this level of accuracy achieved and a superior reproducibility, the tests in the FKFS Thermal Wind Tunnel for brake wet fading

P. E. Pfeffer (ed.), *11th International Munich Chassis Symposium 2020*, Proceedings, https://doi.org/10.1007/978-3-662-63193-5_33

can be considered an improved alternative to on-road tests – in dry and wet conditions.

Keywords: Brake Wet Fading, Wind Tunnel, Test Procedure Development, Friction Coefficient

1 Introduction

Increasing safety requirements and regulations over the last decade have led to a higher development effort concerning the braking system of a passenger car [1, 2].

The main task of a vehicle braking system is to decelerate the vehicle predictably and reliable in any driving condition. Increased vehicle weights and the associated higher braking power requirement lead to high thermal loads on the braking system. Overheating of brake components such as the brake disc and brake pad can reduce the braking power and thus undesirably increase the braking distance - known as "thermal fading". To ensure a sufficient heat rejection and avoid this fading phenomenon concepts such as brake ducts and ventilated brake discs have been introduced and steadily optimized [3]. These measures ensure sufficient heat rejection but can reduce brake performance caused by water contamination in wet conditions [4, 5]. This phenomenon is called "brake wet fading".

As with the development of brake systems, the aerodynamic development of passenger cars has also gained importance. In order to reduce the vehicle emissions, it is vital to minimize aerodynamic drag. A proportion of up to 25% of the total aerodynamic drag can be attributed to the wheels and wheelhouses [6]. To meet the aerodynamic drag targets of today's passenger cars, several components such as chin and wheel spoilers, deflectors and wheel covers have been developed. These measures can improve the overall drag, but can also interact with the airflow through the braking system. This can result in a worsening thermal or wet fading phenomena. Therefore, the evaluation of wet and heat fading phenomena is essential at the late stage of vehicle development.

2 Approaches to Brake Fading Investigations

The following chapter introduces approaches and methods to investigate brake performance in dry and wet conditions. These methods can be divided into optical approaches (qualitative) and direct determination of the friction coefficient between the brake disc and brake pads by measuring the resulting brake force (quantitative).

2.1 Optical Approaches

Optical approaches are used to determine the water contamination of the brake system due to self- and foreign-soiling. Self-soiling describes the water whirl up due to the wheels rotating on a wetted road whereas foreign-soiling is resulting from the spray coming off by a vehicle in front [7]. Spruss et. al [8] published a method to visualize contaminated areas of a brake system e.g. the brake caliper soiling. Puglisevich et. al [4] extended the experimental method and compared it to a numerical approach of the water contamination of the brake disc of an SUV. To simulate a reproducible wet driving condition, the FKFS Thermal Wind Tunnel was used. The open-jet Goettingen type wind tunnel is equipped with two dynamometers to simulate the rotating wheels [9]. For the self-soiling investigations of the vehicle, nozzles are installed in front of the front wheel. For the test, water is mixed with a fluorescent UV dye and applied from the nozzles into the tire contact patch [8]. After a specified testing time, the brake discs are removed and exposed to UV illumination. A photo of each side of the disc is taken and post processed using routines to discretize light intensity. An exemplary evaluation using digital analysis of soiled surfaces developed at FKFS called DiVeAn® is depicted in Fig. 1 [7]. Using this approach, highly contaminated areas are represented in red, less contaminated in green.

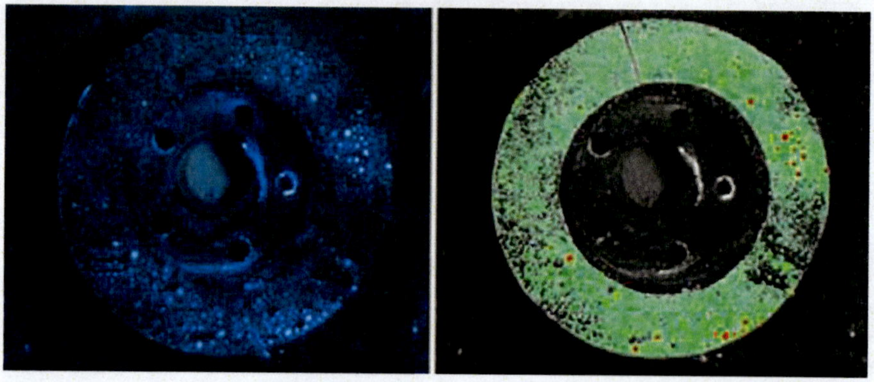

Fig. 1. Optical determination of brake disc contamination with unprocessed UV image with water fluorescence mixture (left) and post processed image (right)

Puglisevich et. al [4] correlated these experimental results with CFD simulations using a Lattice-Boltzmann-Method (LBM) solver. The accumulated water film thickness from CFD results were compared to those of the experimental surface contamination as discussed previously.

Using these optical approaches, surface contamination can be visualized and differences between brake system configurations such as a removing of dust shields can be determined. However, these measures cannot quantify the effects on brake performance under wet conditions and are time-consuming.

2.2 Experimental Approaches

Experimental approaches focus on the boundary layer between the brake pad and brake disc since this boundary layer determines the friction coefficient and therefore the brake performance [10]. The friction coefficient depends on a multitude of parameters such as pad and disc materials, temperature, brake system pressure, wear, dirt and water ingress [3, 11]. To quantify the influence of these parameters different test facilities are used to measure the friction coefficient between the brake pad and the brake discs. The test rigs can be divided into component testing and testing under driving conditions [3]. Component testing can be carried out either on scaled testers or full-size inertia dynamometers. Scaled testers consist of a full-size brake disc and a small sample of the brake pad material. These test benches offer the possibility to determine the friction coefficient as well as comprehensive analyses of the noise, vibration and brake pad topography [3, 10]. These experiments are used for a fundamental evaluation of the friction pairing. In order to verify the results of these measurements and to evaluate the entire brake system, further tests are run on a full size inertia dynamometer [3]. Even in these tests, the real assembly situation in the vehicle and the environmental influences can only be reproduced to limited extent. Therefore, the complete functional range of the braking system in the vehicle must be evaluated by means of road or wind tunnel tests. This involves extreme load cases such as Grossglockner downhill drive, high-speed fading test and the AutoMotorSport test which are focusing on maximum brake system temperature and the above mentioned temperature induced fading [12].

A further extreme load case results from the ingress of water into the brake system especially in the pad and disc. In order to ensure safe operation under wet conditions, a heavy rain scenario can be tested in wind tunnel or on-road tests.

During the on-road tests, the vehicle is operated on a wetted roadway with a defined film of water. A preconditioning phase on this roadway is followed by a braking procedure and evaluated in respect to the friction coefficient of the braking system. This procedure provides the possibility to evaluate the brake system on the vehicle system level. In addition, it provides the opportunity to evaluate measures that have been developed to reduce wet fading, such as brake disc wipers, and quantify their effectiveness [13]. However, environmental conditions such as side wind and ambient temperature rarely remain constant. This, in addition to the limited braking reproducibility of a human driver, makes direct comparisons between different test configurations difficult.

Spruss et. al [14] published an approach to improve the reproducibility of on-road testing and simulated this brake wet fading inside the FKFS Thermal Wind Tunnel. The test vehicle is placed on two dynamometers and mounted with three push/pull rods equipped with load cells. The brake force and corresponding brake system pressure is measured and the friction coefficient is calculated. It has been shown that the restraint of the vehicle and a simultaneous force measurement are complex. In addition, the desired brake pressure cannot be reliably adjusted via the brake pedal. The development of the new test system at FKFS focuses on improving these shortcomings.

3 Test Procedure Development

To enable the evaluation of the brake system performance with regard to wet fading, it is necessary to calculate the friction coefficient μ_B between brake disc and brake pad. The friction coefficient μ_B is defined as the ratio of the tangential force F_t to the clamping force F_{cl} at the brake disc.

$$\mu_B = \left| \frac{F_t}{F_{cl}} \right| \quad (1)$$

From the equilibrium of moments at the wheel (see Fig. 2), the tangential force can be calculated by the time resolved brake force $F_B(t)$, the effective rolling radius r_e and the mean radius of the brake pad r_B [15].

$$F_t(t) = \frac{F_B(t) \cdot r_e}{r_B} \tag{2}$$

In order to calculate the clamping force, the geometric dimensions of the brake and the pressure in the brake system are required. The clamping force results from the measured hydraulic pressure $p_{hyd}(t)$,total piston area for one brake caliper A_p, factor for brake caliper type k (k = 1 for floating caliper, k = 1/2 for fixed caliper).

$$F_{cl}(t) = k \cdot p_{hyd}(t) \cdot A_p \tag{3}$$

Subsequently the friction coefficient for a single wheel can be calculated as

$$\mu_B(t) = \frac{F_B(t) \cdot r_e}{2 \cdot k \cdot p_{hyd}(t) \cdot A_p \cdot r_B} \tag{4}$$

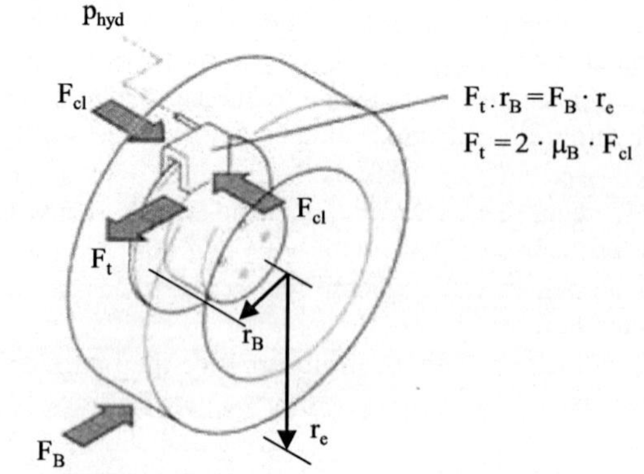

Fig. 2. Torque balance at the wheel to calculate the friction coefficient μ_B [16]

The development of the new brake test procedure at FKFS is based on the system of Spruss et. al [13]. To enhance reproducibility and comparison with the results of on-road tests, the following developments are taken into account. The brake system is decoupled from the vehicle and the hydraulic pressure is applied to the brake system via an external actuator. The actuator controls the desired brake pressure over a specified time. The pressure level can be set reproducibly for each measurement via the closed loop control system of the external hydraulic pressure actuator (see Fig. 3). In addition, this external actuator prevents the intervention of unwanted brake functions, such as brake disc wipers. Instead, this functionality can be evaluated using the FKFS brake system.

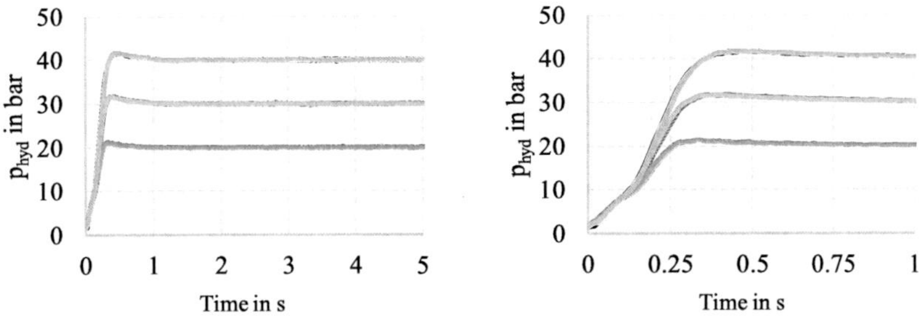

Fig. 3. System pressure build-up for different pressure levels. Each pressure level has been measured three times.

Due to the inertia of the dynamometers and the associated response delay, a direct measurement with them is not feasible. Therefore, the resulting braking force is measured at the trailer coupling via a load cell. The vehicle is restrained by two push/pull rods, which prevent any lateral or yawing movement of the vehicle on the test bench and have no effect on the brake force. The friction coefficient is then determined from the geometric data of the brake and the measured brake force, see equation (4). Fig. 4 depicts the restraint system and the measurement of the brake force. For the determination of the friction coefficient under wet conditions, spray nozzles are installed in front of the wheels to simulate the self-soiling characteristic on a wetted road. A maximum water quantity of 5000 l/h can be injected with the irrigation system of the FKFS Thermal Wind Tunnel. Additional spray nozzles can be installed inside the wind tunnel nozzle for the simulation of foreign-soiling (see Fig. 5). The test procedure involves determining the friction coefficient under dry and wet conditions. Dependent on the aerodynamic vehicle configuration, the influence of different measures e.g. wheel spoilers and brake ducts can be determined.

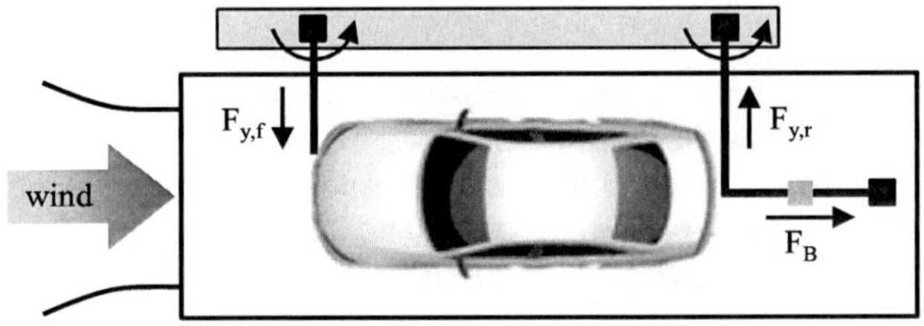

Fig. 4. Experimental setup of the FKFS brake force measurement system in the thermal wind tunnel

In addition, the brake disc temperature is monitored by thermocouples inside the brake disc during the tests. This ensures reproducible initial conditions between measurements and enables the evaluation of the cooling behavior of the brake disc (see Fig. 5).

Further improvements of the new test procedure involve the pre-conditioning of the brake pad before each test run and a brake temperature-controlled test procedure.

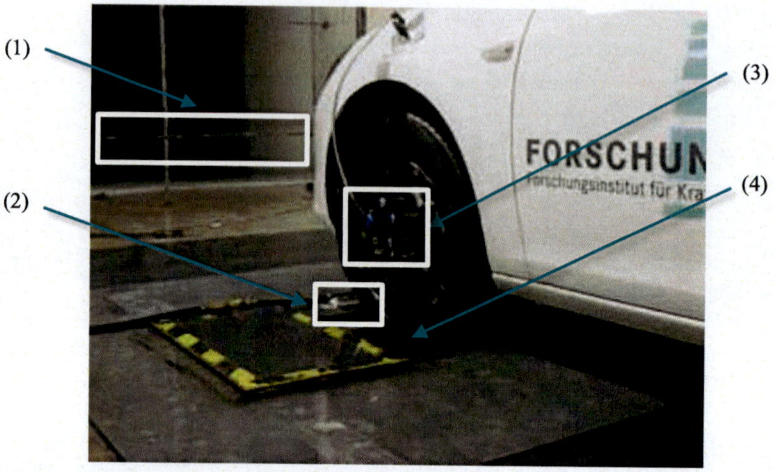

Fig. 5. Test configuration in the FKFS thermal wind tunnel. 1. Nozzle of wind tunnel with spray nozzles for foreign-soiling. 2. Spray nozzles in front of the front wheels for self-soiling. 3. Measurement of brake disc temperature. 4. Dynamometer

The test procedure includes several consecutive braking in dry and wet conditions. In dry conditions, the braking is controlled by a specified initial brake disc temperature. Therefore, every braking is performed at the same start temperature and hydraulic pressure level with the same measurement time. This procedure is used to determine the friction coefficient under dry conditions. In order to simulate driving on a wetted road, the aforementioned spray nozzles are activated. Under these conditions, braking tests at various time intervals can be performed. The results of these tests are used to calculate the friction coefficient under wet conditions and evaluate the system with respect to brake wet fading. If the brake regeneration after the simulated wetted road drive is of interest, an additional test block under dry conditions can be evaluated. After each test procedure, a reconditioning of the brake pad is performed by applying the brakes multiple times until the friction coefficient returns to the one measured in dry conditions. The test procedure described can be used to evaluate brake performance of the front and rear axles separately. In addition, even braking on a single wheel is possible.

4 Experimental Setup

In order to evaluate the described procedure with respect to its comparability to on-road testing, test drives with an equipped test vehicle were conducted. The following chapter provides an overview the vehicle setup and the correlation tests between on-road and wind tunnel measurements.

To calculate the friction coefficient, geometric parameters of the braking systems as well as the test vehicle are necessary (see equation 4).

Effective rolling radius r_e

To calculate the effective rolling radius, it is necessary to drive at constant speed. The radius is then calculated from the vehicle velocity v and the wheel rotation rate n using

$$r_e = \frac{v}{2 \cdot \pi \cdot n} \tag{5}$$

This radius was determined during on-road and wind tunnel tests.

Mean radius of pad r_B

The mean radius of the braking pad can be derived from the position of the braking pad in relation to the wheel axis and the center of gravity of the pad. In the following investigations, this radius was determined by a 3D scan of the braking system and used for further calculations.

4.1 Test Vehicle Setup

For the comparison between the measured friction coefficients, a 2010 Opel Astra was used. For the estimation of the hydraulic brake pressure an additional pressure sensor was fitted to the hydraulic system. In all test runs, the hydraulic system of the rear axle system was decoupled. Therefore, the braking was performed on the front axle only. This approach enables a separate evaluation of the friction coefficient for the front and rear braking systems. For the measurement of single wheels on the front axle, two Kistler RoaDyn P106 wheel torque transducers with a maximum torque of ± 6000 Nm were installed [17]. In addition to the torque measurement, the wheel rotation rate can be determined with a resolution of 1024 measurements per revolution. With the wheel rotation rate and the vehicle speed – measured with Correvit sensor – the wheel effective radius r_e (see equation (5)) can be calculated.

Additionally, the brake disc of the front axle is equipped with two thermocouples on each side. The thermocouples are placed at the mean brake pad radius, one thermocouple on the inside, one on the outside of the brake disc. The temperature of the brake discs are monitored during the on-road tests in order to reproduce the conditions during wind tunnel testing since the friction coefficient can be temperature dependent [11, 18].

During the wind tunnel tests, the vehicle is restrained as described in chapter 3. The braking force is simultaneously measured via the load cell at the trailer coupling and the wheel torque transducers on the front axis (see Fig. 4 and Fig. 6).

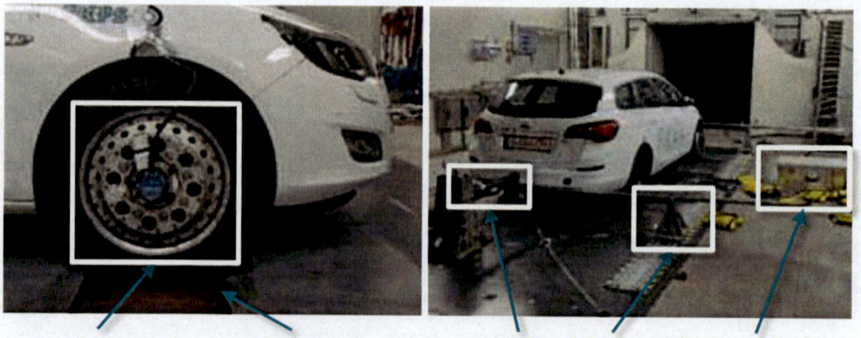

Fig. 6. Experimental setup for correlation tests in the thermal wind tunnel of FKFS. (1) wheel torque transducer, (2) dynamometer, (3) load cell, (4) restraint system and (5) external hydraulic pressure actuator

4.2 On-Road and Thermal Wind Tunnel Investigations

Using the test vehicle described in 4.1, the (time dependent) friction coefficient $\mu_B(t)$ was determined in on-road tests and with the new test procedure in the FKFS Thermal Wind Tunnel. The on-road test was performed on a straight line in dry conditions. After accelerating the vehicle to 80 km/h, a test driver tried to set a pressure of 20 bar as quickly as possible via the brake pedal. Due to limited braking reproducibility of a human driver and the difficulty to set the desired pressure level as a test driver, each pressure signal during on-road measurements must be analyzed. During the decelerating phase, the wheel torque and hydraulic pressure was recorded. For correlation, only the period of quasi-steady hydraulic pressure is evaluated. The pressure level during this period is set as a system pressure during the wind tunnel tests using the external actuator (see Fig. 7). Additionally, the brake disc start temperature is monitored during these tests in order to match it in wind tunnel testing.

The results are first examined with regard to the measured torque and the resulting friction coefficient in comparison between on-road and wind tunnel tests. In the next step, the measurement of the wheel torque transducer during on-road tests is compared to the load cell measurement on the trailer coupling in the wind tunnel.

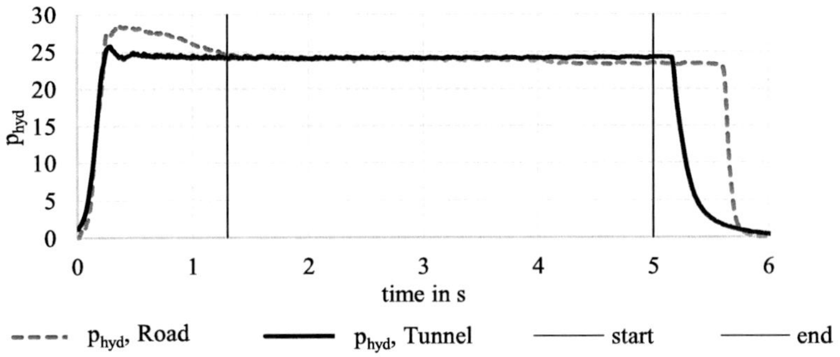

Fig. 7. Comparison of hydraulic pressure during on-road and wind tunnel testing and start and end of quasi-steady pressure level

5 Comparison of Experimental Results between On-Road- and Wind Tunnel Measurements

With the test procedure described in chapter 3, a comparison between on-road and thermal wind tunnel test results was performed. In Table 1, the boundary conditions such as hydraulic pressure p_{hyd} and start temperature T_{start} for each test run on-road and in thermal wind tunnel are listed.

Table 1. Boundary conditions for the comparison of on-road and in wind tunnel test runs

Test run no.	1	2	3	4	5	6
p_{hyd} in bar	25	19	25	24	23	20
T_{start} (Road) in °C	91	83	97	100	109	114
T_{start} (Tunnel) in °C	87	92	96	99	112	115

These initial boundary conditions were obtained from on-road test results as mentioned in chapter 4.2. Fig. 8 shows the time resolved friction coefficient for one test calculated from the wheel torque transducer (WTT) and load cell (LC) in the wind tunnel test.

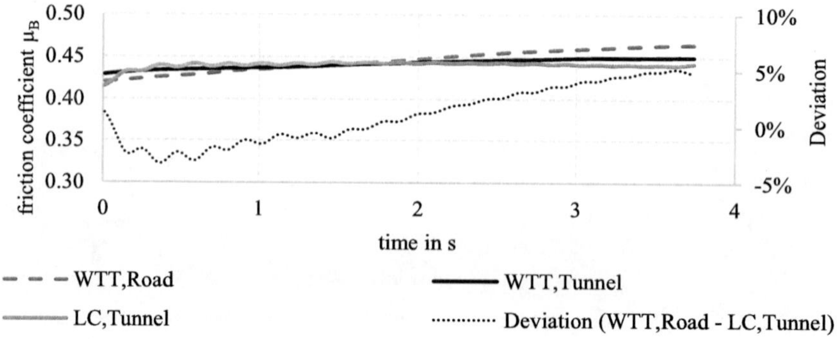

Fig. 8. Time resolved friction coefficient of on-road (Road) test and thermal wind tunnel test (Tunnel) measured with wheel torque transducer (WTT) and with the load cell (LC) in the wind tunnel test.

The comparison of the test runs shows a good agreement over the whole time-period with a deviation of < 5 % between WTT on-road and LC in wind tunnel test. The comparison of the average friction coefficient of all test runs listed in Table 1 is shown in Fig. 9. The maximum deviation of all six test runs is < 3 % for the average friction coefficient of each measurement method. This shows a good agreement in all test runs between the on-road tests and the measurement results in the FKFS Thermal Wind Tunnel with the new test procedure.

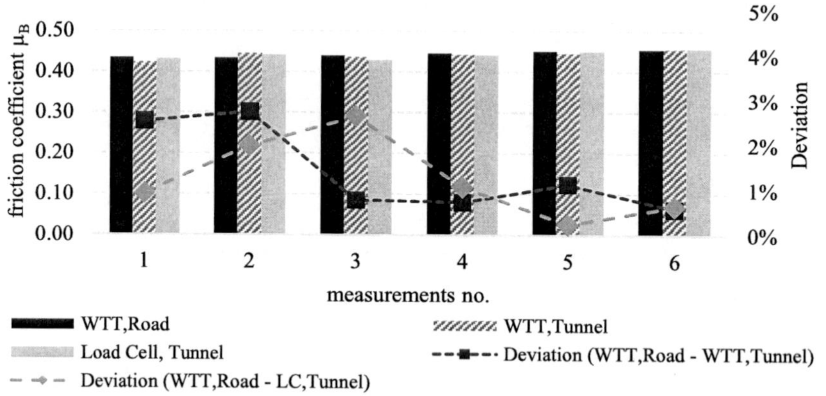

Fig. 9. Averaged friction coefficient μ_B and the averaged deviation with regard to the measurement method of six validation runs.

To evaluate the quality of the time-dependent friction coefficient in both test cases, the root-mean-square error (RMSE) is listed in Table 2. A maximum RMSE of 0.014 between the wheel torque measurements and 0.015 between wheel torque and load cell was observed. This also represents a good agreement between on-road and wind tunnel test results. Overall, the RMSE between WTT on-road and WTT in the thermal wind tunnel is slightly smaller. In summary, the new test procedure is able to reproduce the on-road results of the averaged friction coefficient with a high accuracy.

Table 2. Root-mean-square error of the measurement results in comparison to measurement methods

Test run no.	1	2	3	4	5	6
RMSE $\mu_{B\,(WTT,\,Road\,-\,WTT,\,Tunnel)}$	0.014	0.013	0.008	0.008	0.008	0.007
RMSE $\mu_{B\,(WTT,\,Road\,-\,LC,\,Tunnel)}$	0.013	0.012	0.015	0.013	0.010	0.012

Due to the direct brake force measurement at the wheel, it is assumed that the WTT results are more accurate compared to the load cell measurement method. For later analysis of the brake responsivity in dry and wet conditions, μ_B is evaluated in different time intervals by measuring μ_B using WTT and LC simultaneously. Therefore, a set of 21 runs in the thermal wind tunnel was recorded. In Fig. 10, the percentage deviation of μ_B ((|WTT - LC|) / WTT) is shown in Box-Whisker-Plot split in time intervals.

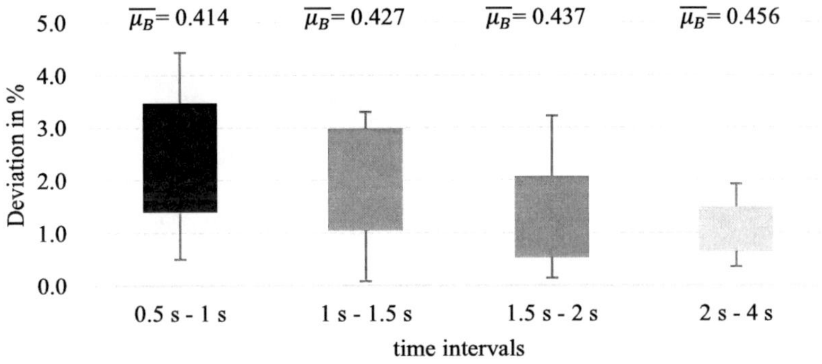

Fig. 10. Box-Plot: Percentage deviation between wheel torque transducer and load cell in thermal wind tunnel divided in intervals of the time-dependent friction coefficient μ_B. Averaged friction coefficient $\overline{\mu_B}$ of 21 test runs is given for each time interval.

The maximum deviation is 4.5 % in the 0.5 s – 1 s time interval within these 21 test runs. This corresponds to a deviation of the friction coefficient $|\Delta\mu_B| = 0.02$ of an average $\overline{\mu_B} = 0.414$ in these 21 runs. For 75 % (third quartile) of all examined runs, a maximum deviation ≤ 3.5 % can be observed, which shows again the high accuracy. In all other time intervals, the third quartile deviation is ≤ 3 %, which indicates a good comparison between the WTT and LC measurement methods. Therefore, the LC measurement method is sufficiently accurate both for steady-state as well as for transient (evaluation of responsivity) results.

6 Evaluation of Brake Wet Fading in Thermal Wind Tunnel

Due to the large amount of water ingress during brake wet fading tests, the wheel torque transducers are removed and replaced by standard rims. In addition, the application of wheel torque transducers affects the flow around the wheels and does not represent a standard vehicle configuration. As mentioned in section 3, spray nozzles in front of the front wheels are installed with a water quantity of 2500 l/h each side corresponding to a heavy rain scenario. The friction coefficient for these tests is determined using the load cell at the trailer coupling, according to the FKFS test procedure. The procedure for the conducted tests includes a brake regeneration phase before each measurement series. One series of measurement consists of three consecutive braking under dry conditions followed by two under wet conditions. A period of 15 minutes with a vehicle speed of 135 km/h before each wet braking has been used for the results presented here to simulate driving on a public highway during rain. In the following, the friction coefficients in wet conditions are designated $\mu_{wet,15}$ and $\mu_{wet,30}$ corresponding to the wetting time. The disc start temperature for the dry friction coefficient is set to 80 °C (see Fig. 11).

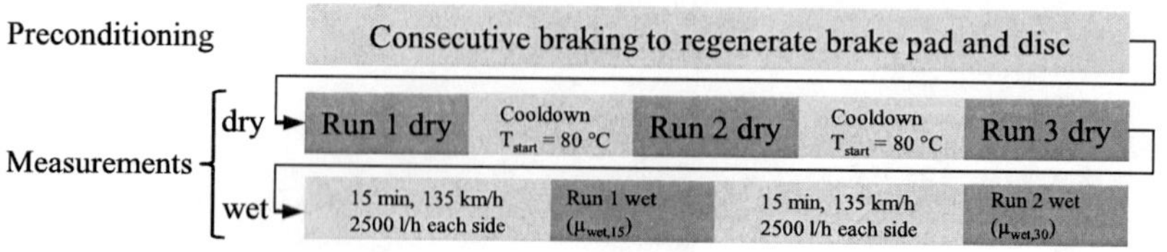

Fig. 11. Test procedure for brake wet fading in the Thermal Wind Tunnel

Using this procedure, two different aerodynamic setups are used to show the impact of aerodynamic measures on brake wet fading. The first configuration describes the standard production setup of the Opel Astra. For the second test run, the aerodynamic features in the front body (chin and wheel spoilers) were removed. The evaluation period for the measurements is four seconds and the brake system pressure is 15 bar. In all configurations, the friction coefficient in dry conditions remains at a similar level. Even in measurement series two (after two times 15 min in wet conditions), the friction coefficient can be stabilized to the absolute value from series one after the brake pad regeneration procedure. Due to the vehicle restraint, a vehicle-characteristic body resonance can be observed, which can be seen in the friction coefficient (see Fig. 12)

The comparison between dry and wet conditions of the first configuration (production setup) shows only a small difference in the time resolved friction coefficient (see Fig. 12). A slight decrease of the friction coefficient can be detected during the first second after brake application. For the interval from 0.5 s to 1 s the friction coefficient in wet conditions is 6 % lower compared to the averaged dry condition coefficient. In addition, the steady-state coefficient in wet conditions at the end of the test run is slightly lower by 5 % in average than in dry conditions in the time interval from 2 s to 4 s. With this standard production configuration, the brake system shows no sensitivity regarding water ingress.

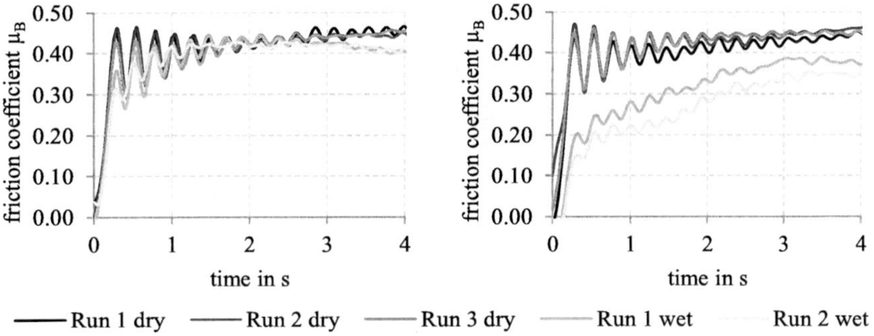

Fig. 12. Comparison of the time resolved friction coefficient for production setup (left) and removed aerodynamic features (right)

With the removal of aerodynamic features in the second vehicle configuration, the sensitivity of brake performance due to water ingress into the brake system can clearly be observed. Both the responsivity as well as the absolute friction coefficient are reduced significantly in wet conditions (see Fig. 12). To compare the responsivity and absolute values, the time resolved signals in dry and wet conditions are compared in time intervals, which is shown in Fig. 13. Evidently, the removal of the aerodynamic features significantly reduces the responsiveness within the first second after the braking after the wetting time of 15 min by 38 % ($\mu_{wet,15}$) and the absolute friction coefficient is reduced by 16 % at the end of the run. This reduction increases even more with an additional wetting time of 15 min after the first braking ($\mu_{wet,30}$). In the first second the friction coefficient decreases by an additional 17 % compared to the first braking in wet condition and the absolute value at the end of the test run shows a further drop by 10 %. These results show the importance of a well-balanced aerodynamic setup not only to meet the aerodynamic drag requirements but also providing a setup for brake system demands.

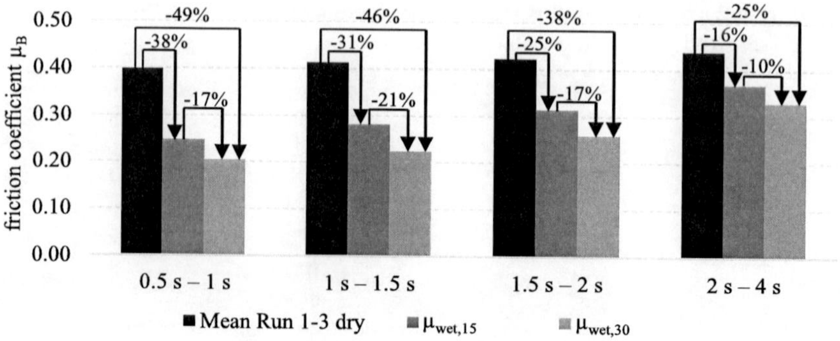

Fig. 13. Time averaged friction coefficient for measurement series two (aerodynamic features removed) in different time intervals for dry and wet conditions

7 Summary and Conclusion

This publication presents a new test procedure to determine brake performance under rainy conditions by means of wind tunnel tests. The main focus is to provide a reproducible test environment to evaluate aerodynamic measures with regard to water ingress into the brake system. Therefore, spray nozzles with a maximum water ejection of 5000 l/h are installed in front of the wheels in a wind tunnel to simulate rainy conditions. In addition, the hydraulic pressure of the brake system is applied by an external actuator. This enables a reliable system pressure built-up. The resulting brake force is measured by a trailer coupling mounted load cell. From the measured braking force, the geometric brake system data and the applied pressure level, the resulting friction coefficient μ_B between brake pad and disc is calculated fully time-resolved. With these time-resolved signals, the average friction coefficient as well as the brake responsivity after the braking can be evaluated. It was shown that aerodynamic measures like wheel and chin spoilers can significantly influence the brake performance in the aforementioned wet conditions.

To validate the results of the new test procedure, a comparison between on-road test and thermal wind tunnel test results is presented. A maximum deviation of < 3 % for the averaged friction coefficient was shown in all test runs. The deviation of the time-resolved μ_B between on-road and wind tunnel tests results in a root-mean-square error of $\mu_B \leq 0.014$, which indicates a high accuracy of the new test procedure. To evaluate responsivity effects measured in the thermal wind tunnel, the force measurement of the load cell used was correlated to the brake force measured by wheel torque transducers. This correlation also shows a high accuracy with a maximum deviation ≤ 4.5%. The results presented show that the new test procedure is capable of simulating on-road tests and providing reliable environmental conditions to evaluate brake friction coefficient in situ like aerodynamic measures regarding brake wet fading or thermal fading.

Abbreviations and Acronyms

A_p	Piston area
CFD	Computational Fluid Dynamics
F_B	Braking force
$F_B(t)$	Time dependent braking force in wheel contact patch
F_{cl}	Clamping force at brake caliper
$F_{cl}(t)$	Time dependent clamping force at brake caliper
FKFS	Forschungsinstitut für Kraftfahrwesen und Fahrzeugantriebe Stuttgart
F_t	Tangential force at brake caliper
$F_t(t)$	Time dependent tangential force at brake caliper
$F_{y,f}$	Yaw force front
$F_{y,r}$	Yaw force rear
k	Factor for type of brake caliper
LBM	Lattice-Boltzmann-Method
LC	Load cell
n	Wheel rotation rate
p_{hyd}	Hydraulic pressure in braking system
$p_{hyd}(t)$	Time dependent hydraulic pressure in braking system
r_B	Mean radius of pad
r_e	Effective rolling radius
RMSE	root-mean-square error
SUV	Sport Utility Vehicle
v	Vehicle velocity
WTT	Wheel torque transducer
μ_B	Friction coefficient between brake pad and brake disc
$\mu_B(t)$	Time dependent friction coefficient between brake pad and brake disc
$\mu_{wet,15}$	Friction coefficient after 15 min of wetting time
$\mu_{wet,30}$	Friction coefficient after 30 min of wetting time

References

[1] United Nations Economic Commission for Europe, "Regulation No 13-H of the Economic Commission for Europe of the United Nations (UN/ECE) — Uniform provisions concerning the approval of passenger cars with regard to braking [2015/2364]," 2015.

[2] "VDA 285-1: Scheibenbremsbeläge der Fahrzeuge bis 3,5 to," vol. 2003.

[3] B. Breuer and K. H. Bill, *Bremsenhandbuch: Grundlagen, Komponenten, Systeme, Fahrdynamik,* 5th ed. Wiesbaden: Springer Fachmedien Wiesbaden, 2017. [Online]. Available: https://ebookcentral.proquest.com/lib/gbv/detail.action?docID=5085047

[4] L. Schembri Puglisevich, A. Gaylard, M. Osborne, J. Jilesen, and A. Gagliardi, "Application of CFD to Predict Brake Disc Contamination in Wet Conditions," *SAE Int. J. Passeng. Cars - Mech. Syst.*, vol. 9, no. 2, pp. 800–807, 2016, doi: 10.4271/2016-01-1619.

[5] T. Schütz, *Ein Beitrag zur Berechnung der Bremsenkühlung an Kraftfahrzeugen*. Zugl.: Stuttgart, Univ., Diss., 2009. Renningen: Expert-Verl., 2009.

[6] G. Wickern, K. Zwicker, and M. Pfadenhauer, "Rotating Wheels - Their Impact on Wind Tunnel Test Techniques and on Vehicle Drag Results," in *SAE Technical Paper Series*, 1997.

[7] T. Landwehr, *Neue Methoden zur Untersuchung der Sichtfreihaltung an Kraftfahrzeugen,* 1st ed., 2020. [Online]. Available: https://doi.org/10.1007/978-3-658-29416-8

[8] I. Spruss, D. Schröck, N. Widdecke, T. Kuthada, and J. Wiedemann, "Investigation of Braking Response Under Wet Condition: 8th MIRA International Vehicle Aerodynamics Conference, 13 - 14 October 2010, RBS Williams F1 Conference Centre," 2010.

[9] T. Kuthada and S. Cyr, "Approaches to vehicle soiling," *Progress in Vehicle Aerodynamics, IV, Numerical Methods*, pp. 111–123, 2006.

[10] G.-P. Ostermeyer, T. Schramm, S. Raczek, F. Bubser, and N. Perzborn, "The automated universal tribotester," in *Proceedings of the EuroBrake Conference, Dresden, Germany*, pp. 4–6.

[11] T. Degenstein, "Kraftmessung in Scheibenbremsen," Dissertation, Technische Universität Darmstadt, 2007. [Online]. Available: http://tuprints.ulb.tu-darmstadt.de/913/

[12] D. Wagner, J. Hoffmann, and M. Lienkamp, "Downsizing potential of wheel brakes in electric vehicles," in *Proceedings, 8th International Munich Chassis Symposium 2017*, P. P. E. Pfeffer, Ed., Wiesbaden: Springer Fachmedien Wiesbaden, 2017, pp. 661–691.

[13] S. Breuer and A. Rohrbach-Kerl, *Fahrzeugdynamik: Mechanik des bewegten Fahrzeugs*. Wiesbaden: Springer Vieweg, 2015. [Online]. Available: http://dx.doi.org/10.1007/978-3-658-09475-1

[14] I. Spruss, D. Schröck, T. Kuthada, and J. Wiedemann, "Aerodynamics as troubleshooting of wet fading," *ATZ worldwide*, vol. 112, no. 10, pp. 22–25, 2010, doi: 10.1007/BF03225146.

[15] P.W.A. ZEGELAAR and H. B. PACEJKA, "Dynamic Tyre Responses to Brake Torque Variations," *Vehicle System Dynamics*, vol. 27, sup001, pp. 65–79, 1997, doi: 10.1080/00423119708969645.

[16] J. Wiedemann and N. Widdecke, *Vorlesungsskript Kraftfahrzeuge II*. Stuttgart: IVK Universität Stuttgart, 2010.

[17] Kistler Instrumente AG, *Data Sheet RoaDyn P106 / P109*. [Online]. Available: https://www.kistler.com/?type=669&fid=106321&model=document (accessed: Mar. 26 2020).

[18] H.-H. Braess and U. Seiffert, *Vieweg Handbuch Kraftfahrzeugtechnik*. Wiesbaden: Springer Fachmedien Wiesbaden, 2013.

On the vehicle state estimation benefits of smart tires

V. Mazzilli[1,2], D. Ivone[3], V. Vidal Muñoz[1,4], S. De Pinto[2] , P. Camocardi[2], L. Pascali[2], A. Doria Cerezo[4], P. Gruber[1], G. Tarquinio[2], A. Sorniotti[1]

[1] University of Surrey, United Kingdom
[2] McLaren Automotive Ltd, United Kingdom
[3] Pirelli & C. S.p.A., Italy
[4] Universidad Politécnica de Cataluña, Spain

Abstract. Smart tires are systems that are able to measure temperature, inflation pressure, footprint dimensions, and, importantly, tire contact forces. The integration of this additional information with the signals obtained from more conventional vehicle sensors, e.g., inertial measurement units, can enhance state estimation in production cars. This paper evaluates the use of smart tires to improve the estimation performance of an Unscented Kalman filter (UKF) based on a nonlinear vehicle dynamics model. Two UKF implementations, excluding and including smart tire information, are compared in terms of estimation accuracy of vehicle speed, sideslip angle and tire-road friction coefficient, using experimental data obtained on a high performance passenger car.

Keywords. Sideslip angle estimation, vehicle speed estimation, tire-road friction coefficient estimation, smart tire system.

1 Introduction

Modern vehicle dynamics controllers are improved by adding more information about vehicle states. The inputs from multiple sources are processed by state estimators to predict vehicle variables that are not easily measurable, such as the sideslip angle at the center of gravity, and the vehicle speed during high tire slip ratio conditions. The recent literature mostly adopts estimation algorithms based on Kalman filters using dynamic models of the vehicle. Despite the good accuracy provided by Extended Kalman filters, nowadays Unscented Kalman filters (UKFs) are widely adopted for their superior robustness with respect to sampling rates and approximation errors, for the same computational effort [1]. UKFs have also been proven to be effective in estimating vehicle parameters, in addition to vehicle motion states, see [2]–[4].

A common feature of the available model based state estimation implementations is to consider the tire as a passive component; the tire-road interaction is usually described by numerical models, which present significant inaccuracies and parameter dependencies that can affect the quality of the estimation. In this context, the introduction of tire sensing systems ([5], [6]) offers the possibility of adding a direct feedback contribution to enhance state estimation robustness with respect to the inaccuracies of the tire force models. These solutions open promising scenarios in terms of vehicle state estimation

P. E. Pfeffer (ed.), *11th International Munich Chassis Symposium 2020*, Proceedings, https://doi.org/10.1007/978-3-662-63193-5_34

and control system development, see [7]– [10].

This paper addresses the benefits of the integration of smart tires in a dynamic model based state estimator, by experimentally comparing two UKF algorithms, including and excluding the information from a smart tire system, during extreme handling maneuvers. The points of novelty are:

- The implementation of smart tire technologies developed by Pirelli, namely the Cyber™ Tyre system, in a high performance passenger car;
- The design of a novel Cyber™ Tyre based UKF, i.e., the UKF-CT, which includes the Cyber™ Tyre signals as feedback contribution in the measurement update phase of the filter;
- The optimization of the estimator parameters to achieve a fair comparison between the two considered estimator designs, including and excluding the Cyber™ Tyre inputs;
- The objective evaluation of the resulting vehicle speed and sideslip angle estimation, together with the road condition factor as additional filter parameter;
- The analysis of the enhanced estimation performance and robustness associated with the longitudinal and vertical tire force information on each vehicle corner, provided by the Cyber™ Tyre system.

The paper is organized as follows: Section 2 is an overview on Cyber™ Tyre system technology; Section 3 describes the state estimation architecture; finally, Section 4 describes the tuning procedure, and shows the comparison results between the UKF-CT and a state-of-the art estimator not using the Cyber™ Tyre inputs, which is referred to as UKF in the remainder.

2 Cyber™ Tyre system technology

In vehicle dynamics, one of the most promising innovations is to use the tires as sensing systems to provide useful information to the vehicle controllers. In fact, the tire is in a very privileged position being the only part of the vehicle that is in contact with the road. The smart tire concept refers to a tire equipped with sensors and digital computing systems for monitoring thermal and mechanical parameters that may be shared with the vehicle control units while driving. The possibility of measuring important variables directly from the tires offers the potential for significantly improving the performances of existing active control systems, and developing new control strategies for enhanced safety and efficiency.

Pirelli Tyre S.p.A. is proposing technologies for embedding sensors in the inner liner of the tire, to make the tire itself an active element of the vehicle. The adopted sensors are tiny and light, and have minimal impact on the production process. They have a wide bandwidth, high reliability, and the robustness required to resist the impulses occurring when the sensor enters and exits the footprint. The Cyber™ Tyre project has the main purpose of developing an innovative sensor able to measure the relevant variables through a tri-axial accelerometer, to condition the signals, and to transmit them to a receiver located on the vehicle. Another crucial aspect is the elaboration of the collected signals during the tire rolling motion, by adopting ad-hoc algorithms to extract useful information.

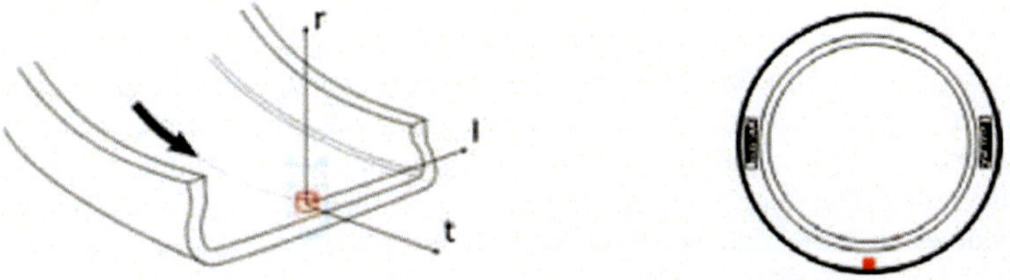

Fig. 1. Sensor reference frame: radial (r), tangential (t) and lateral (l) axes [7].

Among the available Cyber™ Tyre features, this study uses the normal and longitudinal tire force information that is provided once per wheel rotation, as each tire has one sensor placed on the inner liner, as shown in Fig. 1. The generation of these quantities requires specific algorithms developed through preliminary indoor calibration tests, where the extrapolation of relevant parameters is possible in controlled environment.

The Cyber™ Tyre system capability of estimating the normal and longitudinal contact forces is shown in Fig. 2, during a portion of an experimental vehicle test in a handling scenario involving considerable longitudinal and lateral accelerations. The comparison between the forces measured by a commercial dynamometric hub and the corresponding values from the Cyber™ Tyre system highlights the smart tire ability to provide reliable estimation of the contact forces.

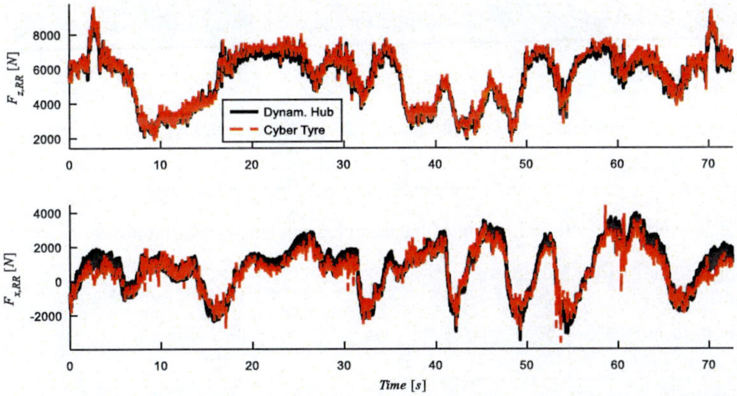

Fig. 2. Comparison between the vertical and longitudinal forces on the rear right wheel, respectively $F_{z,RR}$ and $F_{x,RR}$, measured by a dynamometric hub and generated by the Cyber™ Tyre system during a typical handling maneuver.

3 State estimation architecture

This section describes the vehicle model embedded in the proposed UKF, and the UKF algorithms for state and parameter estimation.

3.1 Vehicle model

The UKF implementations of this study use a 7-degree-of-freedom (7DOF) planar two-track model of the case study rear-wheel-drive high performance passenger car, which considers the longitudinal, lateral and yaw dynamics of the vehicle, as well as the wheel dynamics. The heave and pitch dynamics are not included, whilst the roll angle is considered statically dependent on the lateral acceleration and suspension characteristics.

Fig. 3 shows the vehicle schematic, with positive directions of the vectors, according to the ISO convention [11]. In the following formulations, the subscript i, with $i = F, R$, indicates the axle, while the subscript j, with $j = L, R$, indicates the vehicle side. In the figure, V is the vehicle velocity, having longitudinal and lateral components v_x and v_y; β is the sideslip angle; $F_{x,ij}$ and $F_{y,ij}$ are the longitudinal and lateral tire forces; δ_{Fj} is the steering angle at the front wheels; $\dot{\psi}$ is the vehicle yaw rate; t_i is the track width; and l_i is the longitudinal distance between the center of gravity and the axle. In addition to the steering angle, the model inputs are the engine and braking torque values, M_{ICE} and $M_{B,ij}$.

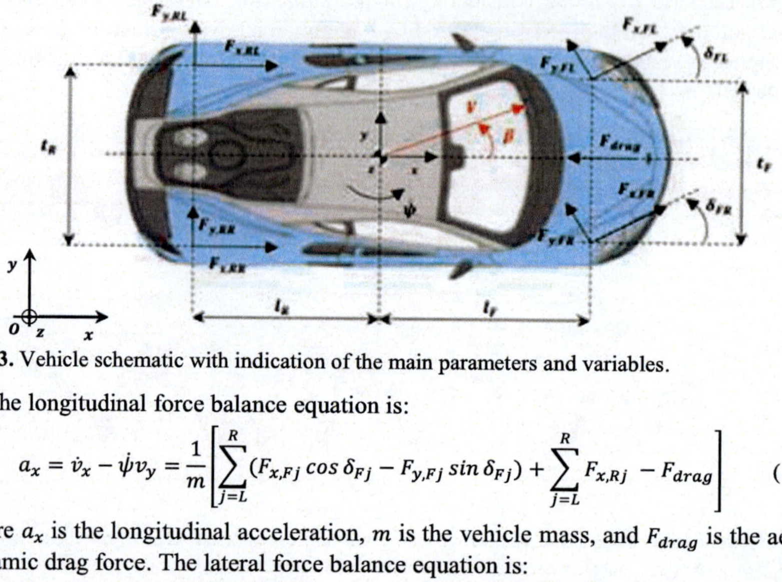

Fig. 3. Vehicle schematic with indication of the main parameters and variables.

The longitudinal force balance equation is:

$$a_x = \dot{v}_x - \dot{\psi} v_y = \frac{1}{m}\left[\sum_{j=L}^{R}(F_{x,Fj}\cos\delta_{Fj} - F_{y,Fj}\sin\delta_{Fj}) + \sum_{j=L}^{R} F_{x,Rj} - F_{drag}\right] \tag{1}$$

where a_x is the longitudinal acceleration, m is the vehicle mass, and F_{drag} is the aerodynamic drag force. The lateral force balance equation is:

$$a_y = \dot{v}_y + \dot{\psi} v_x = \frac{1}{m}\left[\sum_{j=L}^{R}(F_{x,Fj}\sin\delta_{Fj} + F_{y,Fj}\cos\delta_{Fj}) + \sum_{j=L}^{R} F_{y,Rj}\right] \tag{2}$$

where a_y is the lateral acceleration. The yaw moment balance equation is:

$$\ddot{\psi} = \frac{1}{J_z}\left[l_F \sum_{j=L}^{R} (F_{x,Fj} sin\delta_{Fj} + F_{y,Fj} cos\delta_{Fj}) \right.$$
$$- l_R \sum_{j=L}^{R} F_{y,Rj} - \frac{t_F}{2}(F_{x,FL}\cos\delta_{FL} - F_{x,FR}\cos\delta_{FR})$$
$$+ \frac{t_F}{2}\left(F_{y,FL}\sin\delta_{FL} - F_{y,FR}\sin\delta_{FR}\right) - \frac{t_R}{2}(F_{x,RL} - F_{x,RR})$$
$$\left. + \sum M_{z,ij} \right] \tag{3}$$

where $M_{z,ij}$ is the self-aligning moment of the tires. The vehicle speed and sideslip angle are given by:

$$V = \sqrt{v_x^2 + v_y^2}\,; \quad \beta = \tan^{-1}\left(\frac{v_y}{v_x}\right) \tag{4}$$

The Pacejka Magic Formula 2002 calculates the rolling radius, laden radius, rolling resistance, longitudinal force, lateral force, and aligning moment of the tires, namely $R_{e,ij}$, $R_{l,ij}$, $M_{y,ij}$, $F_{x,ij}$, $F_{y,ij}$ and $M_{z,ij}$, as functions of the slip angle α_{ij}, longitudinal slip ratio $\sigma_{x,ij}$, vertical load $F_{z,ij}$, and camber angle γ_{ij}, which depends on the static roll angle. The tire slip ratio formulation resembles the one in [12]:

$$\sigma_{x,ij} = \frac{\omega_{ij} R_{e,ij} - V_{w_{x,ij}}}{\max\left(\left|V_{w_{x,ij}}\right|, \eta v_{x,marg}\right)} \tag{5}$$

where $V_{w_{x,ij}}$ is the longitudinal component of the peripheral wheel speed; ω_{ij} is the angular wheel speed; $v_{x,marg}$ is the marginal speed, namely the minimum speed ensuring system stability when the vehicle model reaches a low speed; and $\eta = 1.1$ is a safety coefficient. This formulation facilitates the removal of the wheel speed oscillations at very low speed, which are typical numerical issues of the conventional slip ratio formulations.

The calculation of the vertical tire loads on each corner, $F_{z,ij}$, includes consideration of the longitudinal and lateral load transfers and aerodynamic downforce contributions. The lateral load transfers are evaluated in static conditions, considering the lateral acceleration and roll moment distribution between the front and rear axles.

Each wheel dynamics are described by a moment balance equation, which, for the rear wheels, is:

$$J_{w_{eq,Rj}} \dot{\omega}_{Rj} = 0.5\, M_{ICE} i_{tr} \eta_{tr} - M_{B,Rj} - F_{x,Rj} R_{l,Rj} - M_{y,Rj} \tag{6}$$

where $J_{w_{eq,Rj}}$ is the equivalent mass moment of inertia of the wheel, including the powertrain inertia; i_{tr} is the transmission gear ratio; and η_{tr} is the transmission efficiency.

3.2 Unscented Kalman filter

The UKF ([1], [13]) is an iterative algorithm including: i) a time update step, in which the set of nonlinear equations describing the process are subject to forward Euler integration; and ii) a measurement update step, in which the data from the available sensors are used to estimate the n states of the system [14].

Within the UKF algorithm, the nonlinear model of Section 3.1 is re-arranged as:

$$\begin{cases} x_k = f(x_{k-1}, u_k, w_{k-1}) \\ y_k = h(x_k, u_k, v_k) \end{cases} \tag{7}$$

where f describes the system dynamics; h is the measurement model; $x_k \in \mathbb{R}^n$ is the state vector at the time step k, assumed to have Gaussian probability distribution $P_k \in \mathbb{R}^n$; $u_k \in \mathbb{R}^r$ and $y_k \in \mathbb{R}^m$ are the input and output vectors; and the random variables $w_{k-1} \in \mathbb{R}^n$ and $v_k \in \mathbb{R}^m$ are the process and measurement noise vectors. The noise vectors are assumed to be uncorrelated with zero-mean Gaussian probability distributions $Q \in \mathbb{R}^{n\times n}$ and $R \in \mathbb{R}^{m\times m}$.

To avoid neglecting any odd-moment information, the UKF algorithm uses an augmented state vector, x_{aug}, with size $N = 2n + m$, and the corresponding augmented error system covariance matrix, P_{aug}, see [1], [2]:

$$x_{aug} = \{x_{k-1} \;\; w_{k-1} \;\; v_k\}^T \; ; \quad P_{aug} = \begin{bmatrix} P_{k-1} & 0 & 0 \\ 0 & Q & 0 \\ 0 & 0 & R \end{bmatrix} \tag{8}$$

As the noise components are uncorrelated, the covariance matrices are diagonal. Ideally, Q and R should change at each iteration because the uncertainties related to the process and sensors might vary. However, unlike P_{k-1}, Q and R are kept constant to reduce the computational load, and aid the estimation of the error covariance P_k and the Kalman gain K_k to quickly converge and stabilize [14].

3.3 Parameter estimation

In vehicle dynamics, the main issue of the model based state estimation approach is its strong dependency on varying vehicle parameters, i.e., vehicle mass, center of gravity position, and tire parameters, including the tire-road friction coefficient. In particular, tire parameter accuracy is crucial in determining the contact forces. In [13], parameter estimation is accomplished by extending the augmented state vector with a new state-space representation during the time update step. The new state, i.e., the estimated parameter, is described by a random walk model, a stationary process only driven by the corresponding process noise.

In this study, similarly to [7], the estimated parameter is the road condition factor, μ_{max}. Its dynamics are described by:

$$\dot{\mu}_{max} = 0 \rightarrow \mu_{max,k} = \mu_{max,k-1} + w_{\mu_{max},k-1} \tag{9}$$

where $w_{\mu_{max},k-1}$ is the associated process noise. This factor constitutes an additional degree of freedom, increasing the vehicle model robustness by scaling the tire-road friction coefficient in the Pacejka tire model formulation [2].

3.4 Unscented Kalman filter implementation

Fig. 4 shows the schematic of the UKF-CT implementation, including the Cyber™ Tyre measurements. The main objective of the UKF-CT is to obtain the values of the most relevant vehicle states, i.e., the vehicle sideslip angle at the center of gravity, β, and the vehicle speed, V, for a wide range of operating conditions. The state and input vectors resemble those of the UKF described in [2], which was implemented on a real car by Bosch. The output vector includes the variables that are available on the controller area network (CAN) bus of the vehicle, see [3], [4]. Moreover, due to the availability of the estimated longitudinal tire forces, $F_{x,ij}$, and vertical loads, $F_{z,ij}$, output by the Cyber™ Tyre system, the vector y is augmented accordingly. Hence, the resulting vectors are:

$$\begin{aligned} x &= \{v_x \quad v_y \quad \dot{\psi} \quad \omega_{ij} \quad \mu_{max}\} \\ u &= \{\delta_{Fj} \quad M_{ICE} \quad M_{B,ij}\} \\ y_{UKF} &= \{a_x \quad a_y \quad \dot{\psi} \quad \omega_{ij}\} \\ y_{UKF-CT} &= \{y_{UKF} \quad F_{z,ij} \quad F_{x,ij}\} \end{aligned} \tag{10}$$

where the subscripts UKF and $UKF-CT$, used here and in the remainder, refer to the estimators excluding and including the smart tire system estimations. The angular wheel speeds, ω_{ij}, allow the estimation of vehicle speed; the additional measurement of a_x, not included the implementation in [2], helps prevent performance degradation in conditions of significant longitudinal tire slip ratios.

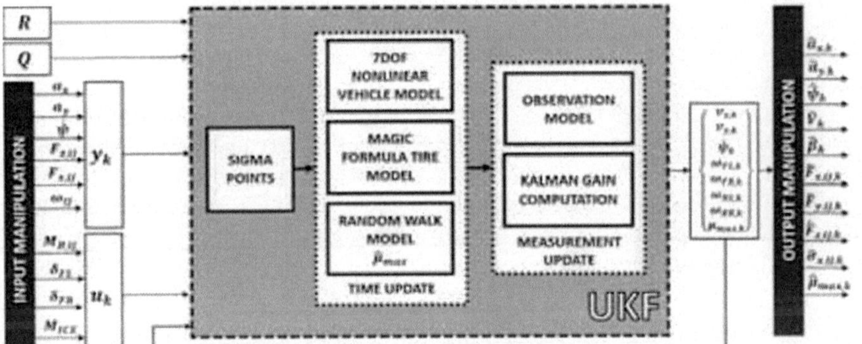

Fig. 4. Schematic of the UKF algorithm for vehicle state estimation.

The measurement noise covariance R, i.e., the variance of the sensors, was determined through the a-priori analysis of sample measurements [14], whilst the matrix Q is a tuning parameter that takes model uncertainties into account. To reduce the number of tuning parameters, the initial error system covariance matrix is set to be equal to Q, as in [4]. The additional parameters are α_{UT}, a constant defining the spread of the sigma points, which is set to 1 in the proposed implementations; β_{UT}, a constant related to the type of probability distribution, equal to 2 for Gaussian distribution; and κ_{UT}, a scaling parameter, set to 1 [1].

4 Results

This section deals with the experimental comparison of the UKF and UKF-CT, and the analysis of the improvements deriving from the addition of the Cyber™ Tyre force outputs. The experimental activity was performed on a McLaren 570s equipped with sensorized Pirelli P Zero Corsa tires, as well as an OxTS RT unit and a 6D IMU, to directly measure the sideslip angle at the center of gravity, vehicle speed, and longitudinal and lateral accelerations.

The two filters were tuned through the optimization routine described in Section 4.1, and are compared by using the experimental measurements acquired during track tests covering a wide range of driving scenarios. The signals to perform the measurement updates within the filters are those available on the CAN bus of the vehicle, whilst the acquired signals from the OxTS RT unit and 6D IMU are used as reference to check the estimation performance.

4.1 Tuning routine

The filter tuning routine uses an optimization process to find the values of the diagonal elements of Q for both filters. The rationale is to obtain two sets of covariances, one for the UKF and one for the UKF-CT, which minimize optimality criteria, to achieve an objective comparison of the estimation performance.

Due to the complexity and important nonlinearities, a multi-objective genetic algorithm has been selected, minimizing three objective functions based on the root mean square error ($RMSE$) between: i) the data acquired from the OxTS RT unit and 6D IMU, in a portion of a handling circuit in dry conditions; and ii) the estimation outputs, obtained by feeding the filters with the required CAN bus signals during the same training maneuver.

The values of the three functions are normalized to build a 3D Pareto frontier. The selected optimum refers to the set of covariances corresponding to the closest point of the Pareto frontier to the origin of the 3D system. The optimization problem is defined as:

$$\begin{aligned}
&\min_{\arg Q} J_\beta \quad s.t. \quad Q_i \in [Q_{i,min}, Q_{i,max}], \quad J_\beta = \sqrt{\frac{1}{N_s}\sum_{i=1}^{N_s}(\beta-\hat{\beta})^2} \\
&\min_{\arg Q} J_{a_x} \quad s.t. \quad Q_i \in [Q_{i,min}, Q_{i,max}], \quad J_{a_x} = \sqrt{\frac{1}{N_s}\sum_{i=1}^{N_s}(a_x-\hat{a}_x)^2} \\
&\min_{\arg Q} J_{a_y} \quad s.t. \quad Q_i \in [Q_{i,min}, Q_{i,max}], \quad J_{a_y} = \sqrt{\frac{1}{N_s}\sum_{i=1}^{N_s}(a_y-\hat{a}_y)^2}
\end{aligned} \tag{11}$$

where J_β, J_{a_x} and J_{a_y} are the objective functions, considering the tracking performance in terms of sideslip angle, longitudinal acceleration, and lateral acceleration; N_s is the number of samples; $Q_{i,min}$ and $Q_{i,max}$ define the range of variation of each element Q_i of Q; and the estimated quantities are indicated with '^'.

The state vector is initialized as:

$$x_0 = \{0.95V \quad 0 \quad 0 \quad 0.95\omega_{FL} \quad 0.95\omega_{FR} \quad 0.95\omega_{RL} \quad 0.95\omega_{RR} \quad 0.7\} \tag{12}$$

where the initial vehicle and wheel speeds are 95% of the acquired CAN signals. The initial conditions for β and $\dot{\psi}$ are set to 0, as the vehicle usually starts from a straight trajectory. Moreover, by considering an initial road condition factor of 0.7, the estimator is brought relatively far from the ideal scenario in terms of speed and parameter estimation, and thus the optimization also tends to improve the convergence features of the estimator.

The optimal covariances computed for the specific dry handling scenario were used to obtain the results in the following sections.

4.2 Selected maneuvers and key performance indicators

The selected maneuvers for state estimation performance comparison are:

- Handling circuit lap in dry conditions. Note that this is a different dataset from the one used for the definition of the optimal covariances in Section 4.1;
- Handling circuit lap in wet conditions, to analyze how the estimators perform in low-grip scenarios;
- Open-loop step steer [15], to evaluate the response of the estimators during extreme cornering transients.

Four key performance indicators (KPIs) are defined for an objective comparison:

- The root mean square value of the estimation error for the considered states:

$$RMSE = \sqrt{\frac{1}{N_s}\sum_{i=1}^{N_s}\left(\Upsilon - \hat{\Upsilon}\right)^2} \tag{13}$$

  where Υ is the real measurement, and $\hat{\Upsilon}$ is the corresponding estimated value;
- The difference (in percentage) between $RMSE_{UKF-CT}$ and $RMSE_{UKF}$:

$$\Delta RMSE_{\%} = \frac{RMSE_{UKF-CT} - RMSE_{UKF}}{RMSE_{UKF}} 100 \tag{14}$$

- The maximum absolute value of the estimation error:

$$e^{max} = max\left|\Upsilon - \hat{\Upsilon}\right| \tag{15}$$

- The difference (in percentage) between e^{max}_{UKF-CT} and e^{max}_{UKF}:

$$\Delta e^{max}_{\%} = \frac{e^{max}_{UKF-CT} - e^{max}_{UKF}}{e^{max}_{UKF}} 100 \tag{16}$$

4.3 Handling circuit lap in dry conditions

The estimation results along a portion of a race track in dry conditions are shown in Fig. 5, while the KPIs are reported in Table 1. The initial μ_{max} was set to 1.1, typical value for the dry tarmac of a vehicle proving ground, see [16].

Both filters provide good performance, with the UKF-CT showing lower $RMSE$ values on the main estimated quantities. The vehicle speed estimation is in line with the target, set by the involved car maker, of not exceeding a $RMSE$ value of 3% of the top speed, with maximum error values of 4.9 km/h and 3.3 km/h for the UKF and UKF-CT. Furthermore, for both estimators, the maximum sideslip angle estimation error remains within the limits mentioned by Grip et al. [16], i.e., within ~1 deg, which allows effective operation of active chassis controllers.

The profiles of the road condition factor are rather steady at reasonable levels for both estimators, even though they converge to different values in the second half of the test, i.e., μ_{max} marginally increases for the UKF, and decreases for the UKF-CT. This kind of difference is normal, and is caused by the fact that the algorithms modify μ_{max} to improve the convergence and overall estimation, since the road condition factor dynamics are not described by a formulation based on the physics of the system, but they are only modeled as process noise [2]. Nevertheless, in the UKF-CT, μ_{max} can be considered closer to the real road condition. In fact, once the road condition factor of the UKF-CT varies, the β estimate matches the measurement very well. On the contrary, in the same condition, the UKF underestimates $|\beta|$, and therefore its μ_{max} can be considered overestimated. The conclusion is that the parameter estimation is enhanced by the additional contribution of the tire forces also in nominal dry tarmac scenarios, close to the one adopted for tuning the estimators. This causes the better sideslip angle estimation performance of the UKF-CT.

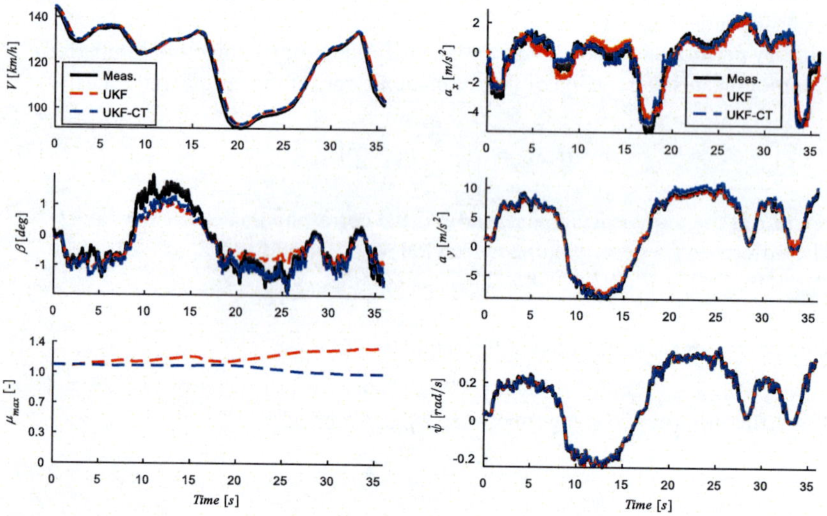

Fig. 5. Estimation results in a handling circuit lap in dry conditions.

Table 1. KPIs for a handling circuit lap in dry conditions.

Variable	Unit	$RMSE_{UKF}$	$RMSE_{UKF-CT}$	$\Delta RMSE_{\%}$	e^{max}_{UKF}	e^{max}_{UKF-CT}	$\Delta e^{max}_{\%}$
V	[km/h]	1.749	1.256	-28.17%	4.910	3.328	-32.22%
β	[deg]	0.363	0.293	-19.34%	1.070	0.957	-10.58%
$\dot{\psi}$	[rad/s]	0.003	0.006	98.36%	0.012	0.029	135.84%
a_x	[m/s^2]	0.526	0.390	-25.76%	2.277	2.790	22.50%
a_y	[m/s^2]	0.433	0.369	-14.80%	1.623	1.176	-27.54%

4.4 Handling circuit lap in wet conditions

The second scenario is a track lap with wet tarmac, in which μ_{max} is expected to be ranging from 0.5 to 0.7. The maneuver was selected to analyze filter robustness when the vehicle operates in low-grip conditions and at high $|\beta|$ values, reaching a maximum of 20 deg (Fig. 6). The initial condition for the estimated μ_{max} was purposely set to 1, to check the estimation convergence.

In such extreme handling conditions, the tire model approximations can lead to inaccuracies that increase the process uncertainty. To cope with this criticality, previous literature implemented adaptive strategies to vary the covariances whenever the process or sensors become unreliable due to the driving conditions [17]. However, as the scope of this research is to analyze the robustness improvements caused by the $F_{x,ij}$ and $F_{z,ij}$ measurements, the filter calibration was kept the same as for the dry handling maneuver.

In Fig. 6, the UKF shows greater oscillation peaks in the a_x and a_y profiles, caused by the overestimation of the road condition factor, which, in the UKF-CT, converges more promptly and to the correct value. Overall, the UKF-CT provides better performance, and is reliable also during the high slip ratio conditions occurring in the time window ~17-19 s, when the vehicle is subject to hard braking after a swift acceleration. Here, the UKF reaches the maximum speed estimation error, due to the loss of reliability of the angular wheel speed signals.

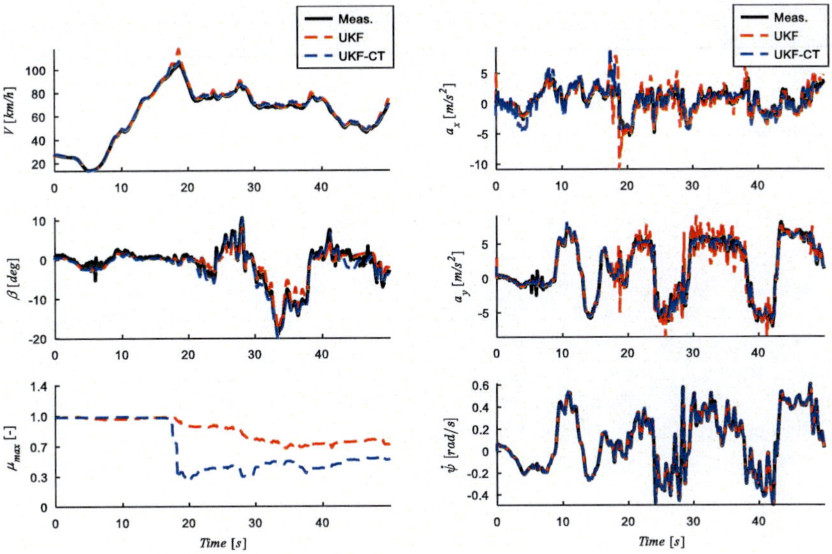

Fig. 6. Estimation results in a handling circuit lap in wet conditions.

In Table 2, the $RMSE_{UKF-CT}$ of the vehicle speed is approximately 1.5 km/h, which corresponds to a -33% $\Delta RMSE_{\%}$. The superiority of the UKF-CT is also noticeable in the sideslip angle estimation, with a $\Delta RMSE_{\%}$ of -11%, and in the longitudinal and lateral accelerations, with $\Delta RMSE_{\%}$ equal to -21.5% and -49%. These estimation errors make the UKF-CT suitable for feedback control applications in large sideslip conditions. In terms of yaw rate, in all considered scenarios the UKF-CT generates higher errors than the UKF; nonetheless, these do not cause concern, as they have limited values with respect to the nature of the maneuver.

Table 2. KPIs for a handling circuit lap in wet conditions.

Variable	Unit	$RMSE_{UKF}$	$RMSE_{UKF-CT}$	$\Delta RMSE_{\%}$	e_{UKF}^{max}	e_{UKF-CT}^{max}	$\Delta e_{\%}^{max}$
V	[km/h]	2.187	1.462	-33.14%	12.946	5.484	-57.64%
β	[deg]	1.397	1.239	-11.28%	4.220	4.715	11.74%
$\dot{\psi}$	[rad/s]	0.007	0.015	110.10%	0.068	0.143	109.35%
a_x	[m/s^2]	1.241	0.974	-21.52%	12.170	7.592	-37.62%
a_y	[m/s^2]	1.055	0.538	-49.02%	8.155	3.069	-62.37%

4.5 Open-loop step steer (ISO 7401, [15])

The last maneuver, i.e., a step steep in dry conditions, allows to evaluate the estimation performance during an extreme cornering transient. The maneuver consists of a first steering wheel angle step with a 100 deg amplitude, followed by an abrupt countersteering action. The initial value of μ_{max} was purposely set to 0.7, which is different from the actual value for the specific experimental scenario.

Grip et al. [16] discussed that when the vehicle is in steady-state conditions, the μ_{max} estimation does not have enough excitation to converge to its final value. Therefore, in this test, the road condition parameter remains constant (Fig. 7) until the step steering input is applied. Fig. 7 shows that the UKF-CT is significantly faster than the UKF in adapting μ_{max}. In fact, the UKF-CT takes 0.72 s to reach 95% of the final μ_{max} value, whilst the UKF takes 2.76 s. This means that the a_x and a_y estimates from the UKF are slower in matching the experimental peak values. For this reason, some authors, such as Piyabongkarn et al. [18], fuse the dynamic model based estimation with the output of kinematic models that are independent from vehicle and tire parameters, and can provide fast state estimation response during transients. However, the important conclusion of this analysis is that good and robust performance in extreme transient conditions can be achieved through the tire force contribution of the UKF-CT, without the addition of any cascade of estimators or adaptive algorithms.

Table 3 shows that the $RMSE$ values for the sideslip angle and the acceleration components are ~30% to ~50% lower for the UKF-CT. On the other hand, the vehicle speed estimation performance is comparable, and the yaw rate trend is well described in both cases, even if the error is smaller for the UKF.

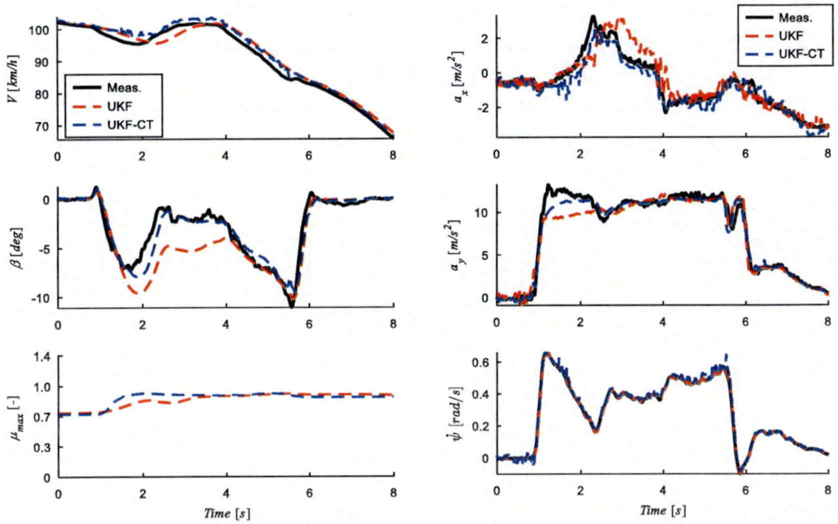

Fig. 7. Estimation results in a step steer maneuver in dry conditions.

Table 3. KPIs for a step steer maneuver in dry conditions.

Variable	Unit	$RMSE_{UKF}$	$RMSE_{UKF-CT}$	$\Delta RMSE_{\%}$	e^{max}_{UKF}	e^{max}_{UKF-CT}	$\Delta e^{max}_{\%}$
V	[km/h]	2.009	1.947	-3.08%	3.733	4.563	22.24%
β	[deg]	2.004	0.916	-54.29%	5.413	2.912	-46.20%
$\dot{\psi}$	[rad/s]	0.007	0.021	184.06%	0.040	0.120	198.66%
a_x	[m/s²]	0.711	0.430	-39.51%	2.723	1.547	-43.18%
a_y	[m/s²]	1.161	0.711	-38.76%	3.566	2.783	-21.97%

Conclusions

The paper described a novel estimator, the UKF-CT, of the key dynamic states for vehicle dynamics control, and discussed a selection of intermediate results in the context of an extensive activity focused on a high performance passenger car. The main feature of the UKF-CT is the inclusion in the measurement vector of the longitudinal and vertical tire forces from the Cyber™ Tyre system.

The performance of the UKF-CT was compared with that of a state-of-the-art estimator, i.e., a UKF not using the Cyber™ Tyre system inputs, during extreme handling maneuvers with different tire-road friction conditions, carried out with a McLaren 570s. For fairness of comparison, the estimators were tuned through an optimization routine based on a genetic algorithm, to find the process covariances minimizing the estimation errors during dry tarmac operation. The covariances were then kept constant for the evaluation.

The experimental results show that the tire force feedback improves the robustness of the estimation and the speed of the adaptation to different road conditions. As a result, the UKF-CT implementation improves the estimation accuracy of the vehicle states, even during extreme transients and in low friction conditions.

Future developments will explore robust tuning methods and adaptive implementations of the process covariances, and involve experimental testing along a wider range of driving scenarios.

References

[1] E. A. Wan and R. Van Der Merwe, "The unscented Kalman filter for nonlinear estimation," *Adapt. Syst. Signal Process. Commun. Control Symp.*, 2000.

[2] S. Antonov, A. Fehn, and A. Kugi, "Unscented Kalman filter for vehicle state estimation," *Veh. Syst. Dyn.*, vol. 49, no. 9, pp. 1497–1520, 2011.

[3] H. Heidfeld, M. Schünemann, and R. Kasper, "UKF-based state and tire slip estimation for a 4WD electric vehicle," *Veh. Syst. Dyn.*, pp. 1–18, 2019 (in press).

[4] M. Wielitzka, A. Busch, M. Dagen, and T. Ortmaier, "Unscented Kalman Filter for State and Parameter Estimation in Vehicle Dynamics," *Kalman Filters - Theory Adv. Appl.*, In Tech, pp. 56–75, 2018.

[5] F. Braghin, M. Brusarosco, F. Cheli, A. Cigada, S. Manzoni, and F. Mancosu, "Measurement of contact forces and patch features by means of accelerometers fixed inside the tire to improve future car active control," *Veh. Syst. Dyn.*, vol. 44, Issue sup1, pp. 3–13, 2006.

[6] S. C. Ergen, A. Sangiovanni-Vincentelli, X. Sun, R. Tebano, S. Alalusi, G. Audisio, and M. Sabatini, "The tire as an intelligent sensor," *IEEE Trans. Comput. Des. Integr. Circ. Syst.*, vol. 28, no. 7, pp. 941–955, 2009.

[7] E. Sabbioni, D. Ivone, F. Braghin, and F. Cheli, "In-tyre sensors induced benefits on sideslip angle and friction coefficient estimation," *SAE Tech. Pap. 2015-01-1510*, 2015.

[8] K. B. Singh and S. Taheri, "Estimation of tire–road friction coefficient and its application in chassis control systems," *Syst. Sci. Control Eng.*, vol. 3, no. 1, pp. 39–61, 2015.

[9] F. Cheli, E. Leo, S. Melzi, and E. Sabbioni, "On the impact of 'smart tyres' on existing ABS/EBD control systems," *Veh. Syst. Dyn.*, vol. 48, Issue sup1, pp. 255–270, 2010.

[10] K. B. Singh, M. A. Arat, and S. Taheri, "Literature review and fundamental approaches for vehicle and tire state estimation," *Veh. Syst. Dyn.*, vol. 57, no. 11, pp. 1643–1665, 2019.

[11] "Road vehicles – Vehicle dynamics and road-holding ability – Vocabulary," *BS ISO 8855*, pp. 1–52, 2011.

[12] T. Y. Kim, S. Jung, and W. S. Yoo, "Advanced slip ratio for ensuring numerical stability of low-speed driving simulation: Part I – longitudinal slip ratio," *Proc. Inst. Mech. Eng. Part D J. Automob. Eng.*, vol. 233, no. 8, pp. 2000-2006, 2019.

[13] S. Haykin, *Kalman Filtering and Neural Networks*, John Wiley & Sons, 2001.

[14] G. Welch and G. Bishop, "An Introduction to the Kalman Filter," SIGGRAPH, pp. 19–33, 2001.
[15] "Road vehicles – Lateral transient response test methods – Open-loop test methods," *ISO 74012011(E)*, pp. 1–67, 2011.
[16] H. F. Grip, L. Imsland, T. Johansen, J. C. Kalkkuhl, and A. Suissa, "Vehicle Sideslip Estimation: Design, Implementation, and Experimental Validation," *IEEE Control Syst. Mag.*, vol. 29, no. 5, pp. 36–52, 2009.
[17] L. Li, G. Jia, X. Ran, J. Song, and K. Wu, "A variable structure extended Kalman filter for vehicle sideslip angle estimation on a low friction road," *Veh. Syst. Dyn.*, vol. 52, no. 2, pp. 280–308, 2014.
[18] D. Piyabongkarn, R. Rajamani, J. A. Grogg, and J. Y. Lew, "Development and experimental evaluation of a slip angle estimator for vehicle stability control," *IEEE Trans. Control Syst. Technol.*, vol. 17, no. 1, pp. 78–88, 2009.

Road-Load Performance Prediction Limitations by RRC Reduction

Ryosuke Okubo[1]

[1] TOYOTA MOTOR CORPORATION, Japan

Abstract. This paper studied the vehicle running resistance prediction limitations based on tyre Rolling Resistance Coefficient (RRC) reduction. Tyre RRC reduction is one of the approaches used to increase fuel consumption efficiency of automobiles. It is known that the RRC value can vary with loading, velocity and temperature, and that the latter has the largest effect. Current tyre developments challenge the trade-off between maximum braking forcee coefficient (μ) and RRC, and present high thermal sensitivity of tyre tread rubber hysteresis loss property. Since different temperature conditions are ordered in ISO28580 tests which is used for Tyre RRC measurements and Road-Load tests which is used for vehicle running resistance measurements, the RRC is rendered less meaningful for estimations of running resistance than intended. The relation between RRC under Road-Load test conditions (RRC_{RL}) and under ISO test conditions (RRCiso) is investigated with the focus on temperature effect. It is found that the estimation of vehicle running resistance improvement by reducing RRCiso on the current tyre development processes is not representative enough of the improvement that can actually be obtained.

Keywords: rolling resistance, Road-load, tyre RRC temperature dependency.

1 Introduction

To protect the earth environment, it is necessary to reduce the vehicle emissions. Tyre rolling resistance coefficient (RRC) is a part of the vehicle running resistance which influences the vehicle emissions. Therefore, reductions of the RRC are contributed to be low running resistance. The RRC grade [1] of tyres which decided by measurement value by ISO-28580 condition are introduced to be informed the customer who purchase the new vehicles and the replace tyres. In Worldwide harmonized Light vehicles Test Procedure (WLTP) [2], Vehicle running resistance and CO2 emissions could be interpolated by the tyre RRC grading class (Equation (1,2), Fig.1,2). The tyre RRC are well known as one of the important characteristics about the vehicle energy efficiency.

$$F=f_0+f_1\cdot V+f_2\cdot V^2 \tag{1}$$

$$f_{0i}=f_{0H}-(f_{0H}-f_{0L})\frac{W_H\cdot RRC_i-W_i\cdot RRC_i}{W_H\cdot RRC_H-W_L\cdot RRC_L} \tag{2}$$

P. E. Pfeffer (ed.), *11th International Munich Chassis Symposium 2020*, Proceedings, https://doi.org/10.1007/978-3-662-63193-5_35

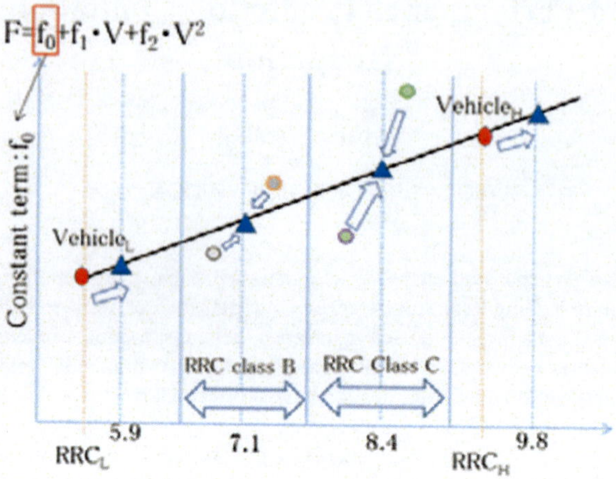

Fig. 1. RRC interpolation method

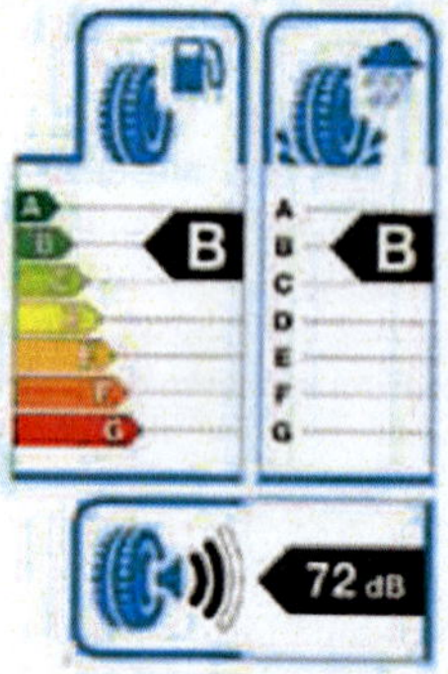

Energy Efficiency Class	C1 class value
A	RRC = 5.9
B	RRC = 7.1
C	RRC = 8.4
D	Empty
E	RRC = 9.8
F	RRC = 11.3
G	RRC = 12.9

Fig. 2. RRC grading and class value [1,2]

To improve an energy loss of vehicle, high efficiency vehicles, such as Electric Vehicle (EV) or Fuel Cell Vehicle (FCV), are getting increased. Contribution of tyre RRC to an energy loss of high efficiency vehicle is larger than internal combustion engines vehicles. Whereby, the reduction of tyre RRC is effective way for the extended electrical cursing distance which is one of the most important characteristics of EV. On the other hands, various tyre technologies such as low hysteresis loss tread compounds and structural change for low mass, are used in order to lower tyre RRC. In particular, the tyre tread compounds has a great influence on the braking and driving characteristics in addition to RRC, so that the tread compound hysteresis loss is designed to keep braking performance and to improve RRC (Fig.3).

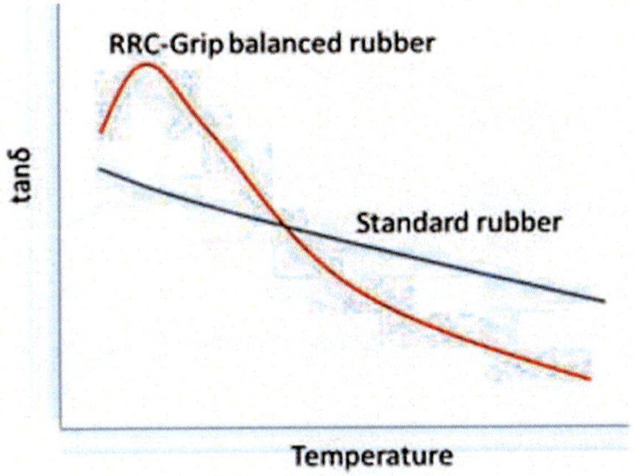

Fig. 3. Tyre tread rubber hysteresis loss properties

The tyre RRC have been reduced for the vehicle low emissions. However, it is necessary to confirm whether this RRC reducing way in the future is effective for the vehicle low emission. If the RRC reducing way is effective, it is important to understand whether there are limitations of the RRC reduction. This paper clarifies these issue by RRC experiments and tyre thermal analysis.

2 Relationship between RRC and Road-load

2.1 Investigation Methods

To survey the relationship between RRC and Road-Load (R/L), some tyres RRC and R/L are measured. There are difficulties from some disturbance, such as ambient temperature and wind speed, to make accurate measurement of actual vehicle. Therefore, R/L measurement accuracy is much worse than RRC bench measurements. So, R/L measured on the bench test machine to except wind and ambient temperature effects and the bench test makes R/L measurement accuracy almost equal the RRC test. The R/L bench test machine are composed of two tyre-wheels and one drum (Fig.4) to reproduce a half of actual vehicle condition. The drum torque is controlled to equal the half of actual vehicle running resistance under installed two tyre-wheels. In this way vehicle running resistance without tyre, such as vehicle internal loss and aerodynamics, are represented as the drum torque. This way is similar to make drum torque control at actual fuel efficiency measurement. Tyre RRC effects on R/L at the same vehicle are confirmed by measured R/L of some tyre under fix the drum torque.

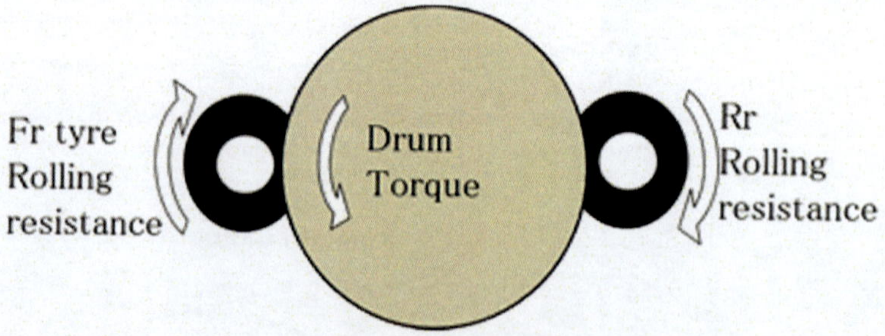

Fig. 4. Bench test machine

2.2 Investigation Results

The vehicle running resistance which is actual measured value of a C segment vehicle is used to the bench test. When drum torque is set, Tyre loads and air inflation pressures are the same condition between actual vehicle test and the bench test to reproduce the actual condition on the bench test. Tyre sizes are used in this measurement from 15inch to 18 inch for the C segment. To confirm the limitation of RRC reduction, tyre inflation pressures were set on 210kPa (ISO standard) to 320kPa. In R/L tests, the tyre inflation Air pressures were set the same as the RRC test condition. The results of these conditions are shown in Fig.5.

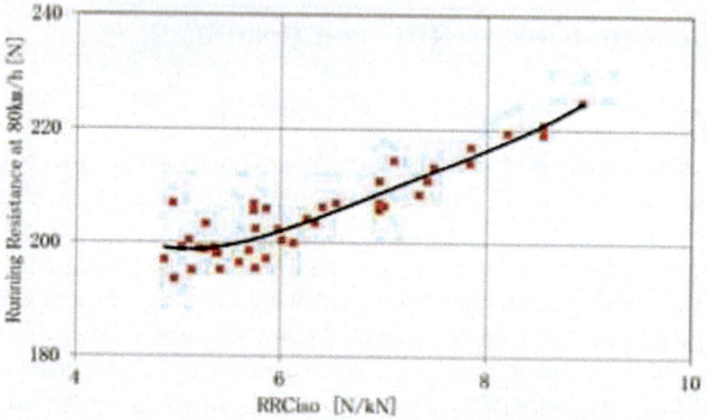

Fig. 5. RRCiso Vs. Running resistance

The lateral axis is RRC which was measured by ISO condition, but different inflation air pressure condition (RRCiso), and the vertical axis is the R/L at 80 km/h which is the same velocity as the RRCiso. The black line in fig.5 is the secondary approximation. In the high RRCiso area, the lower value of RRCiso is, low R/L as known. However, it seems that the effect of RRCiso on R/L became lower in the low RRCiso area. To extract the tyre contribution from the R/L, the similar method is used as in the case of removing parasitic loss at the RRC ISO condition. To compare the same dimension as the RRCiso, the tyre contributed running resistance are divided by the two wheel loads. The tyre RRC during the measured R/L are named as RRC_{RL}, and the relationship between the RRCiso and the RRC_{RL} are shown in Fig.6.

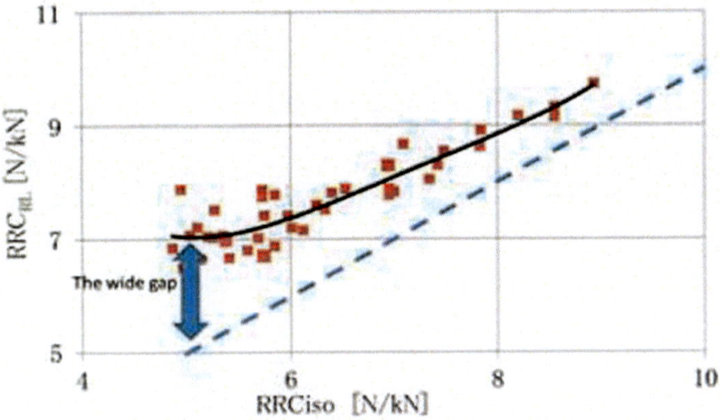

Fig. 6. RRCiso Vs. RRC_{RL}

The vertical axis of Fig.6 is the RRC_{RL}. The black line in Fig.6 is the quadratic approximation, and the blue dash line is the line where the RRCiso equal the RRC_{RL}, and the two lines separate as the RRCiso became lower. This results indicate that the tyre contributions of reduction effects on R/L are reduced even if the RRCiso is lowered.

2.3 Differences between the RRC and the R/L

Theoretically, the RRC reductions should be equivalent for the R/L improvement. In this test, the tyre load and the inflation pressure are the same conditions between the RRC and the R/L test. However, there is a few difference in these protocol. The differences between the RRC and the R/L test are summarized in Table.1.

Table 1. Different terms between RRCiso and R/L

	Ambient Temp	Sequence
RRCiso	25 deg	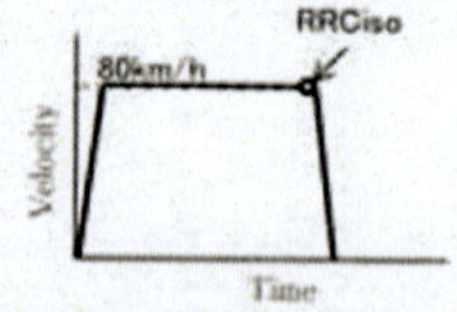 No Interval
R/L (RRC_{RL})	20 deg	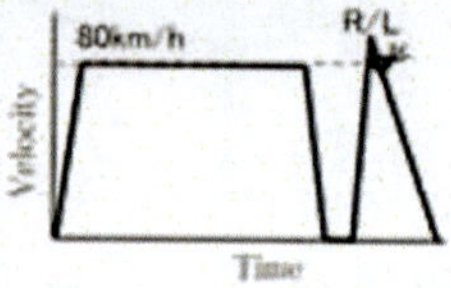 Interval (a few minutes)

First, in the RRC test, the ambient temperature is measured at 25 degree, but the standard temperature of the R/L test is 20 degree. As seen from the above, the RRC tests and the R/L tests have the standard temperature difference of 5 degree. Second, the RRC are measured continuously after warm up, but in the R/L test, generally, there is the sequence of stop, acceleration and coasting after warming up, so that there is an interval of several minutes until it reaches at 80 km/h. The differences cause the tyre temperature differences between the RRC tests and the R/L tests. It is understood that the tyre temperature effected on the RRC characteristics [3-8]. Therefore, it is thought that the RRC gap between the RRCiso and RRC_{RL} is generated from the tyre temperature differences and the RRC temperature dependencies. This RRC gap vary depending on the tyre RRC temperature dependency, but it could not be explained why the gaps grow with lower RRC. Hence, it is investigated whether the gaps could be estimated with a model, depending on the tyre temperatures.

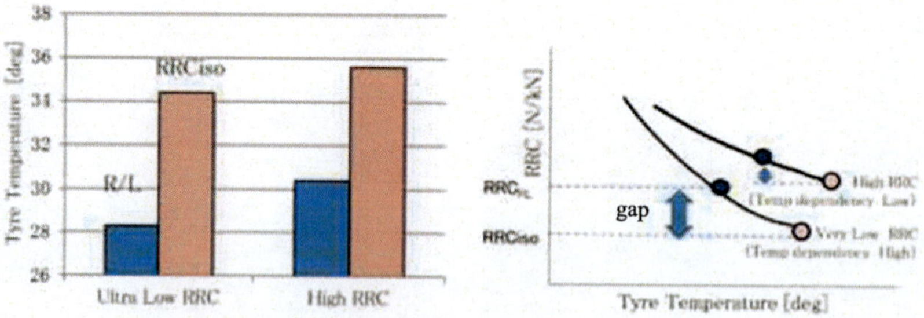

Fig. 7. Tyre surface temperature and RRC variation

3 The Estimated method of RRC$_{RL}$ from RRC Temperature Dependency and RRCiso

3.1 The Estimation Method of Tyre RRC$_{RL}$

It is estimated that the gap between the RRCiso and the RRC$_{RL}$ are caused from the tyre temperature, and a RRC$_{RL}$ estimated method from the RRCiso (Fig.8) was investigated by using the tyre temperature variations and the RRC temperature dependency. Where Ts is a saturated temperature of the RRCiso measurement and the RRC temperature dependency could be expressed as the blak line in Fig.8, and ΔT is the temperature difference between the RRCiso and the RRC$_{RL}$ measurements. And the RRCiso and the RRC$_{RL}$ are represented as the blue point and the red point in Fig.8.

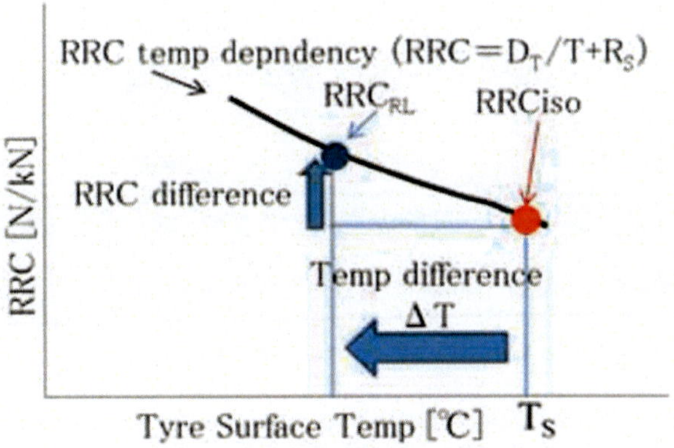

Fig. 8. RRC$_{RL}$ estimating model

Here, it is assumed that RRC temperature dependencies could be expressed as Equation (3).

$$\mathrm{RRC} = \frac{D_T}{T} + R_S \tag{3}$$

Where T is the tyre surface temperature. The larger the coefficient D_T which is represented as the RRC temperature dependency is, the higher RRC temperature dependency is. RRCiso and RRC$_{RL}$ could be described as Equation (4), (5) from Equation (3).

$$\mathrm{RRCiso} = \frac{D_T}{T_S} + R_S \tag{4}$$

$$\mathrm{RRC}_{RL} = \frac{D_T}{(T_S - \Delta T)} + R_S \tag{5}$$

The issues in calculating the RRC_{RL} are the estimation of Ts and ΔT. RRC_{RL} could be transformed as Equation (6) from Equation (4) and (5). So that, calculation of RRC_{RL} needs Ts and ΔT.

$$\mathrm{RRC}_{RL} = \frac{D_T}{(T_S - \Delta T)} - \frac{D_T}{T_S} + RRCiso \tag{6}$$

To calculate the RRC_{RL}, the estimating methods of Ts and ΔT without actual measurements are studied.

3.2 Estimation of Ts and ΔT

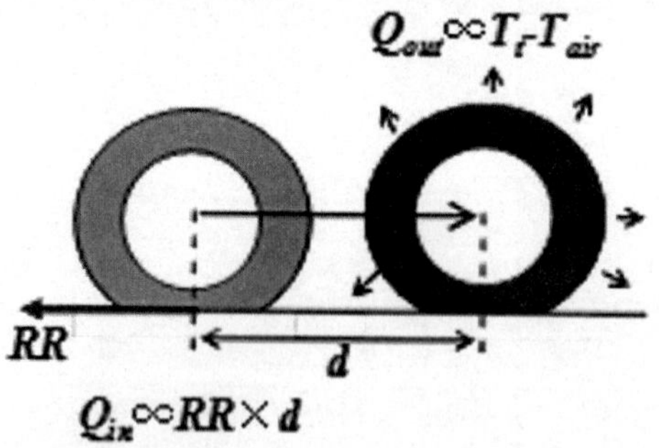

Fig. 9. Simplified Tyre Thermal Model

First, to estimate ΔT, tyre temperature variations are forecasted by used a simplified heat balanced model (Fig.9). In this model, it is assumed that the heat generation during the measurement is only due to the tyre rolling resistance, because braking, driving force and lateral force are not generated during the RRC and the R/L measurements, and also the heat radiation only due to the outside air. In this model, the tyre temperature variation could be expressed by Equation (6), where the first term is expressed the rolling resistance heat generation, and second term is represented the heat radiation to the outside air.

$$\mathrm{dT} = \frac{a \cdot RR}{C} - \frac{b \cdot S \cdot (T - T_{air})}{C} \tag{7}$$

Where, C is tyre heat capacity and S is surface area. Since the temperature difference ΔT between the RRCiso and the RRC_{RL} is composed of the outside temperature difference 5 degree and an effect of the interval of sequence, it is possible to estimate ΔT if the temperature variation of the interval effect is known. The temperature variation during the interval could be described as following equation.

$$\Delta T_{interval} = \frac{Q_{out}^{interval}}{C} \propto \frac{Q_{out}^{interval}}{M} \propto \frac{Q_{out}^{interval}}{RR} \tag{8}$$

Where, $Q^{out}_{interval}$ is the heat radiation of the interval and $\Delta T_{interval}$ is the temperature variation during the interval and M is the tyre mass. The heat capacity C varies depending on tyre structures and tyre tread compounds, etc., but it is assumed that the heat capacity of the tyres are proportional to the mass of the tyre regardless the tyre size, the structure and the tread compound and that the mass is proportional to the tyre rolling resistance (Fig.10).

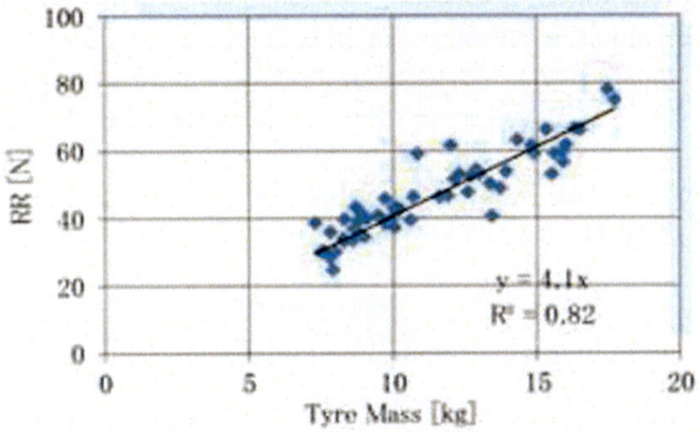

Fig. 10. Relationship between Tyre mass and RR

As the results, it is considered that the temperature variation during the interval is proportional to the -1 power of rolling resistance (RR). Hence, the relationship between the tyre temperature variation and the RR are investigated by some tyres. The results are shown in Fig.11.

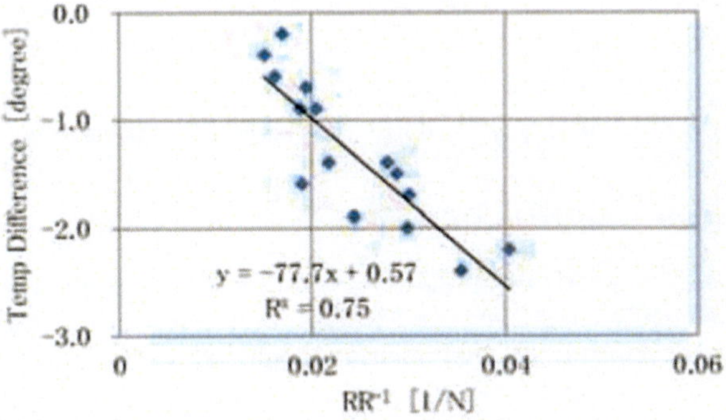

Fig. 11. Relationship between RR and tyre temperature differences

As the results of Fig.11, it is possible to estimate the temperature variation during the interval by used the RR values. The temperature difference, ΔT, could be estimated by adding the interval temperature variation and the 5 degree which is the ambient temperature difference. On the other hands, the saturated tyre temperature during the

RRCiso measurement ,Ts, means that dT,in Equation (7), equal zero, that is, the heat generation and the heat dispersion are balanced. Therefore, Equation (7) could be transformed the following equation:

$$T_S - T_{air} = \frac{a \cdot RR}{b \cdot S} \tag{9}$$

Where it is assumed that S are proportional to the air inflated volumes of tyre, Equation (9) could be described following equation, because of the linear relationship between air inflated volumes and allowed load capacities.

$$T_S - T_{air} = \frac{a \cdot RR}{b \cdot S} \propto \frac{a \cdot RR}{b \cdot W} \propto \frac{a}{b} RRCiso \tag{10}$$

Equation (10) means that Ts are proportional to the RRCiso. The relationship between the Ts and the RRCiso are shown in Fig.12. From this result, it is found that Ts are proportional to the RRCiso.。

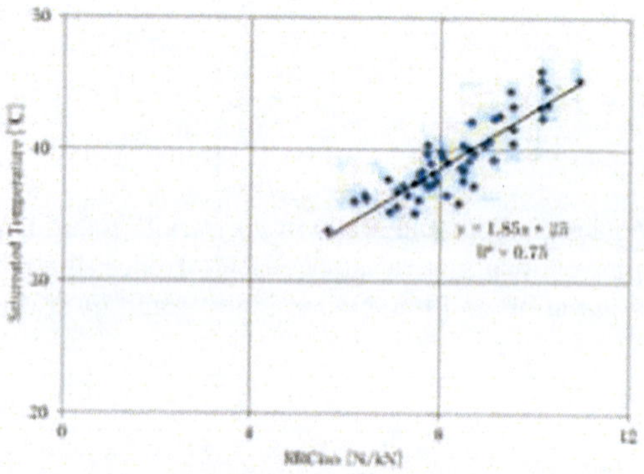

Fig. 12. elationship between RRCiso and saturated temperature

3.3 The comparison between estimated and measured results

From Equation (6), (8) and (10), the RRC_{RL} could be derived from the RRC temperature dependency coefficient D_T and the RRCiso. Therefore, it was investigated how the RRC temperature dependency coefficient D_T could be used to express the relationship between the RRCiso and the RRC_{RL}. Where three levels, high, middle and low, are set from the tyre group whose RRC temperature dependency has been measured, the relationships between the RRCiso and the RRC_{RL} are calculated. The results are shown in Fig.13.

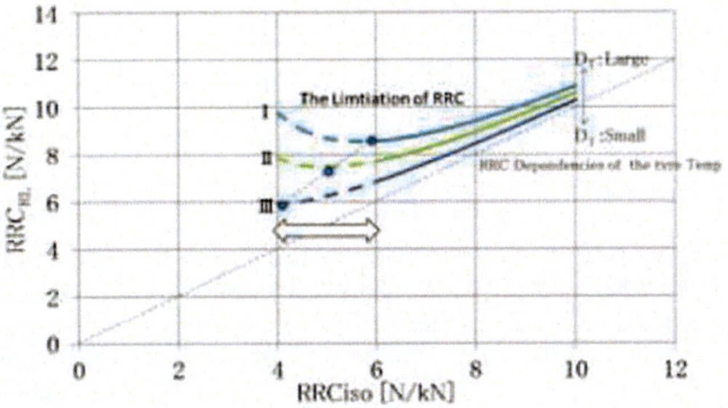

Fig. 13. Estimated RRC_{RL} from RRCiso and RRC dependency of the tyre temperature

In Fig.13, it is recognized that the RRCiso reductions of the high RRC temperature dependency do not improve the RRC_{RL}, but may worsen below RRCiso around 6.0 N/kN. This is because when RRCiso is lower, the saturated temperature decreases (Fig.12) and the RRC temperature dependency coefficient D_T has a large effect. To compare between the estimated results and the measured results on the bench test, RRCiso and RRC_{RL} of the three tyres which have the three levels of the RRC temperature dependency are plotted (Fig.14). Since these plot are on the line which is estimated from the RRC temperature dependencies, thereby the estimated results match the actual measured results. Although the estimated methods have the several assumptions, it has sufficient accuracy to investigate the relationship between tyre characteristics and RRC_{RL}.

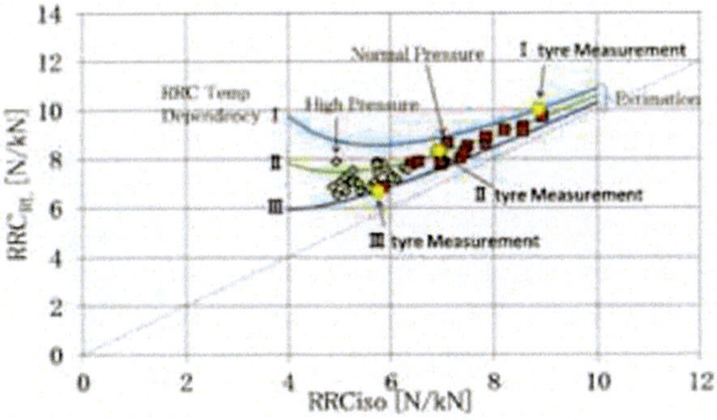

Fig. 14. Comparison between estimated and measured RRC_{RL}

4 Compatibility Limitation of RRC_{RL} and Wet Maximum Braking Force Coefficient

4.1 The conflicted tyre characteristics with the RRC temperature dependency

From the results of Fig. 14, it is found that the RRC_{RL} is effective reduced by the lower RRC temperature dependency and by the RRCiso reduction. Thereby, it is considered that the reducing RRC temperature dependency is essential for improvements of fuel efficiency. On the other hands, the reducing RRC temperature dependency has a bad effect on the tyre maximum braking force coefficient (μ), especially wet condition. The relationship between the RRC temperature dependency and the RRC and wet μ well balanced technology is schematically shown in Fig. 15. In terms of the input frequency, the input in RRC is the low frequency which is tyre rotations, whereas wet μ is the high frequency inputs due to the small unevenness of the road surface. In hysteresis characteristics, the frequency effects could be converted with the temperature effects [9]. Therefore, in terms of the tyre temperature, the input in RRC is higher temperature, whereas the input of the wet μ is the lower temperature. To achieve the well balanced between RRC and wet μ, a current method is the hysteresis loss ($\tan\delta$) curve control, that is, in the RRC (the higher temperature) hysteresis loss, RRCiso, are reduced and in wet μ (the lower temperature) hysteresis loss, $\tan\delta_{wet}$, are maintained or increased. It seems that the RRC temperature dependency is increased. When the RRC temperature dependency is reduced and the RRCiso is maintained, wet μ is decreased because the $\tan\delta_{wet}$ is decreased. Therefore, a compensation technology is required to suppress the decrease of the wet μ.

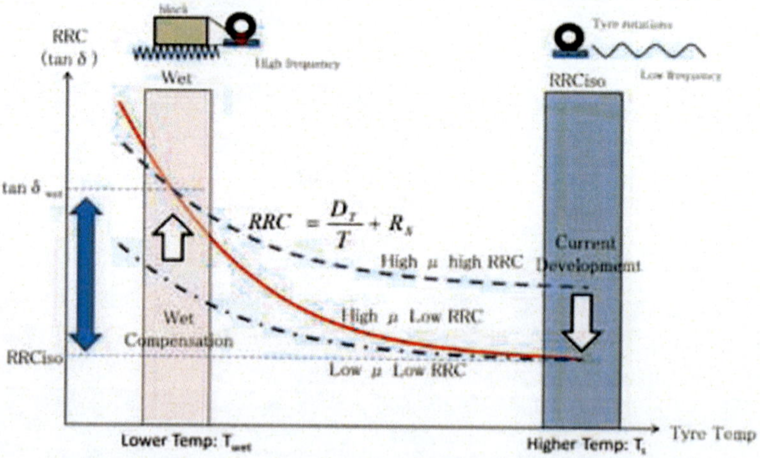

Fig. 15. Effects of RRC thermal dependency on RRCiso and Wet μ

4.2 The Limitation of RRCiso Reduction under Maintaining the Wet μ

Even if the RRCiso unnecessarily lowered, the RRC$_{RL}$ do not always reduce, unless the RRC temperature dependency is taken into the consideration, and in some cases it may worsen. However, the wet μ deterioration due to the reduction of the RRC temperature dependency is inevitable. Therefore, the relationships between the RRC temperature dependency coefficient D_T and the RRCiso and the tan δ_{wet} are examined.

The tan δ_{wet} could be represented as following equation using the wet μ area temperature, T, from the RRC identification equation (2).

$$\tan\delta_{wet} = \frac{D_T}{T_{wet}} + R_S \tag{11}$$

When R_S is deleted from Equation (4) and (11), D_T is expressed as following equation.

$$D_T = \left(\frac{T_S \cdot T_{wet}}{T_S - T_{wet}}\right) \cdot (tan\delta_{wet} - RRCiso) \tag{12}$$

Thus, if the tan δ_{wet} is maintained and the RRCiso is reduced, the RRC$_{RL}$ could be expressed the following equation by plugging in equation (12) for equation (5) .

$$RRC_{RL} = \frac{1}{(T_S - \Delta T)} \cdot \left\{\left(\frac{T_S \cdot T_{wet}}{T_S - T_{wet}}\right) \cdot (tan\delta_{wet} - RRCiso)\right\} + R_S \tag{13}$$

The calculated relationships between the RRC$_{RL}$ and the RRCiso and the tan δ_{wet} by using above equation are shown in Fig.16. From the results of Fig.16, in terms of the RRC$_{RL}$, the limitation of the RRCiso reduction for maintaining the wet μ is around 5.9 N/kN. If RRCiso is reduced below 5.9N/kN, RRC$_{RL}$ might increase. In addition, RRCiso 5.9N/kN matches the class value of grade A.

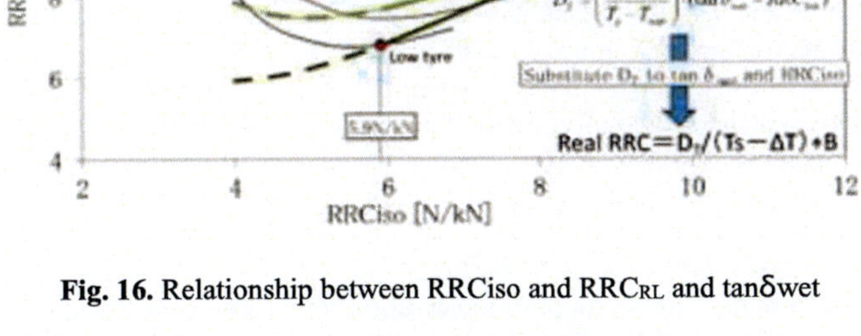

Fig. 16. Relationship between RRCiso and RRC$_{RL}$ and tanδwet

The conventional methods of obtaining the wet μ by tan δ_{wet} result in Fig.16. However, if the wet μ is obtained by another method, the RRC temperature dependency could be further reduced, hence the RRCiso reduction limitation also would be decreased too. Consequently, it is considered that further improvement of fuel efficiency could be achieved by developments of the new technologies which do not depend on the tyre tread compounds hysteresis loss in obtaining the wet μ

5 Conclusion

The following conclusion were obtained from this study.

- Below Tyre RRC 5.9 N/kN, the RRCiso reducing might not be better for Road-load as long as wet μ are maintained.
- RRC temperature dependency improvement is effective for RRC_{RL} reducing but the wet μ deteriorates as a side effect.
- If the wet μ improvement technology which do not use tan δ_{wet} characteristics is established, RRC temperature dependency would be reduced and RRCiso could be also lowered.

Avoiding simplistic reduction of RRCiso and switching to effective reduction of RRC_{RL} could lead to real improvement of fuel efficiency. The next stage of this research is to clarify the effective RRC under actual customer usages that is a wider temperature range.

References

1. REGULATION (EC) No 1222/2009
2. Global technical regulation No. 15 Global technical regulation on Worldwide harmonized Light vehicles Test Procedures (WLTP)
3. Isamu Inoue et al.: Analysis on Influence of Temperature Characteristics of Tire on Rolling Resistance: A Driving on the Road, and Experiment Analysis by New Method Tire Test Equipment. In: Transactions of Society of Automotive Engineers of Japan, 42(6): 193-198 (2011).
4. Kazutaka Yokota at al.: An Estimation of Tire Temperature Distribution and Rolling Resistance Under Running Conditions Including Environmental Factors. In: Proceedings of the Society of Automotive Engineers of Japan, 117(11): 5-8 (2011).
5. Masashi Kitagawa et al.: Study of a Correlation between Tire Temperature and Rolling Resistance during Actual Driving. In: Proceedings of the Society of Automotive Engineers of Japan, 40(11): 11-14 (2011).
6. Takeshi Hotaka at el.: Reasarch on running resistance correction technology focusing on tire temperature. In: Transactions of Society of Automotive Engineers of Japan, 50(2): 411-416 (2019).
7. Takeshi Hotaka at el.: Tire Rolling Resistance Correction Technology Taking Viscoelastics into Consigeration. In: Proceedings of the Society of Automotive Engineers of Japan, No.20195273.

8. Ryosuke Okubo et al.: Effets of Tyre Tread Temperature under Actual Driving on Tyre Characteristics. Transactions of Society of Automotive Engineers of Japan, 44(2): 499-504 (2013).
9. Hideo Sakai.: Tire Engineering. Grand-Prix Pub. Co., Tokyo, Japan (1987).

Extended Evaluation of Tyre Comfort using Wavelet Analysis

Ventseslav Yordanov M.Sc.[1], Univ.-Prof. Dr.-Ing. Lutz Eckstein[1] and Mark Harris M.Sc.[2]

[1] Institute for Automotive Engineering (ika) RWTH Aachen University, Germany
[2] FKA GmbH, Germany

Abstract.
With the emergence of electromobility, demands on driving comfort are changing. Replacing the internal combustion engine with an electric drive train eliminates one of the main sources of vehicle noise and vibration, increasing the influence of those caused by road unevenness. As the only components making contact with the road, tyres filter out or amplify road excitations according to their frequency response. Due to their complex design and multi-material structure, a complete and accurate description of these frequency response characteristics is difficult to achieve. Multibody simulations used in the early phases of a vehicle development process to determine chassis and full vehicle frequency response utilize specialized physical tyre models for this purpose. The most widely used parametric models determine the correct frequency response based on vibration measurements (e.g. cleat tests). State-of-the-art approaches for the evaluation of such measurements are based on the Fourier transform and Power spectral density, which limit the insight into the frequency domain to only the frequency content and its magnitudes. The aim of this paper is to provide a deeper insight into the frequency domain of the tyre vibration by implementing the concept of wavelet analysis. By analyzing real measurement data of tyre cleat tests, some vibration phenomena and the advantages of the proposed methodology will be discussed.

Keywords: Wavelet transform, Synchrosqueezing, Cleat test, Tyre comfort.

1 Introduction

In addition to the power train and the chassis, the tyre is one of the most important components in the virtual vehicle development. As the only component connecting the vehicle with the road surface, tyres influence the driving dynamics and comfort of the vehicle in a wide range. Due to their complex design and multi-material structure, physical and nonlinear mathematical models have been developed to describe and replicate the properties of the majority of these materials. State of the art physical based tyre models [23, 24] use tyre measurements to parameterize the nonlinear behavior under operating conditions without demolishing the specimens. Comfort related aspects of

P. E. Pfeffer (ed.), *11th International Munich Chassis Symposium 2020*, Proceedings, https://doi.org/10.1007/978-3-662-63193-5_36

these models use data from cleat tests in order to parameterize the transient responses. The cleat impulse results in an excitation of a wide frequency band around the tyre natural vibration modes. The model fitting refers to the frequency spectrum obtained from Fourier transform and Power spectral density, while the overall damping behavior is compared in time domain.

The aim of this paper is to introduce a methodology for extended evaluation of tyre comfort data (e.g. cleat tests) with deeper insight into the frequency domain using the wavelet analysis. Researchers in many technical fields [15-22] successfully use time-frequency representations for frequency response investigation and modal parameter identification in multi-layer signals for different purposes. In this work, the synchrosqueezed wavelet transform in combination with ridge reconstruction is proposed for more detailed description of the tyre transient response by separate evaluation of the decomposed tyre natural modes. Section 2 provides an overview of the fundamental mathematical description of both approaches. In Section 3.1, using tyre cleat measurements, the standard Power spectral density approach is compared with the proposed time-frequency based approach. An insight into the evaluation methodology for identifying and extracting tyre comfort features, including a discussion of the vibration phenomena, is given in section 3.2.

2 Frequency- and Time-Frequency Approaches for Data Processing

The tyre comfort characteristics are derived by processing the data recorded during cleat measurements. The following section 2.1 briefly describes the current state of the art frequency based approach used in post-processing of tyre data. A fundamental mathematical description of the proposed time-frequency approach is provided in section 2.2.

2.1 Fourier Transform and Power Spectral Density

This section provides a short overview on the most common mathematical approaches used in frequency based signal analysis. For detailed descriptions on the described approaches, one can refer to [6-10].

The Fourier transform decomposes a signal or time function $x(t)$ to a sum of frequencies represented mathematically by sine and cosine waves. Based on the complex notation of the Fourier series, the transform can be expressed according to Eq. (1).

$$\mathrm{F}(\omega) = \int_{-\infty}^{+\infty} x(t) \cdot \mathrm{e}^{-\mathrm{j}\omega t} dt \tag{1}$$

The value of the transform $|\,\mathrm{F}(\omega)|$ gives the amplitude spectrum, while the phase can be obtained according to Eq. (2).

$$\phi(\omega) = \tan^{-1}\left(\frac{Im(\mathrm{F}(\omega))}{Re(\mathrm{F}(\omega))}\right) \tag{2}$$

If the signal $x(t)$ satisfies the condition given in Eq. (3), the inverse Fourier transformation can be applied for conversion into time domain, as formulated in Eq. (4).

$$\|x\|^2 = \int_{-\infty}^{+\infty} |x(t)|^2 \, dt < \infty \tag{3}$$

$$x(t) = \frac{1}{2\pi} \cdot \int_{-\infty}^{+\infty} \mathrm{F}(\omega) \cdot \mathrm{e}^{\mathrm{j}\omega t} d\omega \tag{4}$$

In practice, $x(t)$ represents a continuous analog signal discretized to a sequence of samples N with sampling period T during measurement. Consequently, the discrete Fourier transform pair can be written as follows

$$F_d(k) = \sum_{n=0}^{N-1} x(n \cdot T) \cdot e^{-\frac{j2\pi kn}{N}}, n = 0,1,2,\ldots N-1, \tag{5}$$

$$x(n) = \sum_{n=0}^{N-1} F_d\left(\frac{k}{T}\right) \cdot e^{\frac{j2\pi kn}{N}}, k = 0,1,2,\ldots N-1. \tag{6}$$

The frequency resolution, defined as the distance between two points in the frequency domain, can be expressed as [11]

$$\Delta f = \frac{1}{N \cdot T}. \tag{7}$$

The Power spectral density describes the power distribution of a signal over a frequency band. This can be formulated using the windowed $F_d(k)$ output, as follows [8]

$$\mathrm{PSD} = \frac{1}{k} \cdot |F_d(k)|^2. \tag{8}$$

The effects of different windowing functions are not discussed here. For more information on the topic, the reader can refer [11, 12].

2.2 Continuous Wavelet Transform and Synchrosqueezed Wavelet Transform

This section provides a brief overview of the fundamentals of wavelet analysis. The idea of the synchronized wavelet technique is explained, starting with the basics of the continuous wavelet transformation. Existing techniques for the reconstruction of signal components from the time-frequency plane are also covered. The reader can refer [1-5] for detailed information on the described mathematical approaches.

In contrast to the Fourier transform, the Wavelet transform decomposes a signal into waves represented by base functions $\psi(t)$ with a mean value of zero. In the literature, various base functions, also called wavelets, are suggested for different purposes, but for the scope of this paper the complex Morlet wavelet is used, given in Eq. (9)

$$\psi(t) = \alpha_{\omega_0} \cdot e^{\left(-\frac{t^2}{2}\right)} \cdot \left[e^{(j\omega_0 t)} - e^{\left(-\frac{\omega_0{}^2}{2}\right)} \right], \tag{9}$$

where α_{ω_0} is a normalizing constant for the wavelet amplitude.

The Continuous wavelet transform in $L^2(\mathbb{R})$ of a signal $x(t)$, satisfying the condition according to Eq. (3), is expressed as

$$W_x(a,b) = \int_{-\infty}^{+\infty} x(t) \cdot \psi_{a,b}^*(t)dt, \quad (10)$$

where $\psi_{a,b}^*(t)$ denotes the complex conjugate of the mother wavelet, given by

$$\psi_{a,b}(t) = \frac{1}{\sqrt{|a|}} \cdot \psi\left(\frac{t-b}{a}\right), \; a, b \in \mathbb{R}, a \neq 0. \quad (11)$$

The frequency content of $\psi_{a,b}(t)$ is controlled by the scaling parameter a. Localization of the frequency content in the time domain is obtained by translating the parameter b. The conversion from the time-scale plane into the time-frequency plane is expressed as

$$a = \frac{\omega_0}{\omega}, \quad (12)$$

where ω_0 represents the mother wavelet frequency. Reconstructions to the time domain are possible, similarly to the inverse Fourier transformation, by integrating $W_x(\omega, t)$ over frequency logarithm.

$$x(t) = \frac{1}{C_\psi} \int_0^{\infty} W_x(\omega, t) \frac{d\omega}{\omega}, \quad (13)$$

where C_ψ represents the admissibility condition as formulated in Eq. (14)

$$C_\psi = \int_{-\infty}^{+\infty} \frac{|\psi(\omega)|^2}{|\omega|} d\omega < \infty. \quad (14)$$

The synchrosqueezing operation intends to increase the resolution of the time-frequency plane obtained by the continuous wavelet transform. This is achieved by reallocating the $W_x(a,b)$ coefficients corresponding to the same phase velocity near the instantaneous frequency, expressed by

$$\omega_x(a,b) = \frac{\frac{\partial W_x(a,b)}{\partial t}}{-i(W_x(a,b))}. \quad (15)$$

The formulation of the synchrosqueezed wavelet transform, with accuracy δ and $W_x(a,b) \neq 0$, is given in Eq. (16).

$$T_x(\omega, b) = \int \delta(\omega_x(a,b) - \omega) W_x(a,b) a^{-\frac{3}{2}} da \quad (16)$$

In contrast to the continuous wavelet transform, there is no direct possibility to reconstruct the signal by integration over the entire frequency range. Instead, integration over each frequency bin is possible and then one can obtain the reconstruction $x^a(t)$ by sum of the results as follows

$$x^a(t) = \sum_k \int_{\sqrt{\omega_{k-1}\omega_k}}^{\sqrt{\omega_k \omega_{k+1}}} T_x(\omega, b) d\omega. \quad (17)$$

Another way to reconstruct components from the time-frequency plane is the ridge reconstruction. The ridge, defined as set of points $(a, b) \in \Omega$ for which $t_s(a, b) = b$, can be extracted from the phase of the wavelet coefficients, as formulated in Eq. (18)

$$\frac{\partial\phi(a,b)}{\partial a} = -\frac{t_s-b}{a^2} \cdot \psi^*_{a,b}\left(\frac{t_s-b}{a}\right), \tag{18}$$

with $\phi(a, b) = arg[T_x(a, b)]$. Once the ridge is obtained, the component can be reconstructed by applying the corresponding inverse wavelet transform.

3 Proposed Approach

The following sections first compare the application of the mathematical approaches described above, followed by an exemplary evaluation of cleat measurements using the proposed approach. In both discussions, data from a standard 205/60 R16 passenger car tyre, measured at the Institute for Automotive Engineering (ika), RWTH Aachen University cleat test bench, is used. The highly rigid rig structure provides a high data fidelity suitable for such investigations.

3.1 Comparison between the Frequency and Time-Frequency Approaches using Tyre Cleat Data

In this section the standard frequency approach based on the Power spectral density (PSD) is graphically compared with the proposed time-frequency representation based on the Synchrosqueezed wavelet transform (WSST). Both approaches use raw data, obtained from cleat measurements. The selected operating conditions produce higher impact energy, which is suitable for representing the capabilities of both approaches.

Starting with the vertical force excited by transversal cleat at 90 [km/h], see Fig. 1, one can clearly recognize the difference in the information amount revealed from both approaches. The time-axis of the WSST time-frequency plane (b) aligns with the x-axis of the force signal (a), while the frequency-axis (b) aligns with the PSD y-axis (c). The amplitude of the PSD x-axis (c) corresponds to the WSST color scale (b). The range between the vertical dashed black lines in (a) and (b) approximate the duration of the cleat event, based on the footprint relative to the measurement environment.

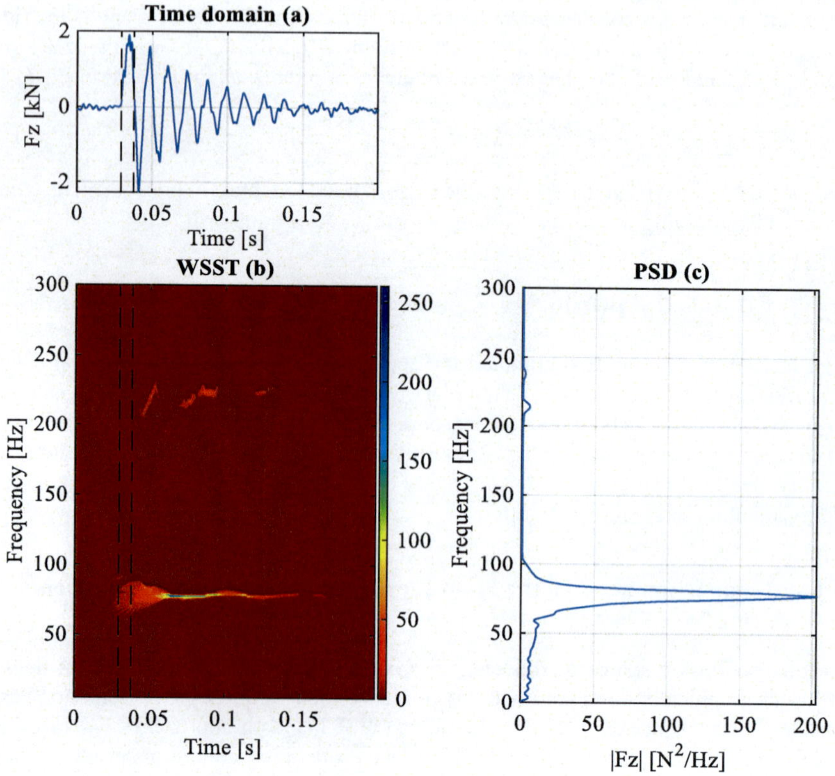

Fig. 1. Vertical force signal – transversal cleat (10 x 20 mm), 100% LI, 90 [km/h]. (a) Raw signal, (b) Synchrosqueezed wavelet transform, (c) Power spectral density.

For the depicted measurement conditions, both approaches indicate the excited vertical vibration mode at ~75 [Hz] as well as the first air cavity mode at 200-250 [Hz], with the corresponding magnitudes. The difference in the magnitude scales comes from the mathematical background of the approaches. The WSST evaluates the energy content of the signal in both time and frequency, while the PSD does so only in frequency perspective. In addition, the WSST offers a time localization of the excited modes with their maximum amplitudes. One can observe the sequence of the excited frequencies during the impact, how they modulate and concentrate together after the impact, as well as their attenuation.

The vertical force frequency content is a rather simple example compared to the longitudinal force under the same measurement conditions. In practice, the extraction of tyre comfort characteristics from measurements at higher speeds, and especially the longitudinal and lateral forces, is a challenging task. The reason for this is the rapid excitation of multiple tyre natural frequencies and the increased power of the air cavity and rim modes. The following Fig. 2 provides a graphical comparison between both approaches applied to a longitudinal force signal measured at 90 [km/h].

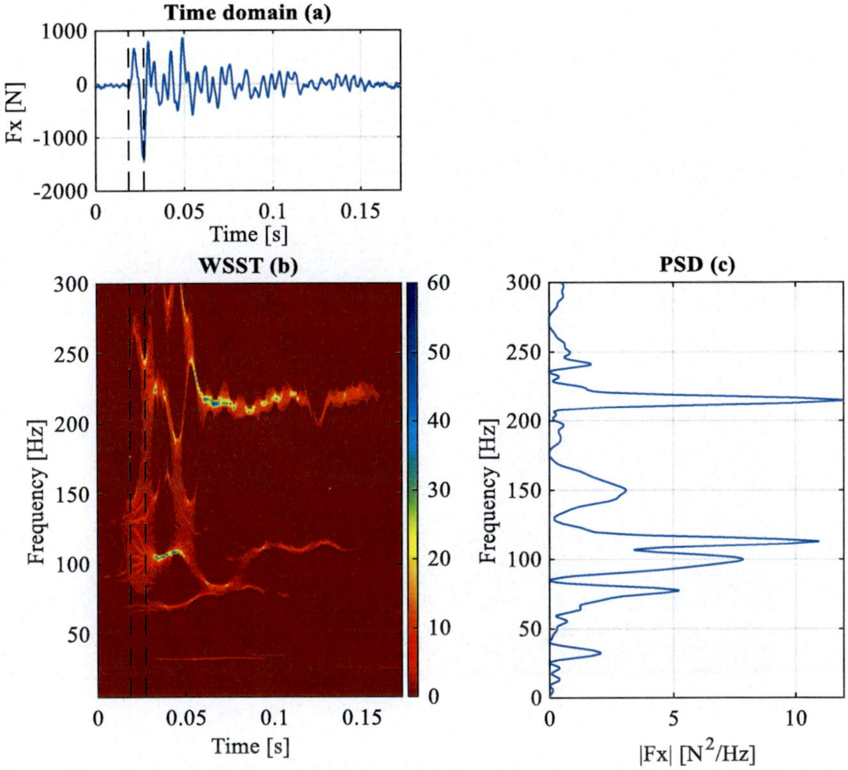

Fig. 2. Longitudinal force signal – transversal cleat (10 x 20 mm), 100% LI, 90 [km/h]. (a) Raw signal, (b) Synchrosqueezed wavelet transform, (c) Power spectral density.

Similar to the previous example, here the frequency content is consistent in both PSD (c) and WSST (b). However, the PSD frequency spectrum indicates multiple peaks around the expected longitudinal and air cavity modes, whose interpretation is rather difficult. In the WSST, starting at the cleat impact, we can observe the excitation of a large frequency band with the highest density before the tyre leaves the cleat. Shortly after the impact, the tyre frequencies fuse together and form the ~100 [Hz] longitudinal mode, which modulates to ~85 [Hz], while the ~30 [Hz] longitudinal mode appears. The sequence of these oscillation modes is discussed further in the next section. The air cavity mode is excited during the impact and reaches its peak shortly after. In addition, the split in the air cavity mode discussed in [14] could be investigated and explained using the wavelet analysis.

The proposed approach is also suitable for the investigation of the wheel vibration profiles with the tyre, the air and the rim as subcomponents. In the following Fig. 3, a comparison of both approaches is presented using the lateral force signal measured under the same conditions as Fig. 2 but on an oblique cleat.

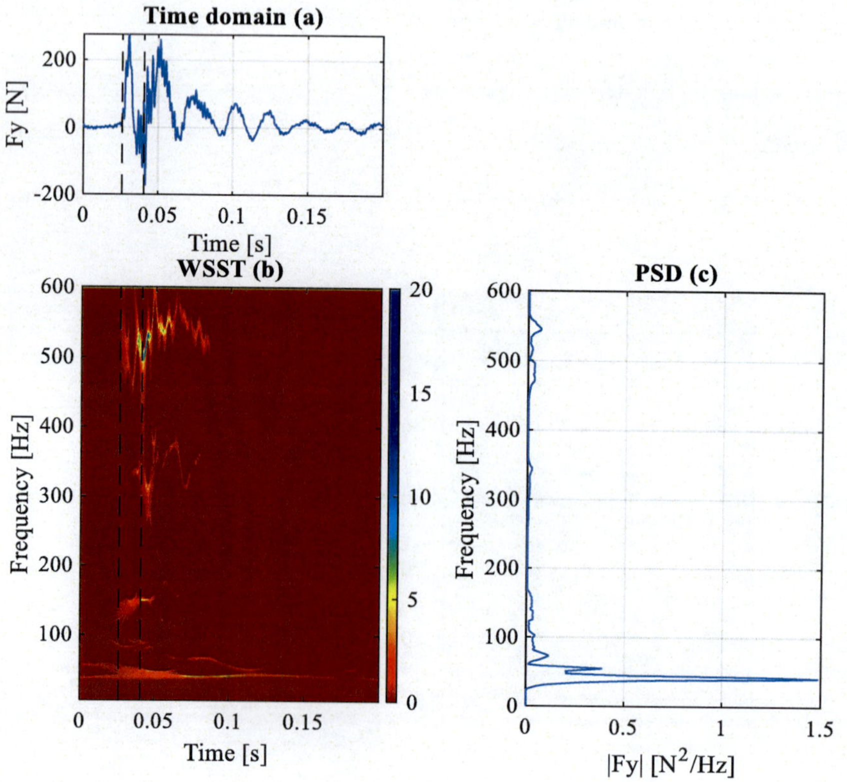

Fig. 3. Lateral force signal – oblique cleat (10 x 20 mm), 100% LI, 90 [km/h]. (a) Raw signal, (b) Synchrosqueezed wavelet transform, (c) Power spectral density.

Here the main lateral mode at ~45 [Hz] is captured with both approaches. Another mode with slightly lower magnitude at ~150 [Hz] can be observed in the time-frequency plane, which is not clearly identified by the PSD. Furthermore, in the frequency range above 300 [Hz], we can observe the natural modes of the rim. This allows a more detailed investigation of the behavior of the tyre-air-rim system in operating conditions. Such information could be used for the further development of high-frequency tyre models where the WSST approach can serve as a validation tool.

3.2 Identification and Extraction of Tyre Comfort Features using the Synchrosqueezed Wavelet Transform and Ridge Reconstruction

The insight provided by the current state of the art frequency approaches used for evaluating and analyzing tyre cleat data is two-dimensional. Comparison between measured and simulated tyres is usually based on the overall frequency and amplitude profiles, and information related to overall tyre damping is extracted from the time domain. The last could be quite challenging if we take the longitudinal force signal from Fig. 2 as an example. This limits the level of detail with which an evaluation can objectify such measurement data and thus influences the application fidelity.

As already presented in the previous section, the time-frequency approaches include an additional dimension, which provides a deeper insight into the vibration behavior of tyres. Revealing the excitation sequence allows tracing and reconstruction of separate vibration modes with the corresponding frequency modulations. The higher WSST time-frequency resolution, combined with the ridge reconstruction technique provides the possibility to estimate the damping behavior of separate modes.

The aim of this section is to demonstrate this by evaluating the longitudinal force measured at 30, 60 and 90 [km/h] under the same loading conditions with a transversal cleat. For each velocity the main two longitudinal tyre modes at ~30 [Hz] and ~100 [Hz] as well as the first air cavity mode at ~225 [Hz] are identified. This is realized by using the maximum amplitude of the WSST output in narrow frequency bands around the selected natural modes. For each mode, the maximum amplitude coordinates in the time-frequency plane serve as a main criterion for tracing of the corresponding ridge. Based on this information, a ridge reconstruction can be applied to convert the modes back into time domain.

In order to validate the reconstructed vibration modes, their sum is compared with the measured force. Fig. 4 provides an example with measurement at 30 [km/h].

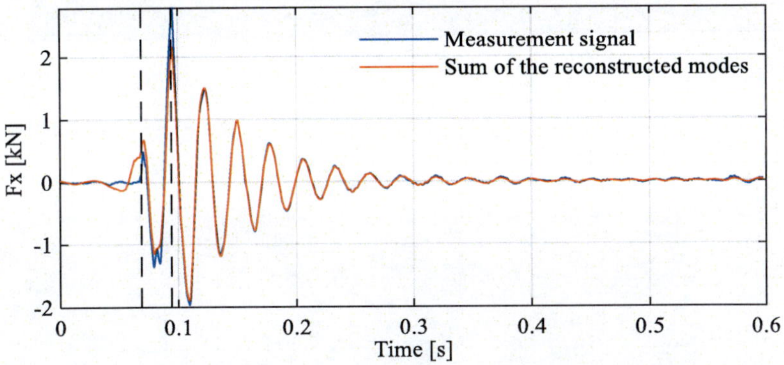

Fig. 4. Comparison between measured force signal and sum of the reconstructed vibration modes. Transversal cleat (10 x 20 mm), 100% LI, 30 [km/h].

The sum of the three reconstructed modes corresponds with high accuracy to the measured force signal after the cleat event. The difference between the two during the impact represents the tyre-enveloping characteristic stored in the lower frequency band.

Fig. 5 provides a visualization of the described methodology, performed at each of the three measurement velocities. The blue dashed lines over the WSST time-frequency plane represent the ridges, while the circles indicate the location of the maximum amplitudes. Similar to section 3.1, the area between the black dashed lines approximate the duration of the cleat event.

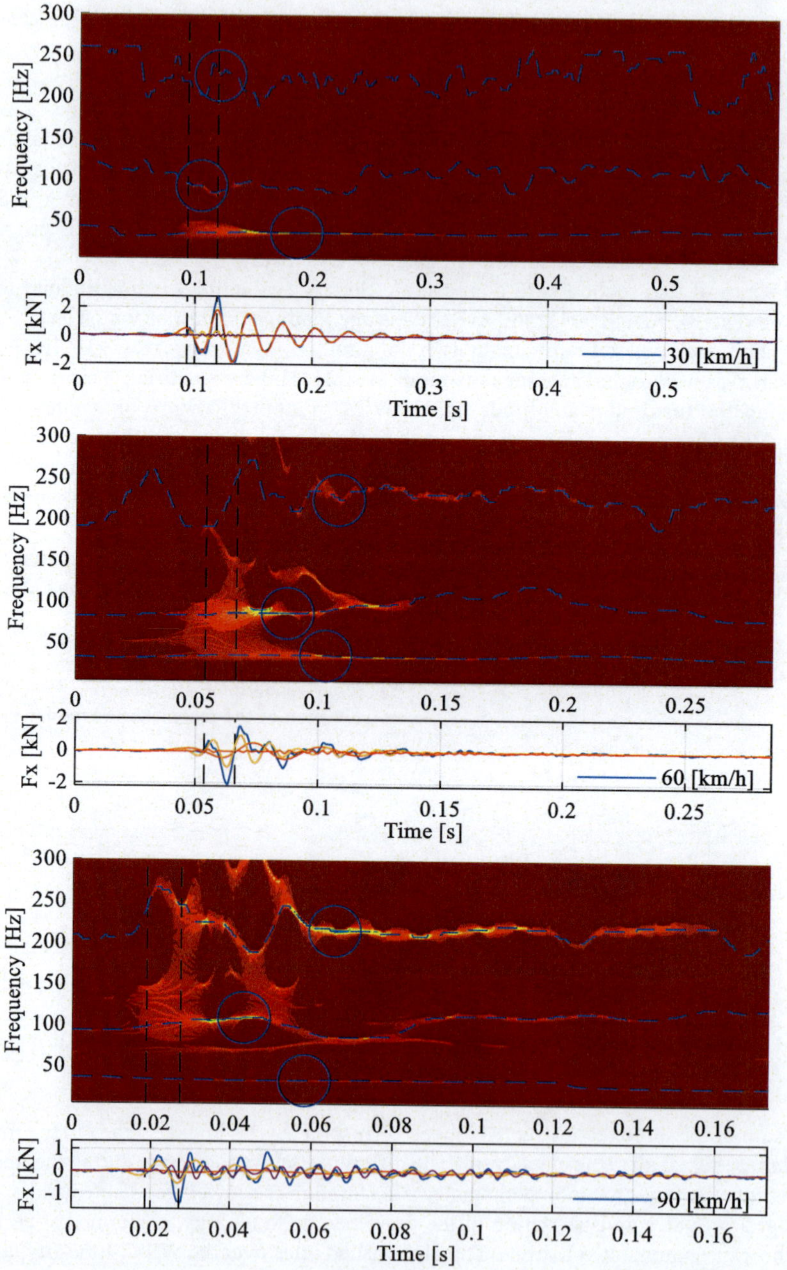

Fig. 5. Reconstruction of the main longitudinal vibration modes at three velocities. Transversal cleat (10 x 20 mm), 100% LI. Red: first mode, Yellow: second mode, Purple: air cavity.

According to [13], the damped natural oscillation of a single mass system is characterized with an exponential decay. Considering each reconstructed mode in a similar way, we can approximate the force decay with a simple exponential function expressed by

$$x(t) = A.e^{-B \cdot t}, \tag{19}$$

where parameter A represents the axis intersection for $x(t = 0)$ and parameter B represents the decay constant. The following Fig. 6 presents an example of fitting Eq. (19) to the oscillation force peaks of the ~30 [Hz] reconstructed mode in time domain.

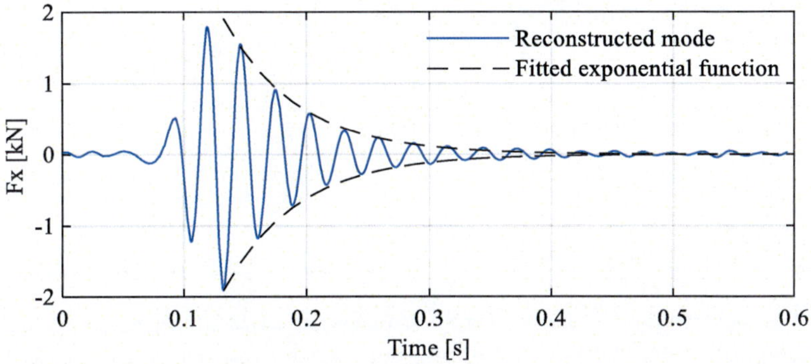

Fig. 6. Example of fitted exponential function on a reconstructed vibration mode

Applying the described methodology, features characterizing the selected vibration modes were derived and summarized in Table 1. In addition, the time delay from the initial impulse caused by the cleat impact until the excited frequency band is concentrated at the maximum WSST amplitude is also considered. We can deduce the distance that the tyre travels until the excited natural mode reaches its maximum amplitude in the frequency domain by integrating the velocity over the time delay. This distance is further referred to in this paper as the concentration length.

Starting with the first mode, the WSST amplitude response dominates the other modes at 30 [km/h] and drastically decreases at higher velocities. The peak frequency remains unchanged for the first two measured velocities. The decay constant increases up to the second measured velocity, which corresponds to increased damping. The concentration length slightly increases with the velocity.

The WSST amplitude of the second mode increases with velocity and indicates a peak at 60 [km/h]. For the first two measured velocities, the instantaneous frequency is below 100 [Hz] and exceeds this threshold at 90 [km/h]. A reduced damping is expressed by the continuous decrease of the decay constant. The concentration length shows that the second mode reaches its peak during the cleat impact at 30 [km/h] and indicates its natural value at higher velocities.

Table 1. Tabular presentation of extracted features using the proposed methodology.

Vibration mode	Feature		Velocity [km/h]		
			30	60	90
First mode	WSST amplitude	[N]	190.5	34.5	11.8
	WSST frequency	[Hz]	35.9	35.86	32.5
	Decay constant	[-]	16.8	20.3	20.14
	Concentration length	[m]	0.78	0.81	0.98
Second mode	WSST amplitude	[N]	35.1	66.6	55.8
	WSST frequency	[Hz]	92.1	87.0	108.4
	Decay constant	[-]	50.9	35.9	23.4
	Concentration length	[m]	0.10	0.55	0.61
Air cavity mode	WSST amplitude	[N]	12.2	13.6	60.0
	WSST frequency	[Hz]	226.3	224.7	214.4
	Decay constant	[-]	10.6	10.3	12.0
	Concentration length	[m]	0.23	0.9	1.16

The frequency and amplitude response of the air cavity mode remains nearly unchanged at the first two velocities. The results for the WSST amplitude at 90 [km/h] show that the air vibrations prevail over the tyre modes. This implies a trend of increasing importance of the air cavity modes at higher velocities. The change of the decay constant is relatively small compared those of the tyre modes. Here it should be noted that the damping approximation given with Eq. (19), is not completely accurate in this case due to the pulse-like shape, illustrated by the reconstruction in time domain (see Fig. 5). The concentration length here describes similar response for the second tyre mode. However, the distance increase between 60 and 90 [km/h] is higher compared to the rest. Taking into account the increase in amplitude, the reduced damping and the higher concentration length, we can conclude that, in this case, air vibrations are the limiting factor for the maximum measuring velocity for a given drum diameter.

4 Summary and outlook

This paper presents a novel methodology for the extended evaluation of tyre comfort data (e.g. cleat tests). Compared to the standard approach, the proposed method based on the synchrosqueezed wavelet transform provides a deeper insight into the frequency domain. The time-frequency plane reveals the excitation sequence of the natural modes, offering the possibility of a more detailed description of the wheel vibrations, which is suitable for the development of future applications, related to comfort assessment.

The selected wavelet base function is appropriate for the investigation of the tyre natural modes in a wide frequency band. The method also offers valid reconstructions of separate wheel vibration modes for selective investigations. The synchrosqueezed wavelet transform could be also a suitable tool for detailed comparison and validation of tyre simulation models with cleat measurements.

As an outlook, the authors are currently working on the implementation of different mother wavelets for the evaluation of tyre enveloping behavior. In addition, further features are being developed to interpret the information from the time-frequency plane more efficiently.

References

1. Iatsenko, D., McClintock, P., Stefanovska, A.: Linear and synchrosqueezed time-frequency representations revisited: Overview, standards of use, resolution, reconstruction, concentration, and algorithms. Digital Signal Processing 42, pp. 1-26. Elsevier Inc (2015).
2. Daubechies, I., Lu, J., Wu, H.-T.: Synchrosqueezed wavelet transforms: An empirical mode decomposition-like tool. Applied and Computational Harmonic Analysis 30, Elsevier Inc, 243-261 (2011).
3. Daubechies, I.: The Wavelet Transform, Time-Frequency Localization and Signal Analysis. IEEE Transactions of Information Theory, vol 36, no.5, pp. 961-1005, (1990)
4. Delprat, N., Escudie, B., Guillemain, P., Kronland-Martinet, R., Tchamitchian, P., Torresani, B.: Asymptotic wavelet and Gabor analysis: extraction of instantaneous frequencies. IEEE Transactions on Information Theory, vol. 38, no. 2, pp. 644-664, (1992)
5. Lilly, J. M., Olhede, S. C.: Higher-Order Properties of Analytic Wavelets. IEEE Transactions on Signal Processing, vol. 57, no. 1, pp. 146-160, (2009)
6. Brigham, E. O., Morrow, R. E.: The fast Fourier transform. IEEE Spectrum, vol. 4, no. 12, pp. 63-70, (1967)
7. Bisina, K., V., Azeez, M., A.: Optimized estimation of power spectral density. International Conference on Intelligent Computing and Control Systems (ICICCS), pp. 871-875, (2017)
8. Howard, R.M.: Power Spectral Density. In Principles of Random Signal Analysis and Low Noise Design, R.M. Howard (Ed.)., doi:10.1002/0471439207.ch3, (2003)
9. Weißberger, W.: Elektrotechnik für Ingenieure. Vieweg & Sohn Verlag GWV Fachverlage GmbH, Weisbaden (2007).
10. Eckstein, L.: Automotive Engineering Practical Course II. Schriftenreihe Automobiltechnik, Institut für Kraftfahrzeuge Aachen, RWTH Aachen University, Aachen (2015).
11. Rapuano, S., Harris, F. J.: An introduction to FFT and time domain windows. IEEE Instrumentation & Measurement Magazine, vol. 10, no. 6, pp. 32-44, (2007)
12. Harris, F. J.: On the use of windows for harmonic analysis with the discrete Fourier transform. Proceedings of the IEEE, vol. 66, no. 1, pp. 51-83, (1978)
13. Mitschke, M., Wallentowitz, H.: Dynamik der Kraftfahrzeuge. Auflage 5, Springer Vieweg, Wiesbaden (2014)
14. Gallrein, A., Baecker, M., Guan, J.: Simulation of Dynamic Gas Cavity Effects of a Tire under Operational Conditions. SAE Technical Paper 2018-01-0682, (2018)
15. Mahato, S., Chakaborty, A.: Sequential clustering of synchrosqueezed wavelet transform coefficients for efficient modal identification. Journal of Civil Structural Health Monitoring vol. 9, pp. 271-291, Springer-Verlag GmbH Germany, (2019)
16. Cherbuliez, M., Jacquot, M.P., Colonna de Lega, X.: Wavelet processing of interferometric signals and fringe patterns. Proc. SPIE 3813, Wavelet Applications in Signal and Image Processing VII, (1999)

17. Park, S.W., Lee, S.K.: Development of a new sound metric for impact sound in a passenger car using the wavelet transform. International Journal of Automotive Technology, vol. 6, pp. 809-818, (2010)
18. Suido, M., Tabaru, T., Shin, S.: Estimating tire condition based on wavelet analysis and frequency modulation. 2007 International Conference on Control, Automation and Systems, pp. 2767-2771, IEEE, (2007)
19. Seki, K., Shin, S., Tabaru, T.: Discrimination of normal and studless tyres by wavelet sound analysis. SICE 2004 Annual Conference, vol. 3, pp. 2312-2315, IEEE, (2004)
20. Lee, S.K., White, P.R.: Application of wavelet analysis to the impact harshness of a vehicle. Proceedings of the Institution of Mechanical Engineers, Part C: Journal of Mechanical Engineering Science, vol. 214, issue 11, pp. 1331–1338, (2000)
21. Seki, K., Shin, S., Tabaru, T.: Analysis of wavelet correlation between tyre sounds and tread patterns. Proceedings. 2005 IEEE Networking, Sensing and Control, pp. 241-246, (2005)
22. Gandetto, M., Guainazzo, M., Regazzoni, C.S.: Use of Time-Frequency Analysis and Neural Networks for Mode Identification in a Wireless Software-Defined Radio Approach. EURASIP Journal on Applied Signal Processing, pp. 1778-1790, (2004)
23. Baecker, M., Gallrein, A., Roller, M.: Noise, vibration, harshness model of a rotating tyre. Vehicle System Dynamics, vol. 54:4, pp. 474-491, (2016)
24. Dorfi, H.: Tire Cleat Impact and Force Transmission: Modeling Based on FTIRE and Correlation to Experimental Data. SAE Technical Paper 2004-01-1575, (2004)

An optimized tire test method on snow

Mauro Martino[1] and Salvatore Danise[2]

[1] FCA Emea Product Development
[2] FCA Emea Product Development

Abstract. For all-season tire development a snow testing session is needed. Snow testing is possible and economically convenient only in a limited period and specific test centers. Therefore, tire development on snow is not achievable out of these boundary conditions. A prediction of snow performances during spring/summer/autumn testing actually is not feasible.
The aim of the study is to identify sets of all-season tire design parameters with same snow performances, rolling resistance and rolling noise but different performances on dry/wet surfaces, of brake, comfort and handling, free to be tuned based on the vehicle target attributes. Pattern geometry (PG), tread compound shore hardness (TSH) and glass transition temperature of tread compound (GTT) have been selected as tire design parameters.
To support the study, the Design for Six Sigma Methodology (DFSS) has been applied. DFSS is a structured approach to obtain the state of maximum robustness in product design. The optimization of the Ideal Function related to the Intended Function (lateral and longitudinal accelerations defined as Output Response, hereafter OR) of the component (tire) is obtained selecting proper Control Factors and Levels (hereafter C.F., which are PG, TSH and GTT) under systematic variation of Noise Factors (dry/wet/snow surfaces, hereafter N.F.) and Input Signals (car speed, hereafter IS). The result of DFSS approach, processed to objective evaluation, has been to identify the optimal combination (sets of all-season tire design parameters) of the above Control Factors and Levels to make the tire minimally sensitive to Noise Factors (dry/wet/snow surfaces) that causing variability in the output responses of component (tire performances).
In summary, the effects of PG, TSH, and GTT on the trade-off between snow performances, rolling resistance, rolling noise and driving dynamics have been identified.

Keywords: All-season tire design, Snow testing, Robust design, DFSS.

1 Introduction

Actually, for the all-season tire development, for an OEM based in Europe, a snow testing session in Scandinavia is possible and economically convenient only from middle of December to middle of March. The target of the study presented is to identify different sets of all-season tire design parameters with same snow performances, rolling resistance and rolling noise. The performances of comfort, brake and handling on different surfaces, dry and wet, are tuned in order to achieve the vehicle target attributes.

P. E. Pfeffer (ed.), *11th International Munich Chassis Symposium 2020*, Proceedings, https://doi.org/10.1007/978-3-662-63193-5_37

During the design phase the pattern geometry, PG, and the rubber compound characteristics, both TSH and GTT, have been identified as the main design parameters [1]. Under these boundary conditions the trade-off among driving dynamics and the desired snow, RR, and Road Noise performances are delimited.
All sets of all-season tire design parameters have defined PG, all-season oriented (APG) and winter oriented (WPG), TSH [2][3] and GTT [4]. Other tire parameters such as cap plies, steel-cord belt plies, textile cord ply, inner liner, sidewall, bead core and vehicle suspension have been considered as out of scope.
PG is based on APG investigated generally. Modification of tread depth, sipe density and number of blocks have been set in order to obtain WPG. Therefore, the modifications balance has been defined in function of the APG and WPG contributions to all the overall tire performances.
The rubber compound constituting the tire tread has a consequential effect on overall tire performance. TSH concurs to rules the tire grip by means of rubber hysteresis [2] [3]. The selected range is the typical all-seasons range for a passenger car.
GTT concurs to define the viscoelastic properties of the rubber compounds. All-seasons/winter tires offer outstanding flexibility and grip at lower temperatures. Summer tire compounds become harder at temperatures below ~ +7°C. Therefore, lower GTT is preferred for all-seasons/winter compounds to maximize the temperature range in which the compound is rubbery [4].
Based on current tire production 205/55-R16, four submission tires (hereafter subm #) have been designed and produced by a premium tire supplier, together with FCA Product Development Department, with different sets of all-season tire design parameters: PG (APG and WPG), TSH and GTT. As per the following L4 Orthogonal Array [5], Submission tires matrix.**Table 1Table 1**, in total four submission tires have been tested and evaluated in terms of rolling resistance, rolling noise, subjective/objective handling and ride, on dry (during summer and winter), wet and snow.

Table 1. Submission tires matrix.

Tire	PG	TSH [ShA]	GTT [°C]
subm #1	all-season pattern	58	-72
subm #2	all-season pattern	65	-61
subm #3	winter pattern	58	-61
subm #4	winter pattern	65	-72

Where:

- 4 tread compounds varying in TSH and GTT.
- 2 patterns (PG) varying only in siping system, APG and WPG.

Pattern Geometry Control Factor details:

- 2 levels: APG and WPG.
- Specific tread depth: APG of (7,5 ; 8,5) mm and WPG of (8,5 ; 9,5) mm. Deep grooves provide additional traction.
- Sipe density: APG lower, WPG higher. Fine sipes for flexible tread blocks bite deeper into the snow. Remove liquid layer on ice.
- Number of blocks: APG lower, WPG higher. The contact area defines the ice performance.
- Optimized Sidewall Design

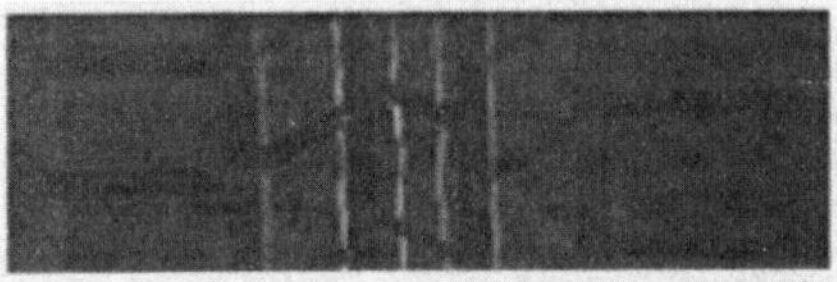

Fig. 1. All-season pattern (APG).

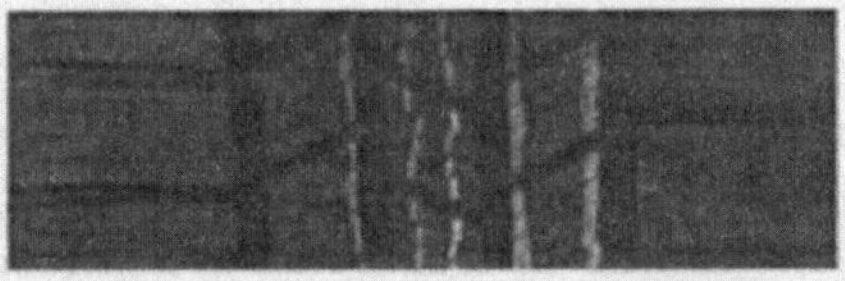

Fig. 2. Winter pattern (WPG).

Tread Compound Control Factor shore hardness details:

- 2 levels: 58 ShA and 65 ShA.
- (58 ; 65) ShA winter/all-seasons range.
- Summer tire 68 ShA with common grip capability.
- UHP summer tire 72 ShA with high grip capability.
- RR optimized adapted Tread Compound.

Tread compound Glass Transition Temperature Control Factor details:

- 2 levels: -72 °C and -61 °C.
- (-72; -61) °C, winter/all-seasons range
- Summer tire -30° C.

In total, on the vehicle, four submission tires have been tested and evaluated through the following study. The driving dynamics have been gauged through handling and ride subjective evaluation and objective evaluation for both handling and braking.

2 Rolling Resistance measurement

Rolling Resistance has been measured by the supplier and by FCA EMEA in accordance with the ISO 28580 [7]. Rolling resistance is almost constant among the four submissions. (9,13 [kg/t] on the average).

Table 2. Weight and Rolling Resistance.

Tire	PG	TSH [ShA]	GTT [°C]	Weight [kg]	RR [kg/t]
subm #1	all-season pattern	58	-72	9,12	9,14
subm #2	all-season pattern	65	-61	9,17	9,09
subm #3	winter pattern	58	-61	9,13	9,18
subm #4	winter pattern	65	-72	9,15	9,11

3 Road Noise measurement

The Road Noise performance, on different tracks and speeds, has been measured for all the submission tires, by FCA EMEA. Namely measurements were done on a cobblestone road, on a rough asphalt road and on a smooth asphalt road. Two microphones were placed near driver's ears (front microphones) and right rear passenger's ears (rear microphones); the noise was acquired in different coast-downs maneuvers. Generally the Road Noise, considering a scattering of ± 0,3 dBA, could be considered constant among the four submissions.

Table 3. Road Noise - Overall front microphones

Overall front [dBA]	COBBLESTONE 50 km/h	ROUGH 50 km/h	ROUGH 75 km/h	ROUGH 100 km/h	SMOOTH 50 km/h	SMOOTH 75 km/h	SMOOTH 100 km/h
subm #1	77,8	68,2	70,4	72,7	61,2	63,4	67,4
subm #2	77,8	68,4	70,2	72,9	61,0	63,4	67,5
subm #3	77,5	68,2	70,3	72,5	60,5	63,0	67,0
subm #4	78,0	68,4	70,3	72,5	61,3	63,6	67,1

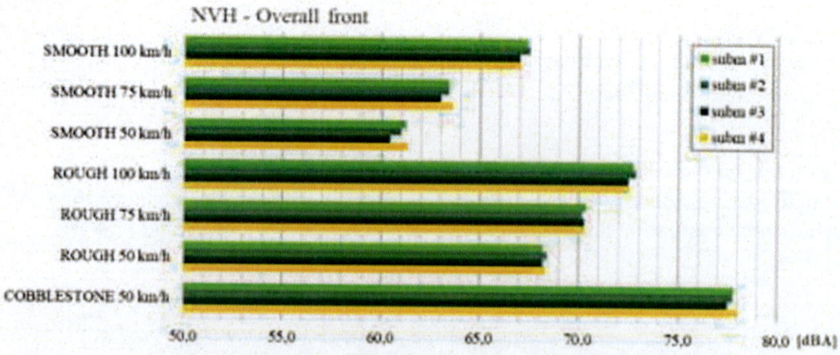

Fig. 2. Road Noise - Overall front microphones.

4 Driving Dynamics

The driving dynamics have been analyzed by means of the following evaluations:

- Objective handling:
  To quantify the effect of the design parameters on different surfaces (dry, wet, snow). In particular, for the dry surface, two seasonal conditions have been tested, during summer and winter in order to explore the whole of the all-season range of use.
- Brake distance:
  To investigate the effects of PG, TSH and GTT trade-off on braking performance.
- Cornering stiffness and relaxation length:
  To analyze the effects of PG, TSH and GTT trade-off on pure cornering stiffness and relaxation length on the flat track.

The primary target of the study is to define Output Responses (OR), by means of the objective handling evaluation. In details OR are:

- $\pm a_y$, lateral acceleration:
  determining the tire lateral traction
- $+a_x$, longitudinal acceleration:
  determining the tire longitudinal traction
- $-a_x$, longitudinal acceleration
  determining the tire longitudinal deceleration
- g-g area:
  lateral and longitudinal accelerations, determining the tire maneuvering area utilized by the driver [8].

The scope of DFFS is to identify the optimum Control Factors and levels combination (PG, TSH and GTT) in order to optimize the OR – maximize the robustness of tire making it minimally sensitive to the effect of factors that cause variability of its output response [5].
Secondarily, subjective handling and ride, brake distance and flat track measurement have been measured. These performances are out of the DFSS analysis.
In order to the test all the subm #x in all the seasonal conditions the following test tracks (N.F.) have been driven at two different levels of car average speed (SL).
In details, N.F.:

- Dry test track on summer (N1: Dry WARM).
- Dry test track on winter (N2: Dry COLD).
- Wet test track, controlled test conditions (N3: WET).
- Snow test track on winter (N4: SNOW).

Then relating to the car speed defined as IS, there are the following SL:

- Normal speed, customer oriented (M1)
- Fast speed, sporty oriented (M2)
- Limit speed, up to the limit (out of scope).

Table 4. Test tracks (N.F.).

NOISE FACTORS TEST TRACKS	N1 DRY WARM	N2 DRY COLD	N3 WET	N4 SNOW
Surface	dry	dry	wet	snow
Ambient Temperature [C°]	+28	+7	+20	-17°
Ground temperature [C°]	+21	+5	+14	-7°
Location	FCA Balocco PG	FCA Balocco PG	Supplier PG	Supplier PG (Sweded)
Month	July	December	May	January

Table 5. Car speeds (IS).

INPUT SIGNALS CAR SPEEDS	M1 Normal speed [km/h]	M2 Fast speed [km/h]
N1 - DRY WARM	85	96
N2 - DRY COLD	86	94
N3 - WET	63	78
N4 - SNOW	32	48

4.1 Objective handling evaluation

All the submission tires have been measured on the instrumented testing vehicle at different car speed. M1 and M2, on all test tracks - configurations N1, N2, N3 and N4. Moreover the lap data acquisition have been used to record and explore both the steady and the transient states analyzing the lateral, longitudinal and g-g accelerations inside linear range up to the limit. Neither evasive nor emergency maneuvers have been investigated in this case study.

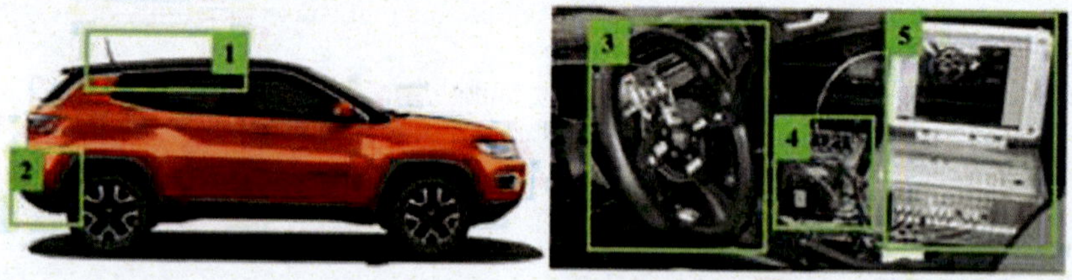

Fig. 3. Instrumented testing vehicle: 1-GPS double antenna, 2-optical sensor, 3-dyno steering wheel, 4-inertia measurement unit, 5-data logger

Table 6. Max of $\pm a_y$, max of $+a_x$, max of $-a_x$ and g-g area (OR).

	N1 DRY WARM		N2 DRY COLD		N3 WET		N4 SNOW	
	M1	M2	M1	M2	M1	M2	M1	M2
	max of ±ay [g]							
subm #1	1,12	1,23	1,04	1,19	1,63	1,88	0,41	0,48
subm #2	1,21	1,20	1,13	1,16	1,51	1,74	0,39	0,45
subm #3	1,22	1,23	1,03	1,18	1,52	1,75	0,42	0,49
subm #4	1,23	1,20	1,05	1,28	1,35	1,55	0,44	0,49

	N1 DRY WARM		N2 DRY COLD		N3 WET		N4 SNOW	
	M1	M2	M1	M2	M1	M2	M1	M2
	max of +ax [g]							
subm #1	0,49	0,47	0,56	0,47	0,50	1,00	0,16	0,17
subm #2	0,38	0,54	0,71	0,63	0,53	1,01	0,18	0,16
subm #3	0,47	0,55	0,54	0,53	0,71	0,75	0,16	0,18
subm #4	0,50	0,51	0,52	0,58	0,53	0,91	0,16	0,17

	N1 DRY WARM		N2 DRY COLD		N3 WET		N4 SNOW	
	M1	M2	M1	M2	M1	M2	M1	M2
	max of -ax [g]							
subm #1	-0,69	-1,18	-0,95	-1,13	-1,01	-1,04	-0,26	-0,31
subm #2	-0,88	-1,05	-1,07	-1,08	-0,81	-1,12	-0,25	-0,29
subm #3	-0,84	-0,96	-0,88	-0,93	-0,99	-0,91	-0,28	-0,32
subm #4	-0,80	-0,96	-0,69	-1,03	-0,87	-0,88	-0,27	-0,32

	N1 DRY WARM		N2 DRY COLD		N3 WET		N4 SNOW	
	M1	M2	M1	M2	M1	M2	M1	M2
	g-g area [g2]							
subm #1	1,87	2,63	1,82	2,62	2,39	3,93	0,31	0,35
subm #2	2,02	2,39	2,10	2,62	2,37	3,57	0,30	0,32
subm #3	1,92	2,34	1,98	2,40	3,02	3,36	0,33	0,36
subm #4	2,03	2,36	1,85	2,39	2,34	3,81	0,32	0,35

Based on the reasonable experience that 5% of measurement scattering on snow surfaces is acceptable, all the submission tires demonstrate same snow performances in terms of longitudinal and lateral accelerations. Besides, on the other surfaces the performances become different inside a common tire tuning range. This means the possibility of submission selection is consistent and feasible.

Thereby, the driver is able to select the best/optimized submission tire preferable to achieve the vehicle target attributes, based on equal snow performances.

Afterwards, in function of the car speed, M1 and M2, the performance ranking is kept up. Hence, there are no saturation phenomena during traction or brake maneuvers.

4.2 Brake distance

Brake distance measurements on dry, wet and snow surface have been completed in accordance with ISO 21944 [9].

Table 7. Brake distances.

Tire	PG	TSH [ShA]	GTT [°C]	brake distance
DRY @ 110 km/h				
subm #1	all-season pattern	58	-72	-
subm #2	all-season pattern	65	-61	-1%
subm #3	winter pattern	58	-61	+3%
subm #4	winter pattern	65	-72	+4%
WET @ 85 km/h				
subm #1	all-season pattern	58	-72	-
subm #2	all-season pattern	65	-61	-1%
subm #3	winter pattern	58	-61	-2%
subm #4	winter pattern	65	-72	+2%
SNOW @ 50 km/h				
subm #1	all-season pattern	58	-72	-
subm #2	all-season pattern	65	-61	-4%
subm #3	winter pattern	58	-61	+5%
subm #4	winter pattern	65	-72	+4%

— Dry brake:
Set up description: +14° C air temperature (+14° C ground), dry surface, μ=0,94.
The dry brake distance seems to be influenced by PG: the WPG tires, subm#3 and subm#4, have longer brake distance with respect to APG tires, subm#1 and subm#2. TSH and GTT combinations are compensated and/or have an influence order lower than PG, both APG and WPG. It must be advised that 2-3% could be considered as scattering.

— Wet brake:
Set up description: +14° C air temperature (+14° C ground), wet surface, 1mm ±0,5 water layer, μ=0,72.
The wet brake distance seems to be influenced by PG: the WPG tires, subm#3 and subm#4, have longer brake distance with respect to APG tires, subm#1 and subm#2. TSH and GTT combinations are compensated and/or have an influence order lower than PG, both APG and WPG. It must be advised that 2-3% could be considered as scattering.

— Snow brake:
Set up description: -13° C air temperature (-7° C ground), compact and scraped snow layer, 1cm fresh snow coating, 83mm snow hardness, μ=0,3.

The dry brake distance seems to be influenced by PG: the WPG tires, subm#3 and subm#4, have longer brake distance with respect to APG tires, subm#1 and subm#2. TSH and GTT combinations are compensated and/or have an influence order lower than PG. It must be advised that 4-5% on snow surfaces could be considered as scattering.

During brake tests, PG, both APG and WPG, have the most influence on the braking performance; in the DFSS section, the effect of TSH and GTT will be discussed.

4.3 Flat track measurement

The flat track analysis has been conducted by the tire supplier, fulfilling FCA requirements. All the submission tires have been characterized on the flat track tire test machine [10].

Since the instrumented test vehicle is constant during all the tests therefore the differences among the cornering stiffness and relaxation length could be considered as differences in terms of vehicle performances [8][11].

Table 8. Cornering stiffness and relaxation length.

Tire	PG	TSH [ShA]	GTT [°C]	Cornering stiffness [N/rad] @5000 N		Relaxation length [m] @5000 N	
subm #1	all-season pattern	58	-72	95242	-	0,68	-
subm #2	all-season pattern	65	-61	101179	+6%	0,71	+4%
subm #3	winter pattern	58	-61	90453	-5%	0,65	-4%
subm #4	winter pattern	65	-72	100365	+5%	0,72	+6%

Focusing on the submission tires PG, TSH and GTT, in details:

— The cornering stiffness is influenced by the TSH mainly.
With respect to high values of TSH, for instance 68 ShA is a common typical summer tire value, parameters PG, both APG and WPG, and GTT have no impact on it. Subm #2 and subm #4 , both with TSH of 65 ShA, (more grip oriented than TSH of 58 ShA), have +5% and +6% higher than subm #1 and subm #3. Typical critical cornering stiffness differences are considered to be about 2%.
With respect to low values of TSH the effect of PG, both APG and WPG, and GTT have a combined influence on the cornering stiffness; to be investigated in the future studies. Subm #1 and subm #2, both with TSH of 58 ShA, (less grip oriented than TSH of 65 ShA), have -6% and -10% lower cornering stiffness than subm #2 and subm #4 respectively.
— The relaxation length is influenced by the TSH mainly.

With respect to high values of TSH, for instance 68 ShA is a common typical summer tire value, the effect of PG, both APG and WPG, and GTT have no impact on it. Subm #2 and subm #4 , both with TSH of 65 ShA, (more grip oriented than TSH of 58 ShA), have +4% and +6% higher than subm #1 and subm #3.

With respect to low values of TSH the effect of PG, both APG and WPG, and GTT have a combined influence on the relaxation length; to be investigated in the future studies. Subm #1 and subm #2, both with TSH of 58 ShA, (less grip oriented than TSH of 65 ShA), have -6% and -10% lower cornering stiffness than subm #2 and subm #4 respectively.

Therefore, within the linear range, TSH influences both cornering stiffness and relaxation length, at high values of the TSH winter/all-seasons range essentially.

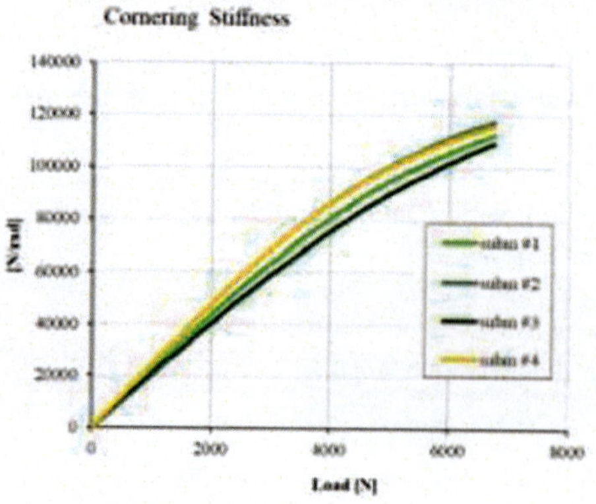

Fig. 4. Cornering stiffness.

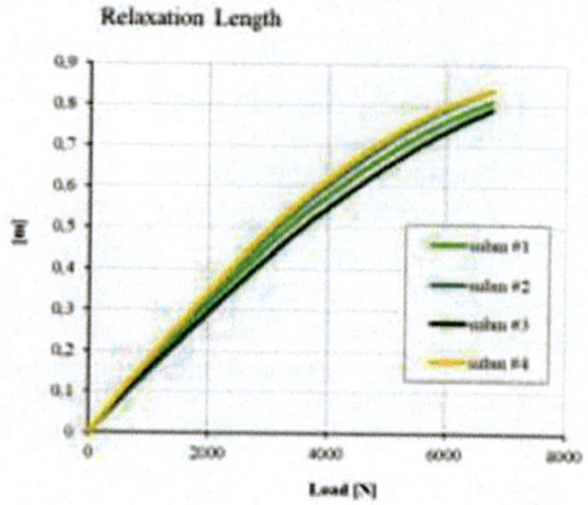

Fig. 5. Relaxation length.

5 DFSS analysis

The DFSS Methodology, Design for Six Sigma – including Taguchi Methods for Robust Optimization [5], is a structured approach to obtain the state of maximum robustness in product/process designs to optimize the intended performance of the system/component for insensitivity to sources of variation (N.F.).

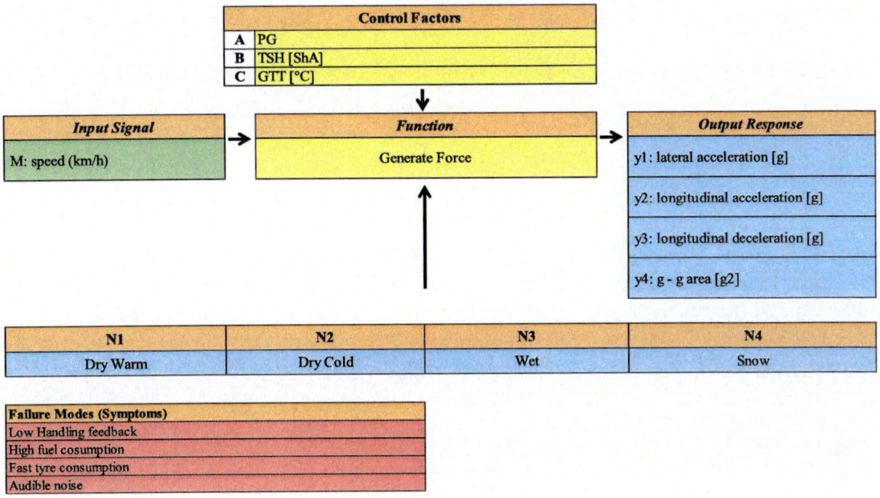

Fig. 6. DFSS approach.

5.1 The 8-steps of Parameter Design and the 6-steps of Tolerance Design

For this study the team applied the Parameter Design to optimize the tire Intended Function - maximizing OR insensitivity to ambient and road conditions N1, N2, N3 and N4, discovering the optimum combination of C.F. levels.

The 8 steps of Parameter Design [5], was conducted to identify the Ideal Function of tire based on energy transformation (lateral and longitudinal accelerations, OR), using systematic variation of N.F. and IS in order to identify the optimum combination of PG, TSH and GTT control factor's levels, using Orthogonal Array.

The optimum combination of C.F.'s levels has been obtained after the "Two Step Optimization" of Parameter Design.

5.2 The 8-Steps of Parameter Design

1 Scope for Optimization.
The Scope of the study is to identify different sets of all-season tire design parameters with same:

— Snow performance.
— Rolling Resistance.
— Rolling Noise.

Moreover different performances, able to be tuned for vehicle target attributes:

— On different dry/wet surface.
— Brake distance.
— Ride and handling.

2 Identify Ideal Function.
The Ideal Function, hereafter IDF, represents an ideal relationship between input (M1 and M2) and the output response of our System/Component (OR) and is related to the Intended Function of the system/component.
For a tire the following 4 energy based OR's were used:

— Speed - $\pm a_y$, lateral acceleration
— Speed - $+a_x$, longitudinal acceleration
— Speed - $-a_x$, longitudinal acceleration
— Speed - g-g area

3 Develop Signal & Noise Factors Strategy.
One of the key aspects of DFSS methodology is to analyze the Design Space (the engineering space where the design parameters, C.F. and levels, can be set) in presence of Customer Space (the downstream conditions where the customers will use the product) using Orthogonal Arrays, to make sure that all the possible customer usage conditions will be cover by the design.

The Input Signal an N.F. represent the Customers Space, the space where the final customer will use product under current study (the drive style and the environment), the PG, TSH and GTT control factors represent the design space (engineering space).
In order to be sure the main customer usage conditions will be covered by the design, a wide range of input signal (car speeds, M1 and M2) and noise conditions (ambient and road surface conditions, N1, N2, N3 and N4) have been considered.

4 Establish Control Factors and Levels, assign they to the Orthogonal Array.
In order to increase tire design knowledge related to compounds formulation and tread design, all sets of all-season tire design parameters (C.F.s) have been definedSubmission tires matrix., as declared in the Introduction paragraph. The experimental layout with Orthogonal Array provides a powerful method of experimentation, with the following advantages:

- Insensitive to test order thanks to Noise Strategy (in the DOE the order of tests need to be randomized to break the effects of uncontrollable factors that can affect the testing)
- Efficiency: allow to explore a wide range of design combinations of control factors and levels using a low number of experiments (for a full factorial experimental approach, it would require 8 sets of tires for this project)
- Fairly balance comparisons between control factors and levels;
- Allow for evaluations for interactions between Control Factors and Noise Factors to uncover how Control Factors and Levels can achieve insensitivity to Noise Factors.

5 Execute test and collect data.
Experimental plan is executed (tests performed), data are collected.

Table 9. OR with Signal-to-Noise Ratio.

	N1 DRY WARM		N2 DRY COLD		N3 WET		N4 SNOW			
	M1	M2	M1	M2	M1	M2	M1	M2		
	max of ±ay [g]								β	S/N
subm #1	1,12	1,23	1,04	1,19	1,63	1,88	0,41	0,48	0,02	-29,95
subm #2	1,21	1,20	1,13	1,16	1,51	1,74	0,39	0,45	0,00	-28,98
subm #3	1,22	1,23	1,03	1,18	1,52	1,75	0,42	0,49	0,00	-29,03
subm #4	1,23	1,20	1,05	1,28	1,35	1,55	0,44	0,49	0,00	-27,93

	N1 DRY WARM		N2 DRY COLD		N3 WET		N4 SNOW			
	M1	M2	M1	M2	M1	M2	M1	M2		
	max of +ax [g]								β	S/N
subm #1	0,49	0,47	0,56	0,47	0,50	1,00	0,16	0,17	0,01	-30,20
subm #2	0,38	0,54	0,71	0,63	0,53	1,01	0,18	0,16	0,00	-29,73
subm #3	0,47	0,55	0,54	0,53	0,71	0,75	0,16	0,18	0,00	-27,46
subm #4	0,50	0,51	0,52	0,58	0,53	0,91	0,16	0,17	0,00	-28,10

	N1 DRY WARM		N2 DRY COLD		N3 WET		N4 SNOW			
	M1	M2	M1	M2	M1	M2	M1	M2		
	max of -ax [g]								β	S/N
subm #1	-0,69	-1,18	-0,95	-1,13	-1,01	-1,04	-0,26	-0,31	-0,01	-25,17
subm #2	-0,88	-1,05	-1,07	-1,08	-0,81	-1,12	-0,25	-0,29	0,00	-23,02
subm #3	-0,84	-0,96	-0,88	-0,93	-0,99	-0,91	-0,28	-0,32	0,00	-23,48
subm #4	-0,80	-0,96	-0,69	-1,03	-0,87	-0,88	-0,27	-0,32	0,00	-23,50

	N1 DRY WARM		N2 DRY COLD		N3 WET		N4 SNOW			
	M1	M2	M1	M2	M1	M2	M1	M2		
	g-g area [g2]								β	S/N
subm #1	1,87	2,63	1,82	2,62	2,39	3,93	0,31	0,35	0,03	-31,81
subm #2	2,02	2,39	2,10	2,62	2,37	3,57	0,30	0,32	0,00	-30,73
subm #3	1,92	2,34	1,98	2,40	3,02	3,36	0,33	0,36	0,00	-30,99
subm #4	2,03	2,36	1,85	2,39	2,34	3,81	0,32	0,35	0,00	-31,48

6 Analyze data and conduct two-step optimization.
Once collected from tests, the data have been analyzed following the procedure below:

a. calculate S/N and β for each test
b. complete and Interpret Response Table
c. perform Two-Step Optimization

The S/N ratio is an index of robustness recommended by Dr. Taguchi. It measures the robustness of energy transformation that occurs within our system/component design. Furthermore, it is the ratio between the energy that is transformed into intended output and the energy that is transformed into unintended output. The response table is the graphical representation of factorial effect [5].
Details of Two-step Optimization:

- First step: reduce variability
- Second step: adjust the response to meet the requirement (to target)

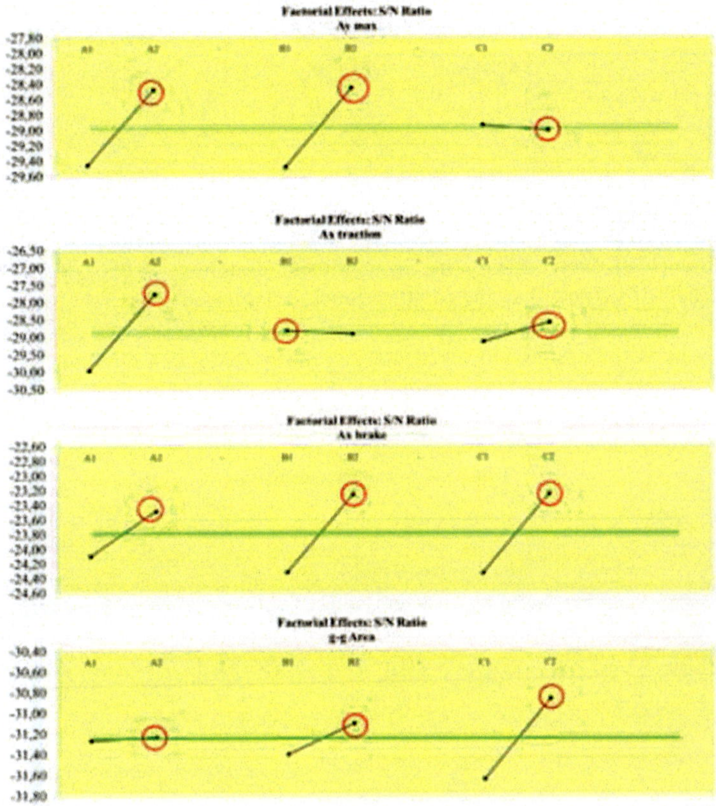

Fig. 7. Response table.

7 Predict and Confirm.

Taguchi's Method requires a confirmation run to check for both the validity of experimentation and the reproducibility of the experimental conclusions. Predictions are made by using the "Additivity of factorial effects" (sum of "gains" from control factors). Good additivity can be defined in case of no strong interaction between control factors. The interaction between C.F. on S/N ratio is an inconsistent effect in the system (the strategy of Robust Design is to eliminate or minimize these interactions so that the main control factor effects are as consistent as possible and the experimental conclusions will be reproducible in downstream environments).

Table 10. Optimum design

Tire	PG	TSH [ShA]	GTT [°C]
subm #1	all-season pattern	58	-72
subm #2	all-season pattern	65	-61
subm #3	winter pattern	58	-61
subm #4	winter pattern	65	-72
Optimum	winter pattern	65	-61

Based on DFSS Methodology the Optimum submission is: WPG, TSH of 65 ShA and GTT of -61° C (note that this combination was not tested in the L4 O.A.).

The Variability Reduction obtained by the Optimum submission is about 28% with respect to the initial design. However, the Variability Reduction for subm #4 is about 20% with respect to the initial design and can be considered more than acceptable. For this reasons subm #4 can be considered as the Best Engineering submission.

Table 11. Variability Reduction.

Prediction	PG	TSH [ShA]	GTT [°C]	S/N	%Variability reduction
max of +ax [g]					
Optimum submission	winter pattern	65	-61	-27,99	25,31
subm #4	winter pattern	65	-72	-27,98	25,50
subm #4 vs Optimum				0,01	0,19
max of +ax [g]					
Optimum submission	winter pattern	65	-61	-27,51	36,30
subm #4	winter pattern	65	-72	-28,19	26,00
subm #4 vs Optimum				-0,68	-10,30
max of -ax [g]					
Optimum submission	winter pattern	65	-61	-22,19	41,03
subm #4	winter pattern	65	-72	-23,55	20,52
subm #4 vs Optimum				-1,36	-20,50
g-g area [g2]					
Optimum submission	winter pattern	65	-61	-30,70	13,66
subm #4	winter pattern	65	-72	-31,48	3,87
subm #4 vs Optimum				-0,78	-9,80

8 Verify.

To make sure that optimized system/component do not harm any surround subsystems and to make sure that all potential benefits from optimized design can be give to customer, verification tests can be execute (specific tests are scheduled).

6 Conclusions

An optimized tire test method on snow is presented. The method has the aim to reduce the number of tests on snow without compromising the general tire performances, indeed granting degrees of freedom during both tire advanced engineering and tuning phase.

Moreover the DFSS Methodology, for the first time applied to tire development, has demonstrated the robustness approach.

These results enable a more effective tire design during the advanced engineering phase and more control during the tuning phase monitoring tire performances trade-off.

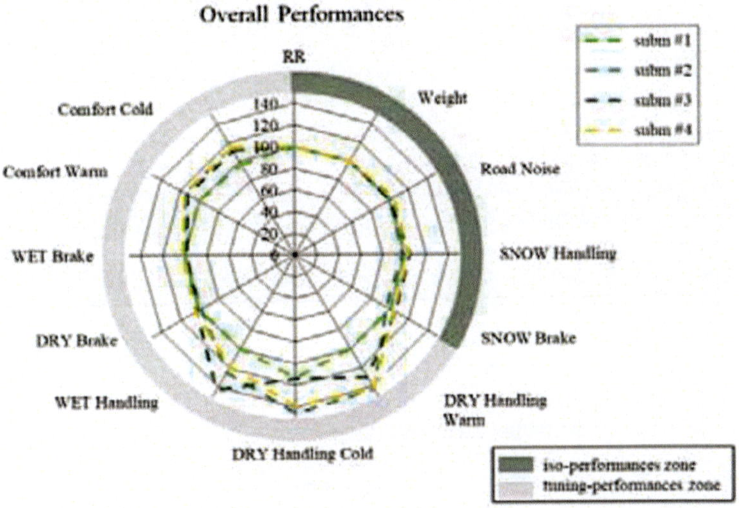

Fig. 8. Overall Performances.

Definitions/Abbreviations

APG	pattern geometry all-season oriented
C.F.	Control Factors
DFSS	Design for Six Sigma
EMEA	Europe, Middle East and Africa
FCA	Fiat Chrysler Automobiles
GTT	tread compound glass transition temperature (°)
IDF	Ideal Fuction
IF	Intendent Fuction
IS	Input Signals
M1	normal car speed Signal Levels
M2	fast car speed Signal Levels
N.F.	Noise Factors
N1	dry warm test track Noise Factors
N2	dry cold test track Noise Factors
N3	wet test track Noise Factors
N4	snow test track Noise Factors
O.A.	Orthogonal Array
OEM	Original Equipment Manufacturer
OR	Output Response
PG	pattern geometry
SL	Signal Levels
S/N	Signal-to-Noise Ratio
subm #x	submission tire #x
TSH	tread compound shore hardness (ShA)
UHP	Ultra High Performance tire
WPG	pattern geometry winter oriented

References

1. Nakajima, Y.: Advanced Tire Mechanics. Springer, Singapore (2019).
2. Haney, P.: The Racing & High-performance. SAE International, Warrendale (2003).
3. Michelin: The tyre. Rolling resistance and fuel savings, Société de Technologie Michelin (2003)
4. Harper, M., Tardiff, J., Haakenson, D., Joandrea, M., and Knych, M.: Tire Tread Performance Modification Utilizing Polymeric Additives. SAE Technical Paper, 2017-01-1501 (2017).
5. Wu, Y., Wu, A.: Taguchi methods for Robust Design. American Society of Mechanical Engineers, New York (2000).
6. Chowdhury, S., Taguchi, S.: Robust Optimization. ASI Consulting Group, LLC, USA , Wiley, UK (2016).

7. The International Organization for Standardization (ISO): Passenger car, truck and bus tires -- Methods of measuring rolling resistance -- Single point test and correlation of measurement results. ISO 28580, (2009).
8. Milliken, W. F., Milliken, D. L.: Race Car Vehicle Dynamics. . SAE International, Warrendale (1995).
9. The International Organization for Standardization (ISO): Passenger cars - Stopping distance at straight-line braking with ABS - Open-loop test method. ISO 21994, (2007).
10. Pacejka, H. B.: Tire and Vehicle Dynamics Third Edition. Butterworth Heinemann, Oxford (2012).
11. Abe, M.: Vehicle handling Dynamics, Second Edition. Butterworth Heinemann, Oxford (2015).

Contact Information

Mauro Martino
R&D - EMEA Product Development – BCI Components
Wheels and Tires - Design Responsible
FCA Italy S.p.A. Gate 8 – Bldg. 5 - 1st floor, Vehicle Suspensions,
C.so Settembrini, 40, 10135 Torino, Italy
Tel. +39 011 0039523
Mob. +39 335 6073868
Email: mauro.martino@fcagroup.com

Salvatore Danise
R&D - EMEA Product Development - Propulsion
DFSS Master Black Belt
Shainin Red-X Black Belt
FCA Italy S.p.A. Gate 8 – Bldg. 5 - Ground Floor
C.so Settembrini, 40, 10135 Torino, Italy
Tel. +39 011 0035875
Mob. +39 334 6900588
Email: salvatore.danise@fcagroup.com

Digital design of chassis and tire: virtual is real!

Patrick PALLOT

Manufacture Française des Pneumatiques MICHELIN, France

Abstract. The digital transformation for vehicle, chassis, systems and tire design is on its way for the sake of vehicle performances, lead-time and cost reduction. It leads to deep changes into automotive industry design processes and Michelin leads these fast-evolving trends and technical strategies with a complete approach from virtual tire design to tire models delivered to the OEM and fast iterations. Based on some on-going application cases, the study will show a range of possible efficient virtualization of the chassis-tire development on different performances such as vehicle dynamics, max handling, interior noise or rolling resistance.

It will be explained how FEA calculations can be connected to tire models' predictions to virtually design different specification letters and pre-screen some design orientation without building any prototype tires.

Virtual is real!

Keywords: Virtual Submission, Virtual Design, Virtual prototyping, TameTire, MF-SWIFT, Magic Formula, F-Tire

P. E. Pfeffer (ed.), *11th International Munich Chassis Symposium 2020*, Proceedings, https://doi.org/10.1007/978-3-662-63193-5_38

1 Simulation: the disruptive solution to automotive revolution

1.1 Context: Automotive revolution reengineering of the OE engineering process

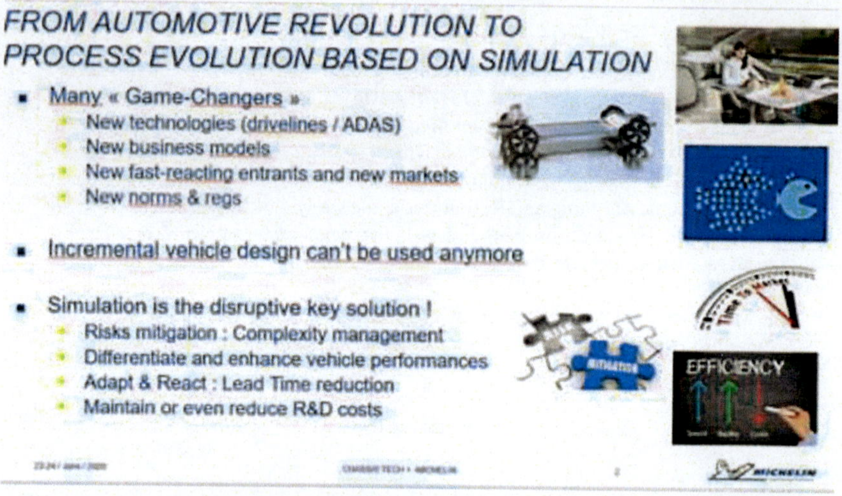

1.2 New mobility solutions to handle

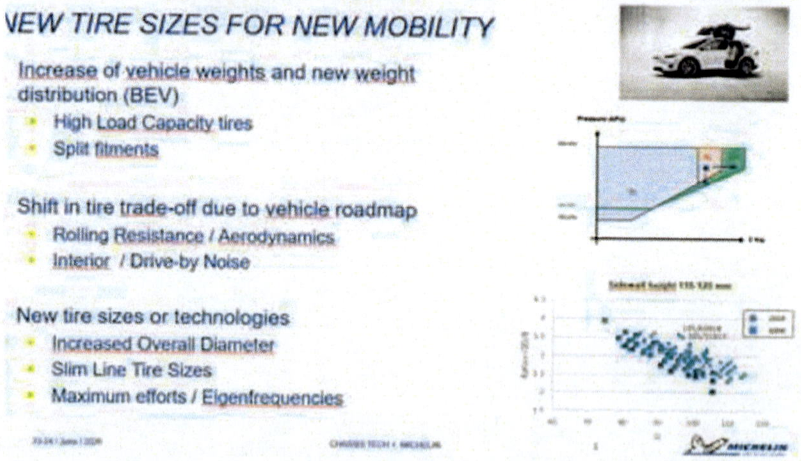

1.3 New engineering process: a mix of physical and digital engineering

Going virtual and digital: how to suppress mule car ?

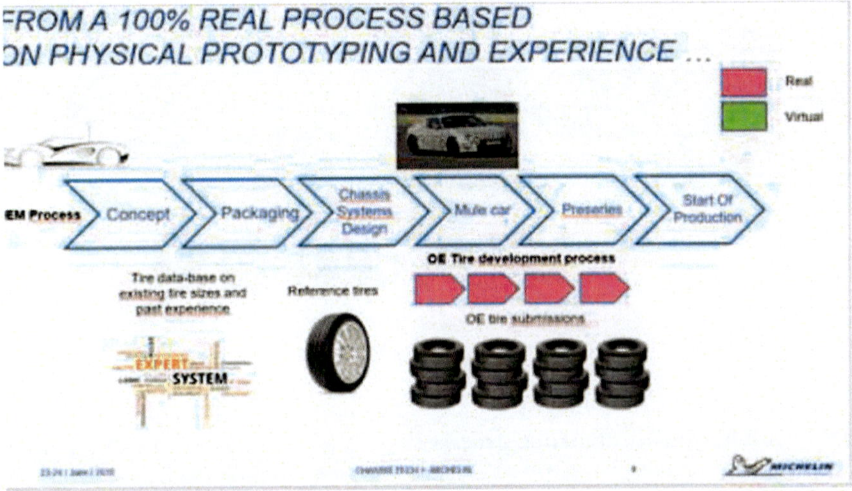

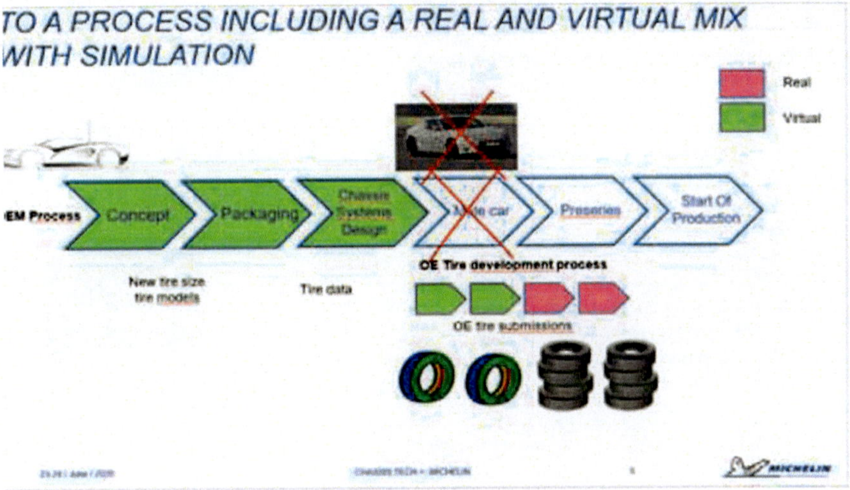

Increasing call for tire data for Virtual Vehicle Design & Virtual Tire Design

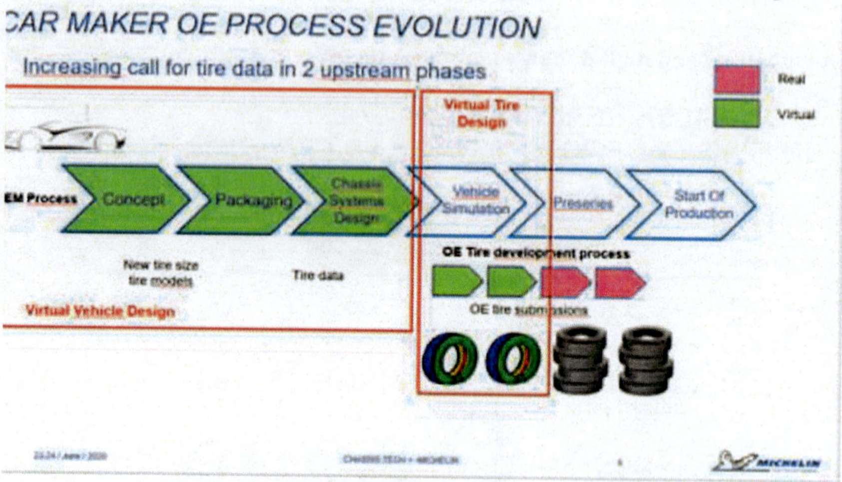

1.4 A tailor-made simulation process

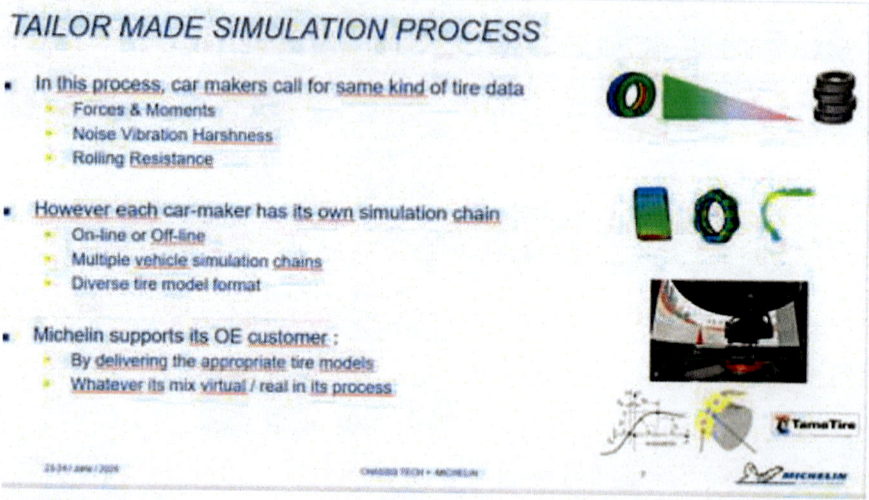

2 Virtual Tire Design: how FEA supports the new OE process

2.1 Virtual Tire Design Principles

From customer expectations to joint tire evaluation by simulation: a new way to design tires

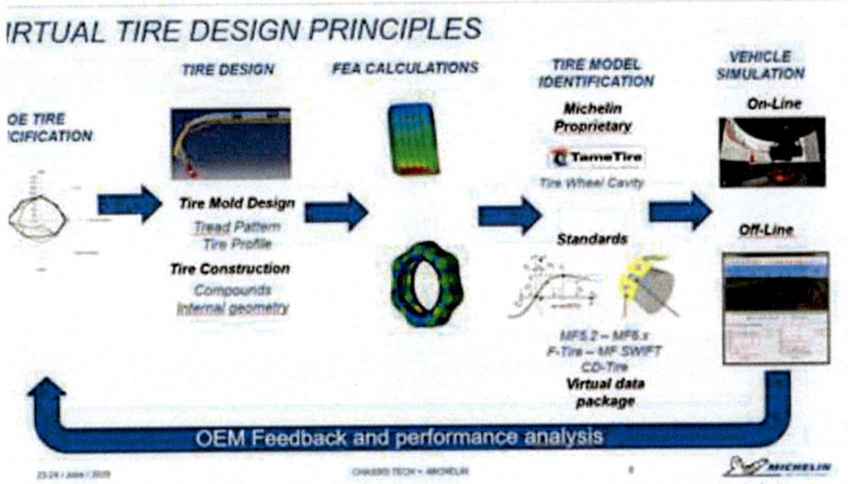

2.2 Michelin assets for virtual submission

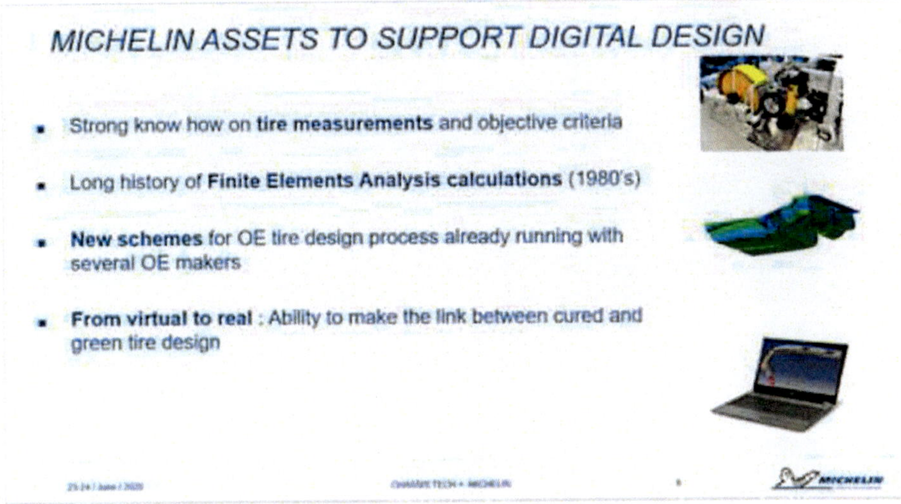

3 Conclusion

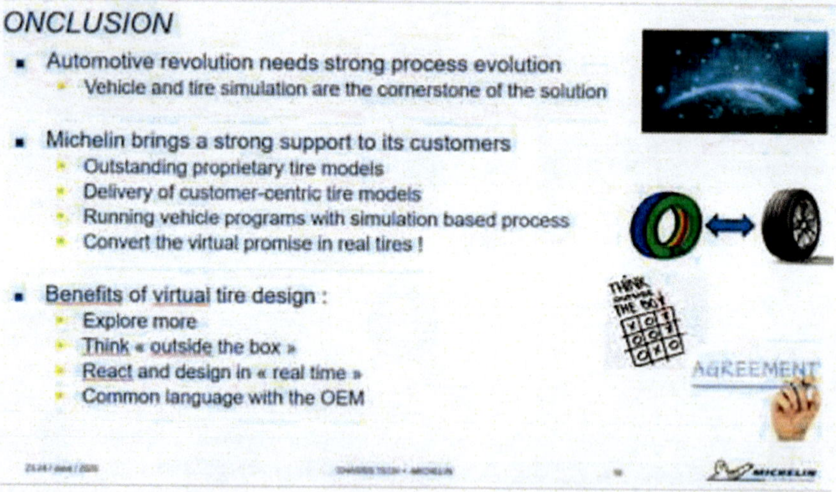

Noise measurement in accordance with UNECE R 51.03: Three years' experience with the new standard

Prof. Gerster Bernhard, BFH-TI and DTC AG, Switzerland

Marcel Strub, DTC AG, Switzerland

Stefan Sempach, DTC AG, Switzerland

1. Introduction

Since 2016, vehicles are tested for noise emissions in accordance with the standard UNECE R 51.03 [1], where the measuring track must be in conformance with ISO 10844:2014. In Vauffelin near Biel (Switzerland), DTC AG has such a track at its disposal, and as an accredited test laboratory to ISO 17025, is able, among others, to measure motor vehicle noise in accordance with the UNECE R 51.03 standard. In order to remain within the scope of this article, the following section will focus on measurements of M1 class vehicles. Serving as a basis for this work are the results of official registration tests and a series of experiments in the context of a student thesis of the Department of Automotive and Vehicle Engineering of the Bern University of Applied Sciences, which demonstrate particular points such as the influence of tyres and powertrain on noise emissions.

Fig. 1. Drive-by measurement (illustrative) on a test track conforming to ISO 10844:2014.

1.1 Why are there limits on the noise emissions of road vehicles?

Noise pollution is a health hazard for humans and thus causes economic costs and human suffering. According to a report by the Swiss Federal Office for the Environment (BAFU) [3], in Switzerland (about 10 times smaller than Germany) approximately one million people are regularly exposed to traffic noise above the legal limits [4]. That means that every seventh person, predominantly in urban regions, is exposed to noise pollution that can cause damage to health. Cardiometabolic diseases such as high blood pressure, heart attacks and strokes (500 heart attacks per year) and diabetes (2500 cases per year) can be caused by noise pollution. Noise can also lead to mental illnesses such as depression or to an impairment of learning and concentration abilities [5]. In 2016 the Federal Office for Spatial Development (ARE) estimated the health costs caused by noise pollution in Switzerland at CHF 1.9 billion. The SiRENE study led by the University of Basel [6] came to the conclusion that the above-mentioned figures are rather conservatively estimated and that road traffic is the most harmful noise source.

1.2 What are the aims of the legislator and UNECE R 51.03?

As before, road vehicles should not emit any noise that is not necessary for their operation. In Switzerland this is required by the VTS (Ordinance on the technical requirements for road vehicles) [2], which is comparable to Article 53 of the StVZo (Road Traffic Licensing Regulations) in Germany. In addition, since 2016 compliance is required with UNECE R 51.03 [7]. This standard envisages a stepwise reduction in vehicle noise and has the goal of mapping the noise level of a vehicle in an urban environment (L_{urban}). It includes constant driving at specific

P. E. Pfeffer (ed.), *11th International Munich Chassis Symposium 2020*, Proceedings, https://doi.org/10.1007/978-3-662-63193-5_39

speeds without engine action, as described in UNECE R 117 [8] for tests on tyres, and accelerated measurement runs, in a gear that is not changed during the test and causes a maximum vehicle acceleration of 2 m/s².

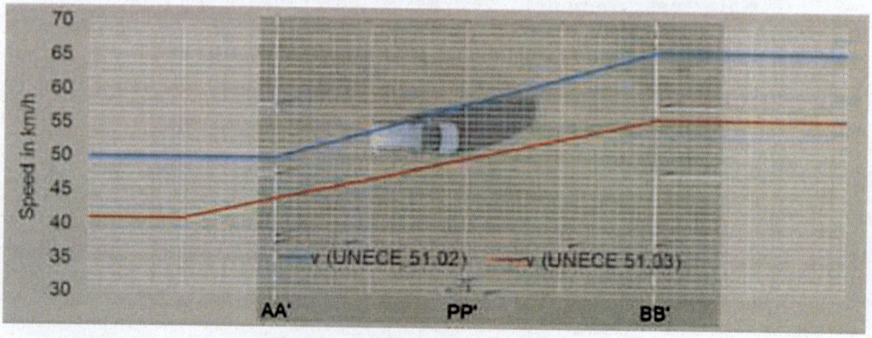

Fig. 2. Velocity curve of measurement run, comparison of UNECE R 51.02 and UNECE R 51.03.

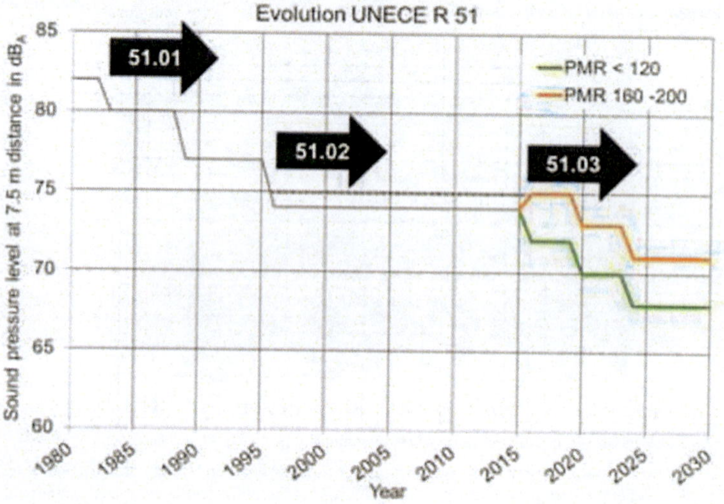

Fig. 3. Development of emission limits at a distance of 7.5 m for power-to-mass ratios (PMR) of under 120 and over 160.

2. Has UNECE R 51.03 made vehicles quieter?

A comparison of vehicles after both series of amendments to the standard shows that a certain reduction has been achieved, but which cannot be observed in every case, especially in high-performance vehicles, due to the shift of the operating point during noise measurements towards lower engine speeds (higher gears). In the case of less powerful vehicles and thus the majority of the sales volume, the goal of noise reduction has essentially been achieved. In addition, various vehicles were measured and compared in accordance with both series of amendments. One observation is that vehicles that have been granted full European approval usually meet the requirements of the standard without any problems, since they have been optimised for the new requirements. Furthermore, vehicles with automatic transmission are rather louder than those with manual transmission, because according to UNECE R 51.03, gear changes are prohibited in the measuring cycle and therefore lower gears are selected. In **Fig. 4** points below the line show vehicles which cause fewer emissions according to the UNECE R 51.03 standard.

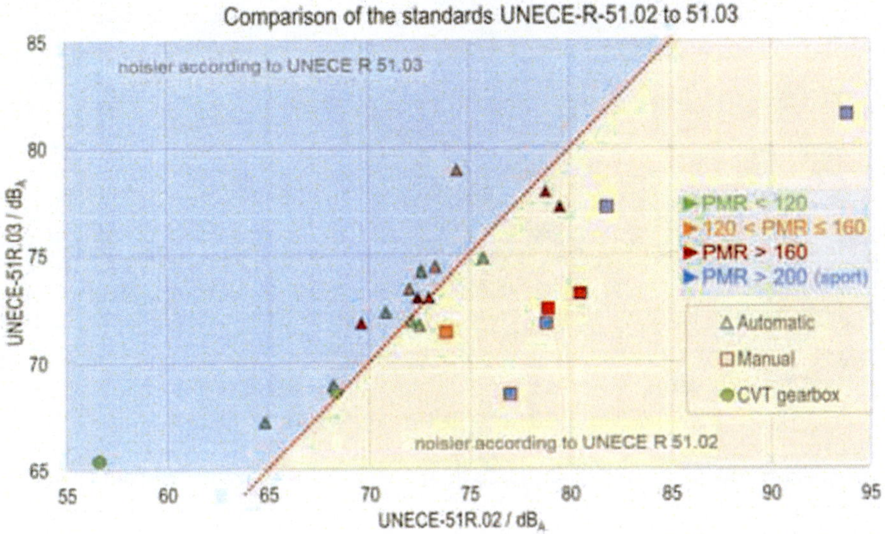

Fig. 4. Comparison of emission values according to UNECE R 51.02 and UNECE R 51.03.

However, not much has changed for the people living in the street environment, because UNECE R 51.03 requires compliance with the limit values neither when changing gears nor during full accelerations in the upper speed range. In addition, for many models, different operating modes can be selected, which have an influence on the acceleration behaviour and possibly the control of exhaust gas flaps. The new ASEPs (additional sound emission provisions)tests brought in under amendment series 03 (accelerated drive-by in the speed range 20 – 80 km/h in various gears), only partially solve the problem, as the manufacturer can currently still declare this himself with a manufacturer's certificate and the particularly annoying thrust bangs due to lack of gear changes are never measured and accelerations above 2 m/s² are only measured at the request of the testing bodies. The influence of emission noise outside the standard conditions indicates values that are significantly too high. In some cases, these have a negative influence on the goodwill towards private transport among the affected people in the inner-city area. One of the reasons for this is because the perception of human hearing is dependent on loudness. The A-weighting is adapted to sound events by 40 phon according to the following **Fig. 5.**

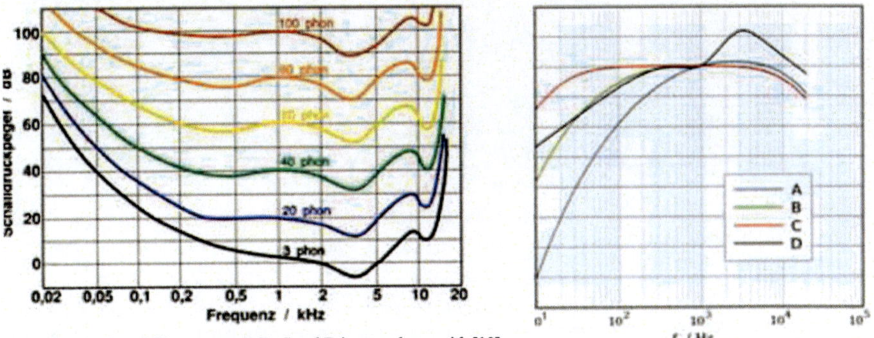

Fig. 5. Isophonic and filter curves A, B, C and D in accordance with [10].

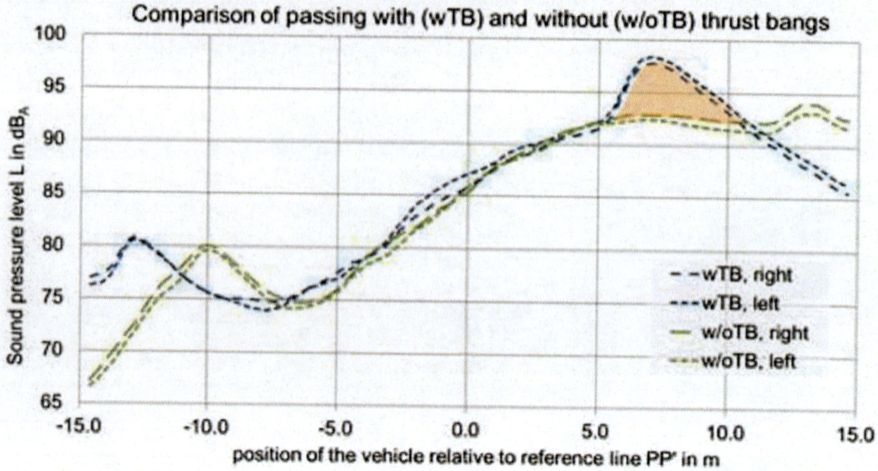

Fig. 6. Comparison of two drive-by measurements with high acceleration and correspondingly open exhaust flap (green curves) plus a gear change (blue curves) at the microphone position (line PP').

The following **Fig. 7** shows how vehicles react to the standard requirements, which indicates the influence of acceleration on the use of the exhaust flap controls.

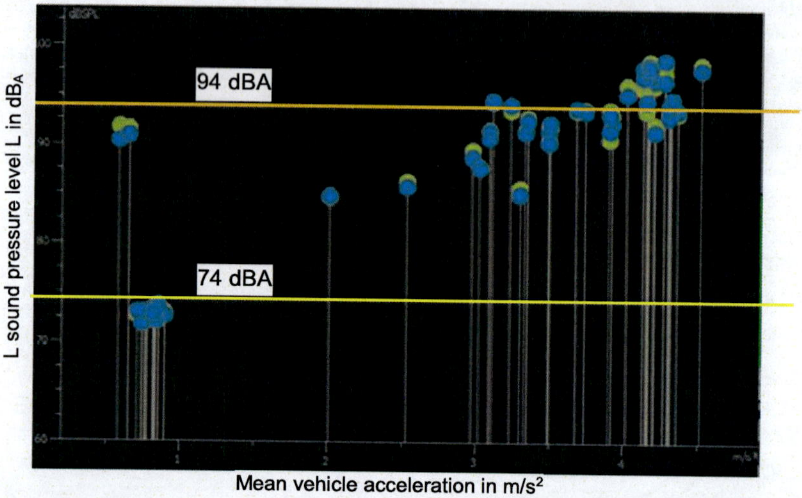

Fig. 7. Overview (screenshot of measuring device) of noise emission with different mean vehicle accelerations. Green dots: left microphone, blue dots: right microphone.

The perception of people in the street environment is also influenced by the effective perception, which, especially at values of 90 to 100 dB, is perceived by our hearing more in accordance with the B-weighting than the A-weighting, which, as mentioned, is designed for the range around 40 phon. The following **Fig. 8** shows the difference between A- and B-weighted sound pressure levels, which for the B-weighting (red bars)) results in a further gain of 4 to 6 dB.

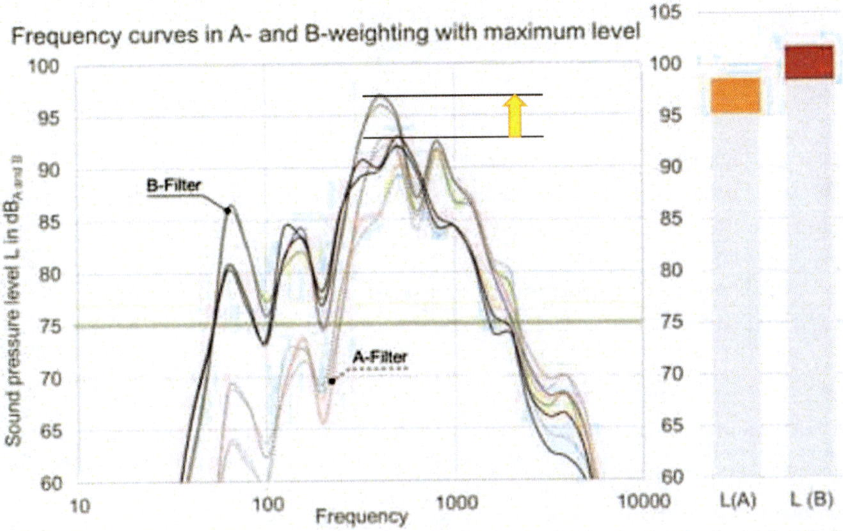

Fig. 8. Comparison of A- and B-weighted spectra of the accelerated drive-by of the left and right microphones with and without thrust bangs (yellow arrow) as an overlay to the open exhaust flaps. The bars on the right-hand side show the change due to thrust bangs in dB_A and dB_B

Both exhaust flaps and thrust bangs could be considered as noise emissions necessary for operation and hence permissible, if the power curve of the engine were to change accordingly, which so far has not been observed in any measured vehicle.

3. What is the contribution of the tyres to the total noise?

The noise levels permissible under the standard will reduce as shown in **Fig. 3**. Because the individual noise sources of the engine, air intake, body and exhaust are reduced by sound insulation measures or aerodynamic developments, the tyres will become even more important in the future. The road surface for the noise measurements will not change, so the entire noise reduction must be achieved by measures on the tyre itself. The noise development on the tyre is just one of several requirements confronting the manufacturer. For this reason, the optimisation of a single point would not be effective overall, because, for example, less abrasion resistance or reduced adhesion would simply solve one problem by creating new problems elsewhere. Moreover, noise measurements are always subject to measurement errors due to varying environmental conditions, which strengthens the case for laboratory tests. The lower the tolerable limit is set, the more important it is to control the variation.

The following measurements were all taken on an open-air noise measurement track in Vauffelin near Biel. Noise was measured from three different vehicles and a total of seven tyres in various configurations in accordance with UNECE R 51.02 and UNECE R 51.03. In addition, two measurements were carried out outside the test requirements (RA_C with road surface temperature at around 0°C and RE_N with a wet road surface).

Designation	Front axle Dimensions	Front axle Tyre label	Rear axle Dimensions	Rear axle Tyre label	remarks	test vehicle	PMR
TA_{cold}	265/30R20	71dBA	285/30R20	73 dBA	OEM / Summer tyres	3	177
TA_{warm}	265/30R20	71dBA	285/30R20	73 dBA	OEM / Summer tyres	3	177
TB	265/30R20	70dBA	285/30R20	70 dBA	Winter tyres	3	177
TC	265/30R20	73dBA	285/30R20	75 dBA	Summer tyres	3	177
TD	265/30R20	69 dBA	285/30R20	69 dBA	Summer tyres	3	177
TE_{dry}	195/55 R 16	72 dBA	195/55 R 16	72 dBA	Summer tyres /dry	1	61
TE_{wet}	195/55 R 16	72 dBA	195/55 R 16	72 dBA	Sommer tyres / wet	1	61
TF	195/55 R 16	68 dBA	195/55 R 16	68 dBA	Summer tyres	1	61
TG	215/50 R 17	70 dBA	215/50 R 17	70 dBA	Summer tyres	2	91

Table 1. Overview of the vehicles and tyres used (noise emission according to tyre label from [9]) The tyres used were exclusively high-quality European products.

For the EU tyre label, the vehicles were measured in accordance with UNECE R 117 in a speed interval from 70 to 90 km/h, where the reference speed is 80 km/h. The officially specified rolling sound pressure level at 80 km/h for the various tyres is also given in Table 1, in the column headed 'Tyre label'.

The following diagrams (**Fig.9** and **Fig.10**) show the separate results for the two measured test vehicles 1 and 3, each with different tyres. The contents of the diagrams represent on one side, the unaccelerated drive-by without an engine at various speeds between 70 and 90 km/h (analogous to the measurements for UNECE R 117) as well as at 50 km/h (measurement for UNECE R 51.03). On the other side, the results of the accelerated drive-by measurements are given as a comparison. This clearly shows what proportion particularly the tyres already contribute to the overall noise during free rolling without circumferential slip, together with a small contribution from air turbulence in the bodywork. For the sake of clarity, the results are presented separately by vehicle. To improve the legibility of the results of accelerated drive-by, they are presented slightly shifted compared to the effectively measured 50 km/h.

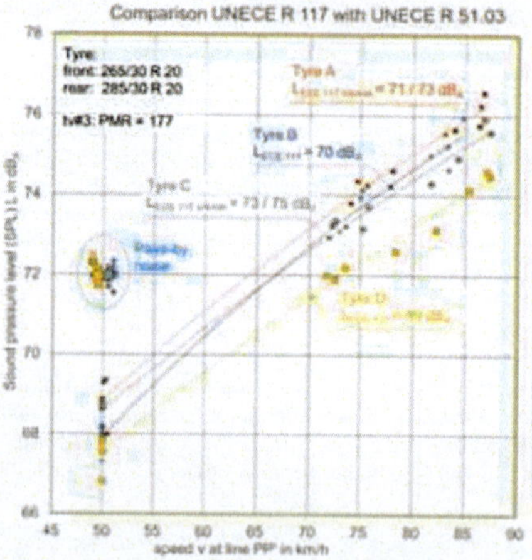

Fig. 9. Tyre noise of Vehicle 3 at constant and accelerated drive-by speeds under standard ambient conditions.

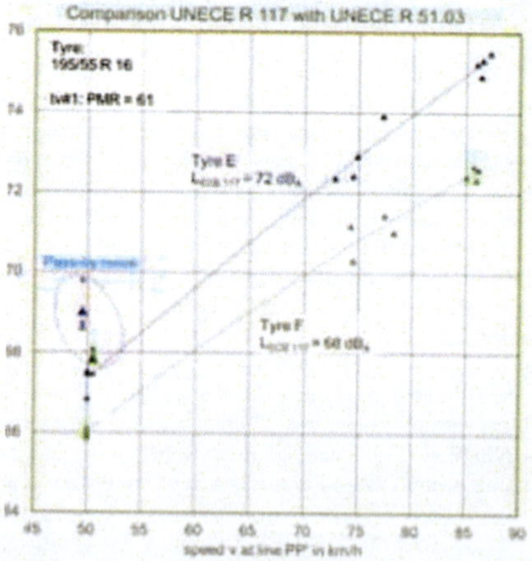

Fig. 10. Tyre noise of Vehicle 1 at constant and accelerated drive-by speeds under standard ambient conditions.

During the measurements it was possible to determine two non-standard conditions: firstly at very low temperatures (**Fig.11**), and secondly, on a wet test track (**Fig. 12**). Depending on the speed, the temperature around freezing point caused the tyres to become louder by 1–2 dB_A which is directly reflected in the results of the accelerated drive-by. On the other hand, when Tyre E with Vehicle 1 was measured under identical conditions once on a wet and once on a dry test track, the resulting noise emissions were over 4 dB_A higher.

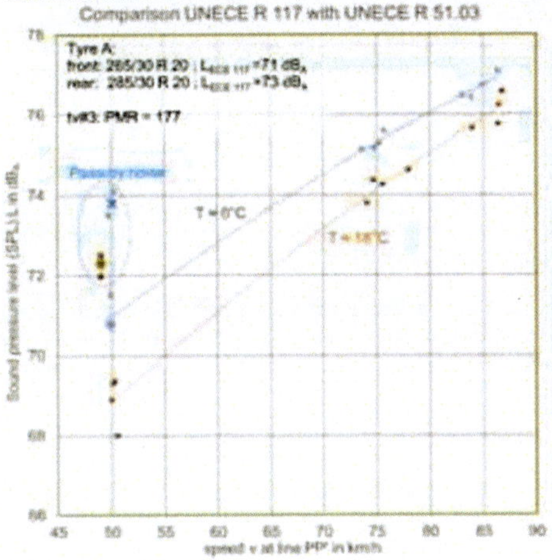

Fig. 11. The effect of tyre temperature on noise emission during constant speed and accelerated drive-by runs with Vehicle 3.

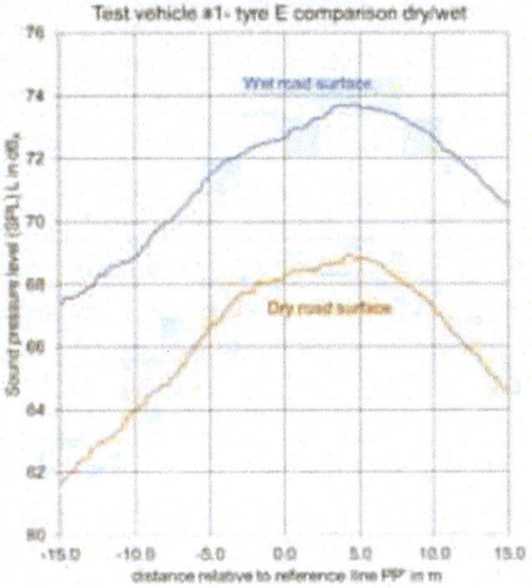

Fig. 12. Noise emission of Vehicle 1 and Tyre E during accelerated drive-by on dry and wet test track surfaces.

In summary, in the following (**Fig. 13**) the mean values of the constant drive-by runs are compared with the mean values for the accelerated drive-by. In each case this is at a speed of 50km/h. With the exception of Tyre C, which produced relatively high noise emissions at 50km/h, but then rose less than other tyres as the speed increased, there is a clear correlation between tyre rolling noise at a constant speed of 50km/h and the total vehicle noise of the accelerated drive-by. This applies even more so for the calculation of L_{Urban}, which is composed of the rolling tyre noise and accelerated drive-by.

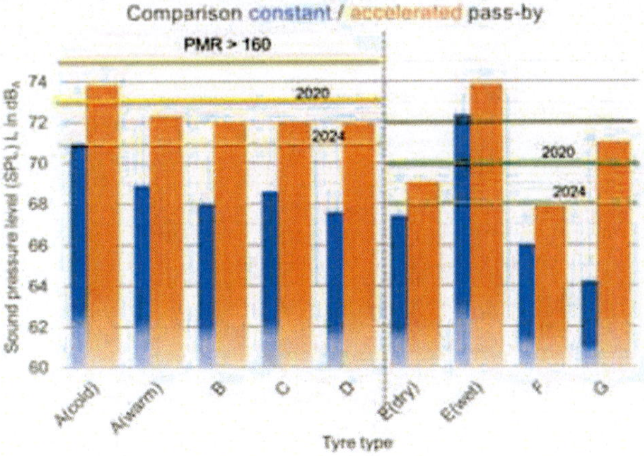

Fig. 13. Comparison of all measurements of the effects of tyres

This clearly shows the influence of the test parameters between A(C) with a cold measured surface of around 0°C, which was significantly under the permissible surface temperature of 5°C, and the measurement A(W) with a warm measured surface. Another influence is the wetness of the road surface, where here Tyre E showed a difference of 4.9 dB_A between a dry (T) and a wet (N) surface. Vehicle 3 with Tyre G is an electric vehicle with relatively high torque. This is not limited to the same extent by the UNECE R 51.03 standard, as is the case with internal combustion engine vehicles. That means the torque force from electric drives has a direct effect via the slippage noises on the measured values of the accelerated drive-by. A further peculiarity of electric vehicles can be the large components at frequencies between 1000 Hz and 5000 Hz. These frequencies have a problematic effect with the A-filter, because the A-filter in this frequency range provides for a gain of up to 1.5 dB compared to the linear measured value (see **Fig. 5**, right diagram).

4. Conclusion

UNECE R 51.03 has certainly improved the situation in terms of the legislator. The noise level has been reduced on a broad front, especially for vehicles with automatic transmission. The effect will be even greater, given the fact that more and more vehicles are likely to be fitted with automatic transmissions.

Somewhat less successful has been the prevention those noises, which are unwanted because they are not operationally necessary, such as those from flap controls or electronically controlled thrust bangs. Neither are wanted by the legislator, but they can be programmed with relatively simple means in such a way that they pose no danger to passing the approval tests.

In principal, almost all noise emitters on the vehicle can be dampened if they are too loud. Tyres are the exception. Therefore, in terms of noise reduction it is the tyre manufacturers who are also faced with the greatest challenge. Over the next four years, the tyre industry must find the ways and means to comply with the very broad and increasingly far-reaching requirements for tyres with relatively low noise levels or a necessary reduction of at least 1 dBA. And they have to do this with increasing electrification of the powertrain, which is typically coupled with an increase in torque and thus slippage during accelerated drive-by measurement. The task is not made any easier by high maximum speeds and, despite CO_2 legislation, vehicles becoming heavier.

This makes it all the more important to keep the scattering in the measurements as small as possible, so as not to jeopardize all the efforts of all those involved through randomness in the measurements. In this respect, as a testing laboratory, we are making every effort to meet this requirement with ever better measurement technology and reliable processes.

List of references:

[1]	UN/ECE, "Regulation No 51 Economic Commission for Europe of the United Nations (UNECE) (03)", *Official Journal of the European Union,* 10 February 2018.
[2]	Verordnung über die technischen Anforderungen an Strassenfahrzeuge, [Ordinance on the Technical Requirements for Road Vehicles], Art. 53 Geräusch, Schalldämpfer, Abs. 1 [Art. 53(1), Noise, Silencers] 15 January 2017.
[3]	Bundesamt für Umwelt BAFU [Swiss Federal Office for the Environment], "Eine Million Menschen sind noch immer übermässigem Verkehrslärm ausgesetzt [A million people are still exposed to excessive vehicle noise]", *Medienmitteilung,* 5 November 2018.
[4]	"Belastungsgrenzwerte für Lärm [Exposure limits for noise]" 25 October 2019.
[5]	Bundesamt für Umwelt BAFU [Swiss Federal Office for the Environment], "Lärm macht krank [Noise makes you ill]", *Faktenblatt,* 15 January 2007.
[6]	Ritter A., "Lärm geht ans Herz [Noise affects the heart]", *Schweizerische Ärztezeitung*, 17 January 2018.
[7]	UN/ECE, "Regulation No 51 Economic Commission for Europe of the United Nations (UN/ ECE) (02)", *Official Journal of the European Union,* 18 June 2007.
[8]	UN/ECE, "Regulation No 117 Economic Commission for Europe of the United Nations (UN/ECE)", *Official Journal of the European Union,* 2 February 2007.
[9]	Allgemeiner Deutscher Automobil-Club e.V. (ADAC) [General German Automobile Club], "EU-Reifenlabel: Das müssen Sie wissen [EU Tyre Labels –What you must know]", (Online) 7 May 2019.
[10]	Lindosland, "Bewertungsfilter A, B, C, D und Empfindlichkeit des Gehörs [Weighting Filters A, B, C, D and Hearing Sensitivity]", wikipedia.org, 21 April 2020.